POULTRY SCIENCE

Editor

Lanzo Mertens

scitus
academics

Poultry Science

Edited by **Lanzo Mertens**

ISBN: 9781681179315
Library of Congress Control Number: 2017948961

SCITUS Academics LLC,
616, Corporate Way, Suite 2, 4766,
Valley Cottage, NY 10989

www.scitusacademics.com

Printed in United States of America.

Preface

Poultry, including meat and eggs, is major agricultural commodity in the world. At some point, almost everyone has eaten chicken, turkey, eggs, or other poultry products. Poultry have been on the earth for over 150 million years, dating back to the original wild jungle fowl. Now we include ducks, geese, turkeys, pheasants, pigeons, peafowl, guinea fowl and chickens in the list of species under the general term poultry. According to the researchers and scientists, 74% of the world's poultry meat, and 68 percent of eggs are produced in ways that are described as 'intensive'. One alternative to intensive poultry farming is free-range farming using lower stocking densities. Poultry producers routinely use nationally approved medications, such as antibiotics, in feed or drinking water, to treat disease or to prevent disease outbreaks. The interest in poultry and poultry products has grown tremendously in the last 20 years. Almost every country in the world has a poultry industry of some kind. Japan is steadily increasing their domestic production of both broilers and egg producing birds. Countries of the former Soviet Union have always been producing poultry and eggs and are continuing to increase their output to meet the new demand. China, the Middle East, and Africa are all areas where increasing demand for poultry has resulted in significant increases in the number of chickens being reared for meat and eggs.

This volume 'Poultry Science' presents an introductory overview to the different fields/branches of poultry science. It covers on special feed resources for poultry, biofilms of salmonella and campylobacter in the poultry production, and mycotoxins in poultry feed. It will review the basics of AI and fertility evaluation in poultry. To better appreciate the biological basis of these techniques, an overview of the reproductive biology of poultry is provided. Anticipation of different contaminants in modern poultry farms is provided as the consumption of poultry meat and products has increased as a consequence of economic crisis, driven by several factors, while people keep away from high priced beef/lamb meat or meat products. Meanwhile, due to this increasing demand in industry resulting strict measures in disease control and environmental factors, these products may involve some chemical and natural compounds with hazardous properties at detectable or even very low concentrations. This book will be of valuable for the graduate students, teachers, researchers, farmers, and other professionals, who are interested to fortify and expand their knowledge about chicken products in fields of poultry science, biotechnology, plant science, and agriculture.

Table of Contents

Chapter 1	Assessment of Maize (*Zea Mays*) as Feed Resource for Poultry	1
Chapter 2	Exploring The Economic Potential of Reducing Broiler Lameness	33
Chapter 3	Behavioural Assessment of Pain in Commercial Turkeys (Meleagris Gallopavo) with Foot Pad Dermatitis	59
Chapter 4	The Effect of Age on Growth Performance and Carcass Quality Parameters in Different Poultry Species	81
Chapter 5	Lipopolysaccharide Reduces Food Passage Rate From The Crop by a Prostaglandin-Independent Mechanism in Chickens	101
Chapter 6	Effects of Dietary Vitamin E on Fertility Functions in Poultry Species	119
Chapter 7	Mycotoxins in Poultry	135
Chapter 8	Precaecal Phosphorus Digestibility of Inorganicphosphate Sources in Male Broilers	157
Chapter 9	Grain Sorghum: A Conundrum for Chicken-Meat Production	175

Chapter 10 **Selenium Requirements and Metabolism in Poultry** 209

Chapter 11 **Phosphorus Bioavailability: A Key Aspect for Conserving this Critical Animal Feed Resource with Reference to Broiler Nutrition** 235

Chapter 12 **The Impacts of Hammer-Mill Screen Size and Grain Particle Size on The Performance of Broiler Chickens Offered Diets Based on Two Red Sorghum Varieties** 257

Chapter 13 **Technology and Poultry Welfare** 277

Chapter 14 **Ranging Behaviour of Commercial Free-Range Laying Hens** 307

Chapter 15 **Effects of Dark Brooders on Behavior and Fearfulness in Layers** 327

Chapter 16 **Distribution, Numbers, and Diversity of ESBL-Producing *E. coli*in the Poultry Farm Environment** 353

Chapter 17 **Assessment of The Effectiveness of Head Only and Back-of-The-Head Electrical Stunning of Chickens** 385

Index 405

CHAPTER 1

Assessment of Maize (*Zea Mays*) as Feed Resource for Poultry

Herbert K. Dei

Department of Animal Science, Faculty of Agriculture, University for Development Studies, Tamale, Ghana

ABSTRACT

Maize, also known as corn (Zea mays L), has been recognised worldwide as a major energy feed ingredient in the diets of poultry. Its major nutritional limitation has been the low protein content and poor protein quality, which necessitates the use of expensive high-protein supplements or synthetic amino acids such as lysine in diets containing large proportion of maize. Therefore, extensive research has been conducted by maize breeders on the world maize germplasms collection with the aim of improving its nutritive value, particularly protein quality for monogastric animals. This chapter assesses the genetic upgrading of the nutritional quality of maize protein that culminated in the development of a new class of maize known as "Quality Protein Maize (QPM)". Various studies on the nutritionally improved maize for poultry as well as future challenges confronting maize utilisation in poultry production are highlighted.

Keywords: maize (*Zea mays*), energy, protein quality, nutritive value, poultry

1. INTRODUCTION

Poultry (avian species) have been recognised as affordable source of high-quality protein worldwide in the forms of meat and eggs. The poultry sector has been shown to become the world's largest meat sector by 2020 (**Figure 1**). Besides, the sector continues to record high global output of eggs (e.g. 70 million metric tonnes in 2014) as additional high-quality protein food.

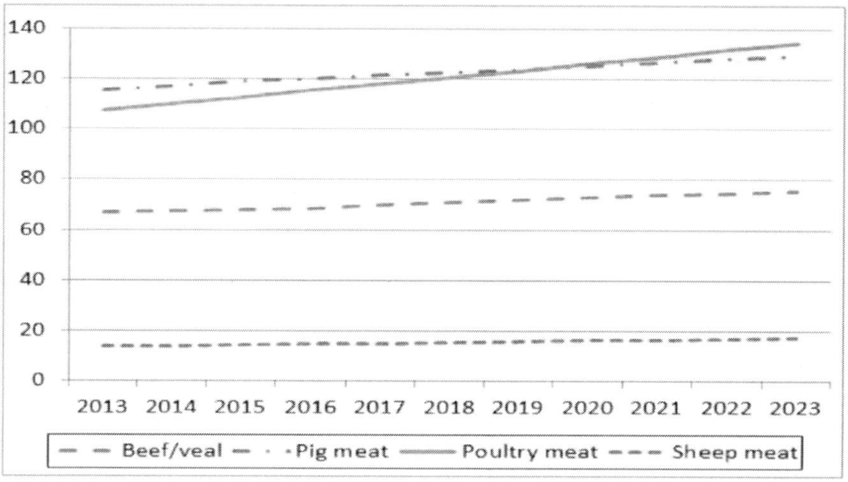

FIGURE 1. Global meat production (million metric tonnes) [1].

The rapid growth of the poultry sector is fuelled by several factors such as an increasing human population, greater purchasing power in developing economies, increased urbanisation and industrialisation in developing countries, development and transfer of feed, relatively short production cycle and advances in poultry breeding, and improved processing technologies. Of these factors, feed has been recognised as the most important factor controlling profitability and product quality [2].

Protein and carbohydrate are by far the two most important nutrients in poultry diets due not only to their marked effect on voluntary feed intake of the bird, but also the fact that they represent approximately 90% of the total cost of the ingredients in a ration [3]. Cereal grains constitute a large proportion (>50%) of poultry diets and contribute largely carbohydrates and to some extent proteins. They are mainly dietary source of energy, but can vary widely between grain types and animal species [4]. The common feed grains for poultry are corn or maize (Zea mays), wheat (Triticum aestivum), barley (Hordeum vulgare) and sorghum or milo (Sorghum bicolor).

Maize is by far the major feed grain grown worldwide, particularly in the United States. Although it is the preferred grain for feeding poultry [5], it is found to be low in protein content as well as protein quality [6], thereby limiting its nutritional value. This has necessitated a search for nutritionally improved maize varieties as well as alternative feed ingredients. The former has resulted in extensive research on the world maize germplasm collection with the aim of improving its nutritive value, particularly the protein quality for poultry.

This chapter discusses maize production and consumption, its genetic upgrading to improve the nutritional value with regard to feeding poultry.

2. IMPORTANCE OF MAIZE IN HUMAN AND ANIMAL NUTRITION

Maize (Zea mays L.) tops other cereals in terms of worldwide production (**Figure 2**), with the United States being the largest producer as well as consumer (**Table 1**).

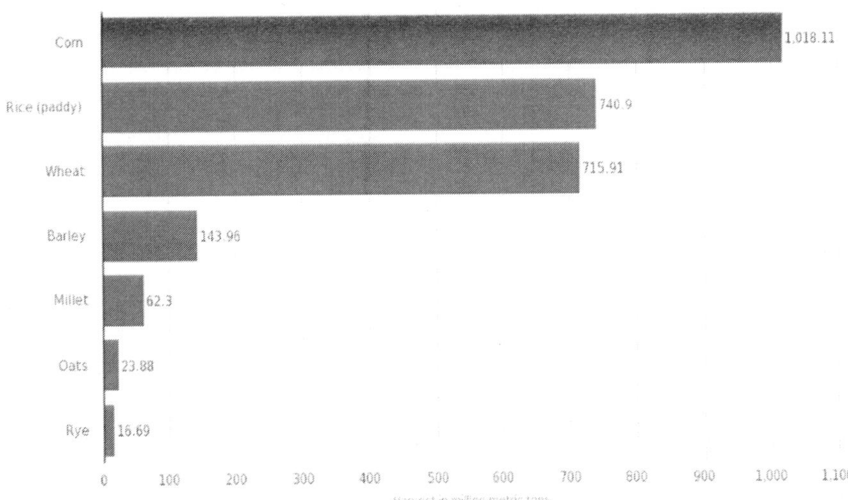

FIGURE 2. Global maize production [7].

On worldwide basis, much of the maize produced is fed to livestock, whereas only a small portion goes directly to human food [9]. The grain provides the world with 19% of its food calories and 15% of its annual production of food crop protein [6]. It is the basic staple cereal grain for large groups of people in Latin America, Africa and Asia [6], where the grain is consumed directly or in modified form as a major item of the diet.

Maize provides more feed for livestock than any other cereal grain [6]. For instance, 65% of the maize grown worldwide is used for livestock feed [10], of which the United States is the highest consumer. Also, rapid increase in poultry production in developing countries in Latin America, Africa and Asia is a major factor contributing to the increased use of maize for livestock feeding. In fact, maize is the preferred grain for feeding domestic birds, because its dietary energy value is the highest among cereals with very low variability between years for a given region [2].

TABLE 1. Worldwide maize production and consumption [8].

Country	Production (million mt)[a]	~Production share (%)	Consumption (million mt)[b]	~Consumption share (%)
United States	345.5	36	301.5	31
China	224.6	23	214.0	22
Brazil	81.5	8	59.0	26
European Union-27	57.8	6	76.0	8
Argentina	25.6	3	–	–
Ukraine	23.5	2	–	–
Mexico	23.5	2	34.3	4
India	21.0	2	21.4	2
Canada	13.6	1	13.4	1
Russia	13.0	1	–	–
Japan	–	–	14.7	2
Egypt	–	–	14.5	2
Indonesia	–	–	12.7	1
Others including sub-Sahara Africa	138.4	14	204.8	21

[a]Total production (967.9 million mt).
[b]Total consumption (966.2 million mt).

Over five hundred products [11] are obtained from industrial processing of maize, particularly from the main end-products of the "wet-milling" process of starch and nutritive sweeteners. The by-products include germ, bran and gluten which are suitable for feeding farm animals [12]. The gluten, in particular, is high in protein and metabolisable energy as well as a concentrated source of xanthophylls pigments, which make it popular in poultry production [2].

Besides, maize has long been an important ingredient in the manufacture of alcoholic beverages [9] including maize beer and whiskey. It is also an essential raw material in the production of industrial alcohols (25.4 kg of maize can yield 9.7 L of anhydrous ethanol plus useful by-products) [9]. The ethanol has a potential use as a partial replacement for gasoline due to increased fuel costs. The main by-product is referred to as "draff" or "distillers dried grains" (DDG), and it is high in protein. The DDG can be added another by-product called "solubles," which comprises the smallest residual particles of maize and yeast. The DDGS is high in protein, trace element and vitamins as well as increased availability of phosphorus, thereby making it popular feed ingredient for poultry production [2].

3. NUTRITIVE VALUE OF NORMAL HYBRID MAIZE GRAIN

The maize grain (**Figure 3**) on dry matter basis is made up of 82.9% endosperm, 11.1% germ, 5.2% pericarp and 0.8% tip cap [13].

Table 2 shows the per cent chemical composition of the maize grain and grain fractions. In general, maize grain is low in protein content (9.1%), oil (4.4%) and ash (1.4%), but very high in starch content (73.4%) when considered on dry matter basis.

3.1. Carbohydrate Content of Maize Grain

The relative proportions of the various carbohydrates are 77% starch, 2% sugars, 5% pentosans [15] and 1.2% crude fibre [16]. The carbohydrate which forms more than 70% of the maize grain [15] is concentrated in two starchy fractions, floury and flinty, of the endosperm. The sugar in the grain is found in the germ and dietary fibre is in the bran [13].

TABLE 2. Chemical composition of normal maize grain and grain fractions (%DM) [14].

	Starch	Protein	Oil	Sugar	Ash (minerals)
Whole grain	73.4	9.1	4.4	1.9	1.4
Endosperm	87.6	8.0	0.8	0.6	0.3
Germ	8.3	18.4	33.2	10.8	10.5
Pericarp	7.3	3.7	1.0	0.3	0.8
Tip cap	5.3	9.1	3.8	1.6	1.6

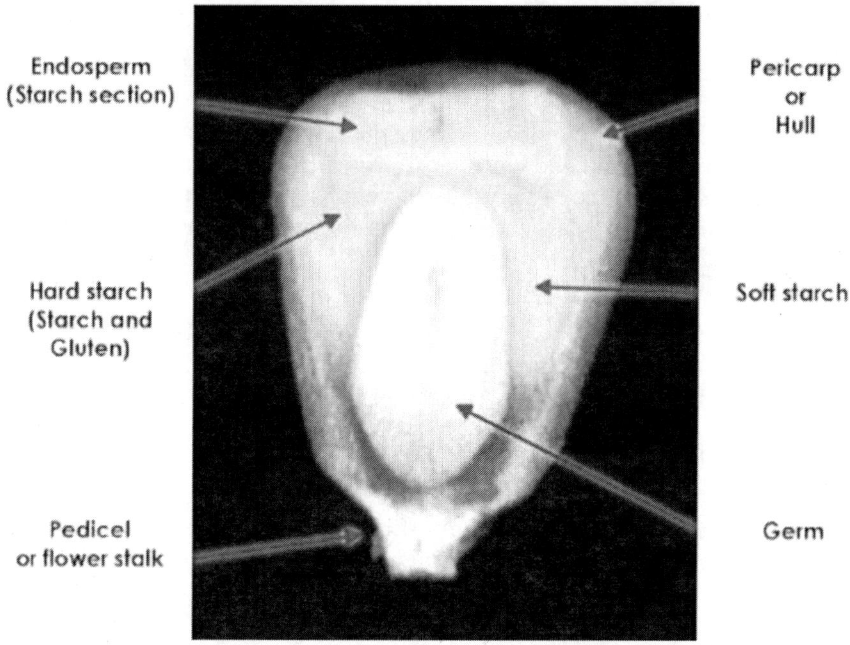

FIGURE 3. Structure of maize kernel (source: www.fao.org).

The endosperm consists of starch granules embedded in a protein matrix. Flinty endosperm has a more rigid protein structure and is also higher in protein content than floury endosperm [13, 17]. The starch in the flinty endosperm consists of 100% amylopectin (large branched molecules), whereas that in the floury endosperm comprises about 27% amylose (linear molecules) and 73% amylopectin [16]. This variation in starch structure does not have any effect on the nutritional value of maize for poultry [2]. The distribution of flinty or floury endosperms in the maize grain determines whether a maize variety is classified as flint or floury (dent) maize. The starch which is the main source of energy in the grain has a digestible energy content ranging from 3.75 to 4.17 kcal/g dry matter [18], thereby making maize one of the highest in energy among the cereal grains (**Table 3**).

TABLE 3. Gross and metabolisable energy of cereals [19].

Cereal	Gross energy (MJ/kg DM)	Metabolisable energy (MJ/kg DM)
Barley	18.3	13.7
Sorghum	18.8	13.4
Maize	19.0	14.2
Millet	18.7	11.3
Wheat	18.4	14.0
Oats	19.0	11.5

The crude fibre content of maize grain averages about 2.7% of dry matter [20]. The crude fibre interferes with nutrient availability of the grain [21]. For instance, the range of protein digestibility of maize is 83–90% [21], while digestibility of carbohydrate is 99% [22]. Nevertheless, the maize grain is highly digestible due to its low crude fibre content [16].

3.2. Lipid Content of Maize Grain

Maize oil (Table 4) is good quality oil both from nutritional standpoint (Table 5) and in terms of cooking quality [13]. Another desirable property of maize oil is its very low concentration of linolenic acid and high level of natural antioxidants [13, 23], thereby making the grain less susceptible to rancidity in storage.

TABLE 4. Concentration of fatty acids in maize oil triglycerides [23].

Fatty acids	Structure[a]	Amount (%)
Palmitic	16:0	11.0
Stearic	18:0	3.0
Oleic	18: 1	43.4
Linoleic	18:2	41.8
Linolenic	18:3	0.6

[a]Number of carbon atoms: number of unsaturated bonds.

TABLE 5. Proportion of chemical constituents contained in each fraction of normal maize grain (% DM basis) [14].

Grain fraction	Starch	Protein	Oil	Sugar	Ash
Endosperm	97.8	73.8	15.4	28.9	17.9
Germ	1.5	26.2	82.6	69.3	78.4
Pericarp	0.6	2.6	1.3	1.2	2.9
Tip cap	0.1	0.9	0.8	0.8	1.0

The maize grain is a fair source of alpha-tocopherol (Vitamin E) which ranges from 0.6 to 2.1 mg/100 g grain [24]. Most of the carotenoids found in maize lipid are xanthophylls which are present only in yellow maize grain, and form about 12,511 mg/100 g grain [25]. Yellow maize, therefore, is one of the best sources of pro-retinal carotenoids [17]. These pigments cause yellow colouration of shanks and skin of broilers and yolks of eggs [26]. However, the pigmented grain tends to colour the carcass fat, which in the United Kingdom is not considered desirable [16].

3.3. Protein Content of Maize Grain

The maize grain is deficient in protein, but its variability is low with standard error of the order 7 g/kg of crude protein [2]. The protein content of maize grain ranges from 8 to 11 g/100 g grain of dry matter [14, 23, 27]. The various fractions of grain vary considerably in protein content (**Table 5**). Even though the majority of protein in the grain occurs in the endosperm, the germ (184 g/kg DM) is considerably higher in protein content than the endosperm (80 g/kg DM) [14].

Generally, the low protein content of the grain limits its nutritive value as the only source of food for both humans and livestock.

The amino acid composition of whole maize grain is determined by both the relative proportions of the various protein fractions and the amino acid composition of each fraction [21]. Maize grain endosperm proteins are usually referred to as albumins, globulins, prolamins and glutelins, depending on their solubility in different solvent systems [21]. Prolamins and glutelins (also referred to as storage proteins) are confined to the endosperm, whereas albumins and globulins (also referred to as water-soluble proteins) are also found in the aleurone layer and the germ. The proportion of each protein fraction is presented in Table 6. In normal maize grain, the prolamin content exceeds that of glutelin and represents about 50–60% of the total protein [21]. Each protein fraction tends to have a characteristic amino acid composition (**Table 7**), and the relative proportion of each fraction strongly affects the level of individual amino acids in the total grain protein [28]. Prolamins are most deficient in lysine, thereby rendering maize protein poor in terms of nutritional quality. The general deficiency of lysine in maize grain is essentially the consequence of its low content of albumin and globulin, which besides having high lysine content exhibit a well-balanced amino acid composition similar to that of animal proteins of superior nutritional value [21]. Moreover, maize prolamins are characterised by larger quantities of leucine than isoleucine, thus causing the typical amino acid imbalance that further reduces the protein quality of maize [29].

TABLE 6. Protein fractions of normal maize [21].

Protein fraction	Amount (g/100 g protein)
Salt-soluble fraction (including NPN), i.e. albumins and globulins	6
Alcohol-soluble fraction (in the presence and absence of reducing agents), i.e. prolamins	64
Acid or alkali-soluble fraction (in the presence of reducing agents and nonextractable N), i.e. glutelins and residue	30

NPN, nonprotein nitrogen; N, Nitrogen.

TABLE 7. Essential amino acid compositions of the endosperm protein fraction of normal maize (g/16 gN) [21].

	Protein fractions		
Amino acid	Albumin+globulin	Prolamin	Glutelin
Isoleucine	3.9–4.6	4.4–4.4	4.9–5.0
Leucine	5.4–6.4	20.3–20.5	9.4–10.1
Lysine	6.1–6.3	0.1–0.2	6.4–7.0
Cystine	0.3–1.6	0.1–0.7	0.2–0.5
Methionine	1.7–2.5	1.9–2.2	2.4–4.0
Tyrosine	2.8–4.5	5.5–6.1	2.4–2.7
Phenylalanine	2.4–4.2	7.8–8.0	5.3–5.6
Threonine	4.2–5.3	3.4–3.4	4.2–5.2
Valine	4.7–6.2	4.1–4.2	7.0–7.1

The real significance of the poor nutritional quality of maize protein, therefore, is that the other food components of the diets of livestock and human may fail to provide adequate amounts of essential amino acids, particularly lysine to offset the nutritional deficiencies of maize protein [30].

3.3.1. Ways Of Improving The Protein Quality Of Maize

Although maize grain is relatively low in total protein and generally low in lysine and tryptophan, these shortcomings can be overcome by appropriate blending with animal products or legumes or oilseed products. The most obvious result of such blending is that the mixture is higher in protein than the maize component alone. Beyond this, animal products, legumes and various oilseed cakes improve the quality of maize protein by supplementing them with limiting amino acids such as lysine and tryptophan. This is called protein supplementation [31]. On the other hand, legumes and some oilseed cakes, which are deficient in methionine, can be supplemented by maize grain, which is not deficient in this amino acid. Such mutual balancing of each other's amino acids is known as protein complementation [31]. Besides, the quality of maize protein can be improved by the addition of synthetic amino acids like lysine. It also appears that a consistent enrichment of the nutritional quality of maize protein can be accomplished by developing new cultivars with a reduced content of prolamins with a parallel increase in glutelins and salt-soluble protein [21].

3.3.1.1. PROTEIN SUPPLEMENTATION OF MAIZE-BASED DIETS

Protein supplementation consists of adding small amounts of proteins which are rich sources of the amino acids deficient in normal maize [31]. Various protein supplements that have been tested on rats [32] include fish protein concentrate, soyabean flour, cottonseed flour, torula yeast, casein and egg protein. The recorded protein efficiency ratio values were normal maize (1.00), fish protein concentrate (2.44), soyabean flour (2.25) cottonseed flour (1.33), torula yeast (1.97), casein (2.21) and egg protein (2.24). This effect on protein quality was attributed to the contribution that the protein supplements made in lysine, tryptophan and protein content.

It has been reported that the supplementation of maize with soyabean flour increased usable protein from about 2.5% for maize alone to 10.6% when 20% of the soyabean flour was included in the diet for children [33, 34].

3.3.1.2. PROTEIN COMPLEMENTATION

Protein complementation comprises various food mixtures of maize with other ingredients of plant origin which are higher in protein quality and protein content [33, 35, 36]. One example of the favourable effect of protein complementation was established in rats [37].

Mixtures of cornflour and soyabean flour were fed to rats, and their weight gains per gram of protein consumed (protein efficiency ratio) were measured. Optimum results were obtained with the 40% cornflour and 60% soyabean flour ratio. With less soyabean flour in the mixture, lysine became limiting, but with more soyabean flour in the mixture, methionine was limiting (**Table 8**).

TABLE 8. Complementation effects in rats fed combinations of soyabean flour and whole normal maize flour at a constant level of dietary protein [37].

	Amino acid content (g/16 gN)			
Maize:soyabean protein ratio	Lysine	Total sulphur amino acids	Tryptophan	Protein efficiency ratio
100:0	2.88	3.15	0.60	1.6
80:20	–	–	–	2.3
60:40	–	–	–	2.7
40:60	4.95	3.14	1.07	2.9
20:80	–	–	–	2.8
0:100	6.32	3.12	1.38	2.6

The nutritive values of vegetable mixtures comprising 38% cottonseed flour plus 58% maize flour, 38% soyabean flour plus 58% maize flour, and 19% cottonseed flour plus 19% soyabean flour plus 58% maize flour with milk and egg proteins were compared in children [37]. The researchers reported biological values of the vegetable mixture which were close to those of high-quality reference proteins, milk and egg. The biological values registered were: milk (69%), egg (64%), soyabean plus maize (63%), cottonseed plus soyabean plus maize (53%) and cottonseed plus maize (50%). For normal maize alone, the biological value was 31%.

3.3.1.3. AMINO ACID SUPPLEMENTATION OF MAIZE-BASED DIETS

The addition of synthetic amino acids such as l-lysine and l-tryptophan either singly or in combination with maize-based diets to improve their protein quality has been demonstrated in feeding studies involving rats [38], pigs [39, 40], humans [34, 41] and poultry [42]. In all these trials, these amino acids have been found to improve protein quality of normal maize-based diets. **Table 9** shows substantial improvement in protein quality of normal maize-based diets supplemented with amino acids.

TABLE 9. Biological confirmation of essential amino acid deficiencies in normal maize [43].

Amino acid added (g/kg)	Protein efficiency ratio of normal maize
None	1.21
Lysine (3), (1)	1.51
Tryptophan (0.5)	1.18
Lysine, tryptophan (3, 0.5); (1, 0.5)	2.66
Lysine, tryptophan, isoleucine (3, 0.5, 2.5)	2.58
Lysine, tryptophan, threonine (3, 0.5, 2)	2.56

3.4. Vitamin Content of Maize Grain

Vitamins in maize grain are concentrated mainly in the aleurone layer and the germ [13]. Analysis of the vitamin content of maize (Table 10) indicates that the grain furnishes significant quantities of riboflavin, panthothenic acid, choline and pyridoxine which are sufficient to satisfy the requirements of most livestock [44]. However, the most significant feature of the vitamin pattern in maize is the low niacin content. Besides, much of niacin that occurs in the grain is in a bound form (niacytin), which is not available to monogastric animals [13].

Furthermore, the high level of the essential amino acid, leucine, in the maize grain increases niacin requirement in humans [45]. Thus, people who live only on a diet of maize suffer from the disease pellagra, associated with niacin deficiency [6]. Nevertheless, niacin shortage alone would not cause pellagra if normal maize were rich in tryptophan [46] or heat-treated with alkali [30]. One approach for improving niacin intake in maize-based diet is complementation with either legumes or animal products [13].

TABLE 10. Vitamin content of normal maize [16].

Vitamin	Concentration (g/kg)
Carotene	4.6
Vitamin E	0.46
Thiamine (B_1)	4.83
Riboflavin (B_2)	1.61
Nicotinic acid	25.29
Pantothenic acid	6.44
Pyridoxine (B_6)	8.74
Choline	655.17

Yellow maize shows vitamin A activity, whereas white maize does not [47]. The vitamin A potency of yellow maize results primarily from the presence of carotenes in the grain. The carotene content of yellow maize is 0.46 mg/100 g of grain [16].

The occurrence of vitamins mainly in the aleurone layer and the germ implies that food preparations that do not retain these parts of the grain further decrease vitamins in the diet.

3.5. Mineral Content of Maize Grain

The inorganic or mineral component (ash) of maize grain constitutes less than 2% [15]. Of this, about 75% is found in the germ. The grain is most abundant in phosphorus and potassium, but deficient in calcium and trace minerals except iron (Table 11). Much of the phosphorus, however, is present in the form of phytic phosphorus which is not digested by monogastric animals [48]. The little calcium that is normally present also has low bioavailability [13] because it forms complexes with phytic phosphorus.

3.6. Moisture Content of Maize Grain

A moisture content of 10–14% is typical of properly ripened and dried maize grain [9]. The grain, therefore, furnishes a very high amount of dry matter. A moisture content of the grain higher than this may enhance the growth of moulds and cause the grain to rot in storage. Some of these moulds produce toxic metabolites like aflatoxins which can cause disease in humans and animals consuming the grain [9].

TABLE 11. Mineral content of normal maize grain [15].

Mineral	Concentration (mg/100 g)
Calcium	6.0
Phosphorus	300.0
Magnesium	160.0
Sodium	50.0
Potassium	400.0
Chlorine	70.0
Sulphur	140.0
Iron	2.5
Manganese	6.8
Copper	4.5

4. FACTORS AFFECTING THE CHEMICAL COMPOSITION OF MAIZE GRAIN

The factors that generally influence the chemical composition of maize grain are either genetic or environmental.

4.1. Genetic Factors

Varieties of maize have been developed through breeding that contain up to 21% crude protein [49, 50–52]. However, maize varieties having more than 12% crude protein do have somewhat lower yields [51, 53], thus not suitable for commercial production.

New varieties of maize collectively called "high-lysine maize" contain nearly double the percentages of lysine and tryptophan of normal maize, even though the two types of maize are similar in overall protein content [6].

Maize varieties also differ in niacin content. For instance, inbred lines of dent maize have niacin content ranging from 13.9 to 53.3 μg/g of grain [44], whereas hybrids tend to be intermediate between their parent lines in concentration of this vitamin. Maize grain with a sugary endosperm had niacin content higher than that of waxy maize grain, which in turn had more niacin than dent grain [54].

The oil content of maize grain is largely characteristic of the particular variety. Breeding for high oil maize gave rise to strains of maize containing up to 20% oil in the grain [24, 55]. The existence of low oil maize varieties with average oil content of 1% [56] has been reported.

4.2. Environmental Factors

Soil nitrogen appears to be the critical environmental factor that affects the protein content of maize [44]. Excess soil nitrogen beyond that required for maximum growth of the plant may increase the protein content of the grain [44]. It is reported that heavy application of nitrogen and phosphate fertilisers increased the thiamine content but decreased the level of niacin in maize grain [49].

Production year (rainfall and temperature) and location are often responsible for variation in the protein content of maize grains of the same variety of maize [52, 57, 58].

5. MAIZE GRAIN AS ANIMAL FEED

As livestock feed, it is the grain that is most important. The stalks, leaves and immature ears are used as forage for ruminants [44]. Maize grain is recognised as giving the highest conversion of dry matter into meat, milk and eggs in relation to other cereal grains [13]. It is used extensively as the main source of calories in the feeding of poultry, pigs and cattle [6].

Maize grain has a digestible energy content of 3.75–4.17 kcal/g [18]. For chicken and pigs, the metabolisable energy values recorded when maize was fed were 3.6 and 3.8 kcal/g, respectively [18, 59], and corresponding gross energy digestibilities were 86% in chickens and 92% in pigs [18]. Maize, therefore, is popular for feeding monogastric animals, particularly poultry. For instance, maize is the basis of the high-energy poultry rations that are recognised throughout the United States whenever "broilers" are fattened [60].

In the feeding of poultry, maize grains are either fed directly or are milled and compounded with other ingredients and thoroughly mixed. The mixture is then fed or converted into forms most desired by specific animals.

The by-products obtained from both wet-milling and dry-milling industrial processes of maize grain are potential feed ingredients for poultry [12, 61–63] as depicted by the favourable nutrient composition of these by-products (Table 12) particularly in terms of protein content. The major by-product ingredients include the germ, bran and gluten [64]. These by-products of maize are usually mixed to produce a feed ingredient called maize gluten feed [62]. Despite the nutritional potential of maize gluten feed as a feedstuff for poultry, its use has been minimal due variously to paucity of research information available [65], perceived low metabolisable energy content [12, 66] and unknown quality of the protein [12] even though the protein content is fairly high.

TABLE 12. Chemical composition (%) of feedstuffs from maize and maize by-products [67].

	Moisture	Ash	CP	CF	EE	NFE	Ca	P
Maize feed meal	10.8	1.9	10.5	2.9	5.5	68.6	0.04	0.38
Maize bran	10.0	2.1	10.0	8.8	6.6	62.5	O. 03	0.14
Maize germ meal	7.0	3.8	20.8	7.3	9.6	51.5	0.05	0.59
Maize gluten meal	8.0	2.2	43.0	3.7	2.7	40.4	0.1	0.47
Maize gluten feed	9.5	6.0	27.6	7.5	3.0	46.4	0.11	0.78
Maize oil meal	8.7	2.2	22.1	10.8	6.8	49.4	0.06	0.62

6. NUTRITIONALLY IMPROVED MAIZE GRAIN

It has been known since 1914 that the quality of maize proteins is poor because they are deficient in the essential amino acids, lysine and tryptophan [38]. These deficiencies were attributed to the high zein fraction of maize protein in the maize grain of most varieties [68, 69]. Results obtained from extensive studies of zein indicated that it contains very low levels of lysine and tryptophan [70, 71]. Several researchers studied the factors that affected the protein quality of maize and reported that both the variety of maize and the environment had in several cases, a significant effect on lysine content [69, 72, 73]. It has been shown that the opaque-2 gene in maize caused a genetic increase in lysine concentration of maize protein [74]. These researchers further reported that the lysine increment in opaque-2 maize was the result of a change in the distribution of endosperm protein fractions, of which the opaque-2 maize contained approximately 22% zein compared with 50% zein in normal maize. Chemical analysis of maize protein for amino acids [74–76] showed that opaque-2 maize contained 60–130% more lysine than did normal maize, plus a 12–40% reduction in leucine as well as elevated level of tryptophan. Since these findings, several other mutant genes of maize have been identified. Collectively designated "high-lysine" genes, all of them control the level of zein accumulation during endosperm development. These "high-lysine" genes include most importantly floury-2 [77]; opaque-7 [78]; opaque-6 and floury-3 [79]. Of these genes, opaque-2 has proven superior in zein reduction [80, 81].

The development of these nutritionally improved maize varieties is of particular significance to those who rely on maize as basic food and animal feed, and can thereby improve such diets nutritionally at no added cost.

6.1. Shortcomings Of Opaque-2 Gene As Tool For Improving Protein Quality In Maize

Although the opaque-2 gene favourably alters the amino acid spectrum in maize, it has several shortcomings that limit its widespread commercial use.

6.1.1. Grain Yield

In general, opaque-2 maize varieties have 10–15% lower grain yields than do normal maize varieties [82, 83]. Besides, opaque-2 maize grain is 10–15% lighter in kernel weight [82, 83]. The lower grain weight can be attributed to loose packing of starch particles in the endosperm [84].

6.1.2. Moisture Content Of Grain

Mature opaque-2 maize grains are higher in moisture content by 1.8–4.2% than their comparable normal maize grains [80, 82]. Higher moisture content of the grains requires additional drying after harvest.

6.1.3. Grain Appearance

Opaque-2 maize kernels are chalky and dull in contrast to the hard and shiny kernels of normal maize varieties [6]. The soft endosperm in opaque-2 grains coupled with the dull appearance restricts acceptance by farmers, millers and consumers.

6.1.4. Susceptibility Of Grains To Pests And Diseases

Opaque-2 maize grains have been found to be more vulnerable to attack by Sitophilus oryzea [85] than normal type grains, in terms of both infestation and loss in weight of grains. Opaque-2 varieties have also been reported to be more susceptible to Chilo zonellus [86]. Besides, opaque-2 grains are more susceptible to seed rot caused by fungi such as Cephalosporium acremonium and Fusarium moniliform [87].

7. DEVELOPMENT OF QUALITY PROTEIN MAIZE (QPM)

Extensive field trials have been carried out at the International Center for Maize and Wheat Improvement (CIMMYT) in Mexico to identify the most productive opaque-2 maize cultivars which are high in lysine and tryptophan contents as well as to change the soft endosperm in opaque-2 grain into a conventional hard vitreous type [88–90]. Through backcrossing and several cycles of recurrent selection of maize, CIMMYT's maize breeders have successfully combined the high-lysine potential of the opaque-2 gene with genetic endosperm modifiers. These new genotypes, collectively called "quality protein maize" (QPM), are becoming of major interest to seed producers, breeders, geneticists and industrialists for their large-scale production and for their potential advantages in human nutrition and animal feeding. The QPM grains are indistinguishable from normal types except by chemical analysis [6].

QPM cultivars retain the protein quality of conventional opaque-2 maize but have improved agronomic traits, notably high yields and hard grain endosperm [6, 23, 81, 90–92].

Several experimental QPM cultivars yielded grains equal to those of the latest experimental releases of normal maize [6, 81, 90–92] in several regions of the world. Mexican QPM cultivars tested at more than twenty locations around the world have shown yields fully comparable to those of normal maize [93].

QPM grains are shiny, transparent and as hard as those of normal maize [6, 23, 94]. The QPM grain now has mostly the same density as that of normal maize, 1.29 g/cm3 [6]. Grain sizes of both QPM and normal maize are similar [6]; however, some of the new QPM hybrids have a grain size greater than that of normal maize [23].

The moisture content of QPM grains at harvest is essentially identical to that of normal maize [6, 91]. On average, the QPM cultivars show higher incidence of ear rots but not disastrously so in humid regions [6]. Apart from ear-rotting organisms, other maize diseases seem to attack both QPM and normal maize with comparable severity [6].

As a result of the hard endosperm of QPM grains, the excessive incidence of broken grains and the accompanying storage damage have been eliminated. Thus, insect infestation is reduced and no worse than in normal maize [6].

7.1. Nutrient Levels In Qpm

The nutrient composition of QPM is similar to that of normal maize with the exception of lysine, tryptophan, leucine and isoleucine contents [6]. The protein content of QPM grains ranges from 7.4% to 10.5% of dry matter [23, 81, 90, 95–97], which is about the same as that of normal maize [6, 81].

The lysine and tryptophan contents of QPM grains are about twice those of normal maize. There is also reduced imbalance between isoleucine and leucine. The grains of most available QPM cultivars contain on the average 3.5–4.5% lysine of total protein in the grain [6, 81]. Also, the lysine content of the gluten of QPM grains is higher than that of normal maize grains [23].

The starch yielded by QPM grains is comparable to normal maize grains [98]. Starch contents of 56.6% grain for QPM and 55.0% grain for normal maize have been reported [23].

The fat content of QPM grains ranges from 3% to 7% [23, 94, 99]. Of the triglycerides, QPM grains contain significantly more palmitic acid and linoleic acid than normal maize grain but less of stearic, oleic and linolenic acids [23]. Yellow grain types of both QPM and normal maize contain similar levels of carotenoids [6].

8. EVALUATION OF PROTEIN QUALITY OF NUTRITIONALLY IMPROVED MAIZE

Many nutritional studies have demonstrated the benefits for monogastric animals of the high protein quality of nutritionally improved maize varieties collectively called "high-lysine maize" (of which opaque-2 maize and QPM are outstanding) as compared with normal maize.

A summary of nitrogen balance studies with weanling rats to evaluate the protein quality of normal maize, opaque-2 maize and QPM is presented in Table 13. The data indicate that opaque-2 maize and QPM show significantly higher nutritional values than normal maize. The nutritional values are similar for opaque-2 maize and QPM. The high efficiency of protein utilisation of high-lysine maize is explained on the basis of its good protein quality [95, 99–101].

TABLE 13. Nutritional evaluation in rats of normal maize, Opaque-2 maize and QPM.

Maize variety	Lysine (g/kg protein)	Biological indicator of nutritional quality			
		True digestibility (%)	Biological value (%)	Net protein utilisation (%)	Protein efficiency ratio
Normal maize[a]	27	–	62	55	1.5
Opaque-2 maize[a]	45	–	87	70	2.8
Normal maize[b]	–	–	–	–	1.6
Opaque-2 (soft)[b]	–	–	–	–	2.8
Opaque-2 (hard)[b]	–	–	–	–	2.9
Normal maize[c]	26	98	63	62	–
Opaque-2 maize	42	96	78	75	–
QPM[c]	40	97	76	74	–
Normal maize[d]	23	–			0.7
QPM[d]	36	–	–	–	1.3

Sources: [a]Bressani et al. [100];
[b]Mertz et al. [101];
[c]Villegas et al. [95];
[d]Ahenkorah et al. [99].

Numerous feeding trials were carried out on rats to study the effect of protein quality of normal maize, opaque-2 maize and QPM on weight gain, feed intake as well as efficiency of feed utilisation. Some of the published work are summarised in Table 14. These growth studies highlighted the superior feed value of QPM over normal maize. One of the nutritional benefits of improved protein quality of maize is a higher food intake from diets made up of the high-lysine maize as compared with normal maize diets [30]. This is because a diet of good protein quality stimulates food intake and a higher diet intake raises calorie intake [36, 100]. Also, efficiency of feed utilisation by rats fed high-lysine maize is better than for normal maize indicating the superior quality of high-lysine maize. The performance of rats fed high-lysine maize was only slightly lower than that observed for their counterparts fed casein (Table 14).

TABLE 14. Comparative average weight gains, average feed intakes and feed conversion efficiencies in rats fed normal maize, high-lysine maize and casein.

Diets	Average gains (g)	Average feed intake (g)	Feed/gain	Sources
Normal maize	25[a]	248[a]	9.92[a]	[100]
Opaque-2 maize	130[b]	455[b]	3.50[b]	
Casein	132[b]	408[b]		
Normal maize	50[a]	211[a]	4.22[a]	[102]
Opaque-2 maize	122[b]	281[b]	2.20[b]	
Normal maize	15[a]	131[a]	9.4[a]	[103]
QPM	27[b]	162[b]	7.0[b]	
Normal maize	9[a]	115[a]	12.78[a]	[99]
QPM	18[b]	154[b]	8.56[b]	
Casein	28[c]	156[b]	5.57[c]	

Values in the same column with different superscripts are statistically different ($P<0.01$).

8.1. Evaluation Of Nutritionally Improved Maize In Poultry Diets

In initial studies with young chicks [104, 105], only slight differences in growth rate and feed conversion efficiency of chicks fed opaque-2 maize in place of normal maize in maize-soyabean meal diet which contained a dietary protein level of 15%. However, when a deficiency of the first limiting amino acid in this type of diet, methionine was corrected by supplementation, significantly better gains and feed conversions were observed with opaque-2 maize as compared with normal maize. Since supplementation of normal maize diets with lysine up to the level in opaque-2 maize diets resulted in equal performance, it was concluded that the beneficial effects of opaque-2 maize over normal maize were mediated solely through the higher lysine content of opaque-2 maize. Analyses of plasma amino acids revealed higher lysine levels in chicks fed opaque-2 maize as compared with chicks fed normal maize. It was reported that chicks given high-lysine maize diets at a dietary protein level of 15% out-performed their counterparts fed normal maize diets at the same dietary protein level [106]. Again, it was found that opaque-2 corn had higher nutritive value than normal corn as determined in modified chick bioassay of protein efficiency ratio [107].

Other reports that demonstrated the superior feeding value of high-lysine maize are summarised in Tables 15–17. Also, it was reported that chicken fed QPM exhibited improved growth rate by 20% as compared with those fed normal maize [92]. The data showed that there was no response to tryptophan supplementation of high-lysine maize. This means that the amount of tryptophan in high-lysine maize is sufficient to meet the requirements of chicks. On the other hand, a significant response is obtained when normal maize is supplemented with tryptophan. Thus, high-lysine maize is superior to normal maize not only for its higher lysine content but also for its higher tryptophan content. Furthermore, there is positive response to methionine supplementation of high-lysine maize as previously reported [104, 105], making the protein quality of the high-lysine maize superior to normal maize for the chick. In addition, the results from the substitution of high-lysine maize protein for soyabean or fishmeal protein indicate that high-lysine maize protein is comparable in quality to that of fishmeal or soyabean meal for the chick. Another interesting observation from these data is that lysine supplementation of high-lysine maize diets resulted in improved performance comparable to that of chicks fed control diets. This indicates that high-lysine maize alone is not adequate as a sole source of protein or lysine for chicks [6].

TABLE 15. Comparative bodyweights and protein efficiency ratios of chickens fed normal maize and high-lysine maize [101].

		Two-week-old chicks	
Source of protein	Protein level (%)	Average body weight (g)	Protein efficiency ratio
Basal protein+SBM	14	124.7^{a}	2.99^{a}
Basal protein+NM	14	77.7^{d}	2.31^{c}
Basal protein+OP (hard)	14	81.6^{c}	2.45^{b}
Basal protein+OP (soft)	14	92.2^{b}	2.81^{a}
Basal protein+Fl-2	14	81.2^{c}	2.46^{b}
BP+SBM+Lys (3 g/kg)	14	122.8^{c}	3.24^{a}
BP+NM+Lys (3.2 g/kg)	14	86.6^{c}	2.74^{c}
BP+OP (hard)+Lys (3 g/kg)	14	114.0^{a}	3.20^{a}
BP+OP (soft)+Lys (2 g/kg)	14	116.0^{a}	3.15^{ab}
BP+Fl-2+Lys (8 g/kg)	14	98.5^{b}	2.94^{b}

BP: basal protein; Fl-2: *floury-2* maize; OP: *opaque-2* maize; SBM: soyabean meal; Lys: lysine. Basal protein: mixture of animal and vegetable proteins of good balance providing 8% protein. Values with the same superscripts are not significantly different from each other (P<0.05).

TABLE 16. Comparative weight gains and feed/gain ratios of chickens fed normal maize and high-lysine maize at 3 weeks or 8 weeks of age.

Source of protein	Protein level (%)	Weight gain (g)	Feed/gain ratio	Source
Normal maize+safflower meal	20	173^{c}	2.74^{c}	[106]
Opaque-2+safflower meal	20	215^{b}	2.25^{b}	
Normal maize+soyabean meal	20	472^{a}	1.46^{a}	[106]
Normal maize+167.3 g/kg soyabean meal	15	300^{c}	1.86^{c}	
Opaque-2+I 67.3 g/kg soyabean meal	15	413^{b}	1.64^{b}	
Opaque-2+77.1 g/kg soyabean meal	15	293^{c}	1.83^{c}	
Opaque-2+214.5 g/kg soyabean meal	20	480^{a}	1.38^{a}	
Normal maize+289.6 g/kg soyabean meal	20	498^{a}	1.40^{a}	
QPM	9.7	708^{b}	4.28^{b}	[108]
Normal maize	9.7	532^{c}	6.55^{c}	
Normal maize+fishmeal	1.8	2017^{a}	2.30^{a}	
Normal maize+I 95 g/kg fishmeal	21	2149	2.60	[108]
QPM+195 g/kg fishmeal	21	2189	2.60	
QPM+175 g/kg fishmeal	20	2229	2.55	
QPM+155 g/kg fishmeal	19	2140	2.70	
QPM+135 g/kg fishmeal	18	2140	2.80	
Opaque-2+soyabean meal	22	390.2^{c}	1.74^{b}	[109, 110]
Opaque-2+soyabean meal	18	373.2^{b}	1.83^{c}	
Opaque-2+soyabean meal	14	320.5^{a}	2.09^{d}	
Normal maize+soyabean meal	22	411.5^{d}	1.64^{a}	
Normal maize+soyabean meal	18	372.8^{b}	1.79^{bc}	
Normal maize+soyabean meal	14	329.7^{a}	2.09^{d}	
NM+SBM	21	465	1.49	
OP+SBM	21	459	1.51	
NM+SBM	19	438	1.57	
OP+SBM	19	422	1.58	
NM+SBM	17	407^{a}	1.67^{a}	
OP+SBM	17	371^{b}	1.72^{b}	
NM+SBM	15	329	1.85^{a}	
OP+SBM	15	332	1.81^{b}	[104]
NM+SBM+Met	21	458	1.40	
OP+SBM+Met	21	468	1.39	
NM+SBM+Met	18	412	1.52	
OP+SBM+Met	18	428	1.48	
NM+SBM+Met	15	286^{a}	1.84^{a}	

OP+SBM+Met	15	369[b]	1.64[b]
NM+SBM	20	451	1.45
OP+SBM	20	463	1.42
NM+SBM+amino acid	20	455	1.41
NM+SBM	16	359[a]	1.76[a]
OP+SBM	16	423[b]	1.58[b]
NM+SBM+amino acid	16	412[b]	1.59[b]
NM+SBM	12	163[a]	2.57[a]
OP+SBM	12	243[b]	2.07[b]
NM+SBM+amino acid	12	255[b]	2.09[b]

NM: normal maize; SBM: Soyabean meal; OP: *Opaque*-2 maize; Met: methionine. Values with the same superscripts are not significantly different from each other (P<0.05). Not significant (P> 0.05).

TABLE 17. Comparative weight gains and feed/gain ratios for chickens fed normal maize and high-lysine maize at 3 weeks of age.

Source of protein 2	Protein level (g/kg)	Weight gain (g)	Feed/gain ratio	Source
Normal maize+soyabean meal+methionine (0.2%)	200	481[a]	1.40[a]	[106]
Normal maize+safflower meal+methionine (0.19%)	200	158[c]	2.92[c]	
Normal maize+safflower meal+Lys+methionine (0.19%)	200	483[a]	1.53[a]	
Opaque-2+safflower mal+methionine (0.44%)	200	214[b]	2.11[b]	
Opaque-2+safflower meal+Lys+methionine (0.14%)	200	487[a]	1.52[a]	
Normal maize+safflower meal+Met (0.26%)	200	145[d]	2.55[a]	[106]
Opaque-2+safflower meal+Met (0.25%)	200	211[c]	2.15[a]	
Opaque-2 safflower meal+Met (0.26%)+Lys (0.15%)	200	364[a]	1.78[b]	
Normal maize+safflower meal+Met (0.25%)+Lys (0.15%)	200	270[b]	1.92[b]	
Normal maize+safflower meal+Met (0.20%)+Lys (0.14%)	150	163[c]	2.44[a]	
Opaque-2+safflower meal+met (0.13%)+Lys (0.18%)	ISO	282[b]	1.99[b]	
Opaque-2(0. 89 g/kg tryptophan)	95	129[a]	3.27[a]	[111]
Opaque-2+0.95 g/kg tryptophan	95	134[a]	3.09[a]	
Opaque-2+1.31 g/kg tryptophan	95	138[a]	2.99[a]	
Opaque-2+2.00 g/kg tryptophan	95	131[a]	3.09[a]	
Normal maize (0.59 g/kg tryptophan)	95	83[d]	4.24c	
Normal maize+0.95 g/kg tryptophan	95	115[bc]	3.54[a]	
Normal maize+1.31 g/kg tryptophan	95	127[a]	3.21[a]	
Normal maize+2.00 g/kg tryptophan	95	103[cd]	3.82[b]	
Opaque-2+sesame meal (0.97 g/kg tryptophan)	120	180*	2.43*	[111]
Opaque-2+sesame meal+1.20 g/kg tryptophan	120	162	2.62	

Opaque-2+sesame meal+1.50 g/kg tryptophan	120	181	2.46	
Opaque-2+sesame meal+2.00 g/kg tryptophan	120	183	2.38	
Normal maize+sesame meal+(0.69 g/kg tryptophan)	120	146	2.84	
Normal maize+sesame meal+120 g/kg tryptophan	120	169	2.59	
Normal maize+sesame meal+1.50 g/kg tryptophan	120	166	2.63	
Normal maize+sesame meal+2.00 g/kg tryptophan	120	163	2.62	
Normal maize replacement with QPM	~220			[112]
Diet 1 (100:0)		634[b]	1.45[a]	
Diet 2 (75:25)		641[b]	1.38[b]	
Diet 3 (50:50)		675[a]	1.34[c]	
Diet 4 (25:75)		686[a]	1.35[c]	
Diet 5 (0:00)		673[a]	1.35[c]	
Diet 6 (100:lys)		676[a]	1.35[c]	

NM: normal maize; SBM: soyabean meal; OP: *opaque*-2 maize; Met: methionine. Values with the same superscripts are not significantly different from each other ($P<0.05$). *Not significant ($P>0.05$).

In summary, the superior performance of chickens fed high-lysine maize has been attributed by all investigators to its higher lysine content and improved balance of essential amino acids.

Table 18 shows that when laying chickens were fed high-lysine maize at suboptimal protein levels, their egg production was significantly higher than that of their counterparts fed normal maize at the same suboptimal protein level. However, at optimal dietary protein levels, similar egg production was recorded. It was suggested that the difference in egg production at suboptimal dietary protein level was due to the higher lysine content of high-lysine maize [109, 110]. An advantage can thus be taken of the high lysine content of high-lysine maize in reducing dietary protein levels, that is, decreasing inclusion level of high protein feedstuffs like soyabean and fishmeal. A study in Ghana [113] found no significant difference in egg production when dietary protein level was reduced from 170 to 140 g/kg in diets in which QPM was incorporated (**Table 18**). Conflicting results have been obtained as to the effect of high-lysine maize on egg weight. Some studies [52, 109, 110, 113] found no significant difference in egg weight of hens fed either high-lysine or normal maize contrary to earlier results reported [106] that hens fed high-lysine maize produced heavier eggs than those fed normal maize. With respect to internal quality of eggs produced, no beneficial effect was reported when high-lysine maize was fed to hens instead of normal maize [52, 109, 110, 113]. Data presented in **Table 18** indicate that at suboptimal dietary protein levels, hens fed high-lysine maize diets utilised feed more efficiently than those fed normal maize diets even though the mean egg weights were similar.

TABLE 18. Comparative performance of laying chickens fed normal maize or high-lysine maize.

Treatment	Protein level (g/kg)	Hen-day egg production (%)	Feed conversion (kg feed/kg eggs)	Egg weight (g)	Haugh units	Specific gravity	Source
QPM+SBM+Met	110	76.1^{b}	1.48^{ab}	53.8	86.0^{c}	1.0826	[110]
QPM+SBM	140	80.6^{bc}	1.43^{b}	54.1	86.0^{c}	1.0806	
QPM+SBM	170	84.2^{c}	1.40^{b}	55.5	86.7^{c}	1.0815	
MN+SBM+Met	110	64.5^{a}	1.60^{a}	53.0	78.9^{a}	1.0838	
NM+SBM	140	79.3^{bc}	1.49^{ab}	54.2	83.9^{b}	1.0829	
NM+SBM	170	82.7^{bc}	1.45^{b}	54.8	79.8^{a}	1.0838	
NM+SBM (90 g/kg)+Met	115.0	62.0^{a}	1.91^{a}	60.3	81.8	1.0760	[110]
QPM+SBM (90 g/kg)+Met	115.0	71.1^{c}	1.65^{c}	60.8	81.8	1.0748	
QPM+SBM (674 g/kg)+Met	107.6	71.1^{c}	1.70^{b}	62.0	82.3	1.0753	
QPM+SBM (45 g/kg)+Met	115.0	68.1 b	1.75^{b}	59.9	83.6	1.0745	
NM+SBM	144	69.1	2.56^{a}	64.1	–	–	[52]
HPC+SBM	158	70.7	2.35^{b}	64.7	–	–	
NM+fishmeal (120 g/kg)	170	57.9	3.28	55.8	–	–	[113]
QPM+fishmeal (120 g/kg)	170	60.2	3.28	56.2	–	–	
QPM+fishmeal (100 g/kg)	160	55.6	3.59	56.4	–	–	
QPM+fishmeal (85 g/kg)	150	63.5	3.29	55.7	–	–	
QPM+fishmeal (65 g/kg)	140	60.2	3.36	55.5	–	–	
Normal maize 1	150	77.4^{a}	2.06	58.9	87.1	–	[114]
QPM1	150	77.3^{a}	2	58.5	89.6	–	
QPM2	140	72.0^{ac}	2	58	89	–	
QPM3	140	64.3^{b}	1.99	56.5	93.2	–	
Normal maize 2	140	66.8^{b}	2.01	57.7	91.9	–	
Normal maize +0.07%Lysine	150	89.6^{b}	2.15	59.2	97.1	–	[115]
QPM	149.4	91.0^{a}	2.16	59.1	97.4	–	

Values with the same superscripts are not significantly different from each other ($P<0.05$).

9. FUTURE CHALLENGES OF MAIZE UTILISATION IN POULTRY DIETS

The major future challenges confronting maize utilisation in poultry production include the following:

- Adverse effects of climate change on maize production have been reported in tropical and subtropical regions. These include frequent droughts, heat, increased temperature and inadequate rainfall during the growing season and water-logging. It has been estimated that one quarter of the global maize areas is affected by drought in any given year [116].

- Competition between humans and animal agriculture. Maize is increasingly being used for human food and other industrial purposes including biofuel production. Thus, in a world where the global population is continually increasing, the argument that producing feed for livestock conflicts with feeding hungry people is likely to continue for some years [117].
- Some challenges for widespread adoption of QPM in developing countries have been described [118], which include lack of profitable markets for commercial producers and lack of government incentive to encourage adoption by subsidising the price of QPM seed.

10. CONCLUSION

Normal maize is the most widely used single grain in poultry feeding due to a combination of desirable nutritional characteristics such as high energy, low fibre and easy digestibility. However, its low protein content and deficiencies of the protein in lysine and tryptophan have been a major nutritional concern for feeding poultry. Therefore, improvement in the protein quality of normal maize through the development of a new class of opaque-2 maize known as "quality protein maize (QPM)" has been a major boost for poultry production, particularly in developing countries, where dietary supplemental protein is either expensive or imported. The benefits of feeding QPM grain to poultry are greater weight gain and more efficient feed conversion as well as less supplemental protein cost.

There are, however, pertinent challenges confronting maize production such as adverse impact of climate change, stiff competition between humans and animal agriculture, and some challenges of widespread adoption of QPM. There is a need to overcome these and other challenges in order to increase cheap meat production to meet the needs of the growing global population.

REFERENCES

1. OECD/FAO, Organisation for Economic Co-operation and Development/Food and Agriculture Organisation. 2014. OECD-FAO Agricultural Outlook 2014–2023. OECD/FAO Agricultural Outlook 2014, OECD Publishing, Paris.
2. Larbier M, Leclercq B. Nutrition and Feeding of Poultry (J. Wiseman, Trans. & Ed.). Loughborough: Nottingham University Press; 1994. 305 p.

3. Summers JD. Factors influencing food intake in practice—Broilers. In: 7th Nutrition Conference for Feed Manufactures (H. Swan and D. Lewis, Eds.); London. 1974. p. 127–140.
4. Black JL. Variation in nutritional value of cereal grains across livestock species. Proceedings of Australian Poultry Science Symposium. 2001; 13: 22–29.
5. Stamen B. Feed preference index on cereal grains for poultry. 2010. Available from: http://hdl.handle.net/1811/45500 [Accessed 2016-06-14].
6. NRC, National Research Council. Quality Protein Maize. Washington: National Academy Press; 1988. 83 p.
7. FAO, Food and Agriculture Organisation. Statistical Yearbook of FAO. Rome: FAO Publications; 2015.
8. USDA, United States Department of Agriculture. World Agriculture Supply and Demand Estimates. Washington: Economic Research Services; 2016.
9. Potter NN, Hotchkiss IH. Food Science, 5th edn. New York: Chapman and Hall; 1995. p. 381–394.
10. FAO, Food and Agriculture Organisation. Statistical Yearbook of FAO. Rome: FAO Publications; 2005.
11. Watson SA. Corn marketing, processing and utilisation. In: Sprague GF, Dudley JW, editors. Corn and corn improvement. Agronomy Journal. 1988; 18: 881–940.
12. Castanon F, Han Y, Parsons CM. Protein quality and metabolisable energy of corn gluten feed. Poultry Science. 1990; 69: 1165–1173.
13. Kling IG. Quality and Nutrition of Maize; IITA Research Guide 33. Ibadan: International Institute of Tropical Agriculture. 1991; p. 9–21.
14. Watson SA, Ramstad PR. Corn Chemistry and Technology. St Paul: American Association of Cereal Chemists. 1987; p.605.
15. IITA, International Institute of Tropical Agriculture. Maize crop. Maize Production Manual. 1982; 1(8), 1–2.
16. MacDonald P, Edwards RA, Greenhalgh J F. Animal Nutrition, 4th edn. Essex: Longman Group Ltd; 1988. 479 p.
17. Lawrence TL. Feeding value of cereals and concentrates. In: Orskov ER, editor. World Animal Science. Amsterdam: Elsevier Science Publishers BV; 1988. p. 129–146.
18. Fetuga BL, Babatunde GM, Oyenuga VA. Comparison of the energy values of some feed ingredients for the chick, rat and pig. Journal of Agricultural Science. 1979; 17: 3–11.
19. MAFF, Ministry of Agriculture, Fisheries and Food. Tables of feed composition and energy allowances for ruminants. Middlesex: Pinner; 1975.
20. Eggum BO. The nutritive quality of cereals. Cereal Research. 1977; 5: 153–157.
21. FAO, Food and Agriculture Organisation. Improvement of nutritional quality of food crops. FAO Plant Production and Protection Paper 34. Rome: FAO; 1981.
22. Shurpalekar KS, Sundaravalli OE, Rao MN. *In vitro* and *in vivo* digestibility of legume carbohydrates. Nutrition Report International. 1979; 19: 111–117.
23. Martinez BF, Sevilla PE, Bjarnason M. Wet milling comparison of quality protein maize and normal maize. Journal of the Science of Food and Agriculture. 1996; 71: 156–162.
24. Morrison WR. Cereal lipids. Proceedings of Nutrition Society. 1977; 36: 143–148.
25. West CE, Pepping F, Temalilwa CR. The Composition of Foods Commonly Eaten in East Africa. Wageningen: Centre for Tropical Agriculture (CTA); 1988. p. 84.
26. Card LE, Nesheim MC. Poultry Production, 11th edn. Philadelphia: Lea Fabiger; 1972. p. 210–243.
27. Blair JC, Harper CD, McNab JM, Mitchell GG, Scougall RK. Analytical Data of Poultry Feedstuffs. 1: Genral Amino Acid Analysis, 1977–1980. Occasional Publication, No.1. Edinburgh: ARC-Poultry Research Centre; 1981.
28. Johnson VA, Lay CL. Genetic improvement of plant proteins. Journal of Agriculture and Food Chemistry. 1974; 22: 558–566.

29. Kies C, Fox HM. Interrelationships of leucine with lysine, tryptophan, niacin as they influence protein value of cereal grains for humans. Cereal Chemistry. 1972; 49: 223–231.
30. Bressani R. Protein quality of high-lysine maize for humans. Cereal Food World. 1991; 36: 806–811.
31. Bressani R. Improving maize diets with amino acid and protein supplements. In: High-quality protein maize. CMMYT-Purdue International Symposium on Protein Quality Maize. Hutchinson and Ross Inc., Stroudsburg. Pennsylvania: CMMYT; 1975. p. 38–57.
32. Bressani R, Marenco E. The enrichment of lime-treated corn flour with proteins, lysine, tryptophan and vitamins. Journal of Agriculture and Food Chemistry. 1963; 6: 517–522.
33. Bressani R, Elias LG. All vegetable mixtures for human feeding. The development of INCAP vegetable mixture based on soyabean flour. Journal of Food Science. 1967; 31: 626–631.
34. Bressani R. The importance of maize for human nutrition in Latin America and other countries. In: Bressani R, Braham JE, Behar M, editors. Nutritional improvement of maize. Guatamala: INCAP; 1972. p. 5–29.
35. Bressani R, Elias EG. Processed vegetable protein mixtures for human consumption in developing countries. Advanced Food Research. 1968; 16: 1–13.
36. Bressani R. Amino acid supplementation of cereal grain flours tested in children. In: Scrimshaw NS, Altschul AM, editors. Amino Acid Fortification of Protein Foods. Cambridge: MIT Press; 1971. p. 184–204.
37. Bressani R, Elias LG, Viteri F. Complementation effects in rats fed combinations of soyabean flour and whole maize flour at a constant level of dietary protein. Journal of Food Science. 1974; 39: 577–580.
38. Osborne TB, Mendel LB. Amino acids in nutrition and growth. Journal of Biology and Chemistry. 1914; 17: 325–349.
39. Gallo JT, Maner JH, Jimenez I. Nutritive value of opaque-2 corn for growing pigs. Journal of Animal Science. 1968; 28: 1152 (Abstr.).
40. Baker DH, Becker DE, Jensen AH, Harmon BG. Protein source and level for pregnant gilts: a comparison of corn, opaque-2 corn and corn-soyabean diets. Journal of Animal Science. 1970; 30: 364–369.
41. Viteri FB, Martinez C, Bressani R. Evaluation of the protein quality of common maize, *opaque-2* maize supplementation with amino acids and other sources of protein. In: Bressani R, Brahman JE, Behar MM, editors. Nutritional improvement of maize. Guatemala: NCAP; 1972.
42. Boomgaart J, Baker DH. The lysine requirements of the chick at five protein levels. Poultry Science. 1970; 49: 1369 (Abstr.).
43. Howe EE, Jansen GR, Gilfillan EW. Amino acid supplementation of cereal grains as related to the world food supply. American Journal of Clinical Nutrition. 1965; 16: 315–321.
44. Leonard WH, Martin IH. Cereal Crops. New York: Macmillan Co.; 1963. p. 131–360.
45. FAO, Food and Agriculture Organisation. 1992. Maize in human nutrition. FAO Report Series 25, Rome, Italy.
46. Bressani R. Protein complementation of foods. In: Karmas E, Harris RS, editors. Nutritional Evaluation of Food Processing, 3rd edn. New York: Van Nostrand Reinhold Company; 1988. p. 627–657.
47. Duffus CM, Slaughter JC. Seeds and Their Uses. New York: John Wiley and Sons Ltd; 1980. p. 95.
48. Maner IH. Quality Protein Maize in Swine Feeding and Nutrition. Arlington: International Agriculture Development Services; 1983.
49. Earley EB, De Turk EE. Corn protein and soil fertility. In: What's New in the Production, Storage and Utilisation of Hybrid Seed Corn. Chicago: America Seed Traders Association; Hybrid Corn Division Annual Report. 1948; 3: 84–95.

50. Pollmer WG, Eberhard D, Klein D, Dhillon BS. Studies on maize hybrids involving inbred lines with varying protein content. Pflanzenzuchtag. 1978; 80: 142–148.
51. Boyat M, Deriux M, Kaan F, Rautou S. Maize breeding for improvement of kernel protein content using Illinois high protein population. In: Pollmer WG, editor. Improvement of Quality Traits of Maize for Grain and Silage. Pflanzenzuchtag. 1980; 80: 173–183.
52. Bond PL, Sullivan TW, Douglas JH, Robeson LG, Baier JG. Composition and nutritive value of experimental high-protein corn in the diets of broiler and laying hens. Poultry Science. 1991; 70: 1578–1574.
53. Sprague GF. Corn and Corn Improvement. New York: Academic Press Inc.; 1955. p. 613–635.
54. Leng ER, Curtis JJ, Shekleton MC. Niacin content of waxy, sugary and dent F segregating kernels in corn. Science. 1950; 111: 665–666.
55. Dudley JW, Lambert RJ. Ninety generations of selection for oil and protein in maize. Maydica. 1992; 37: 81–87.
56. Woodworth CM, Leng ER, Jugenheimer RW. Fifty generations of selection for protein and oil in corn. Agronomy Journal. 1952; 44: 60–64.
57. Hancock 1D, Peo Jr. ER, Lewis AI, Kniep KR, Mason SC. Effects of irrigation and nitrogen utilisation of normal and high lysine corn on protein utilisation by the growing rat. Nutrition Report International. 1988; 38: 413–422.
58. Douglas JH, Sullivan TW, Bond PL, Struwe FJ. Nutrient composition and metabolisable energy value of selected grain sorghum varieties and yellow corn in broiler diets. Poultry Science. 1990; 69: 1147–1155.
59. Nelson TS, May MA, Miles Jr. RD. Digestible and Metabolisable energy content of feed ingredient for rats. Journal of Animal Science. 1974; 38: 555–558.
60. Scott HM, Matterson LD, Singsen EP. Nutritional factors influencing growth and efficiency of feed utilisation. 1. The effect of the source of carbohydrate. Poultry Science. 1947; 26: 554–559.
61. Hillman RI, Kienholz EW, Shroder CD. The effect of corn gluten feed in chicken and turkey breeder diets. Poultry Science. 1973; 52: 2309 (Abstr.).
62. Jones RW. Corn co-products as feed ingredients for swine: effects on growth, carcass composition, fibre and amino acid digestibility [PhD thesis]. Urbana-Champaign: University of Illinois; 1987.
63. Owings WI, Sell IL, Ferket P, Hasiak R1. Growth performance and carcass composition of turkey hens fed corn gluten feed. Poultry Science. 1988; 67: 585–589.
64. Corn Refiners Association. Corn Wet-milled Feed Products, 2nd edn. Washington: Corn Refiners Association; 1982.
65. Castanon F. Nutritional evaluation of corn gluten feed by poultry [MSc thesis]. Urbana-Champaign: University of Illinois; 1988.
66. Sibbald 1R. The T.M.E. System of Feed Evaluation: Methodology, Feed Composition Data and Bibliography. Ottawa: Animal Research Centre Contribution 85–19; 1986.
67. Ellis NR, Kauffman WR, Miller CO. Composition of principal feedstuffs used for livestock. In: Food for Life. Washington: USDA Yearbook of Agriculture; 1939. p. 1065–1074.
68. Forster JF, Yang JT, Yui NH. Extraction and electrophoretic analysis of the proteins of corn. Cereal Chemistry. 1950; 27: 477.
69. Bressani R, Mertz ET. Studies on corn proteins. IV. Protein and amino acid content of different corn varieties. Cereal Chemistry. 1958; 35: 227.
70. Rathman DR. Zein, An Annotated Bibliography (1891–1953). Pittsburg: Mellon Institute; 1954.
71. Mosse J. Alcohol-soluble proteins of cereal grains. Feedstuff Proceedings. 1966; 25: 1663–1669.
72. Bressani R, Valiente AT, Tejada C. All vegetable protein mixtures for human feeding. VI. The value of combinations of lime-treated corn and cooked black beans. Journal of Food Science. 1962; 27: 394.

73. Tello F, Alvarez-Tostado MA, Alvarado G. A study on the improvement of essential amino acid balance of corn proteins. 1. Correlation between racial and varietal characteristics and lysine levels of corn. Cereal Chemistry. 1965; 42: 368–372.
74. Mertz ET, Bates LS, Nelson OE. Mutant gene that changes protein composition and increases lysine content of maize endosperm. Science. 1964; 145: 279–280.
75. Bates LS. Amino acid analysis. In: Proceedings of High-Lysine Corn Conference. Washington: Corn Refiners Association; 1966. p. 61–66.
76. Ahuja VP, Singh J, Naik MS. Amino acid imbalance of proteins of maize and sorghum. Indian Journal of Genetics. 1970; 30:727–731.
77. Nelson OE, Mertz ET, Bates LS. Second mutant gene affecting the amino acid pattern of maize endosperm protein. Science. 1965; 150: 1469–1470.
78. Misra PS, Jambunatha R, Mertz ET, Glover DV, Barosa HM, McWhirter KS. Endosperm synthesis in maize mutants with increased lysine content. Science. 1972; 176: 1425–1427.
79. Ma Y, Nelson OE. Amino acid composition and storage proteins in two new high-lysine mutants in maize. Cereal Chemistry. 1975; 52: 412–418.
80. Paez AV, Helm IL, Zuber MS. Kernel opacity development and moisture content within maize ears segregating for *opaque-2* and floury-2. Agronomy Journal. 1969; 61: 443–445.
81. Zarkadas CG, Yu Z, Hamilton RI, Pattison PL, Rose NGW. Comparison between the protein quality of northern adapted cultivars of common maize and quality protein maize. Journal of Agricultural and Food Chemistry. 1995; 43: 84–93.
82. Lambert RI, Alexander DE, Dudley IW. Relative performance of normal and modified protein (*opaque-2*) maize hybrids. Crop Science. 1969; 9: 242–243.
83. Sreeramulu C, Bauman LF. Yield components and protein quality of *opaque-2* and normal diallels of maize. Crop Science. 1970; 10: 262–265.
84. Dilmer RJ, Report on kernel structure and wet milling of high lysine corn. In: Proceedings of High-Lysine Corn Conference. Washington: Corn Refiners Association; 1966. p. 121–127.
85. Gupta SC, Asnani VL Khare BP. Effects of *opaque-2* gene in maize on the extent of infestation by *Sitophilus oryzae.* Journal of Stored Products Research. 1970; 6: 191–194.
86. Chartterjee SM, Panwar VPS, Siddiqui KH, Young WR, Marwala KK. Field screening of some promising germplasms against Chilo zonellus Swinhoe under artificial infestation. Indian Journal of Entomology. 1970; 32: 167–170.
87. Asnani VL, Gupta SC. Effect of incorporation of opaque-2 gene on yield and yield components in four composites of maize. Indian Journal of Genetics. 1970; 30:377–382.
88. Vasal SK, Villegas E, Bjarnason M, Gelaw B, Goertz P. Genetic modifiers and breeding strategies in developing hard-endosperm *opaque-2*materials. In: Pollmer WG, Phipps RH, editors. Improvement of Quality Traits of Maize for Grain and Silage Use. London: Martinus Nijhoff; 1980. p. 37–73.
89. Ortega EI, Bates LS. Biochemical and Agronomic studies of two modified hard-endosperm *opaque-Z* maize populations. Cereal Chemistry. 1983; 60:107–111.
90. Bjarnason M, Vasal SK. Breeding of quality protein maize (QPM). In: Janick J, editor. Plant Breeding Review 9. New York: Wiley; 1992. p. 181–216.
91. Twumasi-Afriyie S, Badu-Apraku B, Sallah PYK. 1992. The potential of maize as a source of quality protein for Ghana. In: Proceeding of the 12th National Maize and Legume workshop. 25–27 March 1992; Kumasi.
92. Liu ZX, Jie SF, Guo XF, Xu IF, Wang LM. Research on selected high quality protein maize hybrids. Maize Genetics Newsletter (March, 1993). 1993; 67:50–52.
93. CIMMYT, Centro International de Mejoramiento de Maiz Y Trigo. Boosting protein quality in maize. CIMMYT Research Highlights. Mexico City: CIMMYT; 1985.
94. Knabe DA, Sullivan IS, Burgoon KG. QPM as a swine feed. In: Mertz ET, editor. Quality Protein Maize. Washington: The American Association of Cereal Chemists; 1992. p. 225–239.

95. Villegas EM, Eggum BO, Kohli MM, Vasal SK. Progress in nutritional improvements of maize and triticale. Food Nutrition Bulletin. 1980; 2:17–24.

96. Ortega E1, Villegas E, Bjarnason M, Short K. Changes in dry matter and protein fractions during kernel development of quality protein maize. Cereal Chemistry. 1991; 68:482–486.

97. Kniep KR, Mason SC. Lysine and protein content of normal and *opaque-2* maize grain as influenced by irrigation and nitrogen. Crop Science. 1991; 31:177–18l.

98. Gomez, M.H., S.O. Serna-Saldivar, I.Y. Corujo, A.I. Bockholt and L.W. Rooney. Wet milling properties of quality protein maize and regular corns. In: Mertz ET, editor. Quality Protein Maize. Washington: The American Association of Cereal Chemists; 1992. p. 239–260.

99. Ahenkorah K, Twumasi-Afriyie S, Nagai D, Haag W, Asirifi-Yeboah K, Dzah BD. QPM (Obatanpa) as food source in Ghana: nutritional composition, rat growth and protein quality trials. In: Proceeding of the Ghana Animal Science Association Symposium. 1994; 22:25–30.

100. Bressani R, Elias LG, Gomez-Brenes RA. Protein quality of opaque-2 corn evaluation in rats. Journal of Nutrition. 1969; 97:173–180.

101. Mertz ET, Veron OA, Bates LS, Nelson O.E. Growth of rats fed on *opaque-2* maize. Science. 1965; 148:1741–1742.

102. Rivera PH, Peo Jr. ER, Flowerday D, Crenshaw TD, Moger BD, Cunningham PR. Effect of maturity and drying temperature on nutritional quality and amino acid availability of normal and *opaque-2* corn for rats and swine. Journal of Animal Science. 1978; 46:1024–1085.

103. Bosque CM de, Bressani R. Effect of protein quality of corn on liver retinol reserves, food intake, weight gain and feed efficiency in rats. In: Quality Protein Maize. Washington: National Academy Press; 1987. p. 49.

104. Cromwell GL, Rogler JC, Featherston WR, Pickett RA. Nutritional value of opaque-2 corn for the broiler chick. Poultry Science. 1967; 46:704–712.

105. Cromwell GL, Rogler JC, Featherston WR, Pickett RA. A comparison of the nutritive value of opaque-2, floury-2 and normal corn for the chick. Poultry Science. 1968; 47:840–847.

106. Fonseca JB, Featherston WR, Rogler JC, Cline TR. A comparison of the nutritive value of opaque-2 and normal corn for the laying hen. Poultry Science. 1970; 49:532–537.

107. Fernandez R, Lucas E, Ginnis JC. Comparative nutritional value of different cereal grains as protein sources in a modified chic bioassay. Poultry Science. 1974; 53:39–46.

108. Osei SA, Atuahene CC, Donkoh A, Kwateng K, Ahenkorah K, Dzah BD, Haag W, Twumasi-Afriyie S. Further studies on the use of QPM as a feed ingredient for broiler chickens. In: Proceedings of the Ghana Animal Science Association Symposium. 1994; 22:51–55.

109. Chi MS, Speers GM. Nutritional value of high-lysine corn for broiler chick. Poultry Science. 1973; 52:1148–1157.

110. Chi MS, Speers GM. A comparison of the nutritional value of high-lysine, floury-2 corn and normal corn for the laying hen. Poultry Science. 1973; 52:1138–1147.

111. Cuca GM, Pró MA. Tryptophan and Methionine supplementation of opque-2 and normal corn for chicks. Poultry Science. 1972; 51:787–791.

112. Panda AK, Raju MVLN, Rama Rao SV, Lavanya G, Pradeep Kumar Reddy E, Shyam Sunder G. Replacement of normal maize with quality protein maize on performance, immune response and carcass characteristics of broiler chickens. Asia-Australian Journal of Animal Science. 2010; 23:1626–1631.

113. Osei SA, Donkoh A, Atuahene CC, Hagan 1A, Tuah AK. Quality protein maize in the diet of laying chickens. In: Abstracts and Students' Papers Contributed In International Youth Programme. Vo1.IV, World's Poultry Congress (2–5 September 1996); New Delhi: 1996.

114. Osei SA, Dei HK, Tuah AK. Evaluation of quality protein maize as a feed ingredient for layer pullet. Journal of Animal Feed Science. 1999; 8:181–189.

115. Zhai S. Nutritional evaluation and utilisation of quality protein maize (Zhong Dan 9409) in laying hen feed [thesis]. Shaanxi: Northwestern Agricultural and Forestry University of Science and Technology; 2002.
116. Heisey PW, Edmeades GO. Maize production in drought-stressed environments: technical options and research resource allocation. In: Part 1 of CIMMYT 1997/1998 World Facts and Trends. Mexico: CIMMYT; 1999.
117. CAST, Council for Agricultural Science and Technology. Animal Feed vs. Human Food: Challenges and Opportunities in Sustaining Animal Agriculture Toward 2050. Issue Paper 53. Ames: CAST; 2013.
118. Atlin GN, Palacios N, Babu R, Das B, Twumasi-Afriye S, Friesen DK, De Groote H, Vivek B, Pixley KV. Quality protein maize: progress and prospects. In: Janick J, editor. Plant breeding reviews. Hoboken: John Wiley and Sons; 2011. 34:83–130.

CHAPTER 2

Exploring The Economic Potential of Reducing Broiler Lameness

É. Gocsik, A. M. Silvera, H. Hansson, H. W. Saatkamp & H. J. Blokhuis

[a]*Business Economics Group, Wageningen University, Wageningen, the Netherlands;*

[b]*Department of Animal Environment and Health, Swedish University of Agricultural Sciences, Uppsala, Sweden;*

[c]*Department of Economics, Swedish University of Agricultural Sciences, Uppsala, Sweden*

ABSTRACT

1. The present study was designed first to explore the potential economic benefits of adopting management practices to reduce lameness in broiler farms, and second to explore farmers' possible perceptions of this potential in the Swedish context. The likely financial effects were addressed using a normative economic model, whereas a questionnaire-based survey was used to obtain in-depth knowledge about the perceptions of a group of broiler farmers in Sweden.

2. The three alternative practices (out of 6 tested) which realised the greatest improvements in gross margin and net return to management compared to the conventional practice were feeding whole wheat, sequential feeding and meal feeding.

3. The model showed that the negative effect of feeding whole wheat on feed conversion rate was outweighed by the effect of a low feed price and the associated decrease in feed costs. The price of wheat played a major role in the improvement of economic performance, whereas the reduction of lameness itself made a relatively minor contribution.

4. Apparently, the surveyed farmers do not recognise the potential of the positive effects of changing feed or feeding practices on both broiler welfare and farm economics although their implementation can be of great importance in the broiler sector where profit margins are very tight.

Keywords: Animal welfare; broiler production; economic performance; farmers' perception; lameness

INTRODUCTION

Most broiler chickens produced for commercial purposes worldwide have a fast growth rate and are reared in intensive, confined systems with high stocking densities (Knowles et al., 2008). Lameness in broilers is generally considered a consequence of the genetic selection for rapid growth rate (Bradshaw et al., 2002) and is often ssociated with an impaired walking ability. The latter can be measured using agait score (GS) which is expressed in 6 categories ranging from completely normal (0) to immobile (5) (Kestin et al., 1992). Across-sectional study including French, British, Dutch and Italian broiler flocks reared in traditional, indoor farmsystems suggested that the average prevalence of lameness was 15.6%, measured by the percentage of chickens with a GS ≥3 (Bassler et al., 2013). In a study using the Welfare Quality broiler assessment protocol (Welfare Quality®, 2009) in Dutch (majority of the flocks), British, Belgian and Italian flocks, an average 57% of the birds in flocks with standard, fast-growing strains had a GS ≥3, whereas this fell to 17% of birds in flocks with alternative, slower growing strains (De Jong et al., 2011). An earlier study by Sanotra et al. (2003) indicated that in Sweden the prevalence of chickens with a GS ≥3 ranged from 14.1% to 26.1%.

Lameness impairs the welfare of broiler chickens in commercial broiler production (Bradshaw et al., 2002), as lame birds may have difficulty in performing essential behaviours, such as gaining regular access to feeders and drinkers (Danbury et al., 2000; Hall, 2001; Butterworth et al., 2002; Bessei, 2006). These birds may also experience pain, though evidence on the correlations between lameness, pain and underlying pathologies is inconclusive. For example, several studies suggest that lame birds with higher GSs suffer pain when they walk (Mc Geown et al., 1999; Danbury et al., 2000; Buchwalder and Huber-Eicher, 2005; Naas et al., 2009; Caplen et al., 2014; Hothersall et al., 2016), whereas others argue that the relationship between lameness and pain (Skinner-Noble and Teeter, 2009; Siegel et al., 2011) and between lameness and underlying pathologies isweak (McNamee et al., 1998; Sandilands et al., 2011; Fernandes et al., 2012). Regardless of the variation in thefindings of these studies, lameness constitutes a welfare problem because it leads to compromised mobility and probably pain in at least some birds. Furthermore, lameness negatively affects productivity.

Leg abnormalities and lameness, indicated by poor gaits, are associated with higher morbidity and mortality in broilerflocks (Verma, 2007; Wideman et al., 2012) and the level of lameness and the final weight at delivery are negatively related (Butterworth and Haslam, 2009). This negative impact of lameness on production most likely reduces economic performance, i.e. production costs and returns (Bradshaw et al., 2002; Gocsik et al., 2014). Studies of onfarm risk factors and management practices that affect gait health (Butterworth and Haslam, 2009; Bassler et al., 2013) identified opportunities to improve welfare and reduce economic loss.

Steps to reduce lameness may require farmers to change production conditions and management practices that otherwise exacerbate the problem. In this context, the cost and benefits of adopting new practices may be of importance for farmers (Gocsik et al., 2015). Besides the extrinsic characteristics of a new practice, psychological aspects such as farmers' perception of its relative advantage play a significant role in its adoption (Rogers, 1995; Adrian et al., 2005). Hereto though, limited attention has been paid to farmers' perception of the relevance of lameness and the economic benefits of its reduction. In this regard, Butterworth and Haslam (2009) suggested that farmers perceive the impact of lameness on production as relatively small, which may possibly be attributed to difficulties in its detection and assessment.

The objective of this study was twofold. First, it aimed to explore the economic potential of management practices intended to reduce broiler lameness. Evaluating these management practices using real-world experiments is costly and disruptive. Hence, a normative economic model was developed to address the economic potential of these practices under Swedish conditions. This approach is often used to model potential economic performances of different farm systems and farming practices (Acs et al., 2005; Kerselaers et al., 2007). Model results can identify practices that merit further real-world research. Second, farmers' possible perceptions of the management changes were examined by surveying 21 Swedish broiler farmers.

MATERIALS AND METHODS

Normative economic model

Bioeconomic rationale of lameness

Figure 1 outlines the complex relationships between management and housing practices and lameness and the potential effect of lameness on management decisions. Housing and management practices such as lighting programmes, stocking density, diet and feeding regime, and growth rate can all influence the walking ability of broilers (Knowles et al., 2008; Butterworth and Haslam, 2009; Bassler et al., 2013). Walking ability of broilers is commonly assessed using a 6- point gait-scoring system with categories ranging from completely normal (0) to immobile (5) (Kestin et al., 1992; Welfare Quality®, 2009). There is evidence that welfare is poor in birds with a GS ≥3 (Kestin et al., 1992; Mc Geown et al., 1999). Birds with a GS of 3 are considered moderately lame as they have an identifiable abnormality, which impairs their function. Birds with a GS of 4 and 5 are considered severely or completely lame, respectively (Kestin et al., 1992). Hence, in the present study, prevalence of lame birds is defined as the percentage of the flock with a GS of 3, 4 or 5.

Lameness influences not only consumers' perception of bird welfare (Castellini et al., 2008) but also the technical performance of the birds, such as mortality and growth rate (Knowles et al., 2008). Consumers are heterogeneous in terms of their preferences with regard to the various aspects of animal welfare; this causes uncertainty in their willingness to pay for improvements in animal welfare and, hence, in farmers' income. Therefore, this study focused on evaluating the relationships between broiler lameness and the technical and economic performance of the broiler farm; the pivotal role of price premiums in the potential reduction of broiler lameness was not considered.

Technical performance affects economic performance of farms primarily through production costs. From an economic perspective, reducing lameness can be a complex issue. For example, a change in management practice intended to reduce lameness may reduce growth rate (Knowles et al., 2008), hence increases feed costs per kilogram delivered chicken provided that other factors are constant. On the other hand, improved welfare might reduce mortality (Bennett et al., 2002) which in turn can at least partially compensate for the increased feed costs. Moreover, the use of feed differing in composition from the regular compounded feed may reduce the price of feed and thus reduce overall feed costs (Umar Faruk et al., 2010). All in all, it can be ypothesised that extra costs resulting from the implementation of practices designed to reduce lameness could, to a certain extent, be compensated for by other associated gains. A positive evaluation of the studied management practices may encourage adjustment of housing and management practices.

The farm model

To estimate the economic impact of different production practices, a partial budget-based economic farm model was developed from earlier models (Dijkhuizen and Morris, 1997; Gocsik et al., 2014). In the latter study, the approach was used to calculate the economic impact of the most prevalent endemic diseases in Dutch broiler production systems. For present purposes, the technical and economic inputs were adjusted to fit the characteristics of Swedish broiler production. The economic impact of lameness is determined by the prevalence of lame birds (GS ≥3) and by its impact on productivity, and expressed in terms of financial performance indicators, such as production costs, gross margin (i.e. revenues minus variable costs) and net return to management per kilogram of delivered broiler (i.e. revenues minus variable costs minus fixed costs).

In the model, a fixed production period was assumed that corresponded to the time required to reach the average live weight at delivery in the case of healthy chickens. Lame chickens, however, usually grow more slowly than healthy ones (Yalcin et al., 1998), thus implying that given a fixed production period the quantity of kilogram live weight decreases because lame chickens are delivered (when still fit for transport) at a lower weight than healthy ones and because of the culling of severely lame birds. In turn, fixed costs per kilogram live weight delivered increase.

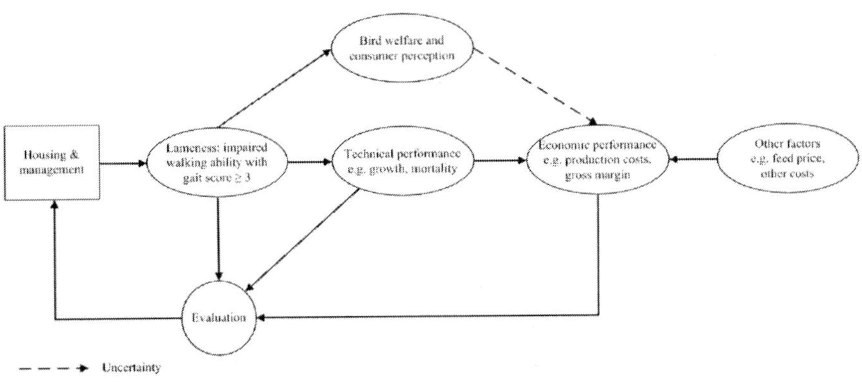

Figure 1. Bioeconomic rationale of lameness in broiler production.

The model includes 4 factors through which lameness may influence farm productivity and, hence financial performance. These factors include increased mortality, increased feed conversion, increased condemnation rate at slaughter and decreased daily weight gain.

Increased mortality due to lameness causes financial loss. Relevant factors include the cost of starting material, i.e. d-old chick, and the lost revenue due to mortality minus the cost of delivery to the slaughterhouse. Accordingly, the cost of mortality was calculated by Equation

$$\textit{Cost of mortality} = \left(\textit{price dayold chick} + \left(\frac{(\textit{producer price} \times \textit{weight at delivery}) - \textit{price dayold chick}}{2}\right) - \textit{cost of delivery}\right) \times \textit{mortality rate} \quad (1)$$

Lame chickens were assumed to have died around the middle of the production period. Fixed costs per delivered broiler and per kilogram live weight may change because fewer chickens are delivered.

Increased feed conversion rate results in higher feed cost when other variables, e.g. final weight at delivery and the price of feed, remain constant. Extra feed costs were calculated using Equation (2).

$$\textit{Cost of condemnation at slaughter} = (\textit{producer price} \times \textit{weight at delivery}) \times \textit{condemnation rate} \quad (3)$$

Like mortality, condemnation also affects the fixed costs per delivered broiler and per kilogram live weight because fewer birds are accepted (and paid for).

Broiler production in weden: conventional practice

The Swedish legislation for broiler production is stricter than the European Directive, which provides minimum rules for the protection of chickens kept for meat production in the European Union (EU) (EC Directive, 2007; Björk, 2012). Moreover, the Swedish Poultry Meat Association introduced the Swedish animal welfare programme to which 98% of broiler producers in Sweden comply (SPMA, 2015). This programme allows a maximum stocking density of 36 kg/m2 (Björk, 2012) that is in contrast to the maximum stocking density of 42 g/m2 set by the EU Directive. This maximum is, however, only allowed if the farms have an adequate ventilation and heating system and if the mortality remains below 3.40% at 40 d. Broilers are on average kept for 32–35 d, and are slaughtered at a weight of 1.6–2.3 kg. Birds are housed indoors, mainly in barns with windows/inlets to provide daylight, which is not required by the EU Directive. In accordance with the EU Directive, broilers must have at least 6 h of darkness per day of which 4 h must be uninterrupted. Ross 308 and Cobb 500 make up 60% and 40% of the broiler population, respectively (Björk, 2012). The cleaning and disinfection period is 7–14 d and wood shavings are usually used as litter. The conventional practice in the present study was defined on the basis of these parameters.

Management practices to reduce lameness

The prevalence of lame birds (percentage of flock with a GS of 3, 4 or 5) and the impact on production performance are influenced by several risk factors, such as lighting programmes, stocking density, diet, feeding regime and growth rate (Knowles et al., 2008; Butterworth and Haslam, 2009; Bassler et al., 2013). Practices targeting these risk factors may reduce the prevalence of lameness and its impact but they might also affect the performance of the healthy birds.

An epidemiological study (Bassler et al., 2013) estimated that increasing the dark period (at 3 weeks of age) from 0 to 6.5 h reduced the prevalence of lame birds from 16.9% to7.4%. Knowles et al. (2008) also found that for every 1 h increase in the daily dark period across a range of 0 to 8.5 h, there was a 0.079 improvement in flock GS. Moreover, Classen (2004) suggested that increasing the dark period from 4 to 8 h decreased final weight but it also led to an improved feed conversion.

When stocking density was decreased from 455 to 625 cm2/bird, the GS at 35 d of age was 0.195 units lower and body weight was positively affected (Sørensen et al., 2000). Hall (2001) then suggested that a stocking density of 40 kg/m2 rather than 34 kg/m2 increased daily mortality and the incidence of leg problems, and negatively affected the birds' behaviour.

A change from ad libitum to meal feeding (2, 3 or 4 times a day while maintaining equal duration of feeding at 240 min/d) reduced the prevalence of leg weakness (Su et al., 1999). Furthermore, Su et al. (1999) also reported that early feed restriction programmes had positive effect on leg weakness but this practice was less effective than meal feeding. The tested feed restriction programmes differed in starting age (5, 7 and 9 d of age), length of restriction (5 and 7 d) and levels of restriction (predetermined amounts of feed calculated to achieve 25%, 50% and 75% of the predicted growth of ad libitum birds during the restriction period).

Sequential feeding using diets varying in energy and crude protein reduced the prevalence of birds with GS ≥3 from 61.3% to 42.0% without impairing growth performance (Leterrier et al., 2008). A low energy, high protein diet and a high energy, low protein diet were rotated every 48 h from d 8 to d 28 or up to no fewer than 8 d before slaughter. The birds were then given a standard finishing diet from d 29 until the last day to compensate for reduced growth. The diet was designed for a production cycle of 38 d, which is 3–5 d longer than the average in Sweden (i.e. 32–35 d).

A fast growth rate increases the birds' susceptibility to lameness (Kestin et al., 2001; Butterworth and Weeks, 2010). Feeding whole wheat to broiler chickens as part of their diet usually slows down digestion, thereby reducing growth rate, and for every percentage increase in dietary wheat (from 0 to 30%), there was a 0.017% improvement in flock GS during the third week of life (Knowles et al., 2008).

Based on the earlier reports, 6 different management practices were selected for further analysis: (1) an increased daily period of darkness (8 h/d), (2) a decreased stocking density (27 kg/m2), (3) meal feeding (2–4 times per day), (4) restricted feeding, (5) sequential feeding and (6) feeding whole wheat (on average 30% of feed during the production period).

Technical inputs

Data on the prevalence of lame birds (percentage of flock with a GS ≥3) and impact on roduction parameters were collected from the literature. The average prevalence of lame birds in the Swedish conventional system is 14.1% according to most recent reports (Sanotra et al., 2003). Prevalence data for alternative practices are scarce. Most studies focused on the improvement in average flock GS achieved by using alternative practices rather than on explicit change in the proportions of broilers in the different gait categories. Hence, the following procedure was used to estimate the prevalence of lame birds in all alternative practices, except sequential feeding for which prevalence data were available.

(1) A continuous lognormal distribution was defined based on the parameters estimated from the discrete prevalence data with respect to the conventional practice described by Sanotra et al. (2003). Figure 2 shows a prevalence of lame birds (GS ≥3) in conventional practice of 14.5%, which roughly corresponds to the 14.1% reported by Sanotra et al. (2003). As a consequence of using a continuous rather than discrete distribution, the reference line indicating the relative frequency of birds with GS ≥3 is at 2 in Figure 2 (because a GS >2 fall in the category of GS ≥3). Note that the available data on prevalence were not suitable to estimate a discrete distribution for the different production practices as it would have required to estimate the frequency of broilers within the 6 different GS categories. Hence, the discrete distribution was transformed into a continuous one to be able to estimate the prevalence of lameness for the different production practices. That transformation resulted in the fact that Figure 2 suggests subscores, although in practice no subscores for GS are given.

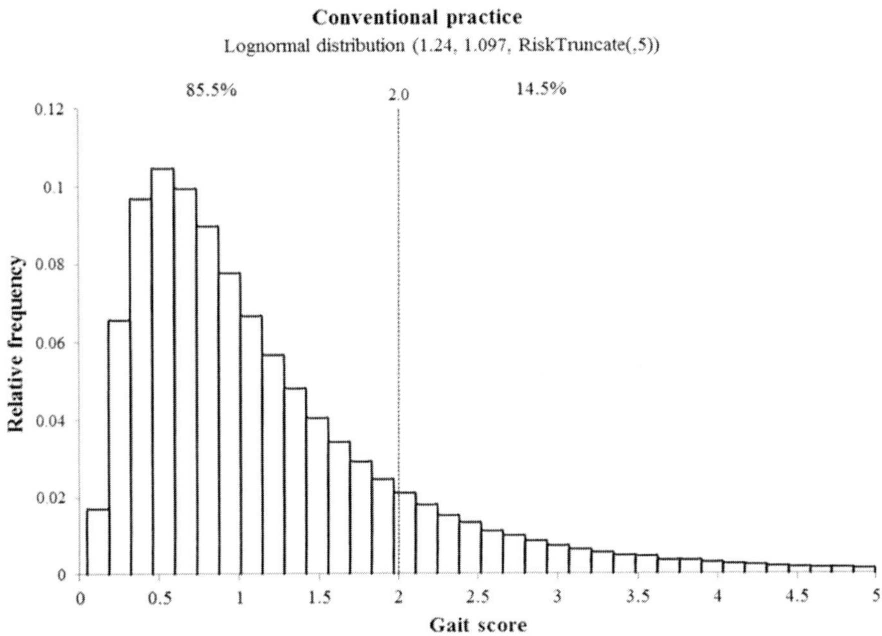

Figure 2. Relative frequency distribution of gait scores in the conventional situation (estimates based on Sanotra et al., 2003).

(2) The improvement in average flock GS in the 5 alternative practices was established from the literature. The mean of the earlier defined lognormal distribution for conventional practice was adjusted with the improvement in the given practice while the standard deviation was held constant. In turn, for each practice, a lognormal distribution was defined with the adjusted mean and standard deviation (Table 1). On the basis of the established distributions, the prevalence of lame birds was estimated for each practice.

Table 2 presents the data on production parameters in healthy and lame broilers.

Economic inputs

Table 3 lists the main economic inputs, such as prices, variable and fixed costs included in the model (Agriwise, 2014). All calculations were made for a farm with a capacity of 80 000 chickens per cycle and it was assumed that one full-time labour equivalent FTE) is available at the farm and can care for the 80 000 birds (Agriwise, 2014). Labour costs associated with 1 FTE were considered as fixed costs. Note that for most management practices no distinction was made in variable costs per chicken. In the case of feeding whole wheat, the price of feed (i.e. 3.46 SEK/kg) was lower than the price of conventional compound feed (i.e. 4.08 SEK/kg) because of the different feed composition. With regard to sequential feeding, no change in feed price was assumed compared to conventional practice.

Sensitivity analysis

Due to scarcity of data on the prevalence of lameness and its effect on production parameters in various management practices, uncertainty about input parameters influences the final model results. Sensitivity analysis was used to evaluate the robustness of the results to the changes in input parameters. Thus, input parameters, such as prevalence of lameness and the daily weight gain, feed conversion rate, mortality and condemnation rate were varied systematically one at a time. Prevalence of lameness was changed by ±10%. By default the daily weight gain of lame birds was assumed to be 10 g/d lower than that of healthy ones. In the sensitivity analysis, the decrease of 10 g/d in daily weight gain was changed by ±5%.

The feed conversion ratio in lame birds was originally assumed to be 0.03 higher than that of healthy ones; this was changed by ±0.1. Compared to healthy chickens, an increase of 0.9% in mortality was, by default, assumed due to lameness (Sullivan 1994; Hall, 2001; Verma, 2007). This increased mortality was changed by ±0.1% resulting in an increase of 0.8% and 1% compared to the healthy situation. In the case of feeding whole wheat, a lower increase in mortality was assumed than with other diets; here, although the level of change was similar at ±0.1%, the resultant increase in mortality amounted to 0.22% and 0.42%. An increase of 0.3% in condemnation due to skeletal problems was originally assumed and in the sensitivity analysis this increase in lame birds was changed by ±0.1%.

Table 1. Parameters of defined distribution to estimate prevalence of lame birds per production practices.

	Lognormal distribution			Estimated prevalence of birds with a gait score ≥3 (%)
Practice	Mean	SD	Upper truncation limit	
Conventional	1.24[a]	1.09[a]	5	14.1[a]
Increased daily period of darkness	1.04[b]	1.09[a]	5	10.7
Decreased stocking density	1.07[c]	1.09[a]	5	11.2
Meal feeding	1.09[d]	1.09[a]	5	11.5
Restricted feeding	1.14[e]	1.09[a]	5	12.5
Sequential feeding	–	–	–	9.66[f]
Feeding whole wheat	0.99[g]	1.09[a]	5	9.8

[a]Prevalence of GS ≥3 is 14.1% in Ross 208 chickens (Sanotra *et al.*, 2003).
[b]An improvement of 16% in GS is assumed compared to the conventional practice (Knowles *et al.*, 2008).
[c]An improvement of 13.6% in GS is assumed compared to the conventional practice (Sørensen *et al.*, 2000).
[d]An improvement of 12.3% in GS is assumed when changing from feeding *ad libitum* to meal feeding (Su *et al.*, 1999).
[e]An improvement of 8% in GS is assumed compared to the conventional practice (Su *et al.*, 1999).
[f]An improvement of 31.5% is assumed in the prevalence of birds with a GS ≥3 (Leterrier *et al.*, 2008).
[g]An improvement of 0.51% in GS by increasing dietary feed with 30% (Knowles *et al.*, 2008).

Table 2. Estimated impact of lameness (lame = gait score ≥3) in each of the alternative practices applied to the broiler farm.

Production system and practices	Daily weight gain (g/d)	Feed conversion rate (g feed/g gain)	Weight at delivery (g)	Mortality (% flock)	Condemnation (% flock)
Conventional production					
Healthy	49.29[a]	1.70[b]	1,725[b]	2[b]	1.5[b]
Lame (gait score ≥3)	39.29[c,d]	1.73[e]	1,375[f]	2.9[g]	1.8[g]
Practices to reduce lameness					
1. Increased daily period of darkness (8 h/d)					
Healthy	48.30[a]	1.67[h]	1,691[h]	2[b]	1.5[b]
Lame	38.30[c,d]	1.70[e]	1,341[f]	2.9[g]	1.8[g]
2. Decreased stocking density (27 kg/m^2)					
Healthy	49.88[a]	1.70[a]	1,746[i]	2[b]	1.5[b]
Lame	39.88[c,d]	1.73[e]	1,396[f]	2.9[g]	1.8[g]
3. Meal feeding (2–4 times per day)					
Healthy	46.82[a]	1.62[j]	1,639[j]	2[b]	1.5[b]
Lame	36.82[c,d]	1.65[e]	1,289[f]	2.9[g]	1.8[g]
4. Restricted feeding					
Healthy	46.82[a]	1.68[k]	1,639[k]	2[b]	1.5[b]
Lame	36.82[c,d]	1.71[e]	1,289[f]	2.9[g]	1.8[g]
5. Sequential feeding					
Healthy	49.29[l]	1.71[a,m]	1,873[b,m]	2[b]	1.5[b]
Lame	39.29[c,d]	1.74[e]	1,493[n]	2.9[g]	1.8[g]
6. Feeding whole wheat (on average 30% of feed)[o]					
Healthy	49.29[a]	1.80[o]	1,725[b,p]	2[b,p]	1.5[b]
Lame	39.29[c,d]	1.83[e]	1,375[f]	2.32[q]	1.8[g]

[a]Daily weight gain (g/d) = weight at delivery (g)/35 d.
[b]Agriwise (2014).
[c]The decrease in daily weight gain due to lameness in conventional system was assumed to be 10 g/d (Gocsik *et al.*, 2014).
[d]Yalcin *et al.* (1998).
[e]The average feed conversion efficiency for chickens with GS4 and GS5 was 0.03 worse than that in the situation without leg problems (Su *et al.*, 1999).
[f]Weight at delivery (g) = daily weight gain (g/d) × 35 d.
[g]Based on Verma (2007), Hall (2001) and Sullivan (1994) an increase of 0.9% in mortality was assumed due to lameness.
[h]Feed conversion was assumed to be improved by 1.7% applying lighting regimen of 16L:8D instead of 20L:4D Classen (2004).
[i]Sørensen *et al.* (2000).
[j]An improvement of 5% in feed conversion and a decrease of 5% in final body weight were assumed compared to the conventional situation (Su *et al.*, 1999).
[k]Feed conversion was assumed to decrease by 1% and the final body weight was assumed to decrease by 5% compared to the conventional situation (Su *et al.*, 1999).
[l]Daily weight gain = weight at delivery (g)/38 d.
[m]Feed conversion rate was assumed to be 1.71 for a production period of 38 d (Leterrier *et al.*, 2008). Final body weight was recalculated for a production period of 38 d, i.e. 49.29 g/d × 38 d = 1873.02.
[n]Weight at delivery = daily weight gain (g/d) × 38 d.
[o]The proportion of whole wheat in the diet was assumed as, on average, 30% during the production period (Bennett *et al.*, 1995, 2002; Knowles *et al.*, 2008).
[p]An increase of 6% in feed conversion rate, as an effect of feeding whole wheat, was assumed. In mortality, no significant differences due to feeding whole wheat were observed (Bennett *et al.*, 1995).
[q]A 20% lower mortality rate was estimated, i.e. 2.32%, compared to the mortality when lameness was present in conventional system (Bennett *et al.*, 2002).

When feeding whole wheat, its price is an important driver of the feed costs. To evaluate the effect of changing feed price on the economic performance of this particular practice, wheat price was changed by ±5%.

Farmers' perceptions of lameness and practices to control lameness

Table 3. Economic inputs (Agriwise, 2014).

Variable	Unit	Value
Prices		
Feed price compound (65%)	SEK/kg	4.08
Feed price own production (35%)	SEK/kg	1.54
Wheat price	SEK/kg	2.00[a]
Price of 1-d-old chick	SEK/chick	3.61
Producer price of broiler chicken	SEK/kg	9.24
Variable costs		
Electricity	SEK/chicken	0.184
Heating	SEK/chicken	0.519
Catching and loading	SEK/chicken	0.531
Litter (straw)	SEK/chicken	0.061
Cleaning	SEK/chicken	0.265
Carrion collecting service	SEK/chicken	0.136
Environmental fee	SEK/chicken	0.020
General costs and insurance	SEK/chicken	0.075
Other	SEK/chicken	0.122
Fixed costs		
Labour own	SEK/h	220
Depreciation	%	3.7
Maintenance	%	1.5
Interest working capital	%	7
Interest buildings	%	3.5
Capital		
Working capital	SEK/m^2/round	32
Buildings and equipment	SEK/m^2/round	400

[a]Applicable in case of the calculations regarding the practice of feeding whole wheat as part of the diet. Average wheat price based on year March 2013–March 2014, No. 1 Hard Red Winter, ordinary protein, FOB Gulf of Mexico.

Questionnaire

A questionnaire (administered using Qualtrics online survey software, Qualtrics Inc., Provo, UT, USA) was sent to Swedish broiler farmers during August to November 2014 to explore their perceptions of problems associated with lameness and of the management practices intended to control it. The questionnaire was pretested with a broiler specialist to ensure the questions were clear and it was edited accordingly. The resulting questionnaire consisted of three distinct parts. The first part contained questions about the farmer and the characteristics of the farm. The second part included questions regarding the farmer's perception of the relevance and importance of lameness in his/ her farm. Thus, as well as questions on the exact figures for prevalence of lameness, mortality and culling due to lameness, which are often difficult for farmers to estimate, they were also asked if they perceived lameness as a minor, moderate or major problem and how they rank lameness in the context of other health problems. Finally, the third part focused on the farmer's thoughts on particular management practices intended to control lameness. Here, they were asked to rank an array of remedial practices in terms of their effectiveness and to add any others that were missing from the predefined list but might also be effective. They were also asked if they believed that a particular management practice could be effective in controlling lameness, and whether it is being used at their farm or if it would be in the future. The farmers were also asked to give reasons for their answers. This combination of quantitative and qualitative questions enabled us to get insight into the farmer's mindset, which primarily refers to their opinions, attitudes and beliefs regarding lameness, and the underlying motivation for lameness control and management in the farm.

Sample

An invitation to participate in the study was sent out by one of the largest slaughterhouses in Sweden to about 60 associated broiler farmers (50% of the broiler farmer population) (Björk, 2012). In total, 21 farmers responded, from which 9 fully and 12 partially completed the questionnaire. Given the relatively small number of returned questionnaires, it is cautioned that our results may not be easily generalised to the whole Swedish broiler farmer population. Therefore, we simply regard the survey results as a source of information that provides a good basis for further investigation and identifies some common viewpoints among the respondents.

The demographic and socio-economic characteristics of the sample are given in Table 4. The mean age (±SD) of respondents was 52 years (±10.1). Most respondents were males who had an agriculture-oriented education. Fourteen respondents had more than 20 years of farming experience. The mean (±SD) farm size was 5231 m2 (±3325) and the stocking density in most farms was 36 kg/m2. Mean (±SD) duration of the growth period was 35 d (±1.2). The dark period averaged, 6 h/d, of which 4 h were uninterrupted.

RESULTS

Normative economic analysis

Gross margin

The financial performance of the broiler farm improved as a result of some of the practices designed to reduce lameness (Table 5). More specifically, the gross margin (return minus variable costs) of all but one alternative practice (i.e. restricted feeding) was higher than that of conventional practice. The greatest improvements in gross margin were realised when the practices of feeding whole wheat, sequential feeding or meal feeding were applied. Indeed, when whole wheat was fed the gross margin more than tripled compared to the conventional practice, this reflects the fact that whole wheat is cheaper than regular compounded feed. Although, the feed conversion rate of chickens fed whole wheat is less efficient, this effect is outweighed by a lower prevalence of lameness, a lower mortality and the lower feed costs. The latter factor contributed the most to the improved economic performance, whereas reduced lameness played a minor role. Improved gross margins found in the other practices can also be primarily attributed to lower feed cost. However, in the case of sequential feeding, the average growth period, i.e. 38 d is longer than that in conventional practice and this results in a higher live weight at delivery and thereby the lower cost of 1-d-old chicks per kilogram live weight.

Net return to management

In terms of net return to management (return minus variable costs minus fixed costs), practices such as feeding whole wheat, sequential feeding, meal feeding and longer dark period performed better than the conventional practice. These improvements reflect the decrease in variable costs in relation to the gross margin (though fixed costs were similar to those of conventional practice). In contrast, decreased stocking density and restricted feeding had lower net return to management than conventional practice. With decreased stocking density, fixed costs related to investments in buildings and equipment were higher per kilogram live weight, which is a direct consequence of keeping fewer chickens per square metre. With restricted feeding, the slower growth rate than in conventional practice resulted in a lower live weight at delivery and, hence poorer returns.

Table 4. Demographic and socio-economic characteristics of the sample.

Variable	*N*	Mean	SD
Farmer characteristics			
Age	15	52	10.1
Sex	17		
Male	*13*		
Female	*4*		
Highest level of education completed	17		
Primary school	*1*		
Secondary school: Agriculture	*2*		
Secondary school: Other	*3*		
Vocational/occupational degree: Agriculture	*3*		
Vocational/occupational degree: Other	*2*		
University/college: Agriculture	*5*		
University/college: Other	*1*		
Experience in farming	17		
Less than 5 years	*0*		
5–10 years	*1*		
11–15 years	*2*		
16–20 years	*0*		
More than 20 years	*14*		
Farm characteristics			
Farm size (m^2)	16	5231	3325
Stocking density (kg/m^2)	11	36	0
Growth period (d)	15	35	1.2
Dark period per day (h)	15	6	0.9
Uninterrupted dark period per day (h)	15	4	0.8

Sensitivity analysis

In case of the various management practices, an increase of 10% in prevalence of lameness resulted in a decrease of about 1% in net return to management per kilogram, whereas a 10% decrease was associated with a 1% increase (Table 6). A 5% higher decrease (compared to the originally estimated 10 g/d) in the daily weight gain of lame birds resulted in a lower net return in all management systems (ranging from −1.70% to −2.40%) with the greatest decrease occurring when whole wheat was fed (−2.40%). In contrast, a 5% decrease in the daily weight gain of lame birds resulted in an increase in net return to management of 1.65–2.36% in the various management systems. When the increase in feed conversion rate of lame birds compared to healthy ones was reduced or increased by 0.1, the net return to management respectively increased or decreased by 0.2–0.3% in the various systems. Changing the increase in mortality or the condemnation rate of lame birds compared to healthy ones by ±0.1% led to a change ranging from −0.1% to +0.1% in net return to management. The sequence of management practices, from highest to lowest net return, was not sensitive to change in any of the input variables. More specifically, feeding whole wheat resulted in the highest net return to management. The sequential feeding, meal feeding and longer dark period practices also had a higher net return to management than did conventional practice. Changes of ±5% in wheat price resulted in a change of ±3% in net return to management.

Table 5. Return, production costs, gross margin and net return to management for different lameness-reducing practices (values in SEK per kg delivered broiler).

	Conventional	Longer dark period	Decreased stocking density	Meal feeding	Restricted feeding	Sequential feeding	Feeding whole wheat
Return	9.24	9.24	9.24	9.24	9.24	9.24	9.24
Variable costs	9.01	8.96	8.94	8.90	9.12	8.74	8.56
1-d-old chick	2.17	2.20	2.13	2.27	2.28	1.97	2.14
Feed	5.44	5.34	5.44	5.16	5.38	5.47	5.03
Electricity	0.11	0.11	0.11	0.12	0.12	0.10	0.11
Heating	0.31	0.32	0.31	0.33	0.33	0.28	0.31
Catching and loading	0.32	0.32	0.31	0.33	0.34	0.29	0.32
Litter (straw)	0.04	0.04	0.04	0.04	0.04	0.03	0.04
Cleaning	0.16	0.16	0.16	0.17	0.17	0.14	0.16
Carrion collecting service	0.08	0.08	0.08	0.09	0.09	0.07	0.08
Environmental fee	0.01	0.01	0.01	0.01	0.01	0.01	0.01
General costs and insurance	0.05	0.05	0.04	0.05	0.05	0.04	0.04
Other	0.07	0.07	0.07	0.08	0.08	0.07	0.07
Mortality	0.11	0.11	0.11	0.11	0.11	0.11	0.11
Condemnations	0.14	0.14	0.14	0.14	0.14	0.14	0.14
Fixed costs	1.72	1.73	2.07	1.76	1.77	1.78	1.71
Labour own	0.50	0.51	0.49	0.54	0.54	0.49	0.50
Depreciation	0.49	0.49	0.63	0.49	0.49	0.52	0.48
Maintenance	0.20	0.20	0.26	0.20	0.20	0.21	0.20
Interest buildings	0.46	0.46	0.60	0.46	0.47	0.49	0.46
Interest working capital	0.07	0.07	0.10	0.07	0.07	0.08	0.07
Total production costs	10.74	10.68	11.02	10.66	10.89	10.52	10.27
Gross margin	0.23	0.28	0.30	0.34	0.12	0.50	0.68
Net return to management	−1.50	−1.44	−1.78	−1.42	−1.65	−1.28	−1.03

Table 6. Sensitivity analysis (% change in net return to management).

Change in variable	Conventional	Longer dark period	Decreased stocking density	Meal feeding	Restricted feeding	Sequential feeding	Feeding whole wheat
Prevalence of lameness							
−10%	1.29%	1.05%	0.89%	1.22%	1.14%	0.99%	1.25%
+10%	−1.29%	−1.05%	−0.89%	−1.22%	−1.14%	−0.99%	−1.25%
Decrease in daily weight gain due to lameness							
−5%	1.18%	1.26%	1.18%	1.36%	1.17%	1.44%	1.69%
+5%	−1.21%	−1.29%	−1.21%	−1.40%	−1.20%	−1.48%	−1.74%
Increase in feed conversion rate due to lameness							
−0.01	0.3%	0.3%	0.2%	0.1%	0.3%	0.2%	0.3%
+0.01	−0.3%	−0.2%	−0.2%	−0.4%	−0.2%	−0.2%	−0.2%
Increase in mortality due to lameness							
−0.1%	0.1%	0.1%	0.1%	−0.1%	0.1%	0.1%	0.1%
+0.1%	−0.1%	0.0%	−0.1%	−0.2%	0.0%	−0.1%	0.0%
Increase in condemnation due to lameness							
−0.1%	0.1%	0.1%	0.1%	0.0%	0.2%	0.1%	0.2%
+0.1%	−0.1%	−0.1%	−0.1%	−0.2%	0.0%	−0.1%	−0.1%
Wheat price							
−5%	0%	0%	0%	0%	0%	0%	+3%
+5%	0%	0%	0%	0%	0%	0%	−3%

Farmers' perceptions of lameness and practices to control lameness

Lameness as a health concern

Thirteen respondents estimated the prevalence of lameness on the last day of production to be lower than the average of 14.1% of Ross 308 chickens and 26.1% of Cobb 500 chickens in Sweden reported by Sanotra et al. (2003) (Table 7). Four respondents regarded lameness as either a minor (N = 1) or a moderate problem (N = 3) in their farm, whereas the other 9 did not regard it as a problem.

Ascites (3), cardiac arrest (2), subcutaneous inflammations (2), small and underdeveloped animals, high mortality and heat stroke were named by 6 respondents as the greatest health problems. Six respondents identified foot pad dermatitis (4) and leg problems (3) as the most frequent health problems associated with lameness.

Adverse consequences of lameness

Culling and impaired growth rate were named as the most frequent adverse consequences of lameness (Table 8).

Eleven respondents further indicated that, on average, 0.1– 0.7% of the flock needs to be culled due to lameness (mean = 0.48 ± 0.58%). Mean condemnation due to breast blisters, hock burns, foot pad lesions and fractures was 0.37 ± 0.40% of the flock; these issues are generally associated with lameness.

Table 7. Respondents' perception of lameness in their farm.

Perception	*N*	*N%*
Prevalence of lameness on the last day of the production round	13	
Lower than 20%	*13*	*100*
Equal to 20%	*0*	*0*
Higher than 20%	*0*	*0*
Prevalence of lameness if different than 20%	5	
Less than 1%	*1*	
5%	*1*	
5–10%	*2*	
10%	*1*	
Lameness is considered as a problem in your farm	13	
Yes, I consider lameness as a problem	*4*	*31*
No, I do not consider lameness as a problem	*9*	*69*
The extent lameness is a problem in your farm	4	
A minor problem	*1*	*25*
A moderate problem	*3*	*75*
A large problem	*0*	*0*

Table 8. Indicated adverse consequences of lameness (N = 10).

About 50% of culling is due to leg problems, it happens early in the rearing period (d 7–20).
Animal suffering, burns (breast-, hock,- foot pad-) due to close contact with the manure.
High culling rate.
It is always a pity to cull healthy animals that for some reason have leg problems.
Low growth rate.
Negative effect on economic performance due to high number of culled animals and lower growth rate.
I have to cull birds that are not sick, but have leg problems.
The chicken is not feeling well and has to be culled.
The animals do not grow as they should.
Chickens need to be culled.

Management practices to control lameness

Respondents generally felt that the suggested practices could not effectively control lameness and some of their remarks are shown in Table 9. More specifically, none of them regarded the use of an 8-h dark period per day or sequential feeding as effective remedial strategies. A main concern about increasing the dark period was that chickens would become inactive and therefore would not eat; it was also expected to cause distress when lights were switched back on. Respondents were unfamiliar with sequential feeding and therefore failed to recognise its potential added value. Some practices, e.g. feeding whole wheat and meal feeding 2–4 times per day were more familiar and had already been applied by some respondents. Results also illustrate the fact that lameness is a complex problem and changing a single management factor may not be sufficient to reduce lameness. For example, respondents believed that changing stocking density would hardly affect lameness because they regarded genetic background and nutrition as more important factors.

DISCUSSION

The main objectives of the present study were to explore the potential economic impact of implementing various alternative practices to reduce lameness in Swedish broiler farms, and to identify farmers' possible perceptions of their potential value in the Swedish context.

Table 9. Perception of management practices to control lameness.

About 50% of culling is due to leg problems, it happens early in the rearing period (d 7–20).
Animal suffering, burns (breast-, hock,- foot pad-) due to close contact with the manure.
High culling rate.
It is always a pity to cull healthy animals that for some reason have leg problems.
Low growth rate.
Negative effect on economic performance due to high number of culled animals and lower growth rate.
I have to cull birds that are not sick, but have leg problems.
The chicken is not feeling well and has to be culled.
The animals do not grow as they should.
Chickens need to be culled.

Results suggest that the application of certain alternative practices can provide better economic performance than conventional practice. Hence, within the Swedish context there seems to be potential to reduce lameness and to consequently improve animal welfare and economic performance. The alternative practices which realised the greatest improvements in gross margin and net return to management in the model were (in decreasing order) feeding whole wheat, sequential feeding and meal feeding. Improved economic performance when feeding whole wheat and meal feeding can mainly be attributed to decreased feed costs which respectively reflect a lower feed price and a more efficient feed conversion rate. In the case of sequential feeding, the decreased cost of d-old chicks (per kilogram delivered broiler) contributed most to the improved economic performance. The complex effect of changing feed composition on technical and economic performance is illustrated well in the case of feeding whole wheat. Feeding whole wheat resulted in a less efficient feed conversion rate, but a reduced mortality in lame chickens compared to the conventional practice. The negative effect of the practice on feed conversion rate was, however, outweighed by the effect of a low feed price (wheat is cheaper than compounded feed), which led to reduced feed costs. Moreover, the lower mortality rate reduced production costs per delivered broiler. It is suggested that farmers can substantially profit from the inclusion of whole wheat in their feeding programmes, but it must be realised that the price of wheat has a pivotal role in determining the level of potential profit. Reduced lameness actually contributed less to the improved economic performance.

Apparently, farmers did not perceive the win-win situation of the remedial practices in terms of economic performance and animal welfare despite the fact that their implementation could be of major importance in the broiler sector where profit margins are very small. There may be various reasons for this oversight. First, for example, our results suggest that farmers may not see lameness as a big problem so any improvements achieved by alternative practices might also be perceived as low. More specifically, 9 of the 13 farmers questioned did not consider lameness as a problem at their farm. Farmers' estimates of the prevalence of lameness at their farms ranged from less than 1% to 10%, which is lower than the 14.1% (for Ross 308 chickens) and 26.1% (for Cobb 500 chickens) reported by Sanotra et al. (2003). Of course, lameness is likely to have become less common since 2003 because improved leg health has been incorporated in the breeding goals (Dawkins and Layton, 2012). However, broiler farmers often fail to detect lameness in their farm (Butterworth and Haslam, 2009) or are less keen to detect it when there are no obvious, clinical signs of the problem. A recent Belgian study (Tuyttens et al., 2014) also indicated that there might be some denial by broiler farmers of welfare problems in their birds. Similarly, dairy cow farmers were found to highly underestimate lameness in their herds (Sárová et al., 2011). Second, the farmers may have more overwhelming problems than lameness, such as negative net returns due to other factors. There is also a possibility of respondent bias, i.e. farmers with less lameness were also more willing to answer the questionnaire, which may account for the difference between our findings and other reports. Third, farmers may primarily focus on the potential negative production effects of an alternative practice, e.g. a higher feed conversion rate, and fail to see the accompanying economic advantages, e.g. lower feed cost when feeding whole wheat. Encouragingly though, some farmers seem to be recognising the win-win potential of alternative practices because some of our respondents were already using the whole wheat and meal feeding strategies.

The present economic model assumed that chickens were kept for a fixed production period. As a consequence,lame broilers (if not culled) were delivered with a lower weight since lameness usually leads to an impaired daily growth (Yalcin et al., 1998). This assumption results in a lower revenue for lame broilers than for healthy ones. Another way to deal with the differences in growth rate between healthy and lame broilers would be to assume that lame birds are kept for longer to reach the same weight as healthy ones. This assumption, however, was not included in the analysis because lameness can become more evere with increasing age (Butterworth and Haslam, 2009) and can negatively impact productivity and financial performance. Detailed input data to model these effects, however, was not available. Keeping lame broilers for longer to increase their final weight is not feasible or desirable for at least three reasons: (1)

the flock would have to be delivered in two batches rather than the normal practice of one batch; (2) more importantly, keeping and selling severely lame broilers raises ethical questions; and (3) transportation of lame birds is not allowed in Sweden.

A normative modelling approach was used to address the economic impact of lameness-reducing management practices. This meant that input data is often not readily available and certain assumptions and estimations had to be made to carry out the analysis. Therefore, we must reflect on the quality of the input data and the potential uncertainties in modelling outcomes. First, input data regarding the production effect of various practices to reduce lameness were usually estimated on the basis of small-scale experiments reported in the scientific literature because there is a lack of commercially based evidence. This might also explain why farmers failed to recognize the estimated impact of the suggested management practices – farmers may have the experience of how the practices may work for them in their commercial settings, or at least they can imagine what the practices may lead to for them. Second, various sources were used to estimate the impact of a particular management practice on production performance and prevalence of lameness. The starting points and settings differed between some of the studies that we based our model on. Nevertheless, for the economic analysis, it was necessary to establish exact parameters by combining the information from these various studies despite their different starting points. For example, in the case of feeding whole wheat, the proportion of whole wheat in the diet varied during the production period in the different studies, but to establish a common ground in the present study it was assumed that the average proportion of whole wheat in the diet is 30% during the production period. Of course, this assumption influences the estimated feed price and hence the resultant feed costs and net returns. Therefore, the extent of economic improvement associated with the alternative feeding practice should be interpreted with care. A sensitivity analysis, which is often used to examine the robustness of results to changes in a certain input parameter when other variables are held constant (Pannell, 1997), was carried out here with regard to input parameters such as prevalence of lameness and performance of lame animals. The outcomes suggest that the model results are robust in a sense that the sequence of management practices, from the highest to the lowest net return to management, remained unchanged when any of the input variables are altered. Thus, the three best performing management practices were feeding whole wheat, sequential feeding and meal feeding. A limitation of such a sensitivity analysis is, however, that interactions between production parameters cannot be taken into account. Hence, the primary task required to increase the accuracy of economic calculations is to improve the accuracy of input parameters. In the case of feeding whole wheat, further research should primarily focus on the relations between the proportion of whole wheat in the diet and feed conversion ratio as well as other production parameters.

Caution must be taken when generalising the results of the present small sample size to a larger population, e.g. the whole Swedish broiler farmer population (Flyvbjerg, 2006). However, the farmers' responses were consistent and offer a wealth of information on their considerations regarding lameness and alternative management practices. Clearly, further research on a larger commercial scale is merited.

In conclusion, the present study indicates that some alternative practices (especially feeding whole wheat, sequential feeding and meal feeding) may not only improve broiler welfare by reducing the prevalence of lameness but may also enhance economic performance. Thus, if these practices are applied in specific situations, a win-win effect in terms of welfare and profitability could be generated. Although reduced lameness contributes to an improved economic performance, the effects of lower feed prices (e.g. when feeding whole wheat) and more efficient feed conversion (e.g. when using meal feeding) on improved economic performance are generally more influential. However, the farmers who participated in the survey did not seem to recognise the potential combined positive effects of changing feed or feeding practices on both broiler welfare and farm economics. There is a clear and pressing need for a wider and more transparent dissemination of the present findings.

ACKNOWLEDGEMENTS

The authors greatly appreciate the assistance of one of the slaughterhouses in Sweden with the collection of data and would like to thank the farmers for their participation in the study. We also thank Dr Bryan Jones for valuable comments on the manuscript and improving the English language.

DISCLOSURE STATEMENT

No potential conflict of interest was reported by the authors.

FUNDING

This work was supported by the Netherlands Organisation for Scientific Research (NWO, The Hague, the Netherlands) and the Dutch Ministry of

Economic Affairs within the programme titled "The Value of Animal Welfare".

ORCID H. J. Blokhuis http://orcid.org/0000-0003-3608-6355

REFERENCES

1. Agriwise (2014). http://www.agriwise.org/.
2. Acs, S., Berentsen, P.B.M. & Huirne, R.B.M. (2005) Modelling conventional and organic farming: a literature review. *Njas-Wageningen Journal of Life Sciences*, 53: 1–18. doi:10.1016/S1573-5214(05)80007-7
3. Adrian, A.M., Norwood, S.H. & Mask, P.L. (2005) Producers' perceptions and attitudes toward precision agriculture technologies. *Computers and Electronics in Agriculture*, 48: 256–271. doi:10.1016/j.compag.2005.04.004
4. Bassler, A.W., Arnould, C., Butterworth, A., Colin, L., De Jong, I.C., Ferrante, V., Ferrari, P., Haslam, S., Wemelsfelder, F. & Blokhuis, H.J. (2013) Potential risk factors associated with contact dermatitis, lameness, negative emotional state, and fear of humans in broiler chicken flocks. *Poultry Science*, 92: 2811–2826. doi:10.3382/ps.2013-03208
5. Bennett, C., Classen, H. & Riddell, C. (2002) Feeding broiler chickens wheat and barley diets containing whole, ground and pelleted grain. *Poultry Science*, 81: 995–1003. doi:10.1093/ps/81.7.995
6. Bennett, C.D., Classen, H.L. & Riddell, C. (1995) Live performance and health of broiler chickens fed diets diluted with whole or crumbled wheat. *Canadian Journal of Animal Science*, 75: 611–614. doi:10.4141/cjas95-090
7. Bessei, W. (2006) Welfare of broilers: a review. *Worlds Poultry Science Journal*, 62: 455–466. doi:10.1079/Wps2005108
8. Björk, K. (2012) *Broiler Production and Welfare in the County of Södermanland*. (Skara, Swedish University of Agricultural Sciences, Department of Animal Environment and Health).
9. Bradshaw, R.H., Kirkden, R.D. & Broom, D.M. (2002) A review of the aetiology and pathology of leg weakness in broilers in relation to welfare. *Avian and Poultry Biology Reviews*, 13: 45–103. doi:10.3184/147020602783698421
10. Buchwalder, T. & Huber-Eicher, B. (2005) Effect of the analgesic butorphanol on activity behaviour in turkeys (Meleagris gallopavo). *Research in Veterinary Science*, 79: 239–244. doi:10.1016/j.rvsc.2004.11.013
11. Butterworth, A. & Haslam, S.M. (2009) A lameness control strategy for broiler fowls, Welfare Quality Reports No. 13, *Welfare Quality Reports* (Cardiff, UK, Cardiff School of City and Regional Planning).
12. Butterworth, A. & Weeks, C. (2010) The impact of disease on welfare, in: Duncan, I.J.H. & Hawkins, P. (Eds) *The Welfare of Domestic Fowl and Other Captive Birds*, Vol. 9, (Dordrecht, Springer Science+Business Media B.V.).
13. Butterworth, A., Weeks, C.A., Crea, P.R. & Kestin, S.C. (2002) Dehydration and lameness in a broiler flock. *Animal Welfare*, 11: 89–94.
14. Caplen, G., Hothersall, B., Nicol, C.J., Parker, R.M.A., Waterman-Pearson, A.E., Weeks, C.A. & Murrell, J.C. (2014) Lameness is consistently better at predicting broiler chicken performance in mobility tests than other broiler characteristics. *Animal Welfare*, 23: 179–187. doi:10.7120/09627286.23.2.179
15. Castellini, C., Berri, C., Le Bihan-Duval, E. & Martino, G. (2008) Qualitative attributes and consumer perception of organic and free-range poultry meat. *World's Poultry Science Journal*, 64: 500–512. doi:10.1017/S0043933908000172

16. Classen, H.L. (2004) Day length affects performance, health and condemnations in broiler chickens. Proceedings of the Proceeding of the Australian Poultry Sci. Soc., Univ. of Sydney, Australia.
17. Danbury, T.C., Weeks, C.A., Waterman-Pearson, A.E., Kestin, S.C. & Chambers, J.P. (2000) Self-selection of the analgesic drug carprofen by lame broiler chickens. *Veterinary Record*, 146: 307–311. doi:10.1136/vr.146.11.307
18. Dawkins, M.S. & Layton, R. (2012) Breeding for better welfare: genetic goals for broiler chickens and their parents. *Animal Welfare*, 21: 147–155. doi:10.7120/09627286.21.2.147
19. De Jong, I., Perez Moya, T., Gunnink, H., Van Den Heuvel, H., Hindle, V.A., Mul, M. & Van Reenen, K. (2011) *Simplifying the Welfare Quality Assessment Protocol for Broilers*. Vol. 533, 74. (Wageningen, Wageningen UR Livestock Research, part of Stichting Dienst Landbouwkundig Onderzoek (DLO Foundation)).
20. Dijkhuizen, A.A. & Morris, R.S. (1997) *Animal Health Economics: Principles and Applications.*. (Sydney, Post Graduate Foundation in Veterinary Science, University of Sydney).
21. EC Directive (2007) Directive, 2007/43/EC, OJ L 182, p. 19 of 12.7. 2007. http://eur-lex.europa.eu/legal-content/EN/ALL/?uri= CELEX% 3A32 007 L0043
22. Fernandes, B.C.D.S., Martins, M.R.F.B., Mendes, A.A., Paz, I.C.D.L.A., Komiyama, C.M., Milbradt, E.L. & Martins, B.B. (2012) Locomotion problems of broiler chickens and its relationship with the gait score. *Revista Brasileira De Zootecnia*, 41: 1951–1955. doi:10.1590/S1516-35982012000800021
23. Flyvbjerg, B. (2006) Five misunderstandings about case-study research. *Qualitative Inquiry*, 12: 219–245. doi:10.1177/1077800405284363
24. Gocsik, É., Kortes, H.E., Oude Lansink, A.G.J.M. & Saatkamp, H.W. (2014) Impact of different broiler production systems on health care costs in the Netherlands. *Poultry Science*, 93: 1301–1317. doi:10.3382/ps.2013-03614
25. Gocsik, É., Van Der Lans, I.A., Oude Lansink, A.G.J.M. & Saatkamp, H.W., (2015) Willingness of Dutch broiler and pig farmers to convert to production systems with improved welfare. *Animal Welfare*, 211–222. doi:10.7120 /09627286.24.2.211
26. Hall, A.L. (2001) The effect of stocking density on the welfare and behaviour of broiler chickens reared commercially. *Animal Welfare*, 10: 23–40.
27. Hothersall, B., Caplen, G., Parker, R.M.A., Nicol, C.J., Waterman-Pearson, A.E., Weeks, C.A. & Murrell, J.C. (2016) Effects of carprofen, meloxicam and butorphanol on broiler chickens' performance in mobility tests. *Animal Welfare*, 25: 55–67. doi:10.7120/09627286.25.1.055
28. Kerselaers, E., De Cock, L., Lauwers, L. & Van Huylenbroeck, G. (2007) Modelling farm-level economic potential for conversion to organic farming. *Agricultural Systems*, 94: 671–682. doi:10.1016/j.agsy.2007.02.007
29. Kestin, S., Knowles, T., Tinch, A. & Gregory, N. (1992) Prevalence of leg weakness in broiler chickens and its relationship with genotype. *Veterinary Record*, 131: 190–194. doi:10.1136/vr.131.9.190
30. Kestin, S.C., Gordon, S., Su, G. & Sorensen, P. (2001) Relationships in broiler chickens between lameness, liveweight, growth rate and age. *Veterinary Record*, 148: 195–197. doi:10.1136/vr.148.7.195
31. Knowles, T.G., Kestin, S.C., Haslam, S.M., Brown, S.N., Green, L.E., Butterworth, A., Pop e, S.J., Pfeiffer, D. & Nicol, C.J. (2008) Leg disorders in broiler chickens: prevalence, risk factors and prevention. *Plos One*, 3: e1545. doi:10.1371/journal.pone.0001545
32. Leterrier, C., Vallee, C., Constantin, R., Chagneau, A.M., Lessire, M., Lescoat, P., Berri, C ., Baeza, E., Bizeray, D. & Bouvarel, I. (2008) Sequential feeding with variations in energy and protein levels improves gait score in meat-type chickens. *Animal*, 2: 1658–1665. doi:10.1017/S1751731108002875

33. Mc Geown, D., Danbury, T.C., Waterman-Pearson, A.E. & Kestin, S.C. (1999) Effect of carprofen on lameness in broiler chickens. *Veterinary Record*, 144: 668–671. doi:10.1136/vr.144.24.668
34. McNamee, P.T., Mccullagh, J.J., Thorp, B.H., Ball, H.J., Graham, D., Mccullough, S.J., Mcc onaghy, D. & Smyth, J.A. (1998) Study of leg weakness in two commercial broiler flocks. *Veterinary Record*, 143: 131–135. doi:10.1136/vr.143.5.131
35. Naas, I.A., Paz, I.C.L.A., Baracho, M.S., Menezes, A.G., Bueno, L.G.F., Almeida, I.C.L. & M oura, D.J. (2009) Impact of lameness on broiler well-being. *Journal of Applied Poultry Research*, 18: 432–439. doi:10.3382/japr.2008-00061
36. Pannell, D.J. (1997) Sensitivity analysis of normative economic models: theoretical framework and practical strategies. *Agricultural Economics*, 16: 139–152. doi:10.1016/S0169-5150(96)01217-0
37. Rogers, E.M. (1995) *Diffusion of Innovations*. (New York, Free Press).
38. Sandilands, V., Brocklehurst, S., Sparks, N., Baker, L., Mcgovern, R., Thorp, B. & Pears on, D. (2011) Assessing leg health in chickens using a force plate and gait scoring: how many birds is enough? *Veterinary Record*, 168: 77. doi:10.1136/vr.c5978
39. Sanotra, G.S., Berg, C. & Lund, J.D. (2003) A comparison between leg problems in Danish and Swedish broiler production. *Animal Welfare*, 12: 677–683.
40. Sárová, R., Stéhulová, I., Kratinová, P., Firla, P. & Pinka, M. (2011) Farm managers underestimate lameness prevalence in Czech dairy herds. *Animal Welfare*, 20: 201–204.
41. Siegel, P.B., Gustin, S.J. & Katanbaf, M.N. (2011) Motor ability and self-selection of an analgesic drug by fast-growing chickens. *Journal of Applied Poultry Research*, 20: 249–252. doi:10.3382/japr.2009-00118
42. Skinner-Noble, D.O. & Teeter, R.G. (2009) An examination of anatomic, physiologic, and metabolic factors associated with well-being of broilers differing in field gait score. *Poultry Science*, 88: 2–9. doi:10.3382/ps.2006-00450
43. Sørensen, P., Su, G. & Kestin, S.C. (2000) Effects of age and stocking density on leg weakness in broiler chickens. *Poultry Science*, 79: 864–870. doi:10.1093 /ps/ 79.6.864
44. SPMA. (2015) http://www.svenskfagel.se/?p=689. Accessed 04-05-2015
45. Su, G., Sørensen, P. & Kestin, S.C. (1999) Meal feeding is more effective than early feed restriction at reducing the prevalence of leg weakness in broiler chickens. *Poultry Science*, 78: 949–955. doi:10.1093/ps/78.7.949
46. Sullivan, T.W. (1994) Skeletal problems in poultry: estimated annual cost and descriptions. *Poultry Science*, 73: 879–882. doi:10.3382/ps.0730879
47. Tuyttens, F., Vanhonacker, F. & Verbeke, W. (2014) Broiler production in Flanders, Belgium: current situation and producers' opinions about animal welfare. *Worlds Poultry Science Journal*, 70: 343–354. doi:10.1017/S004393391400035x
48. Umar Faruk, M., Bouvarel, I., Même, N., Rideau, N., Roffidal, L., Tukur, H.M., Bastianelli, D., Nys, Y. & Lescoat, P. (2010) Sequential feeding using whole wheat and a separate protein-mineral concentrate improved feed efficiency in laying hens. *Poultry Science*, 89: 785–796. doi:10.3382/ps.2009-00360
49. Verma, D.S. (2007) Effects of mycotoxins on the bone structure and leg weakness in poultry: estimation of the economic loss by leg weakness.
50. WELFARE QUALITY®. (2009) *Welfare Quality® Assessment Protocol for Poultry (Broilers, Laying Hens)*. (Lelystad, The Netherlands, Welfare Quality® Consortium).
51. Wideman, R.F., Hamal, K.R., Stark, J.M., Blankenship, J., Lester, H., Mitchell, K.N., Lore nzoni, G. & Pevzner, I. (2012) A wire-flooring model for inducing lameness in broilers: evaluation of probiotics as a prophylactic treatment. *Poultry Science*, 91: 870–883. doi:10.3382/ps.2011-01907
52. Yalcin, S., Settar, P. & Dicle, O. (1998) Influence of dietary protein and sex on walking ability and bone parameters of broilers. *British Poultry Science*, 39: 251–256. doi:10.1080/00071669889213

CHAPTER 3

Behavioural Assessment of Pain in Commercial Turkeys (Meleagris Gallopavo) with Foot Pad Dermatitis

A. Sinclair, C. Weber Wyneken, T. Veldkamp[1], L.J. Vinco[2], And P.M. Hocking

The Roslin Institute and Royal (Dick) School of Veterinary Studies, University of Edinburgh, Midlothian, UK,

[1]Department Animal Nutrition, Wageningen UR Livestock Research, Wageningen, The Netherlands

[2]National Reference Centre for Animal Welfare, Istituto Zooprofilattico della Lombardia e dell'Emilia Romagna, Brescia, Italy

ABSTRACT

1. Two experiments were conducted to investigate the differences in susceptibility to foot pad dermatitis (FPD) of two medium-heavy lines of turkeys, and whether FPD is painful, by detailed analysis of behaviour in birds with and without analgesic treatment (betamethasone).

2. Turkeys housed on dry litter in the first experiment generally had more frequent bouts of different behaviours that were of shorter duration than birds on wet litter. T-patterns (behavioural sequences) were more frequent, varied and complex on dry than on wet litter. Betamethasone-injected birds of line B, but not breed A, had shorter resting and longer standing durations on wet litter than saline-injected birds.

3. In the second experiment, turkeys on wet litter given saline stood less and rested more than all other treatment groups, suggesting that they experienced pain that was alleviated in birds receiving betamethasone.

Turkeys on dry litter had more frequent, varied and complex patterns of behaviour than turkeys on wet litter and birds kept on intermediate litter wetness. Betamethasone provision increased pattern variety regardless of litter treatment.

4. Turkeys with low FPD scores transferred to wet litter and given saline injections had a longer total duration of resting and shorter duration of standing compared to betamethasone-treated birds. Low FPD birds transferred to wet litter had a similar number of patterns and total pattern occurrence as high FPD birds transferred to dry litter. Betamethasone increased pattern variety and frequency compared to saline injections whereas overall pattern complexity was similar.

5. It was concluded that wet litter affects the behaviour of turkey poults independently of FPD and that betamethasone may also change the behaviour of turkeys. There was some evidence from analgesic treatment and T-pattern analyses that FPD was painful. However, there was no evidence of differences in susceptibility to FPD of the two commercial hybrids.

INTRODUCTION

Foot pad dermatitis (FPD) is a highly prevalent and potentially painful contact dermatitis of commercially reared turkeys in which the metatarsal and digital pads become swollen, discoloured and hard. Hyperkeratosis and necrotic lesions form, increasing in size as the condition progresses (Mayne et al., 2007a). These external signs are preceded by histopathological evidence of an inflammatory immune response (Mayne et al., 2007b).

In addition to the potential pain and reduced mobility caused by inflammation and mechanical damage, secondary infections may further reduce welfare in affected turkeys through conditions such as synovitis and subsequent lameness (Clark et al., 2002). Buda et al. (2002) reported the existence of pain receptors and mechanoreceptor sensory nerve endings in the foot pads of turkeys. Additionally, reduced activity and disrupted behavioural sequences are associated with high FPD scores in turkeys kept on wet litter (Hocking and Wu, 2013). However, affected turkeys should be observed both on dry litter and under analgesic intervention to ascertain whether such behavioural alterations are indicative of pain and a major aim of the current research was to critically investigate the hypothesis that FPD in turkeys is painful.

High litter moisture content (LMC) is the main causative factor predisposing poultry to FPD (Mayne et al., 2007a) and an inflammatory response may develop in as little as 48 h (Mayne et al., 2007c). Similarly, fully developed lesions can form under the same conditions over 2–4 d, and heal within 15 d following transfer to dry litter (Mayne et al., 2007a). The relationship with FPD severity is linear when LMC is above a certain proportion depending on the experiment (Wu and Hocking, 2011; Wyneken et al., 2015) consistent with the hypothesis that LMC is the major cause of FPD in turkeys. FPD, however, is multifactorial (Mayne, 2005), and some studies suggest the possibility of genetic factors (Mayne, 2005; Wu and Hocking, 2011). It has been demonstrated that Large White turkey poults are more susceptible to FPD than Broad Breasted Bronze poults reared under the same conditions (Chavez and Kratzer, 1972), and it has been suggested that widely used commercial strains differ in their susceptibility to FPD.

Two questions were therefore addressed in this study: are there differences between two major medium-heavy lines in their susceptibility to FPD; and do turkeys with FPD exhibit behaviours indicative of pain confirmed by analgesic intervention?

MATERIALS AND METHODS

A detailed description of the experimental design of the experiments is presented in the accompanying paper (Wyneken et al., 2015). The research was part of a series of three experiments of which Experiments 1 and 3 were used to assess pain from FPD using analyses of behaviour in floor pens. The results are reported in this paper and the experiment numbers 1 and 3 are retained for ease of comparison with the paper of Wyneken et al. (2015). Experiment 3 consisted of two phases (3.1 and 3.2) in a partial crossover experiment.

Animals and husbandry

In Experiment 1, a total of 240 one-day-old male poults of two medium-heavy commercial lines, 120 of line A and 120 of line B, from breeder flocks that were 33 weeks old were obtained from a commercial hatchery. The poults were beaktrimmed by an infrared method in the hatchery in Experiment 3 but not in Experiment 1. Each pen was sealed with mastic and measured 1.25 × 1.77 m. A brooding lamp was provided until 26 d (Experiment 1) or 28 d (Experiment 3). Each pen was littered with 4.6 kg white wood shaving to a depth of about 50 mm. Feed and water in Experiment 1 were provided by a wallmounted, gridded feeder and a graduated water tank connected to a suspended bell drinker. For Experiment 3, water was supplied in suspended nipple drinkers (10 nipples per drinker). At 64 d in Experiment 3, 72 birds from the larger flock were rehoused in 12 pens (2.5 × 1.77 m) of 6 birds from the same rearing pen and were given feed and water in suspended feeders and bell drinkers. Pens were cleaned and fresh litter provided on d 1 of treatments in Experiment 1 only.

The poults were given 3 h of light (100 lux) followed by 3 h dark alternately until 23:30 h. Thereafter, the photoperiod was 16 h light (07:30–23:30 h), and 8 h dark. Light intensity was 50 lux for d 2 and d 3, and 12 lux from d 4 onward. Ambient temperatures decreased from 28°C at hatch to 16°C at 10 weeks. Feed (Target Feeds Ltd, Whitchurch, UK) was a standard wheatsoya-based commercial turkey starter ration supplied as crumbs in Experiment 1. Birds in Experiment 3 were fed on 4 diets based on maize or wheat and soya or non-soya ingredients as crumbs from 0 to 4 weeks and 3 mm pellets thereafter (5–8 weeks and 9–12 weeks).

Experimental treatments

Experiment 1

The poults were initially housed by line at 10 poults per pen for 26 d. On d 27, litter treatments and birds were randomly allocated across 48 pens (4 poults per pen). Unfamiliar poults were not mixed and excess poults were humanely culled. No water was added to the control (dry, 1D1) pens and in the "wet" treatments water was added to the litter to achieve an initial moisture content of 30%, 40% or 50% (1W1, 1W2, 1W3). The added water was estimated on the basis of the weight of litter in the pen and the previously determined moisture content of dry wood shavings. Water was added daily thereafter for 7 d to maintain the required condition of the litter and induce foot pad lesions as described by Mayne et al. (2007a) except that excreta were not removed.

Between 09:30 h and 10:30 h on d 3 of litter treatment, two randomly selected birds from each pen received a 0.4 ml injection into the breast tissue of either saline (control) or a solution of betamethasone. The concentration of analgesic administered was 0.04 mg/kg of betamethasone sodium phosphate (Betnesol Injection, RPH Pharmaceuticals AB, Haninge, Sweden) as calculated from the average bird weight recorded on d 1 of litter treatments (1162 g). Birds were marked for identification and injected as described for a further 2 d to ensure that an effective concentration of betamethasone was present before filming began. On the third day of injections, aerial-view home-pen recordings were made using 12 monochrome cameras (WV-BP330 Panasonic CCTV, Suzhou Co. Ltd, Suzhou, China) and three 4-channel digital recorders (INS-DVR04V2-5 DVRs, Inspired Security Solutions Ltd, Hastings, UK.) suspended above the centre of each pen. The one hour films were taken twice for each pen at either (i) 11:00– 12:00 h and 15:00–16:00 h or (ii) 13:00–14:00 h and 17:00–18:00 h. Half the pens were filmed on d 3 and half on d 4 starting at 11:00 or 13:00 in alternate pens in each of three blocks.

Experiment 3.1

Between 15:00 and 15:30 h on d 64, water was added to 8 pens to achieve litter scores for treatments 3W1 and 3W2 of 3 and 5 respectively on the Tucker and Walker (1999) scale and no water was added to the 4 dry control pens (3D). Litter scores were maintained as previously described. Daily injections in the morning were conducted on d 10–17 of litter treatments. For birds receiving betamethasone (0.04 mg/kg), doses were calculated using individual bird weights on d 1 of injections and recalculated after 5 d to account for weight gain. In order to minimise potential discomfort from the volume of injected solution, half the dose was injected into each breast. Filming was conducted over one day (injection d 4, litter treatments d 14), with all pens being filmed from 11:00 to 12:00 h and again from 15:00 to 16:00 h.

Experiment 3.2

At 15:00 h on d 16 of litter treatments, the birds were switched from 3D to 3W2 pens and vice versa. Birds from 3W1 pens were not included. Litter and analgesic treatment times were maintained and birds were left to acclimatise overnight (20 h) on the same diets as before. Filming occurred as described for Experiment 3.1 on injection d 7 (litter treatment d 17).

Observations

Individual weights were recorded at the start and end of the experimental periods. LMC in each pen was determined as previously described (Wyneken et al., 2015). Foot pad scores were obtained for every bird at the end of each experiment on an 8-point scale (Mayne et al., 2007a).

For 1D and 1W3 treatments in Experiment 1 and all treatments in Experiments 3.1 and 3.2, home pen recordings were imported into The Observer XT version 11 (Noldus Information Technology, Wageningen, NL) and 25-min clips from each file were randomly selected for analysis. The behaviour of the two injected birds in each clip was scored using focal sampling and the ethogram shown in Table 1. Within each behavioural group (point events, postures and activities), behaviours were mutually exclusive. By definition, point events in Observer have no timescale; therefore, end points are not defined in the ethogram. Inclusion of "reduced activity" was necessary as sets of state events must add up to 100%. Each clip was watched twice as birds were scored individually. Prior to this, 20 h of practice scoring was conducted on non-selected clips to minimise the risk of order effects and to determine optimal software settings. Data were extracted from Observer and processed for THEME version 5 (Noldus Information Technology, Wageningen, NL) using Noldus PatternVision DEP software (PatternVision Ltd, Reykjavik, ISL) for detection and analysis of hidden behavioural patterns (T-patterns). The optimal parameters for pattern detection were identified as a P-value of 0.0001 and a minimum pattern occurrence of 3 (Magnusson, 2000). The simulation filter and randomization testing options were applied (parameters as for real data) to further reduce the chance of type 1 errors.

Bird welfare

All experiments and procedures were conducted after ethical approval under project licence number PPL60/45067. The health of the turkeys was inspected on a daily basis and severely affected birds were humanely killed. All turkeys were killed at the end of each experiment with an intravenous sodium pentobarbital injection (Euthatal, Merial, Toulouse, France).

Statistical analysis

Statistical analyses were conducted in Genstat version 13 (VSN International, Hemel Hempsted, UK). Residual plots were inspected for normality and homogeneity of variance and transformations were used where appropriate (see below). Where residuals violated model assumptions due to outliers (i.e. were 4 or more standard deviations from the mean), the related data points were removed and the analysis repeated; where no notable difference to the output resulted, the original analyses are presented. Post hoc analyses (t-tests) were conducted as necessary to evaluate prior contrasts for significant effects of factors with more than two levels.

Data transformations to achieve normality and homogeneity of variance for Experiment 1 were loge for mean duration of rest and reduced activity; loge + 1 for frequency/h of feeding, drinking, other, peck feeder, peck ground, peck wall, peck others, peck total and also for T-pattern occurrences and original T-patterns. Arc-sine√ transformations were required in Experiment 3.1 for percentage duration of rest, stand and preen standing and in Experiment 3.2 for percentage duration of walk and preen resting. T-pattern occurrences and original T-patterns in Experiment 3.2 were transformed to loge + 1. Where residuals of untransformed data appeared normal and P-values were not notably altered by the transformation, original data are presented; otherwise, transformed data are presented with backtransformed means in parentheses.

Table 1. Ethogram for behavioural observations

Behaviour	Description
Point event behaviours	
Rest together	Bird transitions to rest within one body length's distance of nearest bird
Rest alone	Bird transitions to rest over one body length's distance from nearest bird
Peck and modifiers	Bird's head is thrust forward then retracted quickly, causing the beak to make/almost make contact with the (i) ground (peck ground), (ii) walls (peck wall), (iii) other birds (peck others), (iv) bird itself (allo-peck), (v) drinker or (vi) top/side of feeder (peck feeder). Peck total is the sum of all pecks
State event "posture" behaviours	
Stand	Bird is stationary; body is elevated from the ground/feeder. Behaviour ends when two or more steps are taken in one direction, or body touches the ground/top of feeder for one or more seconds
Rest	Bird is stationary; body is touching the ground/top of feeder. Behaviour ends when body is elevated from the ground/feeder for one or more seconds
Walk	Standing bird takes two or more consecutive steps in one direction. Behaviour ends when bird is stationary for two or more seconds
State event "activities" behaviours	
Feed	Bird pecks inside feeder. Behaviour ends when bird has not pecked into feeder for five or more seconds or takes two consecutive steps away from the feeder
Drink	Bird dips beak into the moat/pecks at the nipple of the drinker. Behaviour ends when bird has removed beak from the moat/has not pecked at the nipple for five or more seconds, or takes two consecutive steps away from drinker
Preen	Bird rubs beak through plumage. Behaviour ends when head is not in contact with feathers for two or more seconds or head remains stationary for five or more seconds
Other	Bird carries out a low-frequency behaviour (leg/wing stretch, fly, flap, dust bathe, fighting/aggression, display)
Reduced activity (RA)	Bird exhibits no behaviours in the "activities" category. Behaviour ends when any behaviour from this category begins
State event behaviours comprising "other"	
Leg/wing stretch	Bird extends one or both leg(s) or wing(s) in a slow, prolonged manner. Behaviour ends when limb returns to original position or is left extended for five or more seconds
Fly	Bird raises and lowers wings repeatedly until no longer in contact with the ground/top of feeder. Behaviour ends when bird is in contact with the ground/top of feeder again
Flap	Bird extends wings, raising and lowering them one or more times but does not take flight. Behaviour ends when wings return to a stationary
Dust bathe	Bird rests on the ground and flicks wood shavings into its plumage. Behaviour ends when bird is motionless for two or more seconds or transitions to standing
Fighting/aggression	Bird displays one or more of the following behaviours: (i) repeatedly presses body/neck against conspecific in a forceful, jerking manner, (ii) runs in tight circles leaning onto conspecific and (iii) faces and pecks at conspecific forcefully possibly retaining hold of flaps of skin/feathers for several seconds. Behaviour ends when birds are no longer in contact for 5 or more seconds
Display	Bird alters appearance and increases size by means of piloerection, lifting its tail and stretching its wings downwards. Behaviour ends when bird's appearance returns to normal.

Experiment 1 was a randomised block design with 48 pens in 6 blocks (replicates) filmed twice. Behaviours were analysed for injected birds in dry and 1W3 litter treatments to give 96 observations (24 pens × 2 birds/pen × 2 times). The mean frequencies (F, converted into frequency/h) and, where applicable, mean bout durations (BD, s) and mean total duration (TD, %) of different behaviours were obtained from the Observer output. General analysis of variances (ANOVAs) were conducted on these variables for all behaviours. Model effects were line, litter (which was confounded with FPD score), provision of analgesia (analgesia vs. saline), time of day (11:00–14:00 h and 15:00–18:00 h) and their interactions. Blocking factors were bird ID nested within pen, nested within block. Analysis of the T-patterns identified by THEME were conducted for the number of unique behavioural patterns (variety), the number of occurrences of all patterns (frequency) and the mean number of behaviours within a pattern (complexity). Effects and blocking factors were the same as described previously for the Observer output analyses.

Experiment 3.1was a completely randomized design with 12 pens (three litter treatments × 4 replicates) filmed twice, giving 24 observations. Each treatment was replicated within each of the 4 diets. Behaviours were analysed for injected birds in all litter treatments (n = 4 for each injection and litter treatment). Behavioural data (as described for Experiment 1) were obtained from Observer and THEME and analysed using general ANOVAs. Effects included for both outputs were litter, analgesia, time of day (11:00–12:00 h and 15:00–16:00 h) and their interactions. Blocking factors were bird ID nested within pen. Experiment 3.2 was a completely randomized design with 8 pens (two litter treatments ("wet moved to dry" and "drymoved to wet") × 4 replicates) filmed twice giving 16 observations. Behavioural data (as described for Experiment 1) were obtained from Observer andTHEME, combinedwith the data from the same birds in Experiment 3.1, and analysed using general ANOVAs. Effects for the Observer and THEME ANOVAs were original litter (confounded with FPD score and referred to as FPD hereafter), analgesia, stage (1 and 2, i.e. pre- and post-relocation), time (11:00–12:00 h and 15:00–16:00 h) and their interactions. Blocking factors were bird ID nested within pen.

RESULTS

Litter scores and production variables are discussed in the companion paper (Wyneken et al., 2015). Mean FPD scores for 1D and 1W3 treatments respectively in Experiment 1 were 0.8 and 6.2 and for 3D and 3W2 in Experiment 3 they were 0.5 and 6.3.

Standing and resting behaviours were scored as occurring on either the ground or near the feeder in Experiment 1; however, as the latter were rarely observed this distinction was removed from the coding scheme for Experiment 3. "Peck ground" and "peck wall" were combined into "peck environment" for Experiment 3. "Peck drinker" in Experiments 1 and 3 and "peck feeder" in Experiment 3 were not analysed because they were rarely observed.

Time was included in the model for all behavioural analyses, but as it was accounted for in the analysis and is not relevant to this study, these results have been omitted. Main and interaction effects have also been omitted when they are superseded by higher-order interaction effects. Finally, effects that were too small to be biologically important are omitted in the presentation of results. For completeness and economy of space, means,

SEDs (standard error of a difference) and P-values of main effects for all behavioural parameters are presented in Supplementary Tables 1–14.

Experiment 1

The mean frequency and duration of different behaviours, SEDs and significance are presented in Table 2. Compared to birds housed on 1W3

Table 2. Effect of dry (1D) and wet (1W3) litter treatments on the number of bouts and total duration of different behaviours in Experiment 1

Behaviour	1D	1W3	SED	*P*-value
Mean bout frequency, n/h				
Stand	57.2	30.7	6.90	0.002
Walk	45.2	23.6	6.15	0.003
Preen	22.8	16.1	2.01	0.005
Peck ground	4.15 (62.4)[a]	1.62 (4.1)	0.274	<0.001
Peck total	5.20 (180.3)	4.47 (86.4)	0.305	0.031
Mean bout length, s				
Feed	40.7	72.1	13.91	0.039
Stand	23.4	52.7	9.66	0.008
Walk	4.87	6.19	0.589	0.041
Total bout duration, %				
Preen	15.79	9.64	2.102	0.010

[a]Backtransformed mean in parenthesis.

litter (high FPD scores), those on dry litter (1D) with lower FPD scores had more frequent bouts of standing, walking, preening, ground-pecking and total pecking, whereas the total duration of bouts of feeding, walking and standing was shorter, and the total duration of preening was higher.

A significant ($t = 2.689$, $P < 0.01$, $P = 0.027$) interaction of litter and analgesic occurred for the total duration of walking that was associated with the longer total duration of walking in salineinjected birds on dry (low FPD) compared to wet litter (high FPD). Means for birds on dry (1D) litter were 5.61% and 6.69% for betamethasone and saline injections respectively compared to 4.77 % and 3.49 % (SED 0.70 %) on wet (1W3) litter.

Significant interactions of line, litter and analgesic occurred for the total duration of resting and the total duration of standing ($P < 0.05$). Betamethasone-injected birds of line B, but not A, had shorter resting ($t = 2.37$, $P < 0.05$) and longer standing ($t = 2.37$, $P < 0.05$) durations than saline injected birds when both were on wet (1W) litter (Table 3).

T-pattern characteristics of birds on dry litter with low FPD scores showed that behavioural sequences were more frequent, varied and complex than in birds on wet litter with high FPD scores (Table 4).

Table 3. Mean total duration of different behaviours with significant line × litter × analgesic interactions in Experiment 1

	Line A		Line B			
Analgesia	1D	1W3	1D	1W3	SED	*P*-value
Resting, %						
Betamethasone	50.5	58.6	65.7	51.9	8.45	0.028
Saline	57.2	55.3	60.8	66.9	(6.34)[a]	
Standing, %						
Betamethasone	29.5	36.5	29.2	43.3	7.12	0.018
Saline	35.6	41.3	32.4	29.5	(5.74)[a]	

[a]SED for same level of line × litter.

Table 4. Effect of dry (1D) and wet (1W3) litter treatments on T-pattern characteristics in Experiment 1

T-pattern characteristic	1D	1W3	SED	*P*-value
Pattern occurrences, *n*	5.73 (307)[a]	4.15 (62)	0.377	<0.001
Original patterns, *n*	3.62 (36)	2.47 (11)	0.262	<0.001
Behaviours/pattern, *n*	4.04	3.25	0.246	0.006

[a]Backtransformed mean in parenthesis.

Experiment 3.1

Means for the frequency and duration of different behaviours in Experiment 3.1 are presented in Table 5. Walking, standing, resting, resting together, environment pecking and total pecking were performed more frequently by low FPD (3D) compared to high FPD birds (3W1) and by 3W1 compared to 3W2 (P < 0.05). Birds on dry litter also had a larger total duration of walking than 3W1, and 3W1 than 3W2 birds (P < 0.01). This pattern was reversed for the duration of resting bouts, with 3D performing the shortest bouts and 3W2 birds performing the longest (P = 0.011). Differences between 3D and 3W1 birds for resting bout duration and between 3W1 and 3W2 birds for environment pecking were not significant.

In contrast to the litter treatments, there were relatively few significant differences between birds given betamethasone or saline. Compared to saline-treated birds, betamethasone-injected poults had longer (P < 0.05) bouts of preenstanding, walking and standing regardless of litter treatment (Table 6). Significant interaction (P < 0.05) effects of litter and analgesic were present for the total duration of standing and resting (Table 6). Turkeys on wet (3W2) litter given saline had a shorter total standing duration compared to birds given betamethasone (P < 0.001) and birds on both 3W1 and 3D litter given saline

Table 5. Mean effect of dry (3D) and wet (3W1 and 3W2) litter treatments on the number of bouts, bout duration and total bout duration of behaviours in Experiment 3.1

Variable	3D	3W1	3W2	SED	*P*-value
Bout frequency, n/h					
Walk	30.9	15.8	5.3	5.32	0.003
Stand	40.2	23.0	9.0	5.85	0.002
Rest	11.1	8.6	4.5	1.05	<0.001
Rest together	7.7	5.1	2.4	1.00	0.002
Peck environment	365	119	49	66.8	0.003
Peck total	413	259	96	71.6	0.005
Mean bout length, s					
Rest	248	336	788	147.8	0.011
Mean total duration, %					
Walk	5.51	3.08	1.06	0.961	0.004

Table 6. Mean effect of analgesia on bout duration of walking, standing and preen standing and analgesia × litter on total duration of standing and resting in Experiment 3.1

Variable		Betamethasone	Saline	SED	*P*-value
Mean bout length, s					
Walk		6.97	4.98	0.715	0.021
Stand		85.5	42.3	17.45	0.035
Preen standing		16.4	9.4	3.1	0.050
Total bout duration, %					
Stand	3D	34.3 (32)[a]	30.9 (26)	7.96	0.016[c]
	3W1	36.7 (36)	31.0 (27)	(4.04)[b]	
	3W2	36.0 (35)	13.4 (5)		
Rest	3D	51.3 (61)	56.5 (70)	8.38	0.024[c]
	3W1	51.0 (60)	57.3 (71)	(4.36)[a]	
	3W2	51.9 (62)	75.8 (94)		

[a]Backtransformed mean in parenthesis.

[b]SED for same level of litter.

[c]Significance of the interaction of analgesia and litter treatment.

Table 7. Effect of dry (3D) and wet (3W1 and 3W2) litter treatments on T-pattern characteristics in Experiment 3.1

T-pattern characteristic	3D	3W1	3W2	SED	*P*-value
Pattern occurrences, *n*	108.2	41.1	16.4	15.77	<0.001
Original patterns, *n*	19.2	7.3	3.6	2.18	<0.001
Behaviours/pattern, *n*	3.82	3.16	1.11	0.306	<0.001

injections ($P < 0.05$). The 3W2 poults given saline had a longer total duration of resting compared to 3W2 birds given betamethasone ($t = 5.48$, $P < 0.001$), 3W1 given saline ($t = 2.21$, $P < 0.05$) and 3D birds ($t = 2.30$, $P < 0.05$).

Results of the THEME analysis are presented in Table 7. There was a significant ($P < 0.001$) effect of litter on pattern frequency, variety and complexity. Birds with low FPD scores (3D) had more frequent ($P < 0.001$), varied ($P < 0.001$) and complex ($P < 0.05$) behaviour patterns compared to 3W1 birds, and also when compared to 3W2 birds ($P < 0.001$) with high scores for FPD. Behavioural patterns in 3W1 birds were more complex than those of 3W2 ($P < 0.001$) and betamethasone provision increased pattern variety regardless of litter treatment (12.8 vs. 7.3, SED 1.81, $P = 0.015$).

Experiment 3.2

Significant ($P < 0.05$) interaction effects of FPD × analgesia × stage were identified for the total duration of resting and standing (Table 8). In low FPD (dry) birds moved to wet litter, those receiving betamethasone had a shorter total resting duration than those receiving saline ($P < 0.01$). The group mean for saline-treatedbirds was virtually unaltered between Experiments 3.1 and 3.2. In Experiment 3.1, high FPD saline birds rested more on wet litter than their betamethasone-injected pen mates,

Table 8. Mean total bout duration of behaviours of turkeys with low (dry litter, 3D) or high (wet litter, 3W2) FPD scores transferred respectively to wet or dry litter with significant interactions of FPD score (wet or dry litter), analgesia and stage (Experiment 3.1 or 3.2)

	Treatment					
Analgesia	3D	3D/3W2	3W2	3W2/3D	SED	*P*-value
Resting, %						
Betamethasone	60.7	47.3	62.9	71.4	11.19[a]	0.036
Saline	68.8	68.4	90.1	74.9		
Standing, %						
Betamethasone	32.1	43.7	35.1	25.3	10.38[b]	0.041
Saline	27.1	28.5	9.6	24.2		

[a]SED except for the same level of original litter and original litter × stage = 7.98; original litter × analgesic = 8.39. [b]SED except for the same level of original litter and original litter × stage = 7.68; original litter × analgesic = 8.15.

whereas when they were moved to dry litter the treatment means were not significantly different. In low FPD turkeys on wet litter, betamethasonetreated birds had a longer total standing duration than saline birds ($t = 1.98$, $P < 0.05$). In high FPD birds on dry litter, there was no significant difference between the betamethasone and saline treatments.

T-pattern characteristics are presented in Table 9 and show significant ($P < 0.01$) interactions between FPD and LMC for the number of patterns and pattern frequency. Low FPD birds transferred from dry to wet litter had a similar number of patterns, pattern frequency and behaviours/ pattern as on dry litter. The T-patterns of high FPD birds transferred from wet to dry litter were similar to those of low FPD birds on wet litter. Compared to saline injection, betamethasone increased pattern frequency (4.06 (57) vs. 3.20 (24), SED 0.230, $P < 0.01$) and number of patterns (i.e. variety) (2.29 (9) vs. 1.88 (6), SED 0.155, $P < 0.05$). A significant ($P = 0.013$) interaction of analgesia and LMC was associated with a single high value for saline-injected turkeys on dry litter and is likely anomalous (data not shown and overall pattern complexity for betamethasone and saline injections respectively were similar (3.34 (28) and 3.38 (28), SED 0.107), not significant).

Table 9. T-pattern characteristics in Experiment 3.2 for turkeys with low (dry litter) or high (wet litter) foot pad scores in Experiment 3.1 (3D and 3W) transferred from dry to wet litter (3D to 2W) or wet to dry litter (3W to 3D)

FPD	Treatment	Occurrence, *n*	Patterns, *n*	Behaviours/ pattern, *n*
Low	3D	4.69 (108)	2.84 (16)[c]	3.82
	3D to 3W2	4.26 (70)	2.39 (10)	3.62
High	3W2	1.93 (6)	1.08 (2)	2.83
	3W2 to 3D	3.62 (36)	2.04 (7)	3.16
SED[a]		0.554	0.293	0.187
SED[b]		0.476	0.301	0.224
Interaction[d], *P* =		0.003	0.002	0.101

[a]SED for different levels of FPD (low, high). [b]SED for the same level of FPD (low, high). [c]Backtransformed mean in parenthesis. [d]Interaction of FPD score (low, high) and litter condition (dry, wet).

DISCUSSION

Birds and other animals in chronic pain may exhibit several behavioural indicators of pain, including self-imposed isolation, changes in appetite and reductions in activity, preening, interest in surroundings and the use of injured body parts (Morton and Griffiths, 1985; Gentle and Corr, 1995; Weary et al., 2006; Lierz and Korbel, 2012). Analgesic intervention can provide additional evidence that these changes are caused by pain. Standing and activity levels were shown to increase in turkeys with severe degenerative hip joint lesions following provision of betamethasone (Duncan et al., 1991); in birds with less extensive joint damage, activity levels were similar to those of unaffected birds and were not increased by provision of betamethasone (Hocking et al., 1999), suggesting that such behavioural changes in turkeys are pain-induced.

Behavioural pattern sequencing analysis potentially reveals more subtle behavioural changes than gross behavioural changes. Reductions in pattern frequency, variety and complexity indicate that either performance of certain behaviours (or activity in general) has decreased, or that durations between behaviours in some patterns have become more variable. Such disruptions have been shown to be associated with stressors such as disease and hunger (Alados et al., 1996; María et al., 2004; MacIntosh et al., 2011).

Experiment 1

Longer feeding bouts without an increase in total feeding duration were observed on wet litter and are reflected in the lower frequency and longer bout duration of standing (Table 2). These changes in feeding behaviour have also been reported in lame broiler chickens (Weeks et al., 2000); however, in the absence of altered feed intake and an effect of analgesia, pain is an unlikely cause of these differences. Birds on wet litter also had lower frequencies but higher durations of walking, possibly due to differences in walking pace and/or reduced interest in stopping to engage with the environment or conspecifics. In saline treatments, birds on wet litter had a shorter total walking duration than those on dry litter, whereas when analgesia was provided, no difference was apparent, suggesting that pain due to the more severe lesions seen on wet litter was the cause.

In line B birds housed on wet litter, shorter resting and longer standing durations were observed when betamethasone was provided, suggesting that these changes were due to pain relief; line A birds showed no such change (Table 3). This may indicate subtle differences in FPD severity missed during scoring, or beed variation in pain perception or responses to pain (Short, 1998). If a variation in pain perception is responsible, in spite of similar breed susceptibility, the negative welfare impact of FPD could differ between breeds.

Frequency, variety and complexity of T-patterns were all lower on wet litter (Table 4), again suggesting that high litter moisture had a negative impact on welfare.

Experiment 3.1

A linear decrease in activity occurred in response to increasing litter moisture as measured by reductions in the frequency of walking, standing, resting, environmental pecking and total pecking as well as reductions in total walking duration and an increase in resting bout duration (Table 5). Lower pecking frequencies in birds kept on wet litter in Experiments 1 and 3.1 compared to birds kept on dry litter are likely to reflect differences in litter conditions such as friability.

The increases in the frequency of standpreening and bout duration of walking, standing and stand-preening observed in betamethasoneinjected birds occurred regardless of litter condition or FPD score (Table 6). These differences may therefore have been due to the presence of other, unknown sources of leg pain that were alleviated with betamethasone injection. Alternatively, this result may reflect a side effect of betamethasone. Betamethasone is a potent glucocorticoid analgesic with anti-inflammatory activity that may also affect behaviour such as increased feeding or foraging (Berthon et al., 2014; Liu et al., 2014). Side effects were not observed in a previous study in a test situation in turkeys (Duncan et al., 1991) or in the companion study on gait analysis in a walkway test (Wyneken et al., 2015). Increased scratching and pecking at the litter, decreased standing and lower mating activity were observed in adult broiler breeders with or without musculoskeletal disease (Hocking, 1994), whereas betamethasone had no effect on activity in adult male turkeys with or without relatively mild musculoskeletal lesions (Hocking et al., 1999). In the current series of experiments, any side effect of betamethasone was controlled for by the use of birds with and without FPD that enabled the effect of the drug on pain perception to be distinguished from that arising from any side effects of the drug.

Birds with high FPD (3W2) receiving saline stood less and rested more than all other treatment groups, suggesting that these were the only birds in pain, which was alleviated in the 3W2 birds receiving betamethasone. Although FPD scores for 3W1 and 3W2 birds were similar, due to the need of the related study their feet were scored 4 d after filming. During this period, 3W1 birds were kept on wet litter for an additional 2 d during which foot pad quality is likely to have deteriorated notably, whereas little, if any, healing would have occurred in the 3W2 birds (Mayne et al., 2007a). The difference between 3W1 and 3W2 lesion scores is therefore likely to have decreased during this period and may, in part, explain discrepancies between FPD scores and the effects of litter × analgesic interactions between 3W1 and 3W2 and this experiment.

Regardless of analgesic or stage, increasing LMC across litter treatments resulted in a gradual reduction in T-pattern frequency, variety and complexity (Table 7). These results are generally consistent with the results of Experiment 1 and support the suggestion that an LMC-associated stressor other than pain (possibly the presence of cold, wet, non-friable litter) was responsible for these behavioural differences.

Experiment 3.2

In birds with high FPD scores housed on dry litter, significant differences between saline and betamethasone treatments were not present for total duration of standing or resting (Table 8). In birds with low FPD scores housed on wet litter, saline birds had a longer total duration of resting and shorter total duration of standing than betamethasone birds. These results appear to suggest that the behavioural differences on wet litter (and lack thereof on dry litter) in Experiments 3.1 and 3.2 were due to LMC rather than pain. However, on wet litter, a motivation to avoid resting on cold, soggy litter may be present. There was a tendency towards standing more and resting less in betamethasone birds transferred from dry to wet litter that is consistent with such an explanation. The means of both behaviours were virtually unchanged by transfer to wet litter in saline birds, suggesting that this motivation was possibly overridden by pain in these birds.

When moved to dry litter, high FPD salineinjected birds showed a tendency to behave more like their betamethasone-injected pen mates (Tables 8 and 9). This could indicate pain suppression caused by an increased interest in their new environment and/or a positive affective state. Conversely, the lack of such an effect in saline birds moved to wet litter may have been caused by an overriding negative affective state due to the introduction of cold, wet, non-friable litter. Cognitive biases are considered good indicators of affective state in both humans and animals (Mendl et al., 2009). Positive cognitive biases (suggestive of positive affective states) occur in animals when environmental improvements are introduced (Matheson et al., 2008; Brydges et al., 2011; Douglas et al., 2012), while animals kept in poor conditions (and those for which good conditions are negatively altered) exhibit negative cognitive biases, suggesting the presence of negative affective states (Bateson and Matheson, 2007; Destrez et al., 2013). It is therefore possible that turkeys moved from dry to wet litter experienced a negative affective state while those moved from wet to dry litter experienced a positive affective state.

In humans, pain perception can be increased or decreased respectively by negative or positive affective states (Villemure and Schweinhardt, 2010). Rats show heightened responses to pain in stressful situations (Andre et al., 2005). Turkeys may experience a similar increase in pain perception due to a negative affective state induced by wet litter. Pain suppression also occurs in humans and chickens when attention is directed away from pain towards other stimuli (Gentle and Corr, 1995; Villemure and Schweinhardt, 2010) and pain perception in turkeys could be subject to similar alterations. If this is indeed the case, turkeys in the earliest stages of FPD onset may experience pain when housed in poor conditions including (though not exclusively) when kept on wet litter. The study of turkey behavior in commercial settings could therefore be informative and yield more pronounced behavioural changes at lower FPD scores than in this experiment. An alternative explanation for the results of Experiments 3.1 and 3.2 is that none of the turkeys were experiencing pain and that differences were due to environmental stress per se. However, this would not explain the role of betamethasone in the litter × analgesia effect observed in Experiment 3.1 and in line B in Experiment 1.

Synthesis and critique

Wet bedding is regularly used to induce depressionlike states in rats (Kompagne et al., 2008). Similar states may have been induced by wet litter in the current experiments and would explain the lower activity levels observed on wet litter. Reduced ground (and total) pecking on wet litter was probably influenced by low litter friability. However, the frequency and total duration of preening, a possible indication of frustration (Duncan and Wood-Gush, 1972), were also lower on wet litter which mplies that reduced welfare was the main cause of activity reductions in wet litter birds.

Pain intensity in humans is known to increase with injury depth (Fruhstorfer et al., 1995). This is also known to be true in rats, where skin-deep incisions to the paw fail to elicit the guarding behaviours that follow deeper incisions (Xu and Brennan, 2009). The scoring system used in this study does not separate superficial tissue damage from deeper lesions in which the dermis is exposed and has previously produced external scores which correlated poorly with histopathological findings (Mayne et al., 2007b). A histologically validated scoring system has recently been formulated for use in both chickens (Michel et al., 2012) and turkeys (Allain et al., 2013) which incorporates lesion depth and may be more suitable for future studies. Attentional biases and affective states are likely to affect pain perception and, even in the earliest stages of FPD, turkeys may be experiencing pain when in negative affective states and barren and/or stressful environments.

CONCLUSIONS

It is clear from these results that wet litter is associated with behavioural changes that suggest a compromised state of welfare in growing turkeys regardless of FPD. T-pattern analysis supplemented the standard analyses of frequency and duration of behaviour and differences between treatments were large and consistent with the conclusion that wet litter adversely affected the welfare of the birds. These results highlight the importance of the environment (wet vs. dry litter) on behaviour and the potential confounding effects of foot pad lesions and litter conditions that will need to be accounted for in future experiments to evaluate pain from FPD.

Betamethasone used as an analgesic also changed behaviour and may indicate that this steroid drug stimulated general activity. There was evidence that FPD was painful in the comparison of analgesia and saline in line B of Experiment 1 and the results of Experiments 3.1 and 3.2. Taken together, these indicate that the welfare of birds with high FPD scores is associated with pain that may be attenuated by attentional shifts.

ACKNOWLEDGEMENTS

This work includes material submitted in partial fulfilment of the MSc in Applied Animal Behaviour and Animal Welfare by CWW and AM. Poults and material support were kindly donated by Aviagen Turkeys Ltd, Chester, UK.

DISCLOSURE STATEMENT

No potential conflict of interest was reported by the authors.

FUNDING

The experiments were funded by the BBSRC (contract number BB/L012022/1) as part of the FP7 ERA-Net ANIHWA, "Coordination of European Research on Animal Health and Welfare" project TURKEYWELFARE [number 182]. Financial support and poults were kindly provided by Aviagen Turkeys, Chester, UK. The Roslin Institute is supported by an Institute Core Strategic Grant from the Biotechnology and Biological Sciences Research Council.

REFERENCES

1. Alados, C.L., Escos, J.M. & Emlen, J.M. (1996) Fractal structure of sequential behaviour patterns: an indicator of stress. *Animal Behaviour*, 51: 437–443. doi:10. 1006/anbe.1996.0040
2. Allain, V., Huonnic, D., Rouina, M. & Michel, V. (2013) Prevalence of skin lesions in turkeys at slaughter. *British Poultry Science*, 54: 33–41. doi:10.1080/0007 1668. 2013.764397
3. Andre, J., Zeau, B., Pohl, M., Cesselin, F., Benoliel, J.J. & Becker, C. (2005)Involvement of cholecystokininergic systems in anxiety-induced hyperalgesia in male rats: behavioral and biochemical studies. *Journal of Neuroscience*, 25: 7896–7904. doi:10.1523/JNEUROSCI.0743-05.2005
4. Bateson, M. & Matheson, S.M. (2007) Performance on a categorisation task suggests that removal of environmental enrichment induces 'pessimism' in captive European starlings (Sturnus vulgaris). *Animal Welfare*, 16: 33–36.
5. Berthon, B.S., MacDonald-Wicks, L.K. & Wood, L.G. (2014) A systematic review of the effect of oral glucocorticoids on energy intake, appetite, and body weight in humans. *Nutrition Research*, 34: 179–190. doi:10.1016/j.nutres.2013.12.006
6. Brydges, N.M., Leach, M., Nicol, K., Wright, R. & Bateson, M. (2011)Environmental enrichment induces optimistic cognitive bias in rats. *Animal Behaviour*, 81: 169–175. doi:10.1016/j.anbehav.2010.09.030
7. Buda, S., Platt, S. & Budras, K.D. (2002) Sensory nerve endings in the foot pads of turkeys, in: Hafez, H.M. (Ed) *Proceedings of the 4th International Symposium on Turkey Diseases*, pp. 78–82 (Berlin, Germany, DVG-Verlag).
8. Chavez, E. & Kratzer, F.H. (1972) Prevention of foot pad dermatitis in poults with methionine. *Poultry Science*, 51: 1545–1548. doi:10.3382/ps.0511545
9. Clark, S., Hansen, G., McLean, P., Bond, P., Wakeman, W., Meadows, R. & Buda, S.(200 2) Pododermatitis in turkeys. *Avian Diseases*, 46: 1038–1044. doi:10.1637/0005-2086(2002)046[1038:PIT]2.0.CO;2
10. Destrez, A., Deiss, V., Lévy, F., Calandreau, L., Lee, C., Chaillou-Sagon, E. & Boissy,A. (2013) Chronic stress induces pessimistic-like judgment and learning deficits in sheep. *Applied Animal Behaviour Science*, 148: 28–36. doi:10.1016/ j.applanim. 2013.07.016
11. Douglas, C., Bateson, M., Walsh, C., Bédué, A. & Edwards, S.A. (2012)Environmental enrichment induces optimistic cognitive biases in pigs. *Applied Animal Behaviour Science*, 139: 65–73. doi:10.1016/j.applanim.2012.02.018
12. Duncan, I.J.H., Beatty, E.R., Hocking, P.M. & Duff, S.R.I. (1991) Assessment of pain associated with degenerative hip disorders in adult male turkeys. *Research in Veterinary Science*, 50: 200–203. doi:10.1016/0034-5288(91)90106-X
13. Duncan, I.J.H. & Wood-Gush, D.G.M. (1972) An analysis of displacement preening in the domestic fowl. *Animal Behaviour*, 20: 68–71. doi:10.1016/S0003-3472 (72)80174-X
14. Fruhstorfer, H., Müller, T. & Scheer, E. (1995) Capillary blood sampling: how much pain is necessary? Part 2: relation between penetration depth and puncture pain. *Practical Diabetes International*, 12: 184–185. doi:10.1002/(ISSN)1528-252X
15. Gentle, M.J. & Corr, S.A. (1995) Endogenous analgesia in the chicken. *Neuroscience Letters*, 201: 211–214. doi:10.1016/0304-3940(95)12181-1

16. Hocking, P.M. (1994) Assessment of the welfare of food restricted male broiler breeder poultry with musculoskeletal disease. *Research in Veterinary Science*, 57:28–34. doi:10.1016/0034-5288(94)90077-9
17. Hocking, P.M., Bernard, R. & Maxwell, M.H. (1999) Assessment of pain during locomotion and the welfare of adult male turkeys with destructive cartilage loss of the hip joint. *British Poultry Science*, 40: 30–34. doi:10.1080/00071669987791
18. Hocking, P.M. & Wu, K. (2013) Traditional and commercial turkeys show similar susceptibility to foot pad dermatitis and behavioural evidence of pain. *British Poultry Science*, 54: 281–288. doi:10.1080/00071668.2013.781265
19. Kompagne, H., Bárdos, G., Szénási, G., Gacsályi, I., Hársing, L.G. & Lévay, G.(2008) Chr onic mild stress generates clear depressive but ambiguous anxiety-like behaviour in rats. *Behavioural Brain Research*, 193: 311–314. doi:10.1016/j.bbr.2008.06.008
20. Lierz, M. & Korbel, R. (2012) Anesthesia and analgesia in birds. *Journal of Exotic Pet Medicine*, 21: 44–58. doi:10.1053/j.jepm.2011.11.008
21. Liu, L., Song, Z., Jiao, H. & Lin, H. (2014) Glucocorticoids increase NPY gene expression via hypothalamic AMPK signaling in broiler chicks. *Endocrinology*, 155:2190–2198. doi:10.1210/en.2013-1632
22. MacIntosh, A.J.J., Alados, C.L. & Huffman, M.A. (2011) Fractal analysis of behaviour in a wild primate: behavioural complexity in health and disease.*Journal of the Royal Society Interface*, 8: 1497–1509.
23. Magnusson, M.S. (2000) Discovering hidden time patterns in behavior: T-patterns and their detection. *Behavior Research Methods, Instruments, & Computers*, 32: 93–110. doi:10.3758/BF03200792
24. María, G.A., Escós, J. & Alados, C.L. (2004) Complexity of behavioural sequences and their relation to stress conditions in chickens (Gallus gallus domesticus): a non-invasive technique to evaluate animal welfare. *Applied Animal Behaviour Science*, 86: 93–104. doi:10.1016/j.applanim.2003.11.012
25. Matheson, S.M., Asher, L. & Bateson, M. (2008) Larger, enriched cages are associated with 'optimistic' response biases in captive European starlings (*Sturnus vulgaris*). *Applied Animal Behaviour Science*, 109: 374–383. doi:10.1016/ j.applanim. 2007.03.007
26. Mayne, R.K. (2005) A review of the aetiology and possible causative factors of foot pad dermatitis in growing turkeys and broilers. *Worlds Poultry Science Journal*, 61: 256–267. doi:10.1079/WPS200458
27. Mayne, R.K., Else, R.W. & Hocking, P.M. (2007a) High litter moisture alone is sufficient to cause footpad dermatitis in growing turkeys. *British Poultry Science*, 48: 538–545. doi:10.1080/00071660701573045
28. Mayne, R.K., Else, R.W. & Hocking, P.M. (2007b) High dietary concentrations of biotin did not prevent foot pad dermatitis in growing turkeys and external scores were poor indicators of histopathological lesions. *British Poultry Science*, 48: 291–298. doi:10.1080/00071660701370509
29. Mayne, R.K., Powell, F., Else, R.W., Kaiser, P. & Hocking, P.M. (2007c) Foot pad dermatitis in growing turkeys is associated with cytokine and cellular changes indicative of an inflammatory immune response. *Avian Pathology*, 36: 453–459. doi:10.1080/03079450701639327
30. Mendl, M., Burman, O.H.P., Parker, R.M.A. & Paul, E.S. (2009) Cognitive bias as an indicator of animal emotion and welfare: emerging evidence and underlying mechanisms. *Applied Animal Behaviour Science*, 118: 161–181. doi:10.1016/ j.applanim.2009.02.023
31. Michel, V., Prampart, E., Mirabito, L., Allain, V., Arnould, C., Huonnic, D., Le Bouquin, S. & Albaric, O. (2012) Histologically-validated footpad dermatitis scoring system for use in chicken processing plants. *British Poultry Science*, 53:275–281. doi:10.1080/00071668.2012.695336
32. Morton, D.B. & Griffiths, P.H.M. (1985) Guidelines on the recognition of pain, distress and discomfort in experimental animals and an hypothesis for assessment. *Veterinary Record*, 116: 431–436. doi:10.1136/vr.116.16.431

33. Short, C.E. (1998) Fundamentals of pain perception in animals. *Applied Animal Behaviour Science*, 59: 125–133. doi:10.1016/S0168-1591(98)00127-0
34. Tucker, S. & Walker, A. (1999) Hock burn in broilers, in: Wiseman, J. &Garnsworthy, P.C. (Eds) *Recent Developments in Poultry Nutrition*, Vol. 2, pp. 107–122 (Nottingham, Nottingham University Press).
35. Villemure, C. & Schweinhardt, P. (2010) Supraspinal pain processing: distinct roles of emotion and attention. *The Neuroscientist*, 16: 276–284. doi:10.1177/ 10738 58409359200
36. Weary, D.M., Niel, L., Flower, F.C. & Fraser, D. (2006) Identifying and preventing pain in animals. *Applied Animal Behaviour Science*, 100: 64–76. doi:10.1016/j. applanim.2006.04.013
37. Weber Wyneken, C., Sinclair, A., Veldkamp, T., Vinco, L.J. & Hocking, P.M. (2015)Foot pad dermatitis and pain assessment in turkey poults using analgesia and objective gait analysis. *British Poultry Science*, 56 (5): 522–530. doi:10.1080/ 0007 1668.2015.1077203
38. Weeks, C.A., Danbury, T.D., Davies, H.C., Hunt, P. & Kestin, S.C. (2000) The behaviour of broiler chickens and its modification by lameness. *Applied Animal Behaviour Science*, 67: 111–125. doi:10.1016/S0168-1591(99)00102-1
39. Wu, K. & Hocking, P.M. (2011) Turkeys are equally susceptible to foot pad dermatitis from 1 to 10 weeks of age and foot pad scores were minimized when litter moisture was less than 30%. *Poultry Science*, 90: 1170–1178. doi:10.3382/ps.2010-01202
40. Xu, J. & Brennan, T.J. (2009) Comparison of skin incision vs. skin plus deep tissue incision on ongoing pain and spontaneous activity in dorsal horn neurons. *Pain*, 144: 329–339. doi:10.1016/j.pain.2009.05.019

CHAPTER 4

The Effect of Age on Growth Performance and Carcass Quality Parameters in Different Poultry Species

Daria Murawska*

Department of Commodity Science and Animal Improvement, Faculty of Animal Bioengineering, University of Warmia and Mazury in Olsztyn, Olsztyn, Poland

ABSTRACT

In recent years, a steady increase in global poultry meat production has been witnessed, accompanied by an increase in a major portion of a poultry carcass, referred to as the inedible portion. In poultry, edible components include meat, skin with subcutaneous fat and giblets (gizzard, liver, and heart) and sometimes also abdominal fat in waterfowl. Age, together with species and environmental conditions, is one of the key factors affecting body growth rate. In four poultry species, chickens, turkeys, Pekin ducks, and geese, an increase in body weight is accompanied by an increase in edible weight and a decrease in inedible weight in the carcass, and more significant age-related changes occur in turkeys and broiler chickens than in ducks and geese. The highest increase in the content of edible components expressed as a percentage of total body weight is noted in turkeys (20% in males, 25% in females), followed by broiler chickens (19.4%), ducks (17.1%), and geese (only 8.2%). Gallinaceous birds have also a higher content of muscle tissue and a lower content of skin (including subcutaneous fat) and bones than waterfowl.

Keywords: tissue components, edible components, non-edible components, poultry species

1. INTRODUCTION

Age, together with species and environmental conditions, belongs to the key factors affecting the growth rate of birds. Age has a significant effect also on carcass tissue composition, and the most profound changes, both qualitative and quantitative, occur at early life stages which are relatively short (compared with the lifespan) and characterized by rapid growth.

Selection progress in meat-type poultry (particularly in broiler chickens and turkeys) has contributed to an increase in their body weight, improved carcass composition, a shorter production period, and a substantial rise in carcass dressing percentage, which results from an increase in the content of edible portions in the total body weight of birds, accompanied by a decrease in the content of inedible components treated as slaughter offal [1]. The total percentage content of edible and inedible components in the carcasses of different poultry species is an important economic consideration since the waste load from meat processing plants has to be effectively managed or disposed of [2–7].

In poultry, edible components include meat, skin with subcutaneous fat, and giblets (gizzard, liver, and heart), sometimes also abdominal fat in waterfowl. Fast-growing birds need concentrated feeds in an adequate amount and appropriate form. Changes in the nutritional regime affect the function of gastrointestinal tract segments and the weight of internal organs [8–10]. The growth rate of muscle tissue—the most valuable edible component—is faster than the growth rates of internal organs and other body parts classified as inedible, such as feathers, blood, head, and others [11–16].

In growing birds, significant age-related changes can be observed not only in carcass tissue composition but also in the location of lean meat, skin with a fat layer, and bones in the carcass [11, 17–20]. Carcass quality is also determined by the distribution of tissue components. Lean meat should be located in the most valuable carcass parts—the breast and legs, rather than in the neck, wings, and back, which are less valuable. The tissue composition of poultry carcasses changes with age because the growth rates of tissue components vary across species.

Carcass parts and tissue components can be classified as edible and nonedible. Methodological differences in carcass preparation (carcass with or without the neck and wing tips) and the classification of abdominal fat as an edible or a nonedible component make it difficult to compare the research findings reported by various authors. The present report summarizes mostly the results of own studies due to the uniform methodological approach to carcass processing and classification of edible and nonedible carcass components in the analyzed poultry species, that is,

Ross 308 broiler chickens [13], Pekin ducks [14], BIG 6 turkeys [15, 20], and Koluda® White geese [16, 21].

2. THE EFFECT OF AGE ON THE GROWTH RATES OF TISSUES AND ORGANS IN BROILER CHICKENS, TURKEYS, DUCKS, AND GEESE

Selective breeding carried out over many generations has resulted in specialized poultry meat and egg production based upon specialized meat stock and egg stock, including single-purpose strains and lines selected separately. Breeds and strains of meat-type poultry are intensely selected for fast growth rate, high body weight, high carcass dressing percentage, and lean tissue deposition. Dual-purpose birds, raised for both meat and eggs, are less cost-effective than specialized layers and broilers in large-scale commercial poultry farming.

2.1. Body Weight and Carcass Weight

Body weight is highly heritable and easy to measure [22, 23]. Body weight variation over time considerably affects production effectiveness in meat-type poultry. The growth rates of birds vary with age and across species. In comparison with ducks and geese, chickens and turkeys grow slower in the first weeks of life. The growth rates of ducks and geese decrease at 7 and 10 weeks of age, respectively. Migratory birds such as ducks and geese have to be capable of flight within a short time after hatching, so they gain weight rapidly during early life stages. Despite domestication and long-term selection aimed at eliminating atavistic features, adjustments enhancing flight capabilities can still be observed in modern ducks and geese [24, 25]. Broiler chickens have been raised since the 1920s, but initially only surplus cockerels from the laying stock were used as a source of poultry meat. In 1923, the average body weight of meat-type cockerels at 16 weeks of age was 1 kg [26]. Today, the average slaughter age of broiler chickens is 35–42 days at a body weight of 2.10–2.80 kg [27]. Between 1 and 10 weeks of age, the body weight and carcass weight of broiler chickens increase nearly 30-fold and over 40-fold, respectively [13] (**Figure 1a**). If the rearing period is longer than 35–42 days, the produced broilers have higher body weight and muscle yield, but also higher carcass fat content.

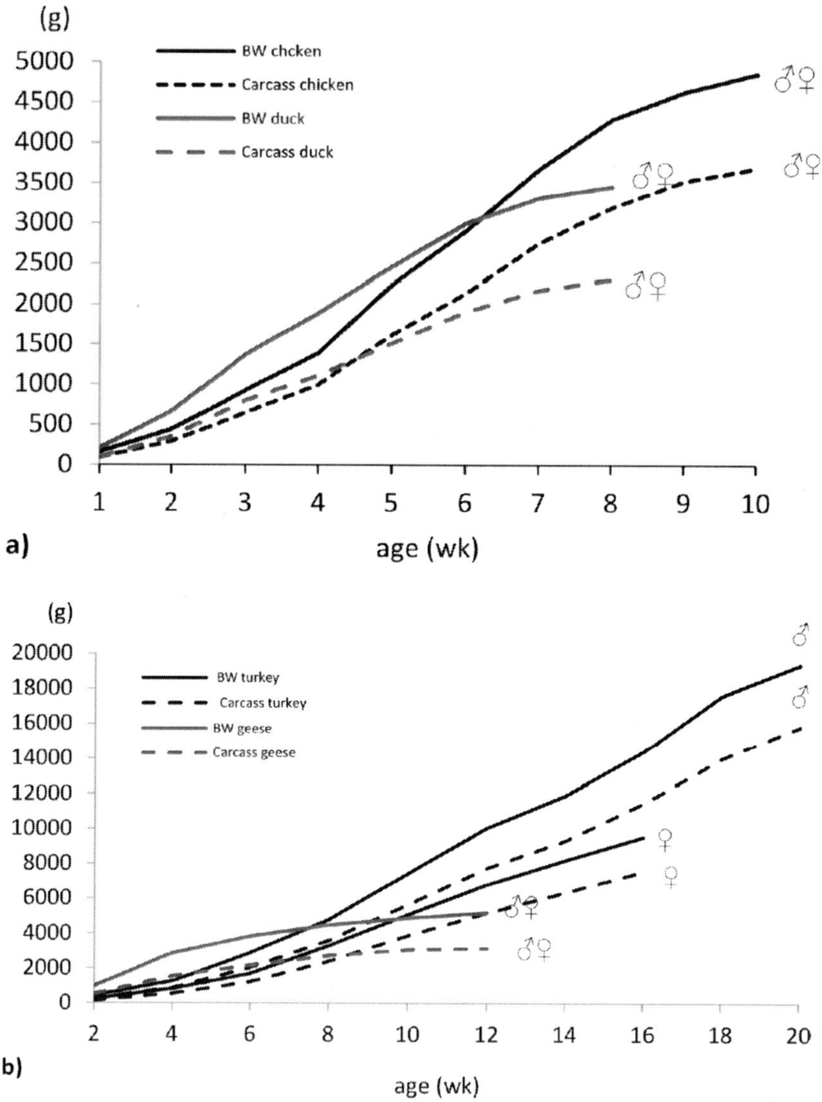

FIGURE 1. Arithmetic means (x) for the body weight and carcass weight of (a) chickens and ducks, (b) turkeys and geese (g).

Turkeys are the second most popular meat-type poultry species, after broiler chickens. Turkeys are heaviest among the four most common poultry species raised for meat. In comparison with other poultry species, turkeys are characterized by the highest carcass dressing percentage and high feed efficiency [4–6]. Turkey carcasses have high lean content and relatively low fat content [28]. Based on their body weight, modern commercial hybrid turkeys can be divided into heavy type, medium-heavy type, and medium type. Depending on the type, the body weight of males and females at 20 weeks of age ranges from 16.8 to 21 kg and from 9.3 to 11 kg, respectively. Due to considerable sexual dimorphism in the body weight of turkeys, males are reared over a longer period of time than females. Turkeys are characterized by the highest carcass dressing percentage among all poultry species, which reaches 81% in 16-week-old females and 84% in 21-week-old males [29].

The average body weight of male and female BIG 6 turkeys increases 45.6-fold between 2 and 20 weeks of age, and 40.8-fold between 2 and 16 weeks of age, respectively. The respective values for carcass weight are 58.3-fold (♂) and 57.7-fold (♀). A nearly 58-fold increase in carcass weight is observed already at 16 weeks of age in females and at 20 weeks of age in males, as compared with initial carcass weight at 2 weeks of age [15, 20] (**Figure 1b**).

Ducks can be divided into meat type, egg-layer type, and dual purpose. Broiler ducks are raised to 7–8 weeks of age and slaughtered at a body weight of approximately 3.5 kg. Hybrid Pekin ducks are used commercially for meat production [9]. Between 1 and 8 weeks of age, the average body weight and carcass weight of Pekin ducks increase 16.5-fold and 23.3-fold, respectively [14] (**Figure 1a**).

Geese are also raised mostly for meat, whereas egg, feather, and down production remains marginal. Geese are slaughtered at different ages, depending on breed and management system [30, 31]. Broiler-type geese are raised intensively to 10–12 weeks of age, and they are slaughtered at a body weight of approximately 5.2 kg. Geese have a fast initial growth rate. From 1 day to 12 weeks of age, their body weight increases over 34-fold [21], compared with a 5.3-fold increase in body weight and a 6.2-fold increase in carcass weight between 2 and 12 weeks of age [16] (**Figure 1b**).

Broiler chickens intended for grilling, with carcass weight of approximately 1 kg, can be slaughtered at 28 days of age. Medium-heavy female turkeys that are to be roasted whole can be slaughtered at 12 weeks of age. However, market demand for such products tends to be seasonal and remains low. Large carcasses that can be divided into retail and market-ready cuts or used for further processing are preferred, and this applies also to ducks and geese.

2.2. Growth Rates of Tissue Components

The whole-body growth rate of birds varies over time, similarly as the growth rates of tissue components, which affects the proportions between carcass parts and tissue components [11, 17–19]. Chicks, poults, ducklings, and goslings are characterized by relatively slow growth of muscle tissue and adipose tissue, and fast growth of bones and internal organs. The content of muscle tissue and skin (including subcutaneous fat) increases significantly with age, whereas the content of giblets (liver, heart, gizzard), bone, and slaughter offal (excluding abdominal fat and feathers) decreases [13–16].

2.2.1. Lean Meat

In birds raised specifically for meat production, both the percentage and the distribution of lean tissue in the carcass are important consideration. In poultry, breast and leg muscles constitute the most valuable carcass portion. The growth rates of muscle groups vary across poultry species. Similar age-related changes in muscle growth can be observed in chickens and turkeys, and in ducks and geese.

In broiler chickens, carcass lean content increases approximately 49-fold between 1 and 10 weeks of age. Over this period, the percentage of muscle tissue increases from 30.9 to 51.3% of total body weight, by 16.5% until week 6, and by approximately 4% between 6 and 10 weeks of age [13] (**Figure 2a**). In the carcasses of 2-week-old broiler chickens, approximately 36 and 35% of lean meat is located in the breast and legs, respectively. As birds grow older, the percentage of lean tissue increases in the breast (to 44%) and decreases insignificantly in the legs (to 32%), relative to the total lean content of the carcass [17].

Turkey carcasses have high lean content [5, 6]. The growth rates of muscle tissue vary between carcass parts. In male BIG 6 turkeys, the weight of breast muscles increases approximately 77.1-fold between 2 and 20 weeks of age, the weight of leg muscles increases 75.4-fold, and the combined weight of the remaining muscles is only 54.4-fold. In female BIG 6 turkeys, the weights of breast muscles and leg muscles increase 66.7-fold and 70.5-fold, respectively, between 2 and 16 weeks of age, whereas the combined weight of the remaining muscles increases 60.5-fold [15, 20]. Muscle tissue accounts for approximately 40 and 34% of the total body weight in 2-week-old males and females, respectively, and increases to approximately 61 and 56%, respectively, at slaughter, that is, at 20 weeks of age in males and 16 weeks in females [15]. Changes in the distribution of lean meat in the carcass can be observed in growing turkeys. In males, lean meat content (relative to total lean weight in the carcass) increases in the breast (by

3.4%) and legs (by 1.8%), and decreases in the wings (by 2.8%), back (by 1.5%), and neck (by 1%). In females, muscle tissue distribution in the carcass is similar, with a higher (by 4.2%) percentage of lean meat located in the breast. The most valuable muscles, that is, breast and leg muscles, account for 72 and 74% of total meat weight in males and females at slaughter age, respectively [20].

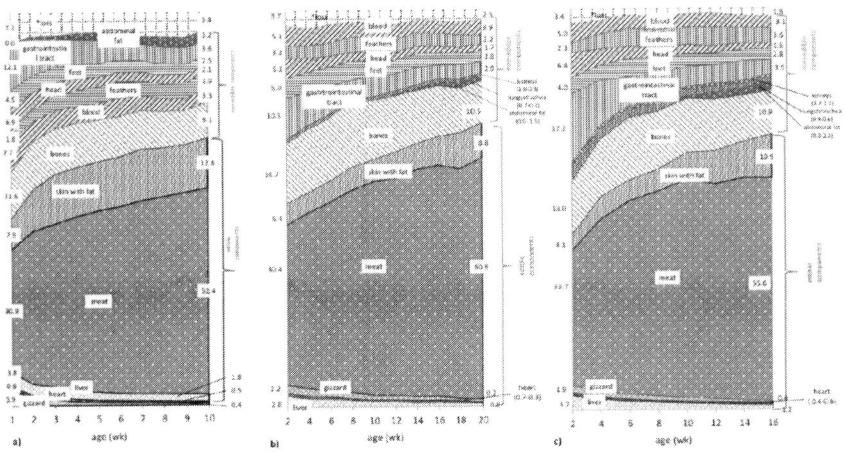

FIGURE 2. Percentage content of particular components in the body weight of (a) broiler chickens and turkeys, (b) males, (c) females. *loss = body weight loss during post-slaughter processing and dissection (%) [13, 15].

Ducklings and goslings have well-developed leg muscles and poorly developed breast muscles. In Pekin ducks, the weight of muscle tissue increases 29.3-fold between 1 and 8 weeks of age. The percentage of lean meat increases from 17.1 to 30.4% of the total body weight over this period. The growth rate of breast muscles increases, and the growth rate of leg muscles decreases with age. Between 1 and 8 weeks, breast muscle weight increases 188-fold, whereas leg muscle weight only 13.6-fold. In 1-week-old ducklings, breast muscles and leg muscles account for approximately 5 and nearly 60% of total muscle weight, respectively. At 8 weeks of age, the proportion of breast muscles increases to approximately 33%, whereas the proportion of leg muscles decreases to 27% of total muscle weight (**Figure 3a** and **c**) [14].

In geese, muscle tissue weight increases 8.5-fold between 2 and 12 weeks of age. Age-related changes in lean meat weight in different carcass parts are similar in geese and ducks. The growth rate of leg muscles decreases, and the growth rate of breast muscles increases between 2 and 10 weeks of age. Over this period, the weights of breast muscles and leg muscles increase 42-

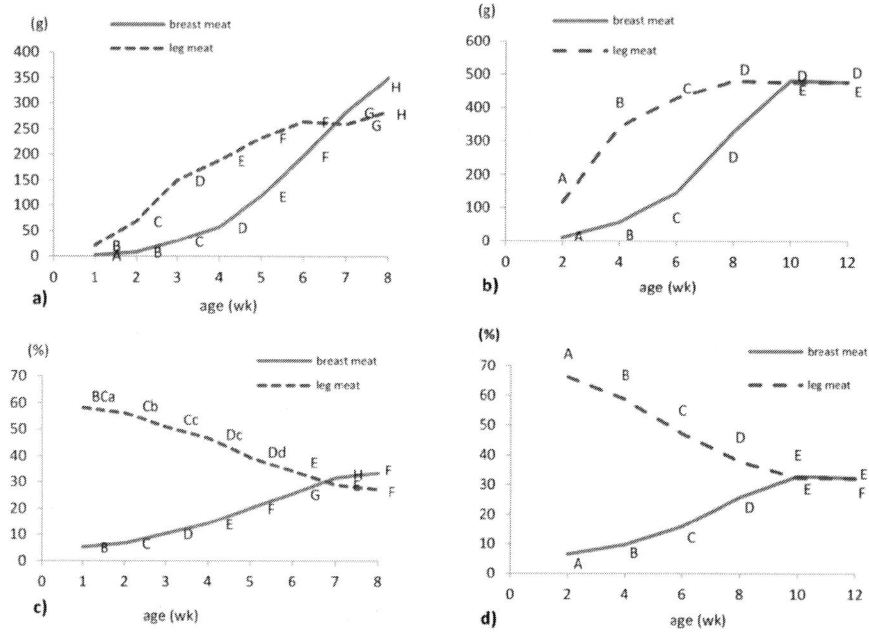

FIGURE 3. Arithmetic means (X‾‾‾X‾) for weight of breast muscles and leg muscles (g): (a) ducks, (b) geese and percentage share of breast muscles and leg muscles in total lean weight, subject to the age of birds (%), (c) ducks, and (d) geese. Values followed by different letters (age) differ significantly: capital letters $\alpha = 0.01$, small letters $\alpha = 0.05$ [14, 16].

fold and only 4-fold, respectively. As a result, the percentage of lean meat increases rapidly in the breast and decreases in the legs [19]. In 2-week-old geese, lean meat accounts for 18.0% of the total body weight, and it increases to nearly 29.0% at 12 weeks of age. In 2-week-old geese, leg muscles and breast muscles account for approximately 66.0% and only 7% of total muscle weight in the carcass, respectively. At 12 weeks of age, the proportion of breast muscles and leg muscles (expressed as a percentage of total lean weight) is comparable, that is, approximately 33 and 32%, respectively (**Figure 3b** and **d**) [16].

2.2.2. Skin With Subcutaneous Fat

Unlike Pekin ducks [9, 18] and geese [30, 31], chickens and turkeys are characterized by relatively low fat content [17, 20, 32]. The weight of fat and skin increases, at a different rate, throughout the growing period of birds. In waterfowl, subcutaneous fat can account for up to 76% of total body fat content [33]. At slaughter age, the percentage share of skin and subcutaneous fat in the total body weight is approximately 13% in broiler chickens, 10 (♂ 21 weeks) to 12% (♀ 16 weeks) in turkeys, 23% in Pekin ducks, and 19% in geese [29]. The results of research investigating different waterfowl species indicate that fat deposition as a reserve of energy is still observed in modern commercial duck and goose lines despite selective breeding aimed at eliminating atavistic features [9, 34].

In broiler chickens, the weight of skin with subcutaneous fat increases 48-fold between 1 and 10 weeks of age. Over this period, the percentage of skin and subcutaneous fat in the total body weight increases from approximately 8% in week 1 to 13% in week 3, and it remains stable until 10 weeks of age. In the analyzed period, the weight of skin and subcutaneous fat increases, but their percentage share remains at a stable level, which results from a fast growth rate of muscle tissue [13]. The distribution of skin fat and subcutaneous fat varies with age. The percentage of skin with a fat layer increases in the back (by 7%) and legs (by approx. 4%), and decreases in the breast (by approx. 4%), wings (by approx. 3%), and neck (by approx. 3.5% [17], relative to the total weight of this tissue component in the carcass. Increased fat deposition in the body cavity can also be observed in growing broiler chickens. Adipose tissue is deposited as abdominal fat, periorgan fat, and peri-intestinal fat, thus increasing the weight of slaughter offal [13].

In BIG 6 turkeys, the weight of skin and subcutaneous fat increases over 75-fold in males (from 2 to 20 weeks of age) and approximately 108-fold in females (from 2 to 20 weeks of age) [15]. Over this period, the percentage content of skin and subcutaneous fat increases by approximately 3.5% in males (from 5.4 to 8.8 %) and by 5.8% in females (from 41 to 10.9%). In males, the percentage share of skin and subcutaneous fat increases in the breast (by 7.6%), legs (by 7.9%), and back (by 1.7%), and decreases in the wings (by 11.6%) and neck (by 5.6%). In females, the percentage share of skin and subcutaneous fat increases in the legs (by 7.5%) and back (by 7.9%), decreases in the wings (by 12.1%) and neck (3.6%), and remains unchanged in the breast [20].

In Pekin ducks, between 1 and 8 weeks of age, the weight of skin with subcutaneous fat increases 25.7-fold, whereas their percentage share of the total body weight increases from 15.5% in week 1 to 24.5% in week 8. Until 3 weeks of age, the content of skin with a fat layer increases by approximately 8%, and between 3 and 8 weeks by only 1% [14]. The distribution of skin with subcutaneous fat in the carcass varies with age, with a rising tendency in the back (5% increase) and wings (4%), and a falling tendency in the neck (7% decrease) and breast (2%) [18].

In geese, the weight of skin and subcutaneous fat increases 5.4-fold from 2 to 12 weeks of age, whereas the percentage of skin with a fat layer in the total body weight remains at a similar level of approximately 19.1–19.6% [16]. The distribution of skin and subcutaneous fat in the carcass undergoes smaller changes with age than the distribution of lean meat. A rising tendency can be observed in the back (4% increase) and wings (2.5%), and a falling tendency in the breast (3.5% decrease), legs (2.5%), and neck (0.5%) [19].

2.2.3. Bones

The growth of bone tissue is completed earlier than the growth of tissue and adipose tissues. Bones are inedible components of the carcass, and the decrease in their proportion observed with age is highly desirable. However, rapid growth of muscles (in particular breast muscles) in the early life stages of birds may negatively affect their health status due to a higher incidence of leg abnormalities and deformities that result in reduced walking ability and feed utilization, followed by increased mortality [35–37].

In Ross 308 broiler chickens, between 1 and 10 weeks of age, bone weight increases approximately 23.4-fold and decreases from 11.5 to 9% when expressed as a percentage of the total body weight [13] (**Figure 2a**). Unlike the growth rates of muscle tissue and adipose tissue, the growth rate of bone tissue in different carcass parts of broiler chickens is uniform and comparable [17].

In BIG 6 turkeys, bone weight increases approximately 33-fold in males (weeks 2–20) and 34-fold in females (weeks 2–16), and the share of bones in the total body weight decreases by 4.2% in males (from 14.7 to 10.5%) and by 2.1% in females (from 13.0 to 10.9%) [15] (**Figure 2b** and **c**). Age-related changes are also observed in bone distribution in the carcass. In males, bone content expressed as a percentage of the total bone weight in the carcass increases significantly in the back portion (by 3.1%), and decreases in the wings (by 1.3%) and neck (by 2.0%). In females, bone content increases only in the back (by approx. 1%) and decreases in the neck (by approx. 1.5%) [20].

In Pekin ducks, between 1 and 8 weeks of age, bone weight increases 15.2-fold. Minor changes are noted in bone content—bones account for 12 and 11% of the total body weight in weeks 1 and 8, respectively [14] (**Figure 3a**). Significant age-related changes are observed in the distribution of bones in carcass parts. Until 7 weeks of age, bone content expressed as a percentage of the total bone weight in the carcass decreases in the legs (from 35.5 to 24.0%), and increases in the neck (from 9.0 to 18%) and breast (from 12% in week 2 to 16% in week 7). The percentage of bones located in the wings, relative to the total bone weight in the carcass, remains stable (approx. 17%) [18].

In Koluda White geese, the patterns of growth in body weight and bone weight are similar until 12 weeks of age (5.3-fold and 5.2-fold increase, respectively). The proportion of bones remains at a stable level (approx. 11.9–11.5%) [16] (**Figure 4b**). Similar to Pekin ducks, also in growing geese bone content expressed as a percentage of the total bone weight in the carcass undergoes greater changes than the content of skin and subcutaneous fat. From 2 to 12 weeks of age, bone content increases rapidly in the wings (from 5.7 to 39.5%) and breast (from 9.0 to 12.9%). Until 8 weeks of age, the proportion of bones decreases rapidly in the legs (from 34.9 to 17.0%) and remains stable in the neck (approx. 9%) [19].

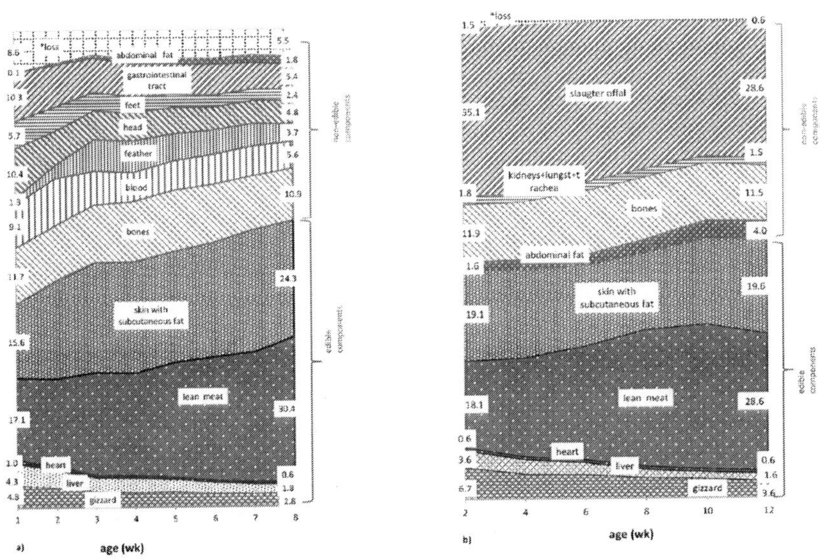

FIGURE 4. Percentage content of particular components in the body weight of (a) ducks and (b) geese (%). *loss = body weight loss during post-slaughter processing and dissection [14, 16].

2.3. Growth Rates of Organs/Giblets

Full expression of the genetic potential of modern poultry lines is largely dependent on feed intake and feed conversion efficiency, which is why the composition and physical form of diets, in particular those offered to gallinaceous birds, are subject to constant modifications. Therefore, the function once performed by gastrointestinal tract segments, in particular the gizzard, has changed. In gallinaceous birds, the storage capacity and grinding activity of the gizzard have been limited. In modern broilers, the gizzard "works" less intensively due to considerable modifications in the composition and structure of feed, which is why this organ gradually diminishes in size [8]. Fast weight gain accompanied by insufficient development of some organs, for example, the heart and lungs, may lead to metabolic diseases, skeletal system diseases, and increased fat deposition [36–39].

In broiler chickens, between 1 and 6 weeks of age, heart weight increases 11.3-fold, liver weight increases 9.3-fold, and gizzard weight only 3.4-fold. In weeks 1, 6, and 10, the total weight of giblets (heart, gizzard, and liver) accounts for 8.2, 3.0, and only 2.4% of the total body weight, respectively. In 1-week-old chickens, the gizzard has a nearly 4.0% share of the total body weight, similar to the liver. Until 10 weeks of age, the percentage content of the gizzard, liver, and heart decreases over 8-fold, 2.5-fold, and 2-fold, respectively. A rapid decrease in the percentage content of the gizzard (from 4.0 to 1.5%) is noted between 1 and 2 weeks of age, and the percentage content of the liver decreases (from 3.8 to 2.1%) from 4 weeks of age. The percentage content of the heart decreases from 0.9% in week 1 to 0.4% in week 10 [13] (**Figure 2a**).

In BIG 6 turkeys, the total weight of giblets increases 13.9-fold in males (weeks 2–20) and 14-fold in females (weeks 2–16). The growth rates of individual organs vary with age. Heart weight increases 22.2-fold in males and 27.2-fold in females. Liver weight increases 15.5-fold in males and 13.5-fold in females. Gizzard weight increases 13.4-fold in males and 14.0-fold in females. At 2 weeks of age, giblets have a 5.7 and 6.1% share of the total body weight in males and females, respectively. The total percentage content of giblets decreases to 1.7% in males at 20 weeks of age, and to 2.2% in females at 16 weeks of age. The share of individual organs decreases as follows in males: gizzard, by 1.5% (from 2.2 to 0.7%); liver, by 1.0% (from 2.8 to 0.8%); heart, by 0.5% (from 0.7 to 0.3%); and in females: gizzard, by 1.3% (from 1.9 to 0.6%); liver, by 2.5% (from 3.7 to 1.2%); heart, by 0.2% (from 0.6 to 0.4%) [15] (**Figure 2b** and **c**).

According to Lilja [40], digestive organs (esophagus, gizzard, intestines, and liver) and leg muscles belong to the group of supplying organs in geese. The supplying organs are characterized by rapid embryonic development and a fast growth after hatching, which is natural since goslings need mature legs to find food and mature digestive organs to process it efficiently.

In Pekin ducks, the total weight of giblets increases 8.9-fold between 1 and 8 weeks of age. Over this period, the weight of the liver, heart, and gizzard increases 6.9-fold, 10.4-fold, and 9.5-fold, respectively. Giblet weight expressed as a percentage of the total body weight decreases from 10.1% in week 1 to 5.3% in week 8. The percentage content of individual organs decreases as follows: liver, by 2.5% (from 4.3 to 1.9%); gizzard, by 2.0% (from 4.8 to 2.8%); heart, by 0.4% (from 1 to 0.7%) [14] (**Figure 4a**).

The growth rates of visceral organs classified as giblets vary with age also in geese. Between 2 and 12 weeks of age, the total weight of giblets increases 2.8-fold, including a 2.3-fold increase in liver weight, a 2.9-fold increase in gizzard weight, and a 4.7-fold increase in heart weight. A statistically significant increase in liver weight, gizzard weight, and heart weight is observed until 4, 6, and 8 weeks of age, respectively. The percentage content of giblets in the total body weight decreases with age, from 10.9% in week 2 to 5.8% in week 12. Over this period, the share of the gizzard and liver decreases by 3.1% (from 6.7 to 3.6%) and 2.0% (from 3.6 to 1.6%), respectively, whereas the share of the heart in the total body weight remains stable at around 0.6% [16] (**Figure 4b**).

2.4. Growth Rate of The Slaughter Offal Portion

During post-slaughter processing, offal is separated from the carcass. In poultry, the slaughter offal portion includes blood, feathers, feet, head, trachea, lungs, intestines (including the contents and peri-intestinal fat), spleen, pancreas, testes, components of the female reproductive system, and abdominal fat. In some regions of the world, abdominal fat and feet are classified as edible components, but they are generally considered as offal. Contemporary consumers prefer low-fat products; therefore, abdominal fat is treated as slaughterhouse waste [41]. However, the fat of waterfowl is used in regional cuisines for the production of specialty foods. Geese have higher abdominal fat content than other poultry species [31]. Fat deposition in the lower parts of the body in waterfowl is a natural adaptation to the environmental conditions. Duck and goose fat is considered to be healthier than pork fat, due to higher concentrations of essential unsaturated fatty acids [33, 42]. However, some researchers emphasize the lower durability of the former. In ducks and geese, the abdominal fat weight increases over the entire growing period, thus increasing both the weight of undesirable offal and overall production costs. Feathers are a by-product of poultry production. Today, the economic importance of the feathers of gallinaceous birds is relatively low [3]. Plumage development is particularly important in ducks and geese because carcass quality deteriorates during the feathering process, which affects the processing suitability of raw material. Feathers account for approximately 4.2, 4.0–4.5, and 5.4% of the total body weight in turkeys, chickens and ducks, and geese at slaughter age, respectively. The rate of feathering is species-specific, affected by nutrition and determined genetically [43].

In broiler chickens, the total weight of slaughter offal increases 17-5-fold between 1 and 10 weeks of age, including a 83-fold increase in feathers weight, a 41-fold increase in abdominal fat weight, a 19-fold increase in the weight of feet, a 14-fold increase in blood weight, an 11-fold increase in the weight of the gastrointestinal tract, and a 10-fold increase in head weight. The growth rates of the above components are slower than the whole-body growth rate, and their percentage content decreases with age, from 33.0% to approximately 19% in broiler chickens. The proportions of most offal components decrease during the growing period, except for abdominal fat and feathers (3.2 and 1.8% increase, respectively). Between 1 and 10 weeks of age, the smallest changes are noted in the percentage content of feet (2% decrease, from 4.5 to 2.5%), and the greatest changes are observed in the percentage content of the gastrointestinal tract (8.5% decrease, from 12.1 to 3.6%). The carcasses of 1-week-old chickens contain low amounts of abdominal fat, whereas at 10 weeks of age abdominal fat content reaches 3.2% and is almost equal to that of gastrointestinal tract [13] (**Figure 2a**).

In male turkeys, the weight of slaughter offal increases 23.5-fold between 2 and 20 weeks of age. Over this period, blood weight increases approximately 35-fold, feathers weight increases approximately 31-fold, and the combined weight of the lungs and trachea nearly 38-fold. The growth rates of the remaining nonedible components are slower; gastrointestinal tract weight increases only 14.5-fold. The percentage content of slaughter offal decreases in males from 30.1% at 2 weeks of age to 15.8% at 20 weeks of age. The greatest decrease is noted in the percentage share of the gastrointestinal tract (by approx. 7%), head (3%), and feet (2%). In female turkeys, the weight of slaughter offal increases 20-fold between 2 and 16 weeks of age. The fastest growth rate is observed in feathers weight (approx. 64-fold increase), followed by the weight of feet (29-fold), lungs and trachea (27-fold), and blood (25-fold). Similar to males, the growth rate of the gastrointestinal tract is relatively slow (approx. 10-fold increase). The percentage content of slaughter offal decreases in females from 38.5% at 2 weeks of age to 18.0% at 16 weeks of age. Age-related changes in the content of individual offal components are similar to those noted in males [15] (**Figure 2b** and **c**).

In Pekin ducks, offal content expresses as a percentage of the total body weight decreases from 37% in week 1 to 24% in week 8. The weights of feathers and abdominal fat change to the greatest extent. Between 1 and 8 weeks of age, feathers weight increases approximately 65-fold. The carcasses of 1-week-old ducks contain low amounts of abdominal fat, but its weight increases nearly 200-fold by 8 weeks of age. The growth rates of the remaining nonedible components are slower; from 1 to 8 weeks of age, the weights of blood, head, feet, and gastrointestinal tract increase 11-fold, 7.7-fold, 7.2-fold, and 9.1-fold, respectively. Changes in the growth rates of the above components and the body weight of birds contribute to an increase in the percentage content of feathers and abdominal fat (by 2.4 and 1.5%, respectively), and a decrease in the percentage content of gastrointestinal fat (5%), blood (3.3%), head (5.4%), and feet (3.3%) [14] (**Figure 4a**).

In geese, between 2 and 12 weeks of age, the total weight of slaughter offal increases 4.2-fold or 4.6-fold, depending on the classification of abdominal fat as an edible or inedible component, respectively. Over this period, abdominal fat weight increases approximately 13.6-fold. The increase in offal weight in geese aged 2–12 weeks is not uniform. The total weight of slaughter offal increases rapidly (2.9-fold) in weeks 2–4, it stabilizes and remains relatively unchanged until week 10, and increases again in weeks 10–12. The main reason for the increase in slaughter offal weight in geese aged 10–12 weeks is enhanced peri-intestinal fat deposition. Despite the increase in slaughter offal weight, its proportion in the total body weight decreases from 35% in week 2 to 29% in week 12 [16] (**Figure 4b**).

3. THE EFFECT OF AGE ON THE PERCENTAGE CONTENT OF EDIBLE AND NONEDIBLE COMPONENTS

The total weights of edible and nonedible components in the carcasses of different poultry species are an important economic consideration since edible carcass parts are used as raw materials for the manufacture of processed meat products, and the waste load from meat-processing plants has to be effectively managed or disposed of. In view of the fact that global poultry meat production continues to increase, the ratio between edible and inedible weights in poultry carcasses deserves particular attention [2, 3, 7]. The whole-body growth rate of birds varies over time, similarly as the growth rates of internal organs and tissue components, which affects the proportions between edible and nonedible components in the carcass. Research results show that the rate of age-related changes in the content of edible and nonedible carcass portions varies across poultry species [13–16].

3.1. Percentage Content of The Edible Portion

In chickens slaughtered at 1 week of age, the ratio between edible and nonedible parts is estimated at 1:1, whereas a more desirable value of 2.2:1 is noted in birds slaughtered at 10 weeks of age. From weeks 1–10, the total weights of edible and nonedible components in broiler carcasses increase 42-fold and only 20-fold, respectively. In 1-week-old broiler chickens, edible components account for approximately 47.0% of the total body weight, and their content increases to 63.0 and 67% at 6 and 10 weeks of age, respectively [13] (**Figure 2a**).

Similar to chickens, age-related changes in the weights and percentages of edible and nonedible components are also observed in turkeys. In males, between 2 and 20 weeks of age, the weights of edible and nonedible portions increase 63.5-fold and approximately 26.5-fold, respectively. In 2-week-old males, edible components account for nearly 52% of the total body weight, compared with 72% in 20-week-old birds. In the group of edible components, muscle tissue is characterized by the highest growth rate (40.4% in week 2, 60.9% in week 20), followed by skin and subcutaneous fat (5.4% in week 2, 8.8% in week 20). The content of giblets, expressed as a percentage of the total body weight, decreases with age (in contrast to the remaining edible components), from 5.7% in week 2 to 1.8% in week 20. A similar trend can be observed in female turkeys. Until 16 weeks of age, edible and inedible weights increase approximately 63.7-fold and 23.7-fold, respectively. The share of muscle tissue in the total body weight increases from 33.7% in week 2 to 55.6% in week 16, the share of skin with subcutaneous fat increases from 4.1% to 10.9%, and the share of giblets decreases from 6 to 2.4% [15] (**Figure 2b** and **c**).

In 1-week-old ducks, the weight of edible components is lower than the weight of nonedible components (0.8:1). In Pekin ducks, between 1 and 8 weeks of age, edible and inedible weights increase 23-fold and 12.1-fold, respectively, and the ratio between edible and nonedible components reaches 1.6:1 in 8-week-old birds. Edible components account for 42.8% of the total body weight in week 1 and for 59.9% in week 8. The most significant changes are observed in the percentage content of muscle tissue (10.5% increase) and giblets (5.1% decrease). The share of skin and subcutaneous fat remains stable at 19.3% throughout the growing period. As a result, the proportions between tissue components change. In 1-week-old ducks, muscle tissue, skin with a fat layer, and giblets account for 40, 37, and 24% of total edible weight, respectively. More desirable values are noted in 8-week-old birds, where muscle tissue, skin with fat, and giblets account for 50, 41, and only 9% of total edible weight, respectively [14] (**Figure 4a**).

In geese, the ratio between edible and inedible weights is 1:1 at 2 weeks of age, and 1.4:1 (variant I, abdominal fat classified as an edible component) or 1.2:1 (variant II, abdominal fat classified as a nonedible component) at 12 weeks of age (Murawska 2013b). The most valuable edible component, muscle tissue, has an 18% share of the total body weight in week 2 and 29% in week 12. Between 2 and 12 weeks of age, the percentage content of skin and subcutaneous fat in the total body weight of geese remains stable at 19.5%, whereas the share of giblets decreases from 10% in week 2 to 6% in week 12 [16] (**Figure 4b**).

3.2. Percentage Content of The Nonedible Portion

The nonedible portion of poultry carcasses comprises slaughter offal and bones. The content of those components, expressed as a percentage of the total body weight of birds, decreases with age.

In 1-week-old broiler chickens, nonedible components account for 45.3% of the total body weight on average, and their content decreases to 34.4 and 30.2% at 6 and 10 weeks of age, respectively. Over this period, a greater decrease is observed in the share of slaughter offal (by 12.5%, from 33.7% in week 1 to 21.2% in week 10), in comparison with bones (by 2.5 %, from 11.6 in week 1 to 9.1% in week 10) [13] (**Figure 2a**).

In 2-week-old turkeys, nonedible components have a 44.7 (♂) and a 49.9% (♀) share of the total body weight, and their content decreases to 26.0 (♂, week 20) and 29% (♀, week 16) at slaughter age. Similar to broiler chickens, slaughter offal content decreases by over 15% (from 29.7 to 14.5%) in males and by 19% (from 36.9 to 17.9%) in females, whereas bone content decreases at a slower rate (♂ 4.2% decrease, from 14.7 to 10.5%; ♀ 2.1% decrease, from 13.0 to 10.9%) [15] (**Figure 2b** and **c**].

In ducks, nonedible components account for 48.6% of the total body weight in week 1, and decrease to 34.6% in week 8. Slaughter offal accounts for 39.7 and 27.2% of the total body weight in weeks 1 and 8, respectively. Bone content decreases from 11.7% in week 1 to 10.9% in week 8 [14] (**Figure 4b**).

In geese, between 2 and 12 weeks of age, the percentage content of inedible components decreases mostly due to a decrease in the share of slaughter offal (approx. 6.5%) because the proportion of bones remains at a stable level (approx. 11.9–11.5%). In variant I (abdominal fat classified as an edible component), the share of nonedible parts decreases from 48.08 to 41.6%, and in variant II (abdominal fat classified as a nonedible component) from 50.2 to 45.4% [16] (**Figure 4b**).

4. CONCLUSION

The research findings presented in this chapter can be used in an analysis of potential health threats resulting from the undesirable differences between the growth rates of individual organs and the whole-body growth rate of birds. The relationship between the growth rates of muscle tissue and adipose tissue is also an important consideration for carcass quality and poultry production efficiency. Due to the specific structure of subcutaneous adipose tissue in poultry, analyses of carcass fat content should include both skin with a subcutaneous fat layer and abdominal fat. Information on age-related changes in the weight and content of edible and inedible carcass components may contribute to the effective processing of poultry meat and the effective management of slaughter offal, including bones separated from muscle tissue in the mechanical deboning process.

A prolonged rearing period contributes to higher final body weight and higher muscle yield, but overall production costs increase because older birds are characterized by a slower growth rate and a higher feed intake per kg of body weight gain. Those factors should be taken into account when determining the optimal slaughter age.

REFERENCES

1. Shahin KA. Sources of shared variability of the carcass and non-carcass components in Pekin ducklings. Annales de Zootechnie. 2000;49:67–72.
2. Meeker DL, Hamilton CR. An overview of the rendering industry. In: Meeker, editor. 16 Essential rendering. National Renderers Association, Virginia; 2006. ISBN: 0–9654660–3–5. p. 1–17. ch1
3. Kornillowicz-Kowalska T, Bohacz J. Biodegradation of keratin waste: Theory and practical aspects. Waste Management. 2011;31:1689–1701.
4. Wood BJ. Calculating economic values for turkeys using a deterministic production model. Canadian Journal Animal Science. 2009;89:201–213.
5. Case LA, Miller SP, Wood BJ. Determination of the optimum slaughter weight to maximize gross profit in a turkey production system. Canadian Journal Animal Science. 2010;90: 349–356.
6. Case LA, Miller SP, Wood BJ. Factors affecting breast meat yield in turkeys. World's Poultry Science Journal. 2010;66:189–201.
7. Plumber HS, Kiepper BH. Impact of poultry processing by-products on wastewater generation, treatment and discharges. In: Proceedings of the Georgia Water Resources Conference (GWRC'11); USA, GA; 11–13 April, 2011. Available from: http://www.gwri.gatech.edu/sites/default/files/files/docs/2011/6.5.3Plumber.pdf
8. Havenstein GB, Ferket PR, Grimes JL, Qureshi MA, Nestor KE. Comparison of the performance of 1966-versus 2003-type turkeys when fed representative 1966 and 2003 turkey diets: growth rate, livability, and feed conversion. Poultry Science. 2007;86:232–240.
9. Fan HP, Xie M, Wang WW, Hou SS, Huang W. Effects of dietary energy on growth performance and carcass quality of white growing Pekin ducks from two to six weeks of age. Poultry Science. 2008;87:1162–1164.
10. Yegani M, Korver DR. Factors affecting intestinal health in poultry. Poultry Science. 2008;87:2052–2063.
11. Murawska D, Bochno R, Michalik D, Janiszewska M. Age-related changes in the carcass tissue composition and distribution of meat and fat with skin in carcasses of laying-type cockerels. Archiv für Geflügelkunde. 2005;69:135–139.
12. Ristic M, Damme K, Freudenreich P. Carcass value of ducks and geese in dependence of breed and age. Fleischwirtschaft. 2006;86:107–110.

13. Murawska D, Kleczek K, Wawro K, Michalik D. Age-related changes in the percentage content of edible and non-edible components in broiler chickens. Asian-Australasian Journal Animal Science. 2011;24:532–539.
14. Murawska D. The effect of age on the growth rate of tissues and organs and the percentage content of edible and non-edible carcass components in Pekin ducks. Poultry Science. 2012;91:2030–2038.
15. Murawska D. Age-related changes in the percentage content of edible and non-edible components in turkeys. Poultry Science. 2013;92:255–264.
16. Murawska D. The effect of age on the growth rate of tissues and organs and the percentage content of edible and inedible components in Polish Koluda White Geese. Poultry Science. 2013;92 :1400–1407.
17. Bochno R, Brzozowski W, Murawska D. Age-related changes in the distribution of meat, fat with skin and bones in broiler chicken carcasses. Polish Journal of Natural Science. 2003;14:335–345.
18. Bochno R, Brzozowski W, Murawska D. Age- related changes in the distribution of lean, fat with skin and bones in duck carcasses. British Poultry Science. 2005;46:199–203.
19. Bochno R, Murawska D, Brzostowska U. Age-related changes in the distribution of lean, fat with skin and bones in goose carcasses. Poultry Science. 2006;85:1987–1991.
20. Murawska D, Kozłowski K, Tomaszewska K, Brzozowski W, Zawacka M, Michalik D. Age-related changes in the tissue composition of carcass parts and in the distribution of lean meat, fat with skin and bones in turkey carcasses. European Poultry Science. 2015;ISSN:1612–9199. DOI: 10.1399/eps.2015.103.
21. Murawska D, Bochno R. Age-related changes in the percentage content of tissue components in geese. Journal Central European Agriculture. 2008;9:211–216.
22. Laughlin K. The evolution of genetics, breeding and production. Temperton Fellowship. University College, Newport, Shropshire. 2007; Report No. 15:1–55.
23. Groeneveld LF, Lenstra JA, Eding H, Toro MA, Scherf B, Pilling D, Negrini R, Finlay EK, Jianlin H, Groeneveld E, Weigend S, GLOBALDIV Consortium. Genetic diversity in farm animals-a review. Animal Genetics. 2010;41(Suppl. 1):6–31. doi: 10.1111/j.1365-2052.2010.02038.x.
24. Lesage L, Gauthier G. Growth and organ development in greater snow goose goslings. The Auk. 1997;114:229–241.
25. Książkiewicz J. Mint: a mallard duck- *Anas platyrhynchos* L., means a flat-billed duck. Wiadomości Zootechniczne. 2006;1:25–30.
26. Gordy JF. Broilers. In Skinner JL, editor. American Poultry History, 1823–1973. Madison, WI: American Poultry Historical Society, 1974, p. 370–432.
27. Aviagen 2016, Parent Stock management manual Ross308. Available from: http:/www.aviagen.com [Accessed: 2016-04-28].
28. Velleman SG, Anderson JW, Coy CS, Nestor KE. Effect of selection for growth rate on muscle damage during turkey breast muscle development. Poultry Science. 2003; 82:1069–1074.
29. Jankowski J, Kozłowski K. Breeding and use of turkeys. In: Jakowski J, editor. Breeding and use of poultry. 1st ed. Warsaw: PWR i L; 2012. ISBN: 978-83-09-01142-2. p. 305–354.
30. Gumulka M, Wojtysiak D, Kapkowska E, Połtowicz K, Rabsztyn A. Microstructure and technological meat quality of geese from conservation flock and commercial hybrids. Annals of Animal Science. 2009;9:205–213.
31. Liu BY, Wang ZY, Yang HM, Wan JM, Xu D, Zhang R, et al. Influence of rearing system on growth performance, carcass traits and meat quality of Yangzhou geese. Poultry Science. 2011;90:653–659.
32. Rivera-Torres V, Ferket PR, Sauvant D. Mechanistic modeling of turkey growth response to genotype and nutrition. Journal Animal Science. 2011;89:3170–3188.
33. Pingel H. Improvement of carcass and meat quality of ducks and geese, Fleischwirtschaft. 2006;86:101–105.

34. Tsuji LJS, Martin ID, Martin ES, LeBlanc A, Dumas P. Spring-harvested game birds in the western James Bay region of northern Ontario, Canada: the amount of organochlorines in matched samples of breast muscle, skin, and abdominal fat. Environmental Monitoring and Assessment. 2008;146:91–104.
35. Bessei W. Welfare of broilers: a review. World's Poultry Science Journal. 2006;62:455–466.
36. Charuta A, Cooper RG, Pierzchała M, Horbańczuk JO. Computed tomographic analysis of tibiotarsal bone mineral density and content in turkeys as influenced by age and sex. Czech Journal Animal Science. 2012;57: 572–578.
37. Charuta A, Dzierzecka M, Komosa M, Kalinowski L, Pierzchała M. Age- and sex-related differences of morphometric, densitometric and geometric parameters of tibiotarsal bone in Ross Broiler chickens. Folia Biologica. 2013;3–4:211–220.
38. McKay JC, Barton NF, Koerhuis ANM, McAdam J. Broiler production around the world. XXI. In: Proceedings of the World's Poultry Congress; 20–24 August 2000; Montreal. Canada: 2000. p. 20–24.
39. Gesek M, Szarek J, Otrocka-Domagała I, Babińska I, Paździor K, Szweda M. et al. Morphological pattern of the livers of different lines of broiler chickens during rearing. Veterinarni Medicina. 2013; 58:16–24.
40. Lilja C. Postnatal growth and organ development in the goose (*Anser anser*). Growth. 1981;45:329–341.
41. Kostecka M. Characterization of poultry fat and rapeseed oil mixture prior to and after the enzymatic interesterification. Żywność. Nauka.Technologia. Jakość. 2008;5:257–272.
42. Chartrin PM, Bernadet D, Guy G, Mourot J, Duclos MJ, Baéza E. The effects of genotype and overfeeding on fat level and composition of adipose and muscle tissues in ducks. Animal Research. 2006;55:231–244.
43. Leeson S, Walsh T. Feathering in commercial poultry II. Factors influencing feather growth and feather loss. World's Poultry Science Journal. 2004;60:52–63.

CHAPTER 5

Lipopolysaccharide Reduces Food Passage Rate From The Crop by a Prostaglandin-Independent Mechanism in Chickens

T. Tachibana[a], M. Ogino[a], R. Makino[a], M. S. I. Khan[b] and M. A. Cline[c]

[a]*Department of Agrobiological Science, Faculty of Agriculture, Ehime University, Matsuyama, Japan;*

[b]*Department of Anatomy and Embryology, Ehime University Graduate School of Medicine, Toon, Japan;*

[c]*Department of Animal and Poultry Sciences, Virginia Polytechnic Institute and State University, Blacksburg, VA, USA*

ABSTRACT

1. We examined the effect of lipopolysaccharide (LPS), a component of Gram-negative bacteria, on food passage in the digestive tract of chickens (Gallus gallus) in order to clarify whether bacterial infection affects food passage in birds.
2. Food passage in the crop was significantly reduced by intraperitoneal (IP) injection of LPS while it did not affect the number of defecations, suggesting that LPS may affect food passage only in the upper digestive tract.
3. Similar to LPS, prostaglandin E2 (PGE2), one of the mediators of LPS, also reduced crop-emptying rate in chickens while it had no effect on the number of defecations.

4. Pretreatment with indomethacin, which is an inhibitor of cyclooxygenase (COX), a prostaglandin synthase, had no effect on LPS-induced inhibition of crop emptying.
5. IP injection of LPS did not affect the mRNA expression of COX2 in the upper digestive tract of chickens.
6. It is therefore likely that LPS and PGE2 reduced food passage rate in the crop by a prostaglandinindependent pathway in chickens.

Keywords: Chick; crop-emptying rate; faeces; lipopolysaccharide; prostaglandin E2

INTRODUCTION

Bacterial infections are associated with non-specific symptoms including fever, somnolence, anorexia and adipsia in animals (Kent et al., 1992). These symptoms are triggered by bacterial components such as lipopolysaccharide (LPS), a component of the outer membrane of Gram-negative bacteria. LPS has been demonstrated to affect the function of the digestive tract. For example, LPS suppresses gastric emptying and/or motility in rats (van Miert and De la Parra, 1970; van Miert and van Duin, 1980; Wirthlin et al., 1996; Collares, 1997; Okumura et al., 1998), mice (Inada, et al., 2006), sheep (Leek and van Miert, 1971), goats (van Miert, 1971), dogs (Cullen et al., 1995) and horses (King and Gerring, 1991). Furthermore, intraperitoneal (IP) injection of LPS inhibits gastrointestinal transit and increases diarrhoea in mice (Liang et al., 2005). However, intravenous injection of LPS accelerates small intestinal transit of liquid diets in rats (Wirthlin et al., 1996) and dogs (Cullen et al., 1997). Moreover, colonic transit is also changed by intravenous injection of LPS in dogs (Cullen et al., 1995; Spates et al., 1998). These results demonstrate that LPS affects food passage and motility in the digestive tract in mammals although the effect is location- dependent within the digestive tract.

The effect of LPS on food passage in the digestive tract of rodents is thought to be mediated by several bioactive molecules such as tumour-necrosis factor-alpha, nitric oxide and epinephrine/norepinephrine (Wirthlin et al., 1996; Takakura et al., 1997; Inada, et al., 2006; Hamano et al., 2007). Prostaglandins (PGs) are also candidates that are proposed to mediate the effect of LPS on food passage in the digestive tract. Prostaglandin E2 (PGE2) is thought to be related to the inhibition of bowel motility by LPS in horses (King and Gerring, 1991). In addition, in rats, carbacholinduced contraction is inhibited by LPS in vitro and the effect is restored by indomethacin (IND), an inhibitor of cyclooxygenase (COX), a PG synthase (Hori et al., 2001). IND also restores LPS-induced inhibition of gastric emptying in rats (Calatayud et al., 2002). Moreover, LPS-induced diarrhoea and suppression of gastrointestinal transit is restored by IND in mice (Liang et al., 2005). There are two isoforms of COX: COX1 is thought to be the constitutively expressed isoform while COX2 is thought to be induced during inflammation. Indeed, IP injection of LPS upregulates mRNA expression of COX2 but not COX1 in the stomach of neonate and adult rats (Martinez et al., 2001). Another study showed that injection of LPS induces COX2 mRNA expression in the stomach of rats (Ferraz et al., 1997). These results suggest that PG, which can be synthesised by the action of COX2, is related to LPSinduced change in the function of the digestive tract in mammals.

In chickens (Gallus gallus), IP injections of LPS have been demonstrated to induce hyperthermia and anorexia (Johnson et al., 1993a, 1993b). In addition, intravenous injection of LPS has been reported to increase plasma PGE2 concentrations in chickens (De Boever et al., 2010), and PG has been demonstrated to be involved in LPSinduced hyperthermia and anorexia in chickens (Johnson et al., 1993b). However, little is known about bacterial infection or the effect of LPS on food passage in the digestive tract in chickens. Furthermore, the effect of PG on food passage in the digestive tract has not been clarified. Thus, the purpose of the present study was to examine whether IP injection of LPS and PGE2 affects food passage in chickens. For this purpose, effects on crop emptying and the number of defecations were investigated. Finally, the relationship between LPS and PG was examined.

MATERIALS AND METHODS

Animals

One-day-old male layer chickens (G gallus, Julia, Nihon Layer, Gifu, Japan) were raised in a room kept at 30°C with continuous lighting. A commercial diet (crude protein: 24%, metabolisable energy: 12.8 MJ/kg, Toyohashi Feed Mills Co. Ltd, Aichi, Japan) and water were available ad libitum to the chickens. Chickens were transferred to their individual cages 1 d before each experiment. They were weighed and distributed into experimental groups so that the average body weight was as uniform as possible between treatment groups. The chickens were maintained in accordance with the recommendations of the National Research Council (1996). This study was approved by the Committee of Animal Care and Use in Ehime University (no. 08- o3-10).

Drugs and injection

All injections were performed between 06:00 and 08:00. LPS (Escherichia coli, O-127:H6, Wako Pure Chemical Industries, Osaka, Japan), PGE2 (Wako Pure Chemical Industries, Osaka, Japan) and IND (Tokyo Chemical Industries, Tokyo Japan) were used for the experiments. LPS was dissolved in phosphate buffered saline (pH 7.4), PGE2 was dissolved in a normal saline solution and IND was dissolved in dimethyl sulfoxide and then diluted 10-fold with sodium carbonate. Each vehicle alone was used for the respective control treatment. These solutions were injected via the IP route at a volume of 0.2 ml per chick except as noted below.

Measurement of crop emptying

Crop-emptying rate was measured based on a method previously reported (Tachibana et al., 2010). Chickens were food-deprived for 15 h to empty residual ingesta within the crop. Feeders were removed from individual cages, while water was available to chickens. Chickens were gavaged into the crop with a feed slurry at a mass of approximately 4% body weight. The feed slurry was made by mixing 40% powdered diet with 60% distilled water on a weight basis. No chickens vomited post gavage in the present study. After gavage, chickens were returned to individual cages and water cups were withheld. One or two hours after the gavage, chickens were deeply anesthetised by inhalation of diethyl ether, after which their crops were exposed, the upper and lower oesophagus clamped and the cropexcised. Total content of the crop was recovered and dried at 55°C for 48 h and further air-dried for 24 h. The air-dried slurry was weighed using a digital balance with a precision of 1 mg. Based on the dry weight, the wet slurry weight was calculated. The weight of the slurry emptied from the crop through the lower oesophagus was calculated by subtracting the weight of slurry within the crop from the weight of administered slurry. Crop-emptying rate was expressed as the percentage of slurry emptied from the crop to the amount gavaged.

Effect of IP injection of LPS on crop emptying

Seven-day-old chickens, which were food-deprived for 15 h, were IP injected with vehicle (control), 2 or 4 mg/kg LPS. The dose of LPS was based on previous studies of chickens (Johnson et al., 1993a, 1993b) and rodents (Okumura et al., 1998; Martinez et al., 2001). Immediately following injection, chickens were gavaged with the feed slurry. Chickens were then returned to individual cages. Two hours after the gavage, the crop content was obtained, dried and weighed, and then crop-emptying rate was calculated as aforementioned.

Time-course change in crop emptying after IP injection of LPS

In the first experiment, 7-d-old chickens, which were fooddeprived for 15 h, were IP injected with vehicle (control) or 2 mg/kg LPS. Immediately following injection, chickens were gavaged with the feed slurry. After gavage, chickens were returned to individual cages. One or two hours after the gavage, the crop content was obtained, dried and weighed, and then crop-emptying rate was calculated as aforementioned.

In the second experiment, 7-d-old chickens, which were food-deprived for 12 h, were IP injected with the vehicle (control) or 2 mg/kg LPS and thereafter food deprivation was continued for 3 h. Then chickens were gavaged with the feed slurry, crops were removed and crop-emptying rate was calculated as described above.

Effect of IP injection of LPS on the number of defecations

Seven-day-old chickens were IP injected with the vehicle (control) or 100 µg LPS under ad libitum feeding conditions. The food and water cups were removed from their cages. At 30 and 60 min post injection, the number of defecations was counted on a white sheet under each cage.

Effect of IP injection of PGE2 on crop emptying

Seven-day-old chickens were food-deprived for 15 h and then IP injected with vehicle (control), 40 or 80 µg/kg PGE2. Immediately after the injection, chickens were gavaged with the feed slurry as described for the other experiments. Chickens were deeply anesthetised with diethyl ether at 2 h after the gavage, and thereafter the crop content was obtained, dried and weighed for the calculation of the crop-emptying rate.

Effect of IP injection of PGE2 on the number of defecations

Eight-day-old chickens were IP injected with the vehicle (control) or 80 µg/kg PGE2 under ad libitum feeding conditions. Food and water cups were removed from cages and the number of defecations counted as for the previous experiment.

Effect of IND on IPS-induced inhibition of crop emptying

This experiment was performed to clarify the involvement of PG on LPS-induced inhibition of crop emptying in chickens using IND, an inhibitor of COX.

Seven-day-old chickens, which were food-deprived for 14.5 h, were divided into 4 groups; vehicle plus vehicle, vehicle plus 2 mg/kg LPS, 10 mg/kg IND plus vehicle and 10 mg/kg μg IND plus 2 mg/kg LPS. Chickens were IP injected with vehicle or 10 mg/kg IND at a volume of0.1 ml. Thirty minutes after the injection, chickens were IP injected with vehicle or 2 mg/kg LPS at a volume of 0.1 ml and then gavaged with the feed slurry. After gavage, chickens were returned to individual cages. Two hours after the gavage, the crop content for each chick was obtained, dried and weighed, and crop-emptying rate was calculated. The second experiment was performed using 20 mg/kg IND and identical procedures.

Effects of LPS on the mRNA expression of COX2 in the digestive tract

Gene expression of COX2 was determined in the digestive tract in order to investigate whether LPS affects the synthesis of PG.

Eight-day-old ad libitum-fed chickens were IP injected with vehicle (control) or 2 mg/kg LPS. At 3 h post injection, crop, lower oesophagus (Es), proventriculus (Pv), gizzard (Gz) and duodenum (Du) were collected, weighed and then frozen with liquid nitrogen and stored at −80°C until analysis.

The expression of COX2 mRNA expression was measured in these tissues. Tissues were homogenised with Sepasol-RNA I Super G (Nacalai Tesque, Kyoto, Japan), and then total RNA was isolated as per the manufacturer's instructions. First-strand cDNA was synthesised from DNase I (Ambion, Austin, TX, USA)-treated total RNA using the reverse transcription kit ReverTra Ace qPCR RT Master Mix with gDNA Remover (Toyobo, Osaka, Japan) with random primers. The cDNA of COX2 was amplified and quantified by quantitative real-time PCR using specific primers for COX2 (product size, 131 bp; forward primer, 5'- ctgctccctcccatgtcaga-3'; reverse primer, 5'-cacgtgaagaattccggtgtt- 3'). As an internal standard, polymerase (RNA) II (DNA directed) polypeptide B (RPII) was also amplified (product size, 172 bp; forward primer, 5'- cagaatttgccgacctcttc-3'; reverse primer, 5'-ggccagcatcacagtctctt- 3'). PCR parameters for assays were as follows: 95°C for 10 min then 45 cycles at 95°C for 10 s, 60°C for 10 s and 72°C for 15 s. The accuracy of these primers was verified by sequencing each PCR product. The Ct value was calculated using the 2−ΔΔCt method.

Statistical analysis

Data for the effects of LPS and PGE2 on the weight of slurry in the crop and crop emptying were analysed with one-way analysis of variance (ANOVA) and the Tukey–Kramer test post hoc. Time-course changes in the effect of LPS on the weight of slurry in the crop and crop emptying were analysed with two-way ANOVA with respect to LPS and time, with t-tests performed at each time. The number of defecations was analysed with two-way repeated-measures ANOVA. Real-time PCR data were analysed with t-tests. Data are expressed as means ± SEM, and statistical significance was set at $P < 0.05$. The numbers of chickens are stated in the figures.

RESULTS

Effect of IP injection of LPS on crop emptying

Chickens were orally administered the slurry at 2.226 ± 0.006 g. IP injection of 2 and 4 mg/kg LPS significantly increased the amount of slurry remaining in the crop at 2 h after the injection and thereby significantly decreased the crop-emptying rate (Figure 1). The effect of LPS was not different among doses.

Time-course change in crop emptying after IP injection of LPS

Chickens were orally administered the slurry at 2.230 ± 0.002 g. Although the effect of LPS was not significant in terms of the amount of slurry remaining in the crop [$F(1,26) = 2.531$, $P = 0.124$] and crop-emptying rate [$F(1,26) = 2.504$, $P = 0.126$], there was a significant interaction between LPS and time for the amount of slurry remaining in the crop [$F(1,26) = 6.489$, $P < 0.05$] and crop-emptying rate [$F(1,26) = 6.535$, $P < 0.05$]. The IP injection of 2 mg/kg LPS did not affect the amount of slurry remaining in the crop and hence the crop-emptying rate at 1 h (Figure 2(a)). However, at 2 h LPS significantly increased the amount of slurry and significantly decreased the crop-emptying rate (Figure 2(a)).

When LPS was injected at 3 h before the gavage, a significant effect of LPS was observed for the amount of slurry remaining the crop [$F(1,29) = 9.526$, $P < 0.01$] and crop-emptying rate [$F(1,29) = 9.789$, $P < 0.01$]. The interaction between LPS and time was also significant for the amount of slurry remaining the crop [$F(1,29) = 7.123$, $P < 0.05$] and crop-emptying rate [$F(1,29) = 7.338$, $P < 0.05$]. There was a significant increase in the amount of slurry remaining the crop and decrease in the crop emptying rate from 1 h that continued to 2 h (Figure 2(b)).

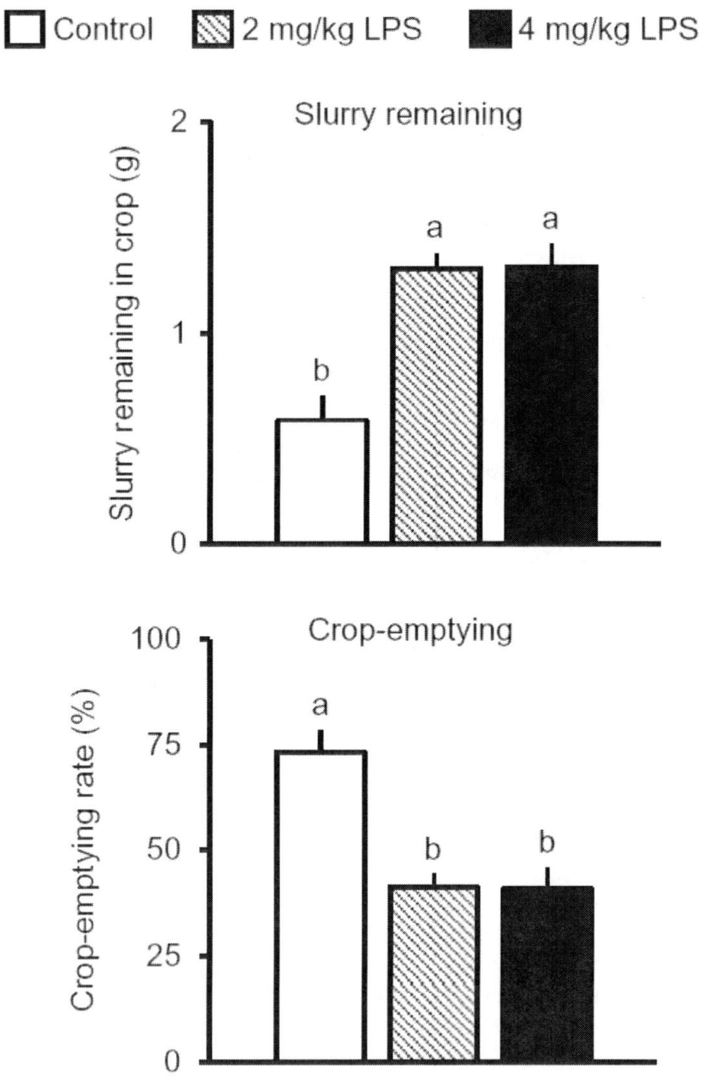

Figure 1. Effect of intraperitoneal injection of lipopolysaccharide (LPS) on slurry remaining in the crop and crop emptying in chickens. Data are expressed as mean ± SEM (n = 7 in each group). Groups with different letters within a time period are significantly different ($P < 0.05$).

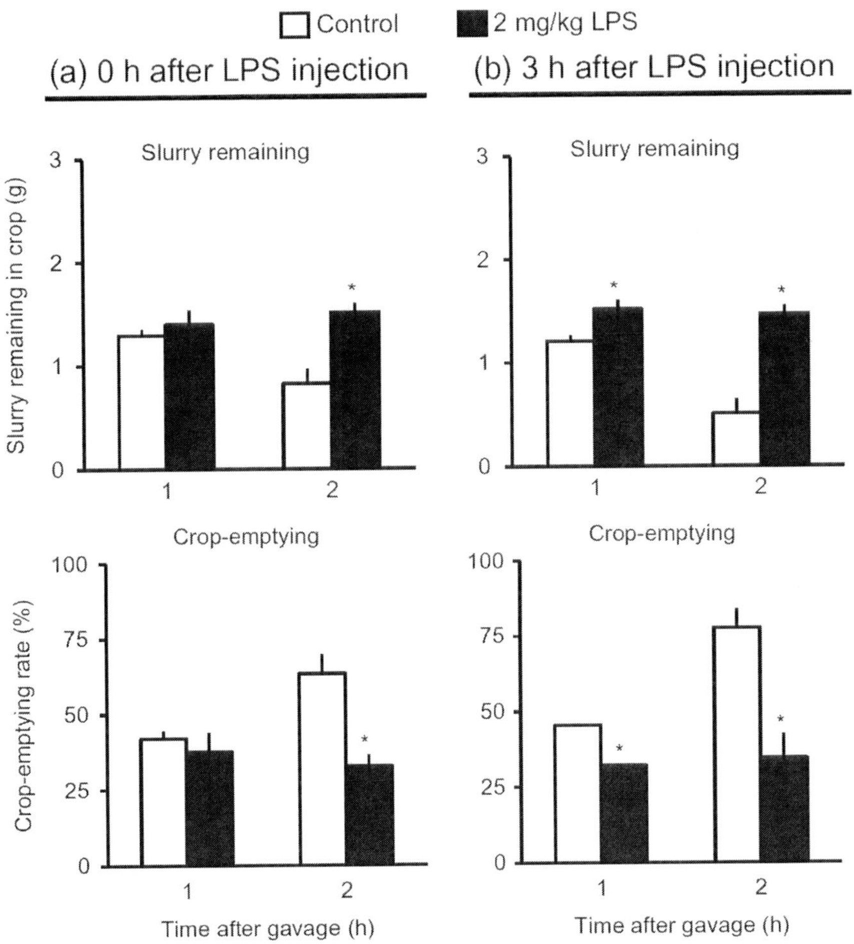

Figure 2. Slurry remaining in the crop and crop emptying after intraperitoneal (IP) injection of lipopolysaccharide (LPS) in chickens. (a) The IP injection was performed immediately before the feed gavage. Data are expressed as mean ± SEM ($n = 7$ in each group for 1 h and $n = 8$ in each group for 2 h). (b) The IP injection was performed 3 h before the feed gavage. Data are expressed as mean ± SEM ($n = 8$–9 for 1 h and $n = 8$ in each group for 2 h). Asterisks indicate a significant difference from the control group at each time ($P < 0.05$).

Effect of IP injection of LPS on the number of defecations

The number of defecations of chickens in the control group was 1.3 ± 0.3 and 2.2 ± 0.4 at 30 and 60 min after the injection, respectively. On the other hand, the number of defecations in the LPS group was 1.7 ± 0.4 and 2.8 ± 0.4 at 30 and 60 min, respectively. The effect of LPS on the number of defecations was not significant [$F(1,23) = 1.223$, $P = 0.280$]. The interaction between LPS and time was also not significant [$F(1,23) = 0.266$, $P = 0.611$].

Effect of IP injection of PGE2 on crop emptying

Chickens were orally administered the slurry at 2.234 ± 0.004 g. IP injection of 40 and 80 µg/kg PGE2 significantly increased the amount of slurry remaining in the crop and thereby decreased the crop-emptying rate at 2 h (Figure 3).

Effect of IP injection of PGE2 on the number of defecations

The number of defecation of chickens in the control group was 1.4 ± 0.3 and 2.2 ± 0.4 at 30 and 60 min after the injection, respectively. On the other hand, the number of defecations in the LPS group was 1.8 ± 0.4 and 2.9 ± 0.4 at 30 and 60 min, respectively. IP injection of 80 µg/kg PGE2 did not affect the number of defecations [$F(1,18) = 1.231$, $P = 0.282$]. The interaction between PGE2 and time was not significant [$F(1,18) = 0.491$, $P = 0.493$].

Effect of IND on IPS-induced inhibition of crop emptying

Chickens were orally administered slurry at 2.227 ± 0.004 g. IP injection of 2 mg/kg LPS alone significantly increased the amount of slurry and decreased crop emptying as noted before, and 10 mg/kg IND treatment had no effect on crop emptying (Figure 4). Pretreatment with 10 mg/kg IND did not affect the amount of LPS-induced slurry in the crop or LPS-induced decreases in crop emptying (Figure 4).

Similarly, LPS-induced increases in the amount of feed slurry and decreased crop emptying were not affected by pretreatment with 20 mg/kg IND (Figure 4).

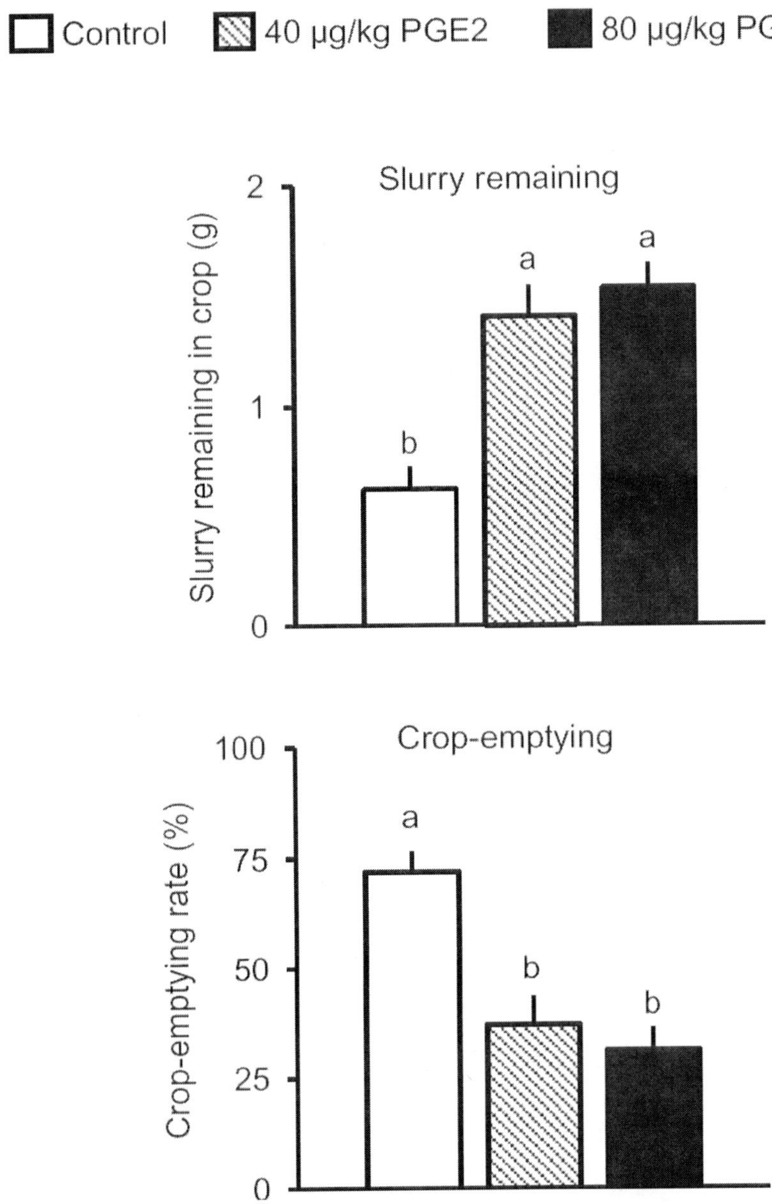

Figure 3. Slurry remaining in the crop and crop emptying after intraperitoneal injection of prostaglandin E2 (PGE2) in chickens. Data are expressed as mean ± SEM (n = 7 in each group). Groups with different letters within a time period are significantly different ($P < 0.05$).

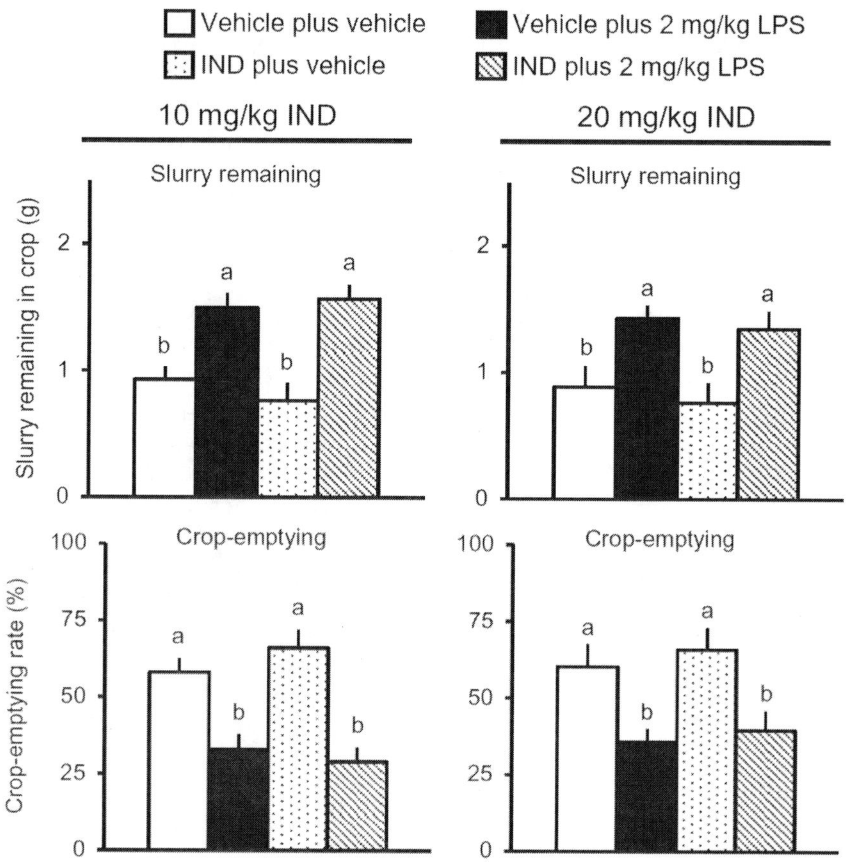

Figure 4. Effect of intraperitoneal (IP) injection of indomethacin (INDO) on lipopolysaccharide (LPS)-induced increase in slurry remaining in the crop and the inhibition of crop emptying. (a) 10 mg/kg INDO was injected at 30 min prior to IP injection of 2 mg/kg LPS. Data are expressed as means ± SEM (n = 6–7 in each group). (a) 20 mg/kg INDO was injected at 30 min prior to IP injection of 2 mg/kg LPS. Data are expressed as mean ± SEM (n = 6–7 in each group). Groups with different letters within a time period are significantly different ($P < 0.05$).

Effects of LPS on the mRNA expression of COX2 in the digestive tract

The IP injection of 100 μg LPS did not affect the mRNA expression of COX2 in any portion of the digestive tract studied (Figure 5).

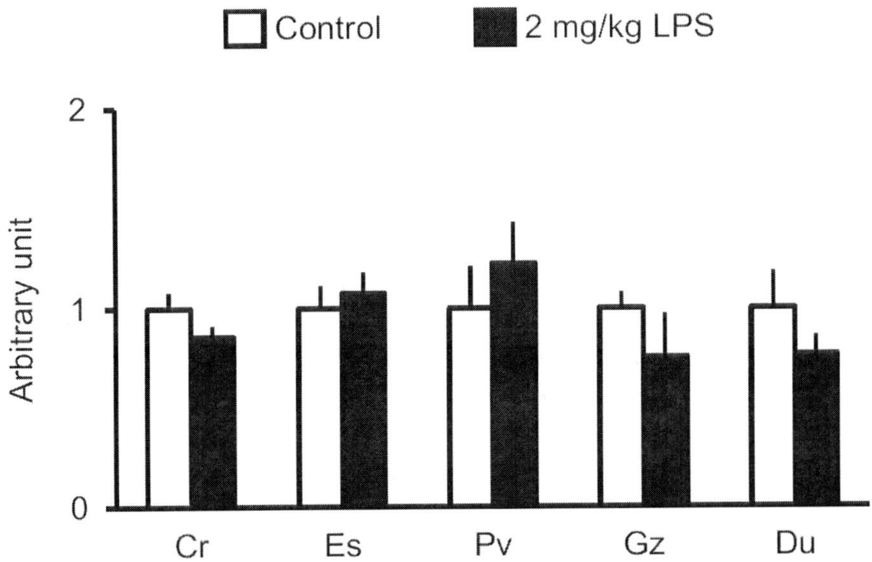

Figure 5. Effect of IP injection of 2 mg/kg LPS on mRNA expression of COX2 in the upper digestive tract. Cr, crop; Es, lower oesophagus; Pv, proventriculus; Gz, gizzard; Du, duodenum. Data are expressed as mean ± SEM (*n* = 7–8 in each group).

DISCUSSION

The present study demonstrated that IP injection of 2 and 4 mg/ kg LPS significantly reduced crop-emptying rate in chickens. Because the crop is located in the middle oesophagus and stores ingesta (Denbow, 2000), Gram-negative bacteria may suppress food passage in the upper digestive tract in chickens, a result similar to mammals. In mammals, many studies have revealed that injection of LPS inhibits gastric emptying and/or motility (van Miert and De la Parra, 1970; Leek and vanMiert, 1971; van Miert, 1971; van Miert and van Duin, 1980; King and Gerring, 1991; Cullen et al., 1995; Wirthlin et al., 1996; Collares, 1997; Okumura et al., 1998; Inada, et al., 2006). For example, IP injection of 4.5 mg/kg LPS significantly decreased gastric emptying in rats (Okumura et al., 1998). Furthermore, IP injection of 2 mg/kg LPS significantly suppressed gastric emptying in mice (Inada, et al., 2006). These results imply that LPS suppresses food passage from the Pv to Gz and thus crop emptying might be reduced in chickens. Indeed, it is likely that whether the diet remains in the crop or is passed on to the Pv is dependent on the contraction of the Gz, while evacuation of the diet from the crop itself is mediated by contraction of the crop wall

(Denbow, 2000). Further studies which investigate the effect of LPS on Pv and Gz will clarify whether LPS directly affects crop emptying. Onthe other hand, IP injection of 2 mg/kg LPS had no effect on the number of defecations in chickens. This result is inconsistent with mammalian studies because LPS affects colonic transit in dogs (Cullen et al., 1995; Spates et al., 1998) and colonic contraction in horses (King and Gerring, 1991). It is thus likely that the effect of LPS on the lower digestive tract is different between chickens and mammals. However, Spates et al. (1998) showed that intravenous injection of 200 μg/kg LPS facilitated colonic transit in dogs. Thus, it remains a possibility that the difference in the experimental conditions affects the present result.

Although IP injection of LPS significantly decreased crop emptying in chickens, the effect is not fast-acting because LPS did not affect the crop-emptying rate at 1 h after the injection. This late onset suggests that LPS might affect crop emptying indirectly and that some mediator might be related to the effect of LPS. PG is one of the candidates thought to mediate LPS-induced changes in motility of and transit in the gastrointestinal tract of mammals (King and Gerring, 1991; Hori et al., 2001; Calatayud et al., 2002; Liang et al., 2005). Among the PG, PGE2 is associated with food passage in the gastrointestinal tract in mammals. For example, it is reported that PGE2 contributes to suppression of gastric emptying induced by intragastric injection of a hyperosmolar NaCl solution in rats (Nishiyama et al., 1992; Mizuguchi et al., 2010). Intragastric injection of enprostil, a PGE2 derivative, also reduces gastric emptying of barium grains in rats (Sakurai et al., 1996). On the other hand, enprostil does not reduce gastric emptying in humans while it increases mouth-to-caecum transit time (Nicholl et al., 1989). In contrast to these results, Berstad et al. (1988) reported that enprostil accelerates gastric emptying in humans. Nompleggi et al. (1980) showed that 15(S)-15-methyl PGE2, which is a PGE2 analogue, facilitates gastric emptying in rhesus monkeys. Rowart and Rush (1984) demonstrated that PGE2 and its analogue 16, 16- dimethyl PGE2 affect gastric emptying and small intestinal transit in rats but that their effects are dependent on dosage and injection route. Thus, the effect of PGE2 on food passage in the digestive tract is complex in mammals. In the present study, on the other hand, IP injection of 40 and 80 μg/kg PGE2 clearly inhibited crop emptying in chickens while it had no effect on the number of defecations. It is therefore likely that PGE2 modifies food passage in the upper but not lower digestive tract in chickens as observed in the LPS study.

However, pretreatment with IND, an inhibitor of PG synthase, did not attenuate the effect of LPS, suggesting that PG might not be associated with LPS-induced inhibition of crop emptying in chickens. In addition, IP injection of LPS had no\ effect on themRNA expression of COX2 in the crop,

Es, Pv, Gz or Du. Thus, it is likely that PG, including PGE2, does not mediate LPS-induced changes in the upper digestive tract in chickens. In contrast to chickens, IND restores LPS-induced inhibition of gastric emptying in rats (Calatayud et al., 2002) and IP injection of LPS upregulates mRNA expression of COX2 but not COX1 in the stomach of neonate and adult rats, whereas expression was not altered in the small and large intestines (Martinez et al., 2001). Thus, the responsiveness of PG synthesis to LPS treatment seems to be different between chickens and rodents. On the other hand, it has been reported that LPSinduced COX2 mRNA expression is dependent on the sources of LPS: COX2 mRNA expression in gastric epithelial cells is increased by E. coli LPS but not Helicobacter pylori LPS in vitro (Smith et al., 2003). Although E. coli LPS did not affect COX2 mRNA expression in the upper digestive tract in the present study, LPS from other sources might affect COX2 transcription.

The factors that are involved in LPS-induced inhibition of crop emptying in chickens are still unknown. Since interleukin 1, nitric oxide, tumour-necrosis factor alpha and the alpha2- adrenergic system are thought to mediate LPS-induced inhibition of gastric emptying in mammals (Takakura et al., 1997; Endo and Kumagai, 1998; DeWinter et al., 2002; Liang et al., 2005; Inada, et al., 2006; Hamano et al., 2007), some of them might be related to the inhibition of crop emptying induced by LPS in chickens as well. Another candidate is corticosterone because IP injection of LPS increases plasma corticosterone (Johnson et al., 1993a) and subcutaneous injection of corticosterone retards crop emptying in chickens (Ogino et al., 2016).

In sum, we showed that LPS injection suppresses food passage in the upper digestive tract in chickens. Further study will clarify the actual mechanism underlying the inhibition of food passage in chickens.

ACKNOWLEDGEMENTS

This work was supported by the Japan Society for the Promotion of Science (JSPS).

FUNDING

This work was supported by the Japan Society for the Promotion of Science (JSPS) KAKENHI [Grant Number: 16K07991]

REFERENCES

1. BERSTAD, K., MASSEY, H. & BERSTAD, A. (1988) Effect of enprostil on basal and meal-stimulated gastric acid and pepsin secretion, serum gastrin and gastric emptying in healthy persons. Alimentary Pharmacology and Therapeutics, 2: 65–72. doi:10.1111/j.1365-2036.1988.tb00673.x
2. CALATAYUD, S., GARCÍA-ZARAGOZÁ, E., HERNÁNDEZ, C., QUINTANA, E., FELIPO, V., ESPLUGUES, J.V. & BARRACHINA, M.D. (2002) Downregulation of nNOS and synthesis of PGs associated with endotoxin-induced delay in gastric emptying. American Journal of Physiology. Gastrointestinal and Liver Physiology, 283: G1360–G1367. doi:10.1152/ajpgi.00168.2002
3. COLLARES, E.F. (1997) Effect of bacterial lipopolysaccharide on gastric emptying of liquids in rats. Brazilian Journal of Medical and Biological Research, 30: 207–211. doi:10.1590/S0100-879X1997000200008
4. CULLEN, J.J., CAROPRESO, D.K. & EPHGRAVE, K.S. (1995) Effect of endotoxin on canine gastrointestinal motility and transit. Journal of Surgical Research, 58: 90–95. doi:10.1006/jsre.1995.1015
5. CULLEN, J.J., CAROPRESO, D.K., EPHGRAVE, K.S., HEMANN, L.L. & HINKHOUSE, M.M. (1997) The effect of endotoxin on canine jejuna motility and transit. Journal of Surgical Research, 67: 54–57. doi:10.1006/jsre.1996.4920
6. DE BOEVER, S., NEIRINCKX, E.A., MEYER, E., DE BAERE, S., BEYAERT, R., DE BACKER, P. & CROUBELS, S. (2010) Pharmacodynamics of tepoxalin, sodium-salicylate and ketoprofen in an intravenous lipopolysaccharide inflammation model in broiler chickens. Journal of Veterinary Pharmacology and Therapeutics, 33: 564–572. doi:10.1111/j.1365-2885.2010.01184.x
7. DE WINTER, B.Y., BREDENOORD, A.J., DE MAN, J.G., MOREELS, T.G., HERMAN, A.G. & PELCKMANS, P.A. (2002) Effect of inhibition of inducible nitric oxide synthase and guanylyl cyclase on endotoxininduced delay in gastric emptying and intestinal transit in mice. Shock, 18: 125–131. doi:10.1097/00024382-200208000-00006
8. DENBOW, D.M. (2000) Gastrointestinal anatomy and physiology, in: Whittow, G.C. (Eds) Sturkey's Avian Physiology. 5th edn., pp. 299–325 (San Diego, Academic Press).
9. ENDO, Y. & KUMAGAI, K. (1998) Induction by interleukin-1, tumor necrosis factor and lipopolysaccharides of histidine decarboxylase in the stomach and prolonged accumulation of gastric acid. British Journal of Pharmacology, 125: 842–848. doi:10.1038/sj.bjp.0702108
10. FERRAZ, J.G., SHARKEY, K.A., REUTER, B.K., ASFAHA, S., TIGLEY, A.W., BROWN, M.L., MCKNIGHT, W. & WALLACE, J.L. (1997) Induction of cyclooxygenase 1 and 2 in the rat stomach during endotoxemia: role in resistance to damage. Gastroenterology, 113: 195–204. doi:10.1016/S0016-5085(97)70095-7
11. HAMANO, N., INADA, T., IWATA, R., ASAI, T. & SHINGU, K. (2007) The alpha2-adrenergic receptor antagonist yohimbine improves endotoxin- induced inhibition of gastrointestinal motility in mice. British Journal of Anaesthesia, 98: 484–490. doi:10.1093/bja/aem011
12. HORI, M., KITA, M., TORIHASHI, S., MIYAMOTO, S., WON, K.J., SATO, K., OZAKI, H. & KARAKI, H. (2001) Upregulation of iNOS by COX-2 in muscularis resident macrophage of rat intestine stimulated with LPS. American Journal of Physiology. Gastrointestinal and Liver Physiology, 280: G930–G938.
13. INADA, T., HAMANO, N., YAMADA, M., SHIRANE, A. & SHINGU, K. (2006) Inducible nitric oxide synthase and tumor necrosis factor-alpha in delayed gastric emptying and gastrointestinal transit induced by lipopolysaccharide in mice. Brazilian Journal of Medical and Biological Research, 39: 1425–1434. doi:10.1590/S0100-879X2006001100006

14. JOHNSON, R.W., CURTIS, S.E., DANTZER, R., BAHR, J.M. & KELLEY, K.W. (1993a) Sickness behavior in birds caused by peripheral or central injection of endotoxin. Physiology & Behavior, 53: 343–348. doi:10.1016/0031-9384(93)90215-2
15. JOHNSON, R.W., CURTIS, S.E., DANTZER, R. & KELLEY, K.W. (1993b) Central and peripheral prostaglandins are involved in sickness behavior in birds. Physiology & Behavior, 53: 127–131. doi:10.1016/0031-9384(93)90020-G
16. KENT, S., BLUTHÉ, R.M., KELLEY, K.W. & DANTZER, R. (1992) Sickness behavior as a new target for drug development. Trends in Pharmacological Sciences, 13: 24–28. doi:10.1016/0165-6147(92)90012-U
17. KING, J.N. & GERRING, E.L. (1991) The action of low dose endotoxin on equine bowel motility. Equine Veterinary Journal, 23: 11–17.doi:10.1111/evj.1991.23.issue-1
18. LEEK, B.F. & VAN MIERT, A.S. (1971) An analysis of the pyrogeninduced inhibition of gastric motility in sheep. Journal of Physiology, 215: 28–29.
19. LIANG, Y.C., LIU, H.J., CHEN, S.H., CHEN, C.C., CHOU, L.S. & TSAI, L.H. (2005) Effect of lipopolysaccharide on diarrhea and gastrointestinal transit in mice: roles of nitric oxide and prostaglandin E2. World Journal of Gastroenterology, 11: 357–361. doi:10.3748/wjg.v11.i3.357
20. MARTINEZ, F.E., RENO, C., TREVENEN, C.L., HART, D.A. & BELIK, J. (2001) Age-dependent changes in the regulation of cyclooxygenases in the gastrointestinal tract after gram-negative endotoxemia. Journal of Pediatric Gastroenterology and Nutrition, 33: 165–170. doi:10.1097/00005176-200108000-00013
21. MIZUGUCHI, S., OHNO, T., HATTORI, Y., AE, T., MINAMINO, T., SATOH, T., ARAI, K., SAEKI, T., HAYASHI, I., SUGIMOTO, Y., NARUMIYA, S., SAIGENJI, K. & MAJIMA, M. (2010) Roles of prostaglandin E2-EP1 receptor signaling in regulation of gastric motor activity and emptying. American Journal of Physiology. Gastrointestinal and Liver Physiology, 299: G1078–G1086. doi:10.1152/ajpgi.00524.2009
22. National Research Council. (1996) Guide for the Care and Use of Laboratory Animals. (Washington, DC, National Academy Press).
23. NICHOLL, C.G., CAROLAN, G., SEVELIUS, H. & BLOOM, S.R. (1989) Effect of enprostil on gastric emptying, intestinal transit time and postprandial release of gastro-intestinal peptides. Digestion, 43: 47–55. doi:10.1159/000199860
24. NISHIYAMA, K., KATORI, M., UENO, A., OHNO, T. & SAIGENJI, K. (1992) Suppression of myoelectrical activity of gastric smooth muscle by endogenous gastric prostaglandin E2. Digestive Diseases and Sciences, 37: 1002–1008. doi:10.1007/BF01300278
25. NOMPLEGGI, D., MYERS, L., CASTELL, D.O. & DUBOIS, A. (1980) Effect of a prostaglandin E2 analog on gastric emptying and secretion in rhesus monkeys. Journal of Pharmacology and Experimental Therapeutics, 212: 491–495.
26. OGINO, M., KHAN, M.S., CLINE, M.A. & TACHIBANA, T. (2016) Acute injections of corticosterone, norepinephrine and epinephrine retards food passage in the crop of chickens. General and Compative Endocrinology, 225: 155–161. doi:10.1016/j.ygcen.2015.10.015
27. OKUMURA, T., SHOJI, E., TAKAHASHI, N., WAKEBE, H., IMAGAWA, K., KIKUCHI, M. & KOHGO, Y. (1998) Delayed gastric emptying by Helicobacter pylori lipopolysaccharide in conscious rats. Digestive Diseases and Sciences, 43: 90–94. doi:10.1023/A:1018828105226
28. ROWART, M.J. & RUSH, B.D. (1984) The effects of PGF2 alpha, PGE2 and 16, 16 dimethyl PGE2 on gastric emptying and small intestinal transit in rat. Prostaglandins, 28: 915–928. doi:10.1016/0090-6980(84)90044-3
29. SAKURAI, T., TORII, A., TANAKA, F., ASAKAWA, H., MATSUOKA, M., NEGISHI, M., KATO, S., HINO, I., INADAMA, E., ARIIZUMI, M. & TODA, G. (1996) A basic study on gastric emptying test using barium grains and effect of drugs on gastric emptying in rats. Nihon Shokakibyo Gakkai Zasshi, 93: 75–82.
30. SMITH, G.V., MORAN, A.P., BAJAJ-ELLIOTT, M. & FARTHING, M.J. (2003) Induction of cyclooxygenase 2 by Escherichia coli but not Helicobacter pylori lipopolysaccharide

in gastric epithelial cells in vitro. Helicobacter, 8: 513–520. doi:10.1046/j.1523-5378.2003.00170.x

31. SPATES, S.T., CULLEN, J.J., EPHGRAVE, K.S. & HINKHOUSE, M.M. (1998) Effect of endotoxin on canine colonic motility and transit. Journal of Gastrointestinal Surgery, 2: 391–398. doi:10.1016/S1091-255X(98)80080-2
32. TACHIBANA, T., MATSUDA, K., SAWA, H., MIKAMI, A., UEDA, H. & CLINE, M.A. (2010) Differential thresholds of neuromedins B-, C-, and bombesin-induced anorexia and crop emptying rate in chickens. General and Comparative Endocrinology, 169: 144–150. doi:10.1016/j.ygcen.2010.08.006
33. TAKAKURA, K., HASEGAWA, K., GOTO, Y. & MURAMATSU, I. (1997) Nitric oxide produced by inducible nitric oxide synthase delays gastric emptying in lipopolysaccharide-treated rats. Anesthesiology, 87: 652–657. doi:10.1097/00000542-199709000-00027
34. VAN MIERT, A.S. (1971) Inhibition of gastric motility by endotoxin (bacterial lipopolysaccharide) in conscious goats and modification of this response by splanchnectomy, adrenalectomy or adrenergic blocking agents. Internationales de Pharmacodynamie et de Thérapie, 193: 405–414.
35. VAN MIERT, A.S. & DE LA PARRA, D.A. (1970) Inhibition of gastric emptying by endotoxin (bacterial lipopolysaccharide) in conscious rats and modification of this response by drugs affecting the autonomic nervous system. Archives Internationales de Pharmacodynamie et de Thérapie, 184: 27–33.
36. VAN MIERT, A.S. & VAN DUIN, C.T. (1980) Endotoxin-induced inhibition of gastric emptying rate in the rat. The effect of repeated administration and the influence of some antipyretic agents. Archives Internationales de Pharmacodynamie et de Thérapie, 246: 19–27.
37. WIRTHLIN, D.J., CULLEN, J.J., SPATES, S.T., CONKLIN, J.L., MURRAY, J., CAROPRESO, D.K. & EPHGRAVE, K.S. (1996) Gastrointestinal transit during endotoxemia: the role of nitric oxide. Journal of Surgical Research, 60: 307–311. doi:10.1006/jsre.1996.0048

CHAPTER 6

Effects of Dietary Vitamin E on Fertility Functions in Poultry Species

Deivendran Rengaraj and Yeong Ho Hong*

Department of Animal Science and Technology, Chung-Ang University, Anseong, Gyeonggi-do 456-756, Korea

ABSTRACT

Vitamin E is found in high quantities in vegetable oils. Although vitamin E has multiple functions in humans and animals, its key function is protecting cells from oxidative damage. Since its discovery, several studies have demonstrated that vitamin E deficiency causes impaired fertility in humans and lab animals. However, the effects of vitamin E deficiency or of its supplementation on the fertility of farm animals, particularly on poultry, are less well studied. Therefore, a comprehensive review of the effects of dietary vitamin E on the fertility of poultry species is needed in order to understand the beneficial role of vitamin E in the maintenance of sperm and egg qualities. Based on the observations reviewed here, we found that a moderate amount of vitamin E in poultry diet significantly protects semen/sperm qualities in male birds and egg qualities in female birds via decreasing the lipid peroxidation in semen/sperms and eggs. This review provides an overall understanding of the effects of dietary vitamin E on fertility functions in poultry species.

Keywords: vitamin E; fertility functions; poultry species

1. INTRODUCTION

Vitamins are vital organic compounds that can be obtained in high quantities from vegetable oils. Vitamin tablets are also commercially available at a reasonable price. To date, thirteen vitamins have been discovered, named alphabetically vitamin A to vitamin K. The fat-soluble vitamin E was discovered in the early 1920s by Evans and Bishop [1]. Vitamin E is a generic term for a group of tocopherols and tocotrienols that have some amount of vitamin activity. Among the four tocopherols (α, β, γ and δ) and four tocotrienols (α, β, γ and δ) discovered, α-tocopherol is the most biologically active form and available in high quantities from vegetable oils, unprocessed cereal grains, and nuts [2,3,4,5]. Since the first discovery of vitamin E, numerous investigators have demonstrated the nutritional value of vitamin E in humans as well as in laboratory and farm animals. Vitamin E is absorbed via the lymphatic pathway and transported into the systemic circulation in association with chylomicrons [5]. After absorption, vitamin E is stored chiefly in the liver. Because of its fat-soluble properties it is incorporated in lipid storage organelles and plasma membranes, therefore it is also widely distributed throughout the body [2,5,6]. Vitamin E interacts with several other dietary components, including selenium, polyunsaturated fatty acids, sulphur-containing amino acids, other vitamins and minerals, and synthetic antioxidants [7,8,9,10,11]. Vitamin E has multiple critical functions in animals. For instance, it acts as an efficient biological antioxidant in protecting cells from the adverse effects of reactive oxygen species or free radical initiators [7,8,10,12]. It is required for the regulation of heme biosynthesis, apparently by controlling induction and repression of aminolevulinic acid synthase and porphobilinogen synthase [12]. Vitamin E plays a specific role in the essential transport of amino acids and possibly lipids in the intestine [7]. Vitamin E is also involved in iron metabolism and steroidogenesis [12], and it stimulates humoral and cellular immune responses against infectious diseases [13].

The symptoms and disorders of vitamin E deficiency vary, depending on the species affected. They include disorders of the nervous system, skeletal system, circulatory system, muscular system, cardiovascular system, immune system, and reproductive system [2,10,13]. Furthermore, vitamin E deficiency may manifest itself in a number of disorders of liver, kidney, and lung, and in adipose tissue [2]. Vitamin E deficiency may increase the risk of ischemic heart disease, breast cancer, and the incidence of infections [14], and it promotes susceptibility to dietary and environmental stress in humans and animals [7,13]. Vitamin E is clinically administered to treat several conditions and diseases in humans, and complications associated with high dose administrations have been reported [4,14]. Compared to observations in humans, data from animal studies show that vitamin E toxicity is low and that the vitamin is not mutagenic, carcinogenic, or teratogenic [4].

The activity of vitamin E was first identified as an essential dietary factor for male and female reproduction in rats [1,15]. Although vitamin E has a broad range of functions in the body, it is primarily crucial for fertility in humans and livestock species, including poultry. Generally, it has been shown that vitamin E deficiency causes abnormal spermatogenesis in males and failure to retain zygotes and fetal resorption in females [5,10,12]. Reproductive functions are crucial for healthy offspring and species survival of all animals, including poultry. Thus, the objective of this paper is to review studies that have examined the effects of dietary vitamin E on male and female fertility in poultry species.

2. VITAMIN E IN POULTRY NUTRITION

Because vitamin contents vary in western diets, it is important to know the recommended allowances of vitamins for humans [16]. However, due to many variables, it is very difficult to estimate the optimum allowances of vitamins for laboratory animal and livestock diets [15]. Dietary supplementation with vitamin E increases the resistance of animals against infectious diseases and is thus recommended for farm animals including poultry, swine, sheep, and cattle, to meet the increasing demand for meat, eggs, and milk [13]. The rearing of domesticated birds including chickens, quail, turkeys, ducks, and geese in poultry farms has been increasing worldwide for the purpose of meat and egg production. Among the poultry species, chickens are reared in huge numbers, *i.e.*, several billions. Nowadays, modern poultry feeds are largely supplemented with proteins, carbohydrates, lipids, minerals, and vitamins to meet the basal nutritional requirements for raising and maintaining healthy birds. The standard dietary supplementation with proteins, carbohydrates, lipids, minerals, and vitamins supports normal growth of birds. Similarly, the standard supplementation with certain minerals and vitamins helps to increase the birds' immune resistance against pathogenic diseases [17,18,19,20]. The dietary requirement for vitamin E in poultry feed is highly variable and depends on the concentration and type of fat in the diet, the concentration of selenium, and the presence of pro-oxidants and antioxidants [21]. The National Research Council's Committee on Animal Nutrition, USA, provided the nutrient requirements for poultry species including chickens, turkeys, geese, ducks, pheasants, and quail. According to its recommendations, poultry feed can be supplemented with 10 IU of vitamin E per kg feed (1 IU = 0.67 mg *dl*-α-tocopheryl acetate) for chickens aged up to six weeks, 5 IU/kg feed for chickens aged over six weeks, 12 IU/kg feed for turkeys aged up to eight weeks, and 10 IU/kg feed for turkeys aged over eight weeks. For ducks and Japanese quail, feed can be supplemented with 10

IU/kg feed and 12 IU/kg feed, respectively, for starting and growing birds. The dietary recommendations of vitamin E for poultry species during laying and breeding vary slightly [21]. Vitamin E is one of the essential nutrients in poultry feed, and its deficiency causes a wide variety of disorders in poultry species (Figure 1). These include nutritional muscular dystrophy that affects striated muscles, erythrocyte hemolysis that affects erythrocytes, and exudative diathesis that affects capillary walls. Furthermore, vitamin E deficiency can lead to membrane lipid peroxidation, affecting hepatic mitochondria and microsomes, as well as to an accumulation of ceroid in adipose tissues, and to cerebellar encephalomalacia in chickens [7,10,12]. In addition, vitamin E deficiency impairs feather development in chickens [10]. Vitamin E deficiency causes gizzard myopathy in turkeys and ducks, and an accumulation of ceroid in turkeys [10].

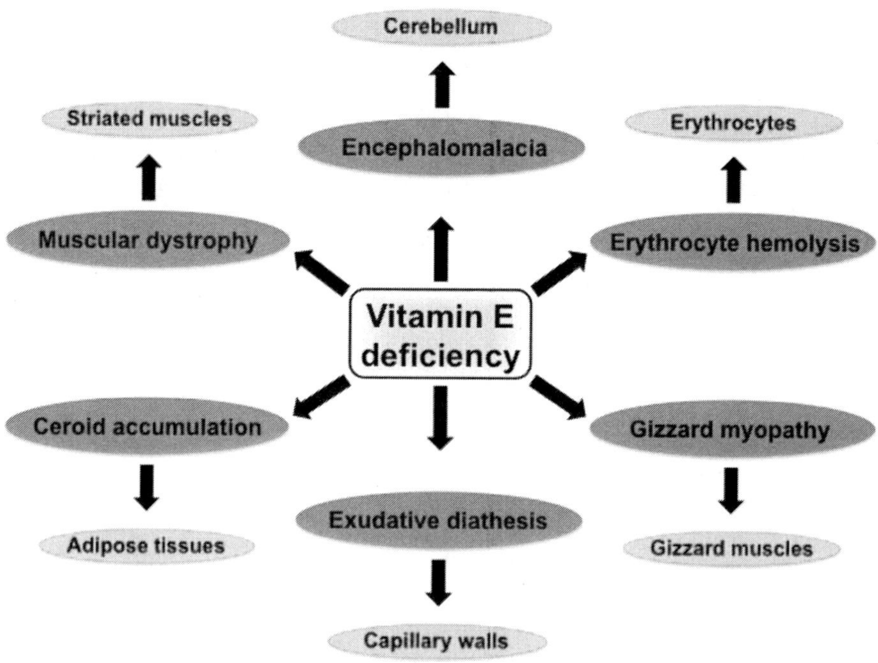

Figure 1. Commonly known vitamin E deficiency diseases and most vulnerable tissues in poultry species.

3. EFFECTS OF DIETARY VITAMIN E ON MALE FERTILITY FUNCTIONS IN POULTRY SPECIES

Since the first study on reproductive functions in rats by Evans and Bishop [1], several investigators have demonstrated the benefits of dietary supplementation with vitamin E on fertility in different animal species. In this chapter, we will review the effects of vitamin E supplementation specifically on fertility functions in male birds, based on evidence provided by several significant studies. Male fertility is principally related to semen and sperm qualities, including the volume of semen, concentration of sperms in the semen, sperm viability, sperm motility, sperm forward progression, and the sperm fertilizing capacity. These qualities can be affected by several environmental factors and endocrine disrupting chemicals, which may enter the body through diet, respiration, and skin contact [22]. The resulting adverse effects may be partially controlled or averted by clinical treatments or by dietary supplementation with certain beneficiary compounds, including minerals, vitamins, and antioxidants. Linoleic acid is one of the essential polyunsaturated fatty acids that cannot be produced *de novo* in vertebrates. If an animal diet is rich in any polyunsaturated fatty acids, the addition of an antioxidant is required in order to avoid peroxidation of the fatty acids [23]. Arscott and colleagues [24] fed adult male White Leghorn (WL) chickens with diets containing high levels of linoleic acid (7.3%) and an either low (4.3 mg/kg feed) or high (166.3 mg/kg feed) amount of vitamin E for 25 weeks. The males fed with diets high in linoleic acid and low in vitamin E showed impaired fertilizing capacity and sperm concentration in the semen. However, diet supplementation with a high amount of vitamin E prevented these adverse effects. The semen volume, hatchability of fertile eggs, and weight of the testes were not affected by any of the above diets. Furthermore, the authors suggested that vitamin E acts as an antioxidant that protects chickens against encephalomalacia by preventing the breakdown of linoleic acid to 12-oxo-*cis*-9-octadecenoic acid (keto acid). Encephalomalacia is a vitamin E deficiency disease, which readily occurs in chickens fed on a diet containing high levels of polyunsaturated fatty acids of the linoleic acid series and low levels of vitamin E [25]. In another study, Arscott and Parker [26] fed male chickens with a diet high in linoleic acid (7.3%) and low in vitamin E (4.3 mg/kg diet) from hatch to 28 weeks of age, and high in both linoleic acid (7.3%) and vitamin E (166.3 mg/kg diet) from 28 to 40 weeks. As a result, the adverse effects on fertility and semen concentration were restored in chickens fed on the diet high in vitamin E. Thus, the adverse effects of linoleic acid on male fertility are not permanent but can be reversed by vitamin E supplementation. Another compound, dilauryl succinate, does not contain peroxide or polyunsaturated fatty acids, however, similar to linoleic acid, it induces encephalomalacia in chickens. In a study by Yoshida and Hoshii [27], fertility of WL roosters fed on a diet containing 12% dilauryl succinate for 16 weeks was significantly low. When the roosters were fed a diet containing 12% dilauryl succinate and 200 mg/kg *dl*-α-tocopheryl acetate, the sperm fertilizing capacity was significantly high. Avian spermatozoa are rich in polyunsaturated fatty acids, in particular in docosatetraenoic acid and arachidonic acid [28,29]. The high proportions of polyunsaturated fatty acids provide membrane flexibility, which is essential for sperm mobility and sperm-egg fusion [30]. However, because of their high levels of polyunsaturated fatty acids, avian sperms are also very sensitive to reactive oxygen species, causing male infertility [28,31,32]. Therefore, an increased antioxidant status in semen or spermatozoa is a prerequisite for the prevention of male infertility. In a study by Surai and colleagues [31], six-month-old male Rhode Island Red chickens were fed with a diet containing 0, 20, 200, or 1000 mg/kg α-tocopheryl acetate for eight weeks. During the final two weeks, they observed that the concentration of vitamin E in semen and sperms had doubled, and the susceptibility of the semen to lipid peroxidation was decreased, in particular in birds fed with 200 mg/kg of vitamin E. In another study [33], 30-week-old WL roosters were fed with a basal diet high in fish/soybean oil and with or without supplementation of vitamin E (30, 200, or 400 mg/kg). After 38 weeks, the roosters fed with fish/soybean oil showed the lowest total antioxidant status in the semen. However, the addition of vitamin E to the fish/soybean oil diet resulted in an increase in semen volume, motility, and sperm potency at 38 weeks. Zaniboni and colleagues [32] allowed male turkeys to feed on a basal diet supplemented with 60 mg/kg of α-tocopheryl acetate from 26 to 39 weeks of age. From 40 to 60 weeks, the turkeys were allowed to feed on the basal diet additionally supplemented with 60 mg α-tocopheryl acetate and 2% fish oil. As a result, the α-tocopheryl status in the turkey semen was increased two-fold. Biswas and colleagues fed Indian native Kadaknath male chicks with a basal diet supplemented with 10, 100, or 200 mg/kg *dl*-α-tocopheryl acetate from hatch to 30 weeks of age. An analysis performed during the final three weeks of the trial revealed that the proportion of abnormal and dead spermatozoa was significantly lower and the fertilizing capacity was significantly higher in birds fed with 100 mg/kg of vitamin E. Moreover, the vitamin E status of semen and spermatozoa was higher in birds fed with 100 mg/kg of vitamin E than that in birds fed with 10 mg/kg of vitamin E [34]. These studies suggest that an increased antioxidant status of semen or sperm is based on an increase in the antioxidant content of the diet. In another study, Biswas and colleagues fed male Japanese quail with a basal diet containing 15 IU, 150 IU, or 300 IU of *dl*-α-tocopheryl acetate per kg feed, from the age of one day for up to 25 weeks. During these dietary trials, plasma testosterone concentration, testicular weight, as well as semen volume were not significantly changed. However, the percentage of abnormal and dead sperms was significantly lower, and the sperm capacity was higher in quail fed with a diet containing 150 IU/kg of vitamin E [35]. In a similar study, Hooda and colleagues fed male Japanese quail with a basal diet supplemented with different concentrations of vitamin E (0–300 IU *dl*-α-tocopheryl acetate/kg feed) from the age of five weeks onwards, for 10 to 13 weeks. Among the different concentrations of vitamin E, 75 IU/kg feed elicited the best result in protecting the male fertility functions in quail [36]. Taken together, these studies, in particular those by Surai *et al.*, Biswas *et al.*, and Hooda *et al.* suggest that a moderate level of vitamin E is best for maintaining the fertility functions in male chickens and quail [31,34,35,36].

Selenium is a trace element and frequently added to animal diet as a supplement for the maintenance of reproductive functions, and a deficiency in dietary selenium causes a decrease in sperm concentration, sperm motility, and sperm capacity in humans, lab animals, and farm animals, including poultry species [37,38]. Selenium is a component of selenoproteins, such as glutathione peroxidase, which protect sperms against oxidative damage [39,40]. It is well known that vitamin E interacts with selenium, and that both play a role in the maintenance of reproductive functions as well as in the reduction of reactive oxygen species and free radical generation [7,8]. Surai and colleagues fed Rhode Island Red males with a balanced diet supplemented with selenium (0.3 mg/kg diet) and vitamin E (20 or 200 mg/kg diet) and observed that the activity of glutathione peroxidase in testes, semen, and sperms was significantly increased [41]. Compared to chickens and quail, geese exhibit lower semen quality, egg production, and fertility and hatchability rates. When three-year-old White Koluda male geese were fed with a control diet and an experimental diet supplemented with 0.3 mg/kg feed of selenium and 100 mg/kg feed of vitamin E, the semen volume and sperm concentrations in the semen were significantly increased (1.5- and 1.7-fold, respectively) during the next reproductive season, in the males fed on the experimental diet. Also, the level of lipid peroxidation was significantly lower in males fed on the experimental diet [42]. The study by Jerysz and Lukaszewicz suggests that combined supplementation with selenium and vitamin E enhances the reproductive functions of certain avian species that naturally show low reproductive behavior. Antioxidant supplementation is not only important for lipid-containing poultry feed but also a prerequisite for any basal diet in order to maintain fertility. In a study by Lin and colleagues [43], freshly hatched Taiwan Native male chicks were fed with maize/soybean diets for up to 23 weeks of age. After 23 weeks, the birds were allowed to feed on maize/soybean diets supplemented with 80 mg/kg feed of *dl*-α-tocopheryl acetate for up to 52 weeks of age. Males fed with the vitamin E supplemented diet showed significantly higher sperm viability, sperm motility, and sperm concentration in the semen. This study suggests that a maize/soybean containing diet is inadequate for maintaining the fertility functions in male birds, thus vitamin E supplementation is required.

Molting in birds is a periodic occurrence, during which, at least once a year, the feathers are replaced. Molting is commonly associated with hens. However, several studies reported that molting in males might lead to an improved semen quality [30,44]. Molting in male birds can be experimentally induced by withdrawal of feed, water, or both, or by dietary manipulation with imbalanced minerals such as sodium, calcium, iodine, and zinc [30,44]. Khan and colleagues [45] induced molting in WL male birds with a basal diet containing 3000 mg zinc oxide per kg feed. After two weeks of molting, the birds were fed with either a basal diet or a diet supplemented with vitamin E (100 IU/kg feed). At least three weeks after the trial, the semen volume, sperm motility, and sperm capacity in fertilizing eggs had significantly increased in birds fed with the vitamin E supplemented diet, compared to the control group. Several investigators have demonstrated the association of vitamin E deficiency with impaired male reproductive functions. It has been identified as an anti-sterility vitamin, and its deficiency or long-term omission in the diet causes abnormal spermatogenesis and affects the semen qualities, including sperm viability, sperm motility, and sperm capacity [5,10,12]. Earlier studies reported that testicular degeneration resulted when Rhode Island Red males were fed a vitamin E deficient diet for two years [23,24]. From the studies reviewed above, it has been understood that dietary supplementation with vitamin E protects semen quality, especially by preventing the breakdown of polyunsaturated fatty acids, which can cause oxidative damage. The recommended amount of vitamin E in poultry diet for maintaining male fertility functions can vary, depending on the investigator. For instance, Khan *et al.* recommended that a poultry diet should contain at least 10 mg vitamin E per kg of feed [45]. However, several studies recommended that a poultry diet should contain a moderate amount (not very low and not very high) of vitamin E [31,34,35,36].

4. EFFECTS OF DIETARY VITAMIN E ON FEMALE FERTILITY FUNCTIONS IN POULTRY SPECIES

This chapter will review the effects of dietary vitamin E supplementation on female fertility functions in different poultry species, based on the evidence provided by several significant studies. The fertility functions of females, like those of males, are crucial for successful production of healthy offspring. More specifically, in poultry species the daily egg production, the egg quality, including egg weight and components of the yolk and albumin, and the egg fertility and hatchability are the most important factors that determine healthy offspring. The number of fertile eggs produced for hatching dictates the ultimate profitability of hens [46]. The nutrients required for embryo development are derived mainly from the yolk and albumin stored in the eggs. A chicken egg contains significant amounts of nutrients, including carbohydrates, proteins, lipids, vitamins, and trace elements, and these nutrients can be increased or decreased in eggs by altering the dietary composition. The concentrations of total lipids, including the polyunsaturated fatty acids of the linoleic acid series, and of antioxidants are relatively stable in the eggs of chickens fed with a standard diet, but they are subject to alteration by major changes in dietary nutrient composition [47,48]. Similar to Arscott's study in males [24], Machlin and colleagues [23] fed WL hens with a basal diet high in linoleic acid (7%) and low in vitamin E (20 IU/pound feed) for eight weeks. By the end of this period, egg production had decreased from 78% to 25%. Only 37% of the eggs were fertile, and none of the fertile eggs was hatched. When hens were fed with a diet containing high amounts of both linoleic acid (7%) and vitamin E (100 IU/pound feed) for eight weeks, the egg production reached an average of 57%. Seventy-six percent of the eggs were fertile and 67% of the fertile eggs were hatched. Together with the results obtained for males [24], this study suggests that vitamin E acts as an antioxidant, in that it protects female fertility by preventing the breakdown of linoleic acid, which would then lead to oxidative damage. Additionally, the study by Machlin and colleagues showed that hens fed with diets low in linoleic acid do not require additional vitamin E or antioxidants for maintenance of egg production, fertility, and hatch of fertile eggs [23]. When Yoshida and Hoshii fed WL hens with a basal diet containing 12% dilauryl succinate, which has similar effects as linoleic acid, the fertility of eggs and the hatchability of fertile eggs were observed to be low. However, these effects were prevented by adding 200 mg/kg feed of *dl*-α-tocopheryl acetate to the same diet, containing 12% dilauryl succinate [27].

Vicine, an alkaloid found in high quantities in faba beans (*Vicia faba* L.), has a marked influence on the metabolism of laying hens. Dietary vicine causes peroxidation of cellular components, resulting in abnormal lipid transport and a decrease in fertility. In a study by Muduuli *et al.*, laying WL hens fed with a vicine containing diet showed decreased egg weight, fertility, and hatchability. However, vitamin E supplementation slightly increased the egg weight and markedly protected fertility and hatchability of eggs [49]. This study clearly indicates that the antioxidant property of vitamin E is very helpful in reducing the adverse effects of toxic compounds like vicine. Similarly to the observations reviewed above, an adequate amount of vitamin E was found necessary in the poultry diet for maintaining female fertility, regardless of whether or not the diet is high in polyunsaturated fatty acids. In a study by Lin *et al.* [48], hatched Taiwan Native female chicks were fed with a maize/soybean diet for up to 17 weeks of age. After 17 weeks, the birds were allowed to feed on a maize/soybean diet supplemented with 80 mg/kg feed of *dl*-α-tocopheryl acetate for up to 46 weeks of age. As a result, hens fed with the vitamin E supplemented diet showed a better performance in egg production and egg mass. Likewise, vitamin E supplemented diet showed a better performance in fertility (over 7%) and hatchability (over 13%). In another study, Hooda and colleagues fed five-week-old male and female Japanese quail with a basal diet containing different concentrations of vitamin E (0, 75, 150, 225, and 300 IU *dl*-α-tocopheryl acetate/kg feed), during a period of 5 to 13 weeks of age. Then, the males and females were allowed to mate in different groups: control males with control females, control males with females on vitamin E supplemented diets, and control females with males on vitamin E supplemented diets. As a result, males paired with females after both had been fed with a vitamin E supplemented diet led to slightly better fertility and hatchability of quail eggs. However, these results were not influenced by an increase in dietary vitamin E [36]. When Japanese quail were maintained on a soybean meal diet containing low levels of vitamin E (2 IU/kg diet) for 35 weeks, the percentage of fertile eggs was significantly decreased and the hatchability of fertile eggs was severely affected. However, no other clinical symptoms were observed [50]. Kling and Soares suggested that this much lower amount of vitamin E in the diet is extremely inadequate for supporting normal reproduction in quail.

In avian species, the sperms are stored for a short period in male reproductive tract (vas deferens). Following natural mating or artificial insemination, the ejaculated sperms undergo a selection process and the selected sperms are stored for a prolonged period in female reproductive tract at two sperm storage sites [51,52,53]. The primary storage site is found in the utero-vaginal junction, and the secondary storage site is found in the lower portion of the infundibulum [52,53]. Studies by Breque and colleagues indicate a positive effect of dietary vitamin E on the antioxidant status of sperm storage sites in hens [53]. In avian species, the proportion of sperms found particularly at utero-vaginal junction significantly correlate with the sperms found at perivitelline layer of eggs [52,54,55]. Moreover, Brillard and Antoine suggest that hen's egg achieves 100% of fertility when the perivitelline layer contains 1440 or more sperms [54]. Taken together, it is well understood that vitamin E deficiency significantly affects the fertility of female poultry species [5,12]. If the hatched chicks show vitamin E deficiency symptoms, the amount of vitamin E in eggs should be increased through supplementation of the maternal diet. In addition, the amount of vitamin E in the diet should be correlated with the amount of dietary selenium and polyunsaturated fatty acids. If the poultry diet contains a sufficient amount of selenium and is not rich in polyunsaturated fatty acids, about 20 mg vitamin E per kg diet is required for maintaining fertilization rate of breeding hens [56]. In addition, Yoshida and Hoshii recommended an amount of more than 500 μg vitamin E per egg to ensure the hatching of healthy chicks [27]. The main effects of dietary vitamin E supplementation on male and female fertility in poultry are shown in Figure 2. From analyses of male and female birds, it has been understood that feeding the birds with at least 100 mg vitamin E per kg diet can prevent vitamin E deficiency problems.

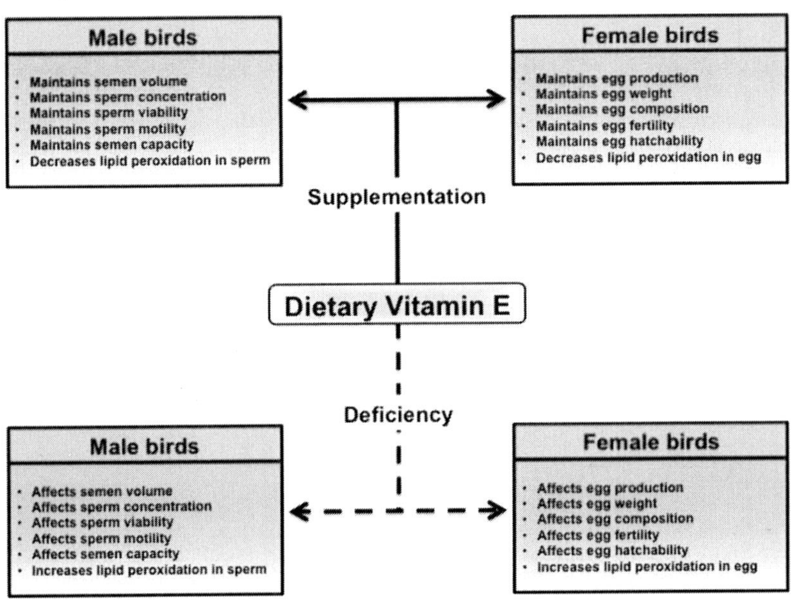

Figure 2. The main effects of dietary vitamin E supplementation on male and female fertility functions in poultry species.

5. CONCLUSIONS AND FUTURE PERSPECTIVES

In this communication, we reviewed the effects of dietary antioxidant vitamin E on fertility of male and female poultry species. The review suggests that a moderate vitamin E supplementation of a balanced poultry diet significantly maintains male fertility functions, including semen volume, sperm concentration, sperm viability, sperm motility, and sperm capacity, in poultry species. In addition, a moderate vitamin E diet supplementation significantly maintains female fertility in poultry species, including egg production, egg fertility, and egg hatchability. It has been understood that the maintenance of fertility in male and female birds by vitamin E is based on its property as a defense mechanism against oxidative damage, which predominantly results from the breakdown of polyunsaturated fatty acids. Furthermore, the deficiency or long-term omission of vitamin E in the diet impairs male and female fertility in birds. Based on the information discussed here, this review suggests a few future perspectives for research on poultry species. (1) Most studies referring to poultry species have focused on the effects of dietary vitamin E on semen and sperm qualities in males, and egg qualities in females. However, the effects on Leydig cells and Sertoli cells in males, and on theca cells and granulosa cells in females need to be elucidated; (2) The effects of vitamin E deficiency or supplementation on male and female steroidogenesis are not well understood; (3) The effects of vitamin E deficiency or supplementation on the functions of fertility-associated genes and proteins need to be elucidated; (4) Most studies recommended a moderate amount of vitamin E (75–100 mg/kg diet) for the maintenance of fertility functions in male and female birds. In agreement with these suggestions, the National Research Council's Committee on Animal Nutrition should establish a recommended allowance; (5) Most studies referring to the fertility of poultry species have focused on the interaction of vitamin E with polyunsaturated fatty acids and selenium, however, the interaction of vitamin E with other vitamins and minerals are less well known. Collectively, this review provides an overall understanding of the effects of dietary vitamin E on fertility of diverse poultry species.

ACKNOWLEDGMENTS

This Project was supported by the National Research Foundation grant (NRF-2010-0009360), Republic of Korea.

REFERENCES

1. Evans, H.M.; Bishop, K.S. On the existence of a hitherto unrecognized dietary factor essential for reproduction. *Science* **1922**, *56*, 650–651.
2. Herting, D.C. Perspective on vitamin E. *Am. J. Clin. Nutr.* **1966**, *19*, 210–218.
3. Horwitt, M.K. The promotion of vitamin E. *J. Nutr.* **1986**, *116*, 1371–1377.
4. Bendich, A.; Machlin, L.J. Safety of oral intake of vitamin E. *Am. J. Clin. Nutr.* **1988**, *48*, 612–619.
5. Bjorneboe, A.; Bjorneboe, G.E.; Drevon, C.A. Absorption, transport and distribution of vitamin E. *J. Nutr.* **1990**, *120*, 233–242.
6. Wang, X.; Quinn, P.J. Vitamin E and its function in membranes. *Prog. Lipid Res.* **1999**, *38*, 309–336.
7. Green, J. Vitamin E and the biological antioxidant theory. *Ann. N. Y. Acad. Sci.* **1972**, *203*, 29–44.
8. Tappel, A.L. Selenium-glutathione peroxidase and vitamin E. *Am. J. Clin. Nutr.* **1974**, *27*, 960–965.
9. Debry, G. Polyunsaturated fatty acids and vitamin E: Their importance in human nutrition. *Ann. Nutr. Aliment.* **1980**, *34*, 337–350.
10. Combs, G.F. Assessment of vitamin E status in animals and man. *Proc. Nutr. Soc.* **1981**, *40*, 187–194.
11. Chen, L.H. Interaction of vitamin E and ascorbic acid (review). *In Vivo* **1989**, *3*, 199–209.
12. Nair, P.P. Vitamin E and metabolic regulation. *Ann. N. Y. Acad. Sci.* **1972**, *203*, 53–61.
13. Tengerdy, R.P. Vitamin E, immune response, and disease resistance. *Ann. N. Y. Acad. Sci.* **1989**, *570*, 335–344.
14. Machlin, L.J. Clinical uses of vitamin E. *Acta Vitaminol. Enzymol.* **1985**, *7*, 33–43.
15. Putnam, M.E.; Comben, N. Vitamin E. *Vet. Rec.* **1987**, *121*, 541–545.
16. Morton, R.A. Vitamins in western diets. *Proc. Nutr. Soc.* **1976**, *35*, 23–29.
17. Tengerdy, R.P.; Heinzerling, R.H.; Nockels, C.F. Effect of vitamin E on the immune response of hypoxic and normal chickens. *Infect. Immun.* **1972**, *5*, 987–989.
18. Lucas, I.A. The use of antibiotics as feed additives for farm animals. *Proc. Nutr. Soc.* **1972**, *31*, 1–8.
19. Wilgus, H.S. Disease, nutrition—Interaction. *Poult. Sci.* **1980**, *59*, 771–781.
20. Latshaw, J.D. Nutrition—Mechanisms of immunosuppression. *Vet. Immunol. Immunopathol.* **1991**, *30*, 111–120.
21. National Research Council. *Nutrient Requirements of Poultry*, 9th revised ed.; The National Academies Press: Washington, DC, USA, 1994; pp. 1–157.
22. Rengaraj, D.; Kwon, W.S.; Pang, M.G. Effects of motor vehicle exhaust on male reproductive function and associated proteins. *J. Proteome Res.* **2015**, *14*, 22–37.
23. Machlin, L.J.; Gordon, R.S.; Marr, J.E.; Pope, C.W. Effect of antioxidants and unsaturated fatty acids on reproduction in the hen. *J. Nutr.* **1962**, *76*, 284–290.
24. Arscott, G.H.; Parker, J.E.; Dickinson, E.M. Effect of dietary linoleic acid, vitamin E and ethoxyguin on fertility of male chickens. *J. Nutr.* **1965**, *87*, 63–68.
25. Mokadi, S.; Budowski, P. Relationship between vitamin E and encephalomalacia in chicks. *Br. J. Nutr.* **1963**, *17*, 347–355.
26. Arscott, G.H.; Parker, J.E. Effectiveness of vitamin E in reversing sterility of male chickens fed a diet high in linoleic acid. *J. Nutr.* **1967**, *91*, 219–222.
27. Yoshida, M.; Hoshii, H. Effect of dilauryl succinate on reproduction of the cock and hen and preventive effect of vitamin E. *J. Nutr.* **1976**, *106*, 1184–1191.
28. Surai, P.F.; Noble, R.C.; Sparks, N.H.C.; Speake, B.K. Effect of long-term supplementation with arachidonic or docosahexaenoic acids on sperm production in the broiler chicken. *J. Reprod. Fertil.* **2000**, *120*, 257–264.

29. Surai, P.F.; Fujihara, N.; Speake, B.K.; Brillard, J.P.; Wishart, G.J.; Sparks, N.H.C. Polyunsaturated fatty acids, lipid peroxidation and antioxidant protection in avian semen. *Asian-Australas. J. Anim. Sci.* **2001**, *14*, 1024–1050.
30. Khan, R.U.; Nikousefat, Z.; Javdani, M.; Tufarelli, V.; Laudadio, V. Zinc-induced moulting: Production and physiology. *World Poult. Sci. J.* **2011**, *67*, 497–505.
31. Surai, P.F.; Kutz, E.; Wishart, G.J.; Noble, R.C.; Speake, B.K. The relationship between the dietary provision of α-tocopherol and the concentration of this vitamin in the semen of chicken: Effects on lipid composition and susceptibility to peroxidation. *J. Reprod. Fertil.* **1997**, *110*, 47–51.
32. Zaniboni, L.; Rizzi, R.; Cerolini, S. Combined effect of DHA and α-tocopherol enrichment on sperm quality and fertility in the turkey. *Theriogenology* **2006**, *65*, 1813–1827.
33. Zanini, S.F.; Torres, C.A.; Bragagnolo, N.; Turatti, J.M.; Silva, M.G.; Zanini, M.S. Evaluation of the ratio of ω6:ω3 fatty acids and vitamin E levels in the diet on the reproductive performance of cockerels. *Arch. Tierernahr.* **2003**, *57*, 429–442.
34. Biswas, A.; Mohan, J.; Sastry, K.V. Effect of higher dietary vitamin E concentrations on physical and biochemical characteristics of semen in Kadaknath cockerels. *Br. Poult. Sci.* **2009**, *50*, 733–738.
35. Biswas, A.; Mohan, J.; Sastry, K.V.; Tyagi, J.S. Effect of dietary vitamin E on the cloacal gland, foam and semen characteristics of male Japanese quail. *Theriogenology* **2007**, *67*, 259–263.
36. Hooda, S.; Tyagi, P.K.; Mohan, J.; Mandal, A.B.; Elangovan, A.V.; Pramod, K.T. Effects of supplemental vitamin E in diet of Japanese quail on male reproduction, fertility and hatchability. *Br. Poult. Sci.* **2007**, *48*, 104–110.
37. Young, V.R.; Nahapetian, A.; Janghorbani, M. Selenium bioavailability with reference to human-nutrition. *Am. J. Clin. Nutr.* **1982**, *35*, 1076–1088.
38. Hansen, J.C.; Deguchi, Y. Selenium and fertility in animals and man—A review. *Acta Vet. Scand.* **1996**, *37*, 19–30.
39. Neve, J.; Vertongen, F.; Molle, L. Selenium deficiency. *Clin. Endocrinol. Metab.* **1985**, *14*, 629–656.
40. Ahsan, U.; Kamran, Z.; Raza, I.; Ahmad, S.; Babar, W.; Riaz, M.H.; Iqbal, Z. Role of selenium in male reproduction—A review. *Anim. Reprod. Sci.* **2014**, *146*, 55–62.
41. Surai, P.; Kostjuk, I.; Wishart, G.; Macpherson, A.; Speake, B.; Noble, R.; Ionov, I.; Kutz, E. Effect of vitamin E and selenium supplementation of cockerel diets on glutathione peroxidase activity and lipid peroxidation susceptibility in sperm, testes, and liver. *Biol. Trace Elem. Res.* **1998**, *64*, 119–132.
42. Jerysz, A.; Lukaszewicz, E. Effect of dietary selenium and vitamin E on ganders' response to semen collection and ejaculate characteristics. *Biol. Trace Elem. Res.* **2013**, *153*, 196–204.
43. Lin, Y.F.; Chang, S.J.; Yang, J.R.; Lee, Y.P.; Hsu, A.L. Effects of supplemental vitamin E during the mature period on the reproduction performance of Taiwan native chicken cockerels. *Br. Poult. Sci.* **2005**, *46*, 366–373.
44. Woodard, A.E.; Ogasawara, F.X.; Snyder, R.L.; Stinnett, V. Effect of forced molting on quantity and quality of semen in turkey breeder males as influenced by diet. *Poult. Sci.* **1975**, *54*, 2094–2101.
45. Khan, R.U.; Rahman, Z.U.; Javed, I.; Muhammad, F. Effect of vitamins, probiotics and protein on semen traits in post-molt male broiler breeders. *Anim. Reprod. Sci.* **2012**, *135*, 85–90.
46. Khan, R.U. Antioxidants and poultry semen quality. *Worlds Poult. Sci. J.* **2011**, *67*, 297–308.
47. Posati, L.P.; Kinsella, J.E.; Watt, B.K. Comprehensive evaluation of fatty acids in foods. III. Eggs and egg products. *J. Am. Diet. Assoc.* **1975**, *67*, 111–115.
48. Lin, Y.F.; Chang, S.J.; Hsu, A.L. Effects of supplemental vitamin E during the laying period on the reproductive performance of Taiwan native chickens. *Br. Poult. Sci.* **2004**, *45*, 807–814.

49. Muduuli, D.S.; Marquardt, R.R.; Guenter, W. Effect of dietary vicine and vitamin E supplementation on the productive performance of growing and laying chickens. *Br. J. Nutr.* **1982**, *47*, 53–60.
50. Kling, L.J.; Soares, J.H., Jr. Vitamin E deficiency in the Japanese quail. *Poult. Sci.* **1980**, *59*, 2352–2354.
51. Bakst, M.R. Sperm recovery from oviducts of turkeys at known intervals after insemination and oviposition. *J. Reprod. Fertil.* **1981**, *62*, 159–164.
52. Brillard, J.P. Sperm storage and transport following natural mating and artificial-insemination. *Poult. Sci.* **1993**, *72*, 923–928.
53. Breque, C.; Surai, P.; Brillard, J.P. Roles of antioxidants on prolonged storage of avian spermatozoa *in vivo* and *in vitro*. *Mol. Reprod. Dev.* **2003**, *66*, 314–323.
54. Brillard, J.P.; Antoine, H. Storage of sperm in the uterovaginal junction and its incidence on the numbers of spermatozoa present in the perivitelline layer of hens eggs. *Br. Poult. Sci.* **1990**, *31*, 635–644.
55. Brillard, J.P.; Bakst, M.R. Quantification of spermatozoa in the sperm-storage tubules of turkey hens and the relation to sperm numbers in the perivitelline layer of eggs. *Biol. Reprod.* **1990**, *43*, 271–275.
56. Hennig, A.; Marckwardt, E.; Richter, G. Relations between vitamin-E supply and the fertility of hens. *Arch. Tierernahr.***1986**, *36*, 519–529.

CHAPTER 7

Mycotoxins in Poultry

Ayhan Filazi, Begum Yurdakok-Dikmen, Ozgur Kuzukiran and Ufuk Tansel Sireli

[1] *Department of Pharmacology and Toxicology, Faculty of Veterinary Medicine, Ankara University, Ankara, Turkey*

2 Veterinary Control Central Research Institute, Ankara, Turkey

3 Department of Food Hygiene and Control, Faculty of Veterinary Medicine, Ankara University, Ankara, Turkey

ABSTRACT

Mycotoxins, the toxic secondary metabolites of fungi, particularly produced by many species of *Aspergillus,Fusarium* and *Penicillium*, have affected animal and human health for over thousand years, whereas little has been discovered so far about these complex substances in poultry, which are generally very sensitive. Even though it varies by species and sex, some common effects are reduced feed intake, weight gain, feed efficiency, growth performance, immunity and hatchability along with increased mortality, organ damages (mainly kidney and liver), carcinogenicity, teratogenicity and decreased egg production. Besides their adverse health effects and the decrease in production rate, concerns over their importance in public health is still under debate. Decontamination approaches to reduce mycotoxins in feed are technologically diverse and based on chemical, biological and physical strategies. Chemical remediation strategies involve the conversion of mycotoxins via chemical reactions. Biological strategies involve various substances such as plant ingredients, enzymes and microorganisms. Physical processes include sorting, milling, dehulling, cleaning, heating, irradiation or combinational approaches. New strategies for the prevention and treatment of mycotoxicosis, including beneficial microorganisms/products, along with alternative treatments, including plant extracts/essential oils, are current hot topics in the poultry industry.

Keywords: Control, mycotoxins, poultry, prevention

1. INTRODUCTION

Mycotoxins, the secondary metabolites of fungi, are a global concern. At aerobic conditions, fungal growth in various feed raw materials is inevitable. There are about 200 species of fungi that produce mycotoxins. Majority of the fungi that form mycotoxin belong to three genuses: *Aspergillus, Penicillium* and *Fusarium*. Although more than 500 mycotoxins produced by these fungi are known, only some of these mycotoxins exert pathogenic characteristics. The poisoning in humans and animals caused by feeds and foods contaminated with mycotoxins may range from a slight reaction to death [1–6].

Fungal growth and mycotoxin production initiate in the cropland, during transportation or storage, and are affected by the environmental conditions including seasons, location of grain cultivation, drought and time of harvest. Long-term analyses show that feed and feedstuffs may be contaminated with mycotoxins, where these contaminated feed materials often include more than one mycotoxin [7]. Also, each of the cereals and oil seeds, available at the poultry feeds are vegetable substances obtained in different climatic conditions during vegetation in the cropland, transport and storage. For this reason, although generally only one mycotoxin is produced in raw feed materials, multiple types of mycotoxins might be found in mixed feeds. Such co-contamination examples in poultry feed are as follows: aflatoxin presence with ochratoxins, T-2 toxin or diacetoxyscirpenol; ochratoxins with T-2 toxin or citrinin and vomitoxin with fumaric acid in the poultry feeds [8].

According to the Food and Agriculture Organization Report, 25% of the world's growing crops are affected by mycotoxins each year, with annual losses of around 1 billion metric tons of food and food products [9]. Generally, there is yield loss or reduced crop value due to diseases induced by toxigenic fungi, and losses in animal productivity and animal or human health costs are due to mycotoxin contamination. Apart from these, the extra costs include the management of mycotoxin, such as prevention, control, sampling, mitigation, labor loss and research costs. Thus, the economic problems related to mycotoxins concern all sections of society [10].

The reasons for mycotoxins to receive this particular attention are their undesirable health effects, decrease in the production rate due to the spoilage of feed, and overall economic effects which are reflected in international trade of food and food products. Therefore, control of the fungal development and mycotoxin production are crucial for feed and animal producers [6].

Mycotoxins are metabolized in the alimentary canal, liver or kidneys of the poultry in accordance with their chemical properties. Their transfer to poultry meat and eggs leads to undesirable health effects in humans, leading to major concerns in public health. Contamination of the feeds with fungi both damages their organoleptic properties and increases poisoning risk by decreasing their nutritional value. Toxicity of the mycotoxins depends on the amount of absorption, number of the metabolites that are formed, exposure period and sensitivity of the animal [1].

Some mycotoxins like aflatoxins (AF), ochratoxin A (OTA), fumonisins (FUM), deoxynivalenol (DON) and T-2 toxin significantly affect the health and productivity of poultry species [11]. The aim of this review is to discuss in detail the important mycotoxins for poultry and their effects, along with the recent developments in prevention strategies.

2. SELECTED MYCOTOXINS IN POULTRY PRODUCTION

2.1. Aflatoxins

Aflatoxins, a group of harmful secondary metabolites characterized by polyketide-derived furanocoumarins, are produced mainly by Aspergillus fungi, such as *A. flavus* and *A. parasiticus.* Aflatoxins B1 (AFB1), G1 (AFG1) and their dihydroxy derivatives B2 (AFB2) and G2 (AFG2) naturally contaminate feeds. The presence of Aflatoxin M1 and M2 (AFM1 and AFM2), the 4-hydroxy metabolites of AFB1 and B2 in biological fluids including milk and tissues, is related to the exposure of the contaminated feed. International Agency for Research in Cancer (IARC) classified these highly toxic compounds as highly carcinogenic to humans (Group 1) [12].

AFB1 was explored in the early 1960s as the main etiological agent of "Turkey X Disease" responsible for the death of young turkeys in England as a result of contaminated peanut-based feed [13]. It is a widespread dietary hepatotoxin and hepatocarcinogen, and a major public health concern throughout the world. There are substantial species-specific differences with regard to susceptibility to the toxic effects of AFB1, and domestic turkeys (*Meleagris gallopavo*) are among the most susceptible species known so far [14, 15].

Aflatoxins are usually found in feed ingredients used for poultry rations. Most extensive forms of AF include B1, B2, G1 and G2, with AFB1 being the most widespread and biologically active form [16]. In fact, AFB1 is a "pro-carcinogen" that is activated to a reactive form by the enzyme hepatic microsomal cytochrome P450 (CYP450), whereas electrophilic AFB1-8,9-epoxide (AFBO) is required for carcinogenic and toxic activity [13]. This compound forms AFB-N7-guanine adduct with DNA, which is not stable and is transformed into formamidopyrimidine. DNA adducts and repair activities through modulation are considered as important markers in carcinogenesis susceptibility. AFB-N7-guanine adduct in urine is also a potential biomarker of AFB1 exposure in animals and humans, and is vital for estimating exposure conditions and potential risk in individuals consuming AFB1 [12].

Major AFB1 detoxification route is via conjugation of the AFBO to endogenous glutathione (GSH) catalyzed by the classical detoxification enzymes glutathione S-transferases (GSTs) in mammals. Xenobiotics, including chemical carcinogens and environmental contaminants, are metabolized through detoxification processes in phase-II metabolism through these proteins [17]. Due to the expression of A3 subunit (mGSTA3), mice bioactivate AFB1 and are assumed as AFB1-resistant with great catalytic activity for AFBO. The present approach is that efficiency of GST conjugation is a major "rate-limiting" determinant for AFB1 action in individuals and species, irrespective of the efficiency of AFB1 bioactivation [14].

Aflatoxins cause a variety of effects in poultry, including decreased weight gain; poor feed efficiency; reduced egg production and egg weight; increased liver fat; changes in organ weights; reduction in serum protein levels; carcass bruising; poor pigmentation; liver damage; decreased activities of several enzymes involved in the digestion of starch, protein, lipids, and nucleic acids; and induction of immunosuppression. Evidence suggests that immunosuppression caused by AF results in many disease outbreaks, vaccination failures and poor antibody titers [9, 11]. At necropsy, livers are usually pale and enlarged, as a result of aflatoxicosis. Histologically, liver lesions include congestion of the hepatic sinusoids, focal hemorrhages, centrilobular fatty cytoplasmic vacuolation and/or necrosis, biliary hyperplasia, and nodular lymphoid infiltration. AF produces a malabsorption syndrome characterized by steatorrhea, hypocarotenoidemia, and decreased concentrations of bile salts and pancreatic lipase, trypsin, amylase and RNase at levels that do not affect growth [11].

Broiler chicken fed with 1.0 mg of AFB_1/kg of diet were found to show decreased hepatic gene expression of superoxide dismutase, GST, and epoxide hydrolase and increased gene expression of interleukin-6 and CYP1A1 and 2H1 at cellular level [18].

Ingestion of 2 ppm AFB1 in male broiler chicks was found to alter various hepatic genes causing up-down regulation. For instance, enzymes having role in the production of energy and metabolism of fatty acids (carnitine palmitoyl transferase), development and growth (insulin-like growth factor 1), coagulation (coagulation factors IX and X), protection of immune system (interleukins), antioxidant protection (GST), detoxification (epoxide hydrolase) were found to be downregulated; while cell-proliferation enzymes (ornithine decarboxylase) were upregulated [19].

A study reported that wild turkeys are significantly more resistant to AFB1 compared to domestic turkeys. Intensive breeding technologies and industrial alliance to produce modern domestic turkey led to the unintentional loss of AFB1-protective GST alleles directing a relative resistance. Actually, it has been shown that similar breeding pressures have eventuated in a remarkable loss of rare alleles and genetic diversity of single-nucleotide polymorphisms in commercial breeds of chickens [14].

As mentioned previously, mycotoxins not only lead to the aforementioned economic and health problems in poultry, but also cause public health concerns due to their residues in food for human consumption. Major metabolites of AFB1 formed in chicken liver are AFM1 and AFB2a. AFB1 and B2 are then degraded to cyclopentanol and aflatoxicol through NADP. Both AFB1 and aflatoxicol are known to accumulate at the layer of the egg. While AFB1 and AFM1 are present in chicken muscle and blood, the levels are found much higher in turkeys; the aflatoxicol levels were found to be less prominent in these animals. As a comparison, 1/1200 of AFB1 taken with feeds was found to accumulate in poultry meats, while 1/2200 of AFB1 was found to accumulate in the eggs [20].

2.2. Ochratoxins

Ochratoxins are a family of structurally related metabolites that are produced by Aspergillus and Penicillium species, including *A. ochraceus*, *A. niger* and *P. verrucosum* [21]. The most prevalent form is ochratoxin A (OTA) followed by its non-chlorinated metabolite ochratoxin B (OTB) and the ethyl ester form ochratoxin C (OTC). OTA is the most frequent and relevant form of this family, while OTB and OTC are generally counted to be of lesser importance [22]. IARC classified ochratoxin A as a compound possibly carcinogenic to humans (Group 2B) [23].

Aspergillus species can generate OTA and OTB in parallel, and experiments with *A. ochraceus* have ascertained growth-associated production of OTA and OTB, in which the yield and the ratio were dependent on the current culture terms. Mostly, the amount of OTB generated was quite lower than that of OTA, but under some situations, the level of OTB production was comparable to that of OTA. The informed generation ratios (OTA:OTB) ranged from 2:1 to 34:1 [22]. Herein, it was reported that a complex interaction of various carbon sources, basal media and nitrogen sources seems to be considerable. High OTA production was related to an induction of OTA polyketide synthase expression, whereas OTB production is not connected with transcription of the polyketide synthase gene. Laboratory fermentation experiments with *A. ochraceus* result in production of OTA at high yields (by 10 mg/g), OTB and temporarily also ochracin [24]. The intermediate metabolite OTβ was determined to be biotransformed in an effective manner into both OTA and OTB (14% and 19%, respectively), whereas OTα was biotransformed only into OTA (4.9%). In addition, OTB is inadequately converted (1.5%) into OTA, whereas some OTB may be produced by dechlorination of OTA [22].

OTA is hepatotoxic, nephrotoxic, neurotoxic, teratogenic and immunotoxic as confirmed by *in vivo* experiments with different animal species and various *in vitro* methods; its adverse effects include renal toxicity and carcinogenesis. Molecular studies with OTA revealed a non-DNA-reactive genotoxic mechanism, which includes various epigenetic mechanisms principally connected to oxidative stress, compensatory cell proliferation and disruption of cell signaling and division [25]. However, a direct genotoxic mechanism including OTA bioactivation and DNA adduct formation was also suggested [23] and this mechanism was found to be in accordance with some *in vivo* gene expression results [25]. Overall, the mode of action of OTA for renal carcinogenesis is yet under discussion.

Ochratoxins cause significant health problems and economic losses in poultry [26] and cause mycotoxic porcine nephropathy (MPN) [27]. Ochratoxin-related diseases are characterized by severe kidney damage, which could be overtly related to the exposure to ochratoxins, sometimes in combination with different mycotoxins [27]. Likewise, a slow, progressive renal disease (endemic nephropathy, EN), characterized by cellular interstitial fibrosis, tubular atrophy, and karyomegaly predominately in proximal convoluted tubules was described in humans. The etiology of this disease is still unknown, but researchers agree that the causative agent is of natural origin. The most common causes of the multiethiologic disease, EN, were the aristolochic acid from the plant birthwort (*Aristolochia clematitis*) and mycotoxins (OTA and citrinin) [28].

OTA consists of an isocoumarin moiety linked through the 7-carboxy group to the amino acid L-β-phenylalanine. OTA interferes with DNA, RNA and protein synthesis by inhibiting the enzyme phenylalanine-tRNA synthetase at a cellular level. It also affects renal carbohydrate metabolism through the reduction of the renal mRNA coding for phosphoenolpyruvate carboxykinase, a key enzyme in gluconeogenesis [11]. The effects of OTA on DNA, RNA and protein synthesis are thought to be due to the phenylalanine moiety of the toxin competing with phenylalanine in the enzyme-catalyzed reaction. OTA also causes hypocarotenidemia which has more severe effects in broilers than AF [29].

Signs of OTA toxicity in poultry include weakness, anemia, decreased feed consumption, reduced growth rate and egg production, poor feathering and excessive mortality at high dietary concentrations [21]. Pathophysiological changes include decreased urine concentration and glomerular filtration rate, impairment of proximal tubular function, and degeneration and ultrastructural alterations in renal integrity [30]. Increases in the relative weights of liver, spleen, pancreas, proventriculus, gizzard and testes have also been reported in poultry fed OTA [21]. A study found that the expression of *Eimeria tenella* and its pathological effects were maximum in the presence of OTA compared to the incidence of coccidiosis alone in broiler chicks [31].

Comparative toxicity studies of OTA and OTB have shown that *in vivo* and *in vitro* effects are very different. OTB is overtly less toxic *in vivo* compared to OTA as indicated in different models. LD50 values, found in a comparative study using 1-day-old chicks, were 120 μg for OTA (about 3.5 mg/kg) and 1890 μg for OTB (54 mg/kg) [32]. OTB is more easily excreted and has a lower affinity for plasma proteins, which may partly elucidate its lower toxicity. Both OTA and OTB toxins induce acute cytotoxic effects *in vitro,* ensuring similar amounts are taken up and are intracellularly bound, while other complex molecular mechanisms were introduced for chronic cytotoxicity studies. Moreover, it can be supposed that the small structural difference, although not responsible for the toxicity, may be crucial for the differential uptake and binding in cells. Furthermore, OTC seems to be similarly acute toxic *in vivo* and *in vitro* compared to OTA; however, the mode of action of OTC and OTA remains to be explained. In a study, oral LD50 values were reported for OTC (216 mg animal^{-1}) and OTA (166 mg animal^{-1}) in day-old chicks. Other ochratoxin ethyl or methyl esters showed lower toxicity compared to OTA. In comparison to OTA, the methyl ester of OTA was less toxic than OTA in day-old chicks, while OTB methyl and ethyl esters were found to be non-lethal to orally exposed day-old ducklings [22]. OTα is much less toxic (approximately 100 fold) than OTA as indicated in different studies [33]. It is obvious that the isocoumarin moiety alone is not effective but must be bound to phenylalanine to show toxic effects. With the current knowledge, no clear general toxicity ranking can be drawn; after all, OTA seems to be overall the most toxic, followed by OTC, OTB and OTα [33].

2.3. Fumonisins

Fumonisins (FUM) are a group of mycotoxins that were first isolated from cultures of *Fusarium moniliforme* and chemically characterized in 1988 by Gelderblom and colleagues [34]. Six different FUM have been identified (A1,

A2, B1, B2, B3, B4) and their structures elucidated. However, fumonisin B1 (FB1) has been reported to be the predominant form produced by *Fusarium moniliforme*. Several other Fusarium species and a species of Alternaria have also been found to produce FB1 [35]. Based on all these animal studies, FB1 is classified by IARC as possibly carcinogenic to humans (Group 2B) [36].

The metabolism of fumonisin is yet to be elucidated. FB1 is metabolized into partially hydrolyzed FB1 and then to the hydrolyzed form (HFB1) in both gastrointestinal tract and liver, where it persists at low concentration for few more days in pigs [37]. FB1 was found to be more toxic than HFB1 in piglets [38]. Even though N-acylation of FB1 and the formation of HFB1 are shown in human cell lines and in rats, the metabolism in the avian species still remains uncertain and yet it is not possible to generalize the metabolic pathways in all animal species [39].

The mechanism that causes toxicity of fumonisins in animals seems to be due to the disruption of sphingolipid metabolism. Present evidence shows that the FUM are specific inhibitors of ceramide synthase (sphinganine/sphingosine N-acyltransferase), a key enzyme needed for the synthesis of ceramide and more complex sphingolipids. Inhibition of this enzyme system causes an increase in tissue concentrations of the sphingolipids sphingosine (SO) and sphinganine (SA), and a change in the SA:SO ratio. An increase in the SA:SO ratio has been demonstrated in tissues of broilers, turkeys, and ducklings fed FB1 [40].

In comparison to horses and swine, two susceptible species, chicks and turkeys, are relatively resistant to the toxic effects of FB1. Mild to moderate toxicity was reported in chicks, ducks and turkeys fed rations containing 75–400 mg FB1/kg for 21 days. The primary changes in chicks, ducks and turkeys were decreased body weight gain and liver pathology [41–43]. Hepatic changes in chicks were multifocal hepatic necrosis and biliary hyperplasia. Hepatocellular hyperplasia and increased extramedullary hematopoiesis were also noted in one study [44]. The primary liver pathology observed in turkeys fed with 150–300 mg FB1/kg [43] and ducklings fed with 400 mg FB1/kg [41] were diffuse hepatocellular hyperplasia, with biliary hyperplasia (more evident in turkeys). In studies designed to evaluate the chronic effects of FB1, chick performance up to 7 weeks was not affected by up to 50 mg FB1/kg diet, whereas turkeys fed with 50 mg FB1/kg diet had lower feed intakes than birds fed 0 or 25 mg FB1/kg diet [45].

2.4. Trichothecenes

Trichothecene mycotoxins are a group of fungal metabolites with the same basic backbone structure and include T-2 toxin, HT-2 toxin, diacetoxyscirpenol (DAS), monoacetoxyscirpenol (MAS), neosolaniol, 8-acetoxyneosolaniol, 4-deacetylneosolaniol, nivalenol, 4-acetoxynivalenol (Fusarenone-X), DON (vomitoxin) and 3-acetyldeoxynivalenol. They are known as the most potent small molecule inhibitors of protein synthesis and the main toxic effect at the cellular level appears to be the primary inhibition of protein synthesis followed by a secondary disruption of DNA and RNA synthesis [11]. The overall conclusion by IARC was that toxins derived from *Fusarium sporotrichioides* are not classifiable as to their carcinogenicity to humans (Group 3) [46].

For livestock, the most important trichothecene mycotoxin is DON, which is commonly a contaminant of corn, wheat and other commodity grains. Lesser amounts of T-2 toxin and DAS are found sporadically in the same sources. Poultry and cattle are more tolerant of trichothecenes than are pigs. Compared to the related DON, T-2 toxin is less frequent in crops. Some reports indicate that trichothecenes such as DON, nivalenol and fusarenon X are more frequent (57%, 16% and 10% of tested grain samples) in European grain samples than other trichothecenes like T-2 toxin (20%), HT-2 toxin (14%), T-2 tetraol (6%), neosolaniol (1%), DAS (4%), MAS (1%) [47].

Trichothecene poisoning in poultry is acute or chronic. Acute poisoning has a characteristic clinical picture and can be readily diagnosed, while chronic poisoning shows unspecific clinical symptoms [48].

Toxic effects of trichothecenes include oral lesions, growth retardation, abnormal feathering, decreased egg production and egg shell quality, regression of the bursa of Fabricius, peroxidative changes in liver, abnormal blood coagulation, leucopoenia and proteinemia, and immunosuppression [49]. Concentrations of T-2 that cause oral lesions are lower (0.4 mg/kg) than concentrations reported to decrease chick performance (3–4 mg/kg) [11]. In a comprehensive review, Danicke [49] concluded that broiler performance is affected at dietary concentrations of 3–4 mg/kg of T-2 toxin, whereas ducks were affected when the dietary concentration was as low as 0.4 mg/kg.

T2 toxin was found to decrease the immune response, represented by the decrease of lymphoid cells in the bone marrow, thymus and spleen causing resistance to infectious diseases including salmonellosis and *Escherichia coli* and cause resistance to treatments of these diseases in poultry [50]. In broilers, T-2 toxin may cause a decrease in body weight and relative weights of bursa of Fabricius, thymus, and spleen, enlarged liver, friable, and yellowish discoloration with distended gall bladder during *Mycoplasma gallisepticum* infections. Microscopical findings include vacuolar degeneration along with augmented hyperplasia in bile duct epithelia; Kupffer cell activity and infiltration of inflammatory cells in liver; vacuolar degeneration with pyknotic nuclei in kidney; lymphocytolysis and reduction of prominent reticuloepithelial cells in lymphoid organs; desquamation of villous-type epithelial cells and lymphoid intrusion in the submucosa of proventriculus; mild hemorrhage along with inflammatory cells in the heart; desquamation and erosion of the mucosa in trachea and the thickening of the air sacs along with edema and the presence of inflammatory cells in air sacs [51]. The toxic effect also manifests as reduced proliferation of lymphocytes stimulated by phytohemagglutinin and lipopolysaccharide in Pekin duck broilers [52].

DON was found to be less toxic than T-2 toxin, and the level of DON that affects chick performance is still disputed. Some researchers [53, 54] reported toxic effects at 16 mg/kg diet, whereas others [55] report no toxic effect until dietary concentrations exceeded 116 mg/kg of DON. A review paper summarizing results of 49 studies with DON concluded that a dietary concentration of 5 mg/kg had no negative effects on performance [56]. DON has also been demonstrated to have both immunosuppressive and immunomodulating effects in poultry [49]. Recent studies indicate that DON at concentrations ranging from 1 to 7 mg/kg diet significantly alters several key functions of the intestinal tract including decaying villus surface area available for absorption and altering the permeability of the alimentary canal [57].

3. INTERACTIONS AMONG MYCOTOXINS

In nature, co-occurrence of mycotoxins is generally observed. Meanwhile, for many years the research community focused on the occurrence of singular mycotoxins. Nowadays, scientific interest is shifted to studies involving multiple mycotoxins using various co-occurrence scenarios. One fungus may produce many different mycotoxins, and the same mycotoxin may be produced by several species. A paper conducted a meta-analysis of publications (> 100) describing toxicological interactions among mycotoxins. Results indicated that most of the studies showed a synergistic or additive interaction on animal performance. However, results with respect to other response variables indicated that there were many types of interactions ranging from synergistic to antagonistic for the same association [58]. They also observed from their review that a combination of mycotoxins, at concentrations that individually should not cause negative effects, may negatively affect animals.

The individual and combined effects of dietary AFB1 and FB1 on liver pathology, serum levels of aspartate amino-transferase (AST) and plasma total protein (TP) of broilers were quantified from 8 to 41 days of age with the dietary treatments of AFB1 (0, 50 and 200 μg AFB1/kg), and FB1 (0, 50 and 200 mg FB1/kg). Following treatment, AST levels were found to be higher in all treatment groups (except 50 mg FB1) compared to controls at day 33. TP levels were found to be reduced at 6 days post feeding in AFB1-treated group (200 μg) and in FB1 combination group. At 33 days post feeding, the combination group (200 μg AFB1 and 200 mg FB1) were found to have higher plasma TP, proliferation of bile duct and trabecular disorders in liver tissue compared to control; while the changes in other groups were insignificant compared to controls. Overall AFB1 alone and in combination caused damage in liver at varying degrees and an increase of serum AST levels [59].

Aflatoxicosis causes a reduction in the production of egg and a decrease in egg weight in laying hens. Meanwhile, the antagonistic effects of AF and FB on egg production were reported in quails, where the decrease in egg production in FB-only treated group was much evident compared to AF+FB combination [60].

Few studies examined the combined effect of AF and FB on immunity. AF and FB co-contaminated feed was found to reduce lymphocyte proliferation by mitogenic stimulation as less than additive [61]or as additive [62]compared to single contamination. A study indicated a synergistic decrease of the antibody titers against Newcastle disease [59].On the contrary, another study demonstrated an unexpected increase and an additive effect of the two toxins when looking at the hemagglutination titers against sheep red blood cells in turkey poults. However, the phytohemagglutinin delayed hypersensitivity response was not affected by dietary treatment. These results indicate that FB1 and AF, alone and in combination, can adversely affect poult performance and health [63].

As is known, both AF and OTA reduce egg production and hatchability. The combined effects of these two toxins were studied in laying hens [64, 65]. An additive interaction of AF and OTA was observed on egg production and on the feed efficiency (consumption for egg production) [64]; meanwhile this interaction was dependent on the concentration, resulting from synergistic to slightly lower than additive effect and also modulates the protein and energy usage [65].

AF- and OTA-contaminated feed resulted in microscopic lesions in the liver and kidneys, along with respective target organs in chicken, while contradictory results are presented in different studies. As such, OTA in the diet was found to prevent the hepatic fatty infiltration caused by AF in chicken [66]. Pigs fed the co-contaminated diet offer the same hepatic lesions as those fed with the diet contaminated with AF alone [67]. On the contrary, a study recorded more severe hepatic lesions in chickens taking the co-contaminated diet, with granular and vacuolar degenerative changes, necrosis of liver parenchyma and areas of hemorrhages [68]. The same conflict was realized for the histology of the kidney. In pigs, less severe renal lesions and lower creatinine and blood urea nitrogen concentrations were observed in animals fed the co-contaminated diet compared to animals fed the OTA-contaminated diet [67]. In contrast, a study observed that renal injuries appeared earlier and were more developed in chickens fed a multi-contaminated diet than in animals taking the mono-contaminated diets, which caused destruction of tubular epithelium, with detachment of tubular cells from basement membrane [68]. The species used may explain these conflicts. Apart from that, chronic DON exposure did not induce any effect on FB1 toxicokinetics in broilers [69].

The results on combination toxicity are yet quite limited and occasionally conflicting. Nowadays, very little is known about mycotoxin interactions although combined exposure is clearly more relevant to real-life conditions. It is known that the combined effects of mycotoxins are mostly additive or synergistic; whereas depending on the concentrations and the *in vitro* model employed, antagonistic interactions have also been determined. The results on multiple mixtures are still inadequate [22].

4. PREVENTION AND CONTROL

Prevention of fungi production in the feeds may be achieved by always keeping the feeds fresh, keeping the humidity low and equipment clean and also adding fungistatic substances. Humidity exceeding 11% promotes fungal production in cereals and feed. Storage conditions that afford high relative humidity also significantly affect humidity content of the feed. Good ventilation of the storehouse removes humidity from the raw material of the feed and storehouse. Physically damaged cereals are more prone to fungus production compared to the healthy ones. Changing the raw materials at the places where they are stored at short intervals decreases mycotoxin formation [2,3,6].

Research efforts progressively increase to develop mitigation strategies based on risk monitoring, risk characterization, prevention, intervention, and remediation strategies for multiple mycotoxins, initiating from critical points along the production chain comprising field, storage, processing and transportation. However, monitoring and good agricultural, storage, and transportation practices along with an effective Hazard Analysis and Critical Control Point (HACCP) approach do not completely prevent mycotoxin presence in the food or feed chain. Decontamination strategies then offer a last resort to salvage contaminated batches along the production chain [70].

Considering the variation of mycotoxin structures, it could be inferred that there is no single method which can be used to deactivate mycotoxins in feed. Therefore, different strategies have to be combined in order to specifically target individual mycotoxins without affecting the quality of feed [11].

Decontamination strategies to reduce mycotoxins in food and feed commodities are technologically diverse and based on chemical, biological and physical approaches. Chemical remediation strategies involve the conversion of mycotoxins via chemical reactions. Ammoniation, alkaline hydrolysis, peroxidation, ozonation and the use of bisulphites are reported to be effective on one or more mycotoxins but a detailed insight into the toxicity of eventual end products or the impact on palatability and nutritive quality is questionable [71].

Biological approach in treatment strategies involves various substances (algae, plant ingredients, etc.) that protect critical organs such as the liver and strengthen the immune system of animals. Enzymatic or microbial detoxification, also referred to as "biotransformation" or "biodetoxification", uses microorganisms or purified enzymes thereof to catabolize the entire mycotoxin or transform or cleave it to less or non-toxic compounds [11]. Some microorganism such as *Rhodococcus erythropolis*[72],*Armillariella tabescens*[73] and*Myxococcus fulvus* [16] have been suggested to have different AF-degrading ability. *Rhizopus oryzae*[74], *Bacillus licheniformis* [75] and Pseudomonas sp [76] were searched for their abilities to degrade ZEA. Some studies found that *B. subtilis* had protective effects against aflatoxicosis in layers and broilers fed naturally AF-contaminated diets and also healed ZEA toxicosis in pre-pubertal gilts when fed diets including ZEA. Therefore, *B. subtilis*, as a new feed additive for biodegradation of AF and ZEA may have promising potential in feed industrial applications [77].

Some physical processes aim to remove highly contaminated fractions from bulk material through sorting, milling, dehulling, cleaning, heating, irradiation or combinational approaches [78, 79]. Another physical removal strategy is the use of inorganic or organic mycotoxin binders [80]. Due to low feed inclusion requirements and easy management of AF enterosorbents, the widespread acceptance of these products by the farm animal industry has led to the introduction of a variety of diverse materials and/or complex mixtures for AF binding. These have been labeled as mycotoxin enterosorbents, binders, sequestrants, interceptor molecules, trapping agents, adsorbents, toxin sorbents, and so on. These materials (and/or mixtures) are reported to contain smectite clays, zeolites, kaolinite, mica, silica, charcoal, sodium bentonite and various biological constituents including chlorophyllins, yeast products, lactic acid bacteria, plant extracts and algae. Some contain smectite or zeolite minerals that have been amended with natural or synthetic surfactants resulting in hydrophobic organoclays or organozeolites [81–84]. There is considerable evidence indicating that smectite clays are the most effective AF enterosorbents. Although these adsorbing binders have some promising features, some may have adverse nutritional effects due to binding of vitamins and minerals or reducing the efficacy pharmacokinetics of antibiotics [85]. Also, possible dioxin contamination might pose a risk for using natural clays in case of forest and trash fire near the sources [86]. Furthermore, the adsorption efficacy of binding agents is limited to only a few mycotoxins, such as AF, ergot alkaloids, and some other fungal toxins, while binders have been shown to be ineffective for trichothecenes [87]. Therefore, alternative approaches for efficient detoxification of mycotoxins are required.

Use of microorganisms and their specific products such as enzymes to detoxify specific mycotoxins not only work for non-adsorbable mycotoxins, but for all other toxins for which respective microbes can be isolated from nature. This approach has been known for a long time, even longer than the binder concept. Within few years after the discovery of AF, the first report on a bacterium capable of detoxifying AF by catabolization was published [88]. Since then, many microorganisms were isolated from different habitats such as the gastrointestinal tract (GIT) of animals, soil, mycotoxin-contaminated materials (e.g., grains) and insects feeding on such materials. The ability of various bacteria, yeast, fungi and enzymes in detoxifying mycotoxins by transformation, cleavage and catabolization has been recently reviewed [89]. However, only a few of these organisms were useful or further investigated for practical applications in animal nutrition. Such microorganisms or enzymes need to fulfill many different requirements before they can be used for gastrointestinal detoxification of mycotoxin in animals, such as:

- The microorganism and its reaction products need to be non-toxic and safe.
- High detoxification reactivity.
- Good technological properties (fermentation, downstream processing, stabilization).
- High stability in feed and during feed processing.
- No negative impact on feed (ingredients).
- Compatibility and stability in the GIT.

- Detoxification reaction in the GIT needs to be fast and as complete as possible.

One of the microorganisms which has been further developed for practical application is *Trichosporon mycotoxinivorans*, a yeast strain capable of detoxifying OTA and ZEN [90]. Application of this yeast in poultry diets has been proven to detoxify OTA. Another organism is an anaerobic rumen bacterium BBSH 797 (Genus Novus of family Coriobacteriacae, formerly Eubacterium) which was isolated and developed as a trichothecene-detoxifying feed additive [91]. BBSH 797 detoxifies trichothecenes by cleavage of the 12, 13 epoxide ring resulting in deepoxy trichothecenes. Several microorganisms, mainly aerobic bacteria and also yeasts, with FUM degradation properties were also explored and isolated in order to detoxify FUM. However, for various reasons, none of these microorganisms were found to be useful as a mycotoxin-deactivating feed additive. Therefore, the catabolic pathway of FUM degradation was investigated and the gene coding for the key enzyme of FUM detoxification (FUMzyme) was identified, cloned and expressed in a yeast strain [92]. FUMzyme (carboxyl-esterase) was further developed and tested in swine for gastrointestinal detoxification of FUM by cleaving the tricarballylic side chains of FUM leading to the non-toxic metabolite hydrolyzed FUM (HFB1) [93].

One of the common approaches to overcome mycotoxicosis in poultry is using herbal products including essential oils as plant-based fumigants in feed storage [94]. Essential oils are complex compounds, and their chemical composition and concentrations of various compounds are variable. Essential oils basically consist of two classes of compounds, the terpenes and phenylpropenes, depending on the number of 5-carbon building blocks. For example, 500 ppm of the ethanolic extract of *Thymus vulgaris* could partially restore the negative impact of AFB1 (600 ppb) in commercial broilers [9]. They suggested that this herb can be used as natural non-antibiotic feed additive on broilers in the prevention of aflatoxicosis. As a result of the change in diet (change in nutrients, phytochemicals, contamination, xenobiotics), the levels of the drug-metabolizing enzymes (phase-I and phase-II) are expected to change, which would eventually lead to a change in AFB1 adducts. On the other hand, as phenolic phytochemicals have antioxidant effects at varying degrees due to their various chemical structures, they are assumed to have a protective role in the cellular components against free radical–induced damage caused by aflatoxicosis [95]. Apart from that, a herbal mycotoxin binder comprising of a combination of minerals (extra purified clay containing diatomaceous earth minerals), antioxidants (curcuminoids extracted from turmeric) and enzymes (Epoxidases and Esterases) in proportions of 15, 10 and 75%, respectively, partially restored feed consumption and egg production, alleviating some side effects of AFB1 (500 ppb in feed) in broiler breeders [96].

5. CONCLUSION

Understanding the occurrence and prevalence of mycotoxins and their individual as well as additive negative effects on poultry has become imperative. New insights on actual microbial detoxification routes are needed in the field, which could be based on the biodegradation metabolisms of non-mycotoxins found in diverse microbial communities. Indeed, many hazardous, undesirable, deleterious or recalcitrant molecules in other research fields share structural analogies with diverse mycotoxins and are reported to be successfully degraded by microorganisms. These unexplored worlds may serve as resource for cutting-edge research in the field of mycotoxin remediation or in the field of metagenomics screening surveys in search for new microbial degraders of mycotoxins. The usage of latest analytical techniques such as liquid chromatography tandem-mass spectrometry will increase the precision in determination of the concentrations of multiple mycotoxins present in agricultural commodities, at once. Latest enzymatic deactivation technologies help to eliminate the mycotoxins that cannot be bound using binder products. Overall, mycotoxins still impose a great risk for the poultry sector and alternative approaches for the prevention are still being sought by researches around the world.

REFERENCES

1. Basalan M, Hismiogullari SE, Hismiogullari AA, Filazi A. Fungi and aflatoxin B1 in horse and dog feeds in Western Turkey. Revue Med Vet. 2004;156:248–252.
2. Becer UK, Filazi A. Aflatoxins, nitrates and nitrites analysis in the commercial cat and dog foods. Fresen Environ Bull. 2010;18:2523–2527.
3. Demircioglu S, Filazi A. Detection of aflatoxin levels in red pepper produced in Turkey.Vet Hekim Der Derg. 2010;81:63–66.
4. Filazi A, Ince S, Temamogullari F. Survey of the occurrence of aflatoxin M1 in cheeses produced by dairy ewe's milk in Urfa city, Turkey. Ankara Univ Vet Fak Derg. 2010;57:197–199.
5. Gundinc U, Filazi A. Detection of aflatoxin M1 concentrations in UHT milk consumed inTurkey markets by ELISA. Pak J Biol Sci. 2009;12:653–656.
6. Kaya S. Mycotoxins. In: Kaya S, editor. Veterinary Toxicology, 3rd ed. Ankara-Turkey:Medisan Publisher; 2014, pp. 393–433.
7. Streit E, Schatzmayr G, Tassis P, Tzika E, Marin D, Taranu I, Tabuc C, Nicolau A,Aprodu I, Puel O, Oswald IP. Current situation of mycotoxin contamination and co-occurrence in animal feed--focus on Europe. Toxins. 2012;4:788–809. doi:10.3390/ toxins4100788.

8. Gentles A, Smith EE, Kubena LF, Duffus E, Johnson P, Thompson J, Harvey RB, EdringtonTS. Toxicological evaluations of cyclopiazonic acid and ochratoxin A in broilers. PoultSci. 1999;78:1380–1384. doi:10.1093/ps/78.10.1380.
9. Manafi M, Hedayati M, Yari M. Aflatoxicosis and herbal detoxification: the effectivenessof thyme essence on performance parameters and antibody titers of commercial broilers fed aflatoxin B1. Res Zool. 2014;4:43–50. doi:10.5923/j.zoology.20140402.02.
10. Filazi A, Sireli UT. Occurrence of aflatoxins in food. In: Razzaghi-Abyaneh M, editor.Aflatoxins-Recent Advances and Future Prospects. Rijeka-Croatia; InTech; 2013,pp.143–170. DOI:10.5772/51031.
11. Murugesan GR, Ledoux DR, Naehrer K, Berthiller F, Applegate TJ, Grenier B, Phillips TD,Schatzmayr G. Prevalence and effects of mycotoxins on poultry health and performance,and recent development in mycotoxin counteracting strategies. Poult Sci. 2015;94: 1298–1315. doi:10.3382/ps/pev075.
12. Dohnal V, Wu Q, Kuca K. Metabolism of aflatoxins: key enzymes and interindividualas well as interspecies differences. Arch Toxicol. 2014;88:1635–1644. doi:10.1007/s00204-014-1312-9.
13. Rawal S, Kim JE, Coulombe R. Aflatoxin B1 in poultry: toxicology, metabolism and prevention.Res Vet Sci. 2010;89:325–331.
14. Kim JE, Bunderson BR, Croasdell A, Reed KM, Coulombe RA. Alpha-class glutathioneS-transferases in wild Turkeys (*Meleagris gallopavo*): characterization and role in resistance to the carcinogenic mycotoxin aflatoxin B1. PLoS One. 2013;8: e60662. doi:10.1371/ journal.pone.0060662.
15. Rawal S, Coulombe RA. Metabolism of aflatox in B1 in Turkey liver microsomes: the relative roles of cytochromes P450 1A5 and 3A37. Toxicol Appl Pharmacol. 2011;254:349–354. doi:10.1016/j.taap.2011.05.010.
16. Zhao LH, Guan S, Gao X, Ma QG, Lei YP, Bai XM, Ji C. Preparation, purification and haracteristics of an aflatox in degradation enzyme from Myxococcus fulvus ANSM068.
17. Hayes JD, Flanagan JU, Jowsey IR. Glutathione transferases. Annu Rev PharmacolToxicol. 2005;45:51–88. doi:10.1146/annurev.pharmtox.45. 120403. 095857.
18. Yarru LP, Settivari RS, Gowda NKS, Antoniou E, Ledoux DR, Rottinghaus GE. Effectsof turmeric (*Curcuma longa*) on the expression of hepatic genes associated with biotransformation,antioxidant, and immune systems in broiler chicks fed aflatoxin. Poult Sci.2009;88:2620–2627. doi:10.3382/ps.2009-00204.
19. Yarru LP, Settivari RS, Antoniou E, Ledoux DR, Rottinghaus GE. Toxicological and geneexpression analysis of the impact of aflatoxin B1 on hepatic function of male broilerchicks.Poult Sci. 2009;88:360–371. doi:10.3382/ps.2008-00258.
20. Hossain SA, Haque N, Kumar M, Sontakke UB, Tyagi AK. Mycotoxin residues in poultryproduct: their effect on human health and control. Wayamba J Anim Sci. 2011; ISSN:2012-578X: P92–P96.
21. Gibson RM, Bailey CA, Kubena LF, Huff WE, Harvey RB. Ochratoxin A and dietary protein.1. Effects on body weight, feed conversion, relative organ weight, and mortality inthree-week-old broilers. Poult Sci. 1989;68:1658–1663. doi:10.3382/ps.0681658.
22. Heussner AH, Bingle LE. Comparative ochratoxin toxicity: a review of the availabledata. Toxins. 2015;7:4253–4282. doi:10.3390/toxins7104253.
23. Pfohl-Leszkowicz A, Manderville RA. An update on direct genotoxicity as a molecularmechanism of ochratoxin A carcinogenicity. Chem Res Toxicol. 2012;25:252–262.doi:10.1021/tx200430f.
24. Harris J, Mantle P. Biosynthesis of ochratoxins by *Aspergillus ochraceus*. Phytochemistry. 2001;58:709–716. doi:10.1016/S0031-9422(01)00316-8.
25. Vettorazzi A, van Delft J, López de Cerain A. A review on ochratoxin A transcriptomicstudies. Food Chem Toxicol. 2013;59:766–783. doi:10.1016/ j.fct.2013.05.043.

26. Elaroussi MA, Mohamed FR, el Barkouky EM, Atta A, Abdou AM, Hatab MH.Experimental ochratoxicosis in broiler chickens. Avian Pathol. 2006;35:263–269.doi:10.1080/03079450600817115.
27. Stoev S, Denev S. Porcine/chicken or human nephropathy as the result of joint mycotoxinsinteraction. Toxins. 2013;5:1503–1530. doi:10.3390/toxins5091503.
28. Pepeljnjak S, Klarić M. "Suspects" in etiology of endemic nephropathy: aristolochic acidversus mycotoxins. Toxins. 2010;2:1414–1427. doi:10.3390/toxins2061414.
29. chaeffer JL, Tyczkowski JJ, Hamilton PB. Alterations in carotenoid metabolism duringochratoxicosis in young broiler chickens. Poult Sci. 1987;66:318–324.
30. Glahn RP, Shapiro RS, Vena VE, Wideman RF, Huff WE. Effects of chronic ochratoxinA and citrinin toxicosis on kidney function of single comb white leghorn pullets. PoultSci. 1989;68:1205–1212.
31. Manafi M, Mohan K, Noor Ali M. Effect of ochratoxin A on coccidiosis-challenged broilerchicks. World Mycotoxin J. 2011;4:177–181. doi:10.3920/WMJ2010.1234.
32. Peckham J, Doupnik BJ, Jones OJ. Acute toxicity of ochratoxins A and B in chicks. ApplMicrobiol. 1971;21:492–494.
33. Xiao H, Madhyastha S, Marquardt RR, Li S, Vodela JK, Frohlich AA, Kemppainen BW.Toxicity of ochratox in A, its opened lactone form and several of its analogs:structure-activity relationships. Toxicol Appl Pharmacol. 1996;137:182–192. doi:10.1006/taap.1996.0071.
34. Gelderblom WC, Jaskiewicz K, Marasas WF, Thiel PG, Horak RM, Vleggaar R, Kriek NP.Fumonisins- novel mycotoxins with cancer-promoting activity produced by *Fusariummoniliforme*. Appl Environ Microbiol. 1988;54:1806–1811.
35. Chen J, Mirocha CJ, Xie W, Hogge L, Olson D. Production of the mycotoxin fumonisin B(1)by Alternaria alternata f. sp. Lycopersici. Appl Environ Microbiol. 1992;58:3928–3931.
36. Domijan AM. Fumonisin B(1): a neurotoxic mycotoxin. Arh Hig Rada Toksikol.2012;63:531–544. doi:10.2478/10004-1254-63-2012-2239.
37. Fodor J, Balogh K, Weber M, Miklós M, Kametler L, Pósa R, Mamet R, Bauer J, Horn P,Kovács F, Kovács M. Absorption, distribution and elimination of fumonisin B(1) metabolitesin weaned piglets. Food Addit Contam Part A Chem Anal Control Expo RiskAssess. 2008;25:88–96. doi:10.1080/02652030701546180.
38. Grenier B, Bracarense AP, Schwartz HE, Trumel C, Cossalter AM, Schatzmayr G, Kolf-Clauw M, Moll WD, Oswald IP. The low intestinal and hepatic toxicity of hydrolyzedfumonisin B1 correlates with its inability to alter the metabolism of sphingolipids.Biochem Pharmacol. 2012;83:1465–1473. doi:10.1016/j.bcp.2012.02.007.
39. Guerre P. Fusariotoxins in avian species: Toxicokinetics, metabolism and persistence intissues. Toxins. 2015;7:2289–2305. doi:10.3390/toxins7062289.
40. Tran ST, Auvergne A, Benard G, Bailly JD, Tardieu D, Babile R, Guerre P. Chronic effectsof fumonisin B1 on ducks. Poult Sci. 2005;84:22–28. doi:10.1093/ps/84.1.22.
41. Bermudez AJ, Ledoux DR, Rottinghaus GE. Effects of *Fusarium moniliforme* culture materialcontaining known levels of fumonisin B1 in ducklings. Avian Dis. 1995;39:879–886.doi:10.2307/1592427.
42. Brown TP, Rottinghaus GE, Williams ME. Fumonisin mycotoxicosis in broilers: performanceand pathology. Avian Dis. 1992;36:450–454. doi:10.2307/1591528.
43. Weibking TS, Ledoux DR, Bermudez AJ, Turk JR, Rottinghaus GE. Effects on turkeypoults of feeding *Fusarium moniliforme* M-1325 culture material grown under differentenvironmental conditions. Avian Dis. 1995;39:32–38. doi:10.2307/1591979.
44. Weibking TS, Ledoux DR, Bermudez AJ, Turk JR, Rottinghaus GE, Wang E, MerrillAH. Effects of feeding *Fusarium moniliforme* culture material, containing known levelsof fumonisin B1, on the young broiler chick. Poult Sci. 1993;72:456–466. doi:10.3382/ps.0720456.

45. Broomhead JN, Ledoux DR, Bermudez AJ, Rottinghaus GE. Chronic effects of fumonisinB1 in broilers and turkeys fed dietary treatments to market age. Poult Sci. 2002;81:56–61.doi:10.1093/ps/81.1.56.
46. International Agency for Research on Cancer. WHO IARC monographs on the evaluationof carcinogenic risks to humans. Some naturally occuring substances: food itemsand constituents, heterocyclic aromatic amines and mycotoxins. Toxins derived from *Fusarium graminearum*, *F. culmorum* and *F. crookwellense*: zearalenone, deoxynivalenol, nivalenol and fusarenone X, vol. 56. IARC, Lyon; 1993, pp. 397–444.
47. Sokolovij M, Garaj-Vrhovac V, Simpraga B. T-2 toxin: incidence and toxicity in poultry.Arh Hig Rada Toksikol. 2008;59:43–52. doi:10.2478/ 10004-1254-59-2008-1843.
48. Resanovic RM, Nešic KD, Nesic VD, Palic TD, Jacevic VM. Mycotoxins in poultry production.Proc Nat Sci. Matica Srpska Novi Sad. 2009;116:7–14.
49. Danicke S. Prevention and control of mycotoxins in the poultry production chain:a European view. World Poult Sci J. 2002;58:451–474. doi:10.1079/WPS20020033.
50. Boonchuvit B, Hamilton PB, Burmeister HR. Interaction of T-2 toxin with Salmonellainfections of chicken. Poult Sci. 1975;54:1693–1696. doi:10.3382 /ps.0541693.
51. Manafi M, Pirany N, Noor Ali M, Hedayati M, Khalaji S, Yari M. Experimental pathology of T-2 toxicosis and mycoplasma infection on performance and hepatic functions of broiler chickens. Poult Sci. 2015;94:1483–1492. doi:10.3382 /ps/pev115.
52. Rafai P, Pettersson H, Bata A, Papp Z, Glávits R, Tuboly S, Ványi A, Soós P. Effect ofdietary T-2 fusariotoxin concentrations on the health and production of white Pekinduck broilers. Poult Sci. 2000;79:1548–1556. doi:10.1093/ps/79.11.1548.
53. Huff WE, Kubena LF, Harvey RB, Hagler WM, Swanson SP, Phillips TD, Creger CR.Individual and combined effects of aflatoxin and deoxynivalenol (DON, vomitoxin) inbroiler chickens. Poult Sci. 1986;65:1291–1298. doi:10.3382/ ps.0651291.
54. Kubena LF, Huff WE, Harvey RB, Phillips TD, Rottinghaus GE. Individual and combinedtoxicity of deoxynivalenol and T-2 toxin in broiler chicks. Poult Sci. 1989;68:622–626. doi:10.3382/ps.0680622.
55. Moran ET, Hunter B, Ferket P, Young LG, McGirr LG. High tolerance of broilers tovomitoxin from corn infected with *Fusarium graminearum*. Poult Sci. 1982;61:1828–1831.doi:10.3382/ps.0611828.
56. Danicke S, Gareis M, Bauer J. Orientation values for critical concentrations of deoxynivalenol and zearalenone in diets for pigs, ruminants and gallinaceous poultry. Proc SocNutr Physiol. 2001;10:171–174.
57. Osselaere A, Devreese M, Goossens J, Vandenbroucke V, de Baere S, de Backer P,Croubels S. Toxicokinetic study and absolute oral bioavailability of deoxynivalenol,T-2 toxin and zearalenone in broiler chickens. Food Chem Toxicol. 2013;51:350–355.doi:10.1016/j.fct.2012.10.006.
58. Grenier B, Oswald IP. Mycotoxin co-contamination of food and feed: meta-analysis of publications describing toxicological interactions. World Mycotoxin J. 2011;4:285–313.doi:10.3920/WMJ2011.1281.
59. Tessari EN, Kobashigawa E, Cardoso AL, Ledoux DR, Rottinghaus GE, Oliveira CA.Effects of aflatoxin B1 and fumonisin B1 on blood biochemical parameters in broilers.Toxins. 2010;2:453–460. doi:10.3390/toxins2040453.
60. Ogido R, Oliveira CAF, Ledoux DR, Rottinghaus GE, Correa B, Butkeraitis P, Reis TA,Goncales E, Albuquerque R. Effects of prolonged administration of aflatoxin B1 and fumonisin B1 in laying Japanese quail. Poult Sci. 2004;83:1953–1958. doi:10.1093/ps/83.12.1953.
61. Effects of feeding fumonisin B1 present in *Fusarium moniliforme* culture material and aflatox in singly and in combination to turkey poults. Poult Sci. 1995;74:1295–1303.doi:10.3382/ps.0741295.

62. Harvey RB, Edrington TS, Kubena LF, Elissalde MH, Rottinghaus GE. Influence of aflatoxin and fumonisin B1-containing culture material on growing barrows. Am J Vet Res.1995;56:1668–1672.
63. Weibking TS, Ledoux DR, Bermudez AJ, Rottinghaus GE. Individual and combined effects of feeding *Fusarium moniliforme* culture material, containing known levels offumonisin B1, and aflatoxin B1 in the young turkey poult. Poult Sci. 1994;73:1517–1525. doi:10.3382/ps.0731517.
64. Verma J, Johri TS, Swain BK. Effect of varying levels of aflatoxin, ochratoxin and theircombinations on the performance and egg quality characteristics in laying hens. Asian-Australas J Anim Sci. 2003;16:1015–1019. doi:10.5713/ajas.2003.1015.
65. Verma J, Johri TS, Swain BK. Effect of aflatoxin, ochratoxin and their combination on proteinand energy utilization in white leghorn laying hens. J Sci Food Agric. 2007;87:760–764. doi:10.1002/jsfa.2655.
66. Huff WE, Doerr JA. Synergism between aflatoxin and ochratoxin A in broiler chickens.Poult Sci. 1981;60:550–555. doi:10.3382/ps.0600550.
67. Tapia MO, Seawright AA. Experimental combined aflatoxin B1 and ochratoxin A intoxicationin pigs. Aust Vet J. 1985;62:33–37. doi:10.1111/j.1751- 0813. 1985.tb14229.x.
68. Sakhare PS, Harne SD, Kalorey DR, Warke SR, Bhandarkar AG, Kurkure NV. Effect ofToxiroak® polyherbal feed supplement during induced aflatoxicosis, ochratoxicosis and combined mycotoxicoses in broilers. Vet Arhiv. 2007;2:129–146.
69. Antonissen G, Devreese M, Immerseel FV, De Baere S, Hessenberger S, Martel A,Croubels S. Chronic exposure to Deoxynivalenol has no influence on the oral bioavailability of Fumonisin B1 in broiler chickens. Toxins. 2015;7:560–571. doi:10.3390/ toxins7020560.
70. Bhat R, Rai RV, Karim AA. Mycotoxins in food and feed: present status and future concerns.Compr Rev Food Sci Food Safety. 2010;9:57–81. doi:10.1111/ j.1541-4337.2009.00094.x.
71. Vanhoutte I, Audenaert K, De Gelder L. Biodegradation of mycotoxins: tales from knownand unexplored worlds. Front Microbiol. 2016;7:561. doi:10.3389/ fmicb.2016.00561.
72. Alberts JF, Engelbrecht Y, Steyn PS, Holzapfel W, van Zyl W. Biological degradationof aflatoxin B-1 by Rhodococcus erythropolis cultures. Int J Food Microbiol. 2006;109: 121–126. doi:10.1016/j.ijfoodmicro.2006.01.019.
73. Cao J, Zhang H, Yang Q, Ren R. Efficacy of Pichia caribbica in controlling blue mold rotand patulin degradation in apples. Int J Food Microbiol. 2013;162:167–173. doi:10.1016/j.ijfoodmicro.2013.01.007.
74. arga J, Peteri Z, Tabori K, Teren J, Vagvolgyi C. Degradation of ochratoxin A and othermycotoxins by Rhizopus isolates. Int J Food Microbiol. 2005;99:321–328. doi:10.1016/j.ijfoodmicro.2004.10.034.
75. Yi PJ, Pai CK, Liu JR. Isolation and characterization of a *Bacillus licheniformis* strain capable of degrading zearalenone. World J Microb Biot. 2011;27:1035–1043.
76. Altalhi AD, El-Deeb B. Localization of zearalenone detoxification gene(s) in pZEA-1plasmid of *Pseudomonas putida* ZEA-1 and expressed in *Escherichia coli*. J Hazard Mater.161:1166–1172.
77. Jia R, Ma Q, Fan Y, Ji C, Zhang J, Liu T, Zhao L. The toxic effects of combined aflatoxinsand zearalenone in naturally contaminated diets on laying performance, egg quality andmycotoxins residues in eggs of layers and the protective effect of Bacillus subtilis biodegradationproduct. Food Chem Toxicol. 2016;90:142–150. doi:10.1016/j.fct.2016.02.010.
78. Kaushik G. Effect of processing on mycotoxin content in grains. Crit Rev Food Sci Nutr.2015;55:1672–1683. doi:10.1080/10408398.2012.701254.
79. Matumba L, Van Poucke C, Njumbe Ediage E, Jacobs B, De Saeger S. Effectiveness ofhand sorting, flotation/washing, dehulling and combinations thereof on the decontamination of mycotoxin-contaminated white maize. Food Addit Contam Part A Chem Anal

80. Kolosova A, Stroka J. Substances for reduction of the contamination of feed by mycotoxins:a review. World Mycotoxin J. 2001;4:225–256. doi:10.3920/WMJ2011.1288.
81. Eraslan G, Akdogan M, Arsan E, Essiz D, Sahindokuyucu F, Hismiogullari SE, Altintas L.Effects of aflatoxin and sodium bentonite administered in feed alone or combinedon lipid peroxidation in the liver and kidneys of broilers. Bull Vet Inst Pulawy. 2004; 48:301–304.
82. Manafi M, Umakantha B, Swamy HDN, Mohan K. Evaluation of high-grade sodiumbentonite on performance and immune status of broilers, fed ochratoxin and aflatoxin.World Mycotoxin J. 2009;2:435–440. doi:10.3920/WMJ2009.1136.
83. Manafi M. Counteracting effect of high grade sodium bentonite during aflatoxicosis inbroilers. J Agric Sci Tech. 2012;14:539–547.
84. Ortatatli M, Oguz, H. Ameliorative effects of dietary clinoptilolite on pathologicalchanges in broiler chickens during aflatoxicosis. Res Vet Sci. 2001;71:59–66. doi:10.1053/ rvsc.2001.0487.
85. De Mil T, Devreese M, Broekaert N, Fraeyman S, DeBacker P, Croubels S. In vitroadsorption and in vivo pharmacokinetic interaction between doxycycline and frequentlyused mycotoxin binders in broiler chickens. J Agric Food Chem. 2015;63:4370 4375.doi:10.1021/acs.jafc.5b00832.
86. Rosa CA, Miazzo R, Magnoli C, Salvano M, Chiacchiera SM, Ferrero S, Saenz M, CarvalhoEC, Dalcero A. Evaluation of the efficacy of bentonite from the south of Argentina to ameliorate the toxic effects of AF in broilers. Poult Sci. 2001;80:139–144. doi:10.1093/ps/80.2.139.
87. Vekiru E, Fruhauf S, Sahin M, Ottner F, Schatzmayr G, Krska R. Investigation of variousadsorbents for their ability to bind Aflatoxin B1. Mycotoxin Res. 2007;23:27–33.doi:10.1007/BF02946021.
88. Ciegler A, Lillehoj B, Peterson E, Hall HH. Microbial detoxification of aflatoxin. Appl Microbiol. 1966;14:934–939.
89. Mc Cormick SP. Microbial detoxification of mycotoxins. J Chem Ecol. 2013;39:907–918.
90. Molnar O, Schatzmayr G, Fuchs E, Prillinger H. Trichosporon mycotoxinivorans sp.nov., a new yeast species useful in biological detoxification of various mycotoxins. SystAppl Microbiol. 2004;27:661–671. doi:10.1078/0723202042369947.
91. Schatzmayr G, Zehner F, Taubel M, Schatzmayr D, Klimitsch A, Loibner AP, BinderEM. Microbiologicals for deactivating mycotoxins. Mol Nutr Food Res. 2006;50:543–551.doi:10.1002/mnfr.200500181.
92. Hartinger D, Moll WD. Fumonisin elimination and prospects for detoxification by enzymatictransformation. World Mycotoxin J. 2011;4:271–283. doi:10.3920/WMJ2011.1285.
93. Grenier B, Applegate TJ. Modulation of intestinal functions upon mycotoxin ingestion:meta-analysis of published experiments in animals. Toxins. 2013;5:396–430. doi:10.3390/toxins5020396.
94. Prakash B, Kedia A, Kumar Mishra P, Dubey NK. Plant essential oils as food preservativesto control moulds, mycotoxin contamination and oxidative deterioration ofagri-food commodities – potentials and challenges. Food Control. 2015; 47:381–391.doi:10.1016/j.foodcont.2014.07.023.
95. Abdel-Wahhab MA, Hassan NS, El-Kady AA, Khadrawy YA, El-Nekeety AA, MohamedSR, Sharaf HA, Mannaa FA. Red ginseng extract protects against aflatoxin B1 and fumonisins- induced hepatic pre-cancerous lesions in rats. Food Chem Toxicol. 2010;48:733–742. doi:10.1016/j.fct.2009.12.006.
96. Manafi M, Khosravinia H. Effects of aflatoxin on the performance of broiler breeders andits alleviation through herbal mycotoxin binder. J Agr Sci Tech. 2013;15:55–63.

CHAPTER 8

Precaecal Phosphorus Digestibility of Inorganicphosphate Sources in Male Broilers

P. Bikker, J. W. Spek, R. A. Van Emous & M. M. Van Krimpen

Wageningen Livestock Research, Department of Animal Nutrition, Wageningen, The Netherlands

Abstract

ABSTRACT

The aim of this study, comprising two experiments, was (1) to determine in Experiment 1 the relationship of incremental dietary P (phosphorus) content on precaecal digestible P in male broilers and (2) to determine in Experiment 2 the precaecal P digestibility of various inorganic P sources at marginal levels of P supply.

In Experiment 1, a total of 260 male Ross 308 broilers were divided into groups of 10 birds per pen resulting in 8 replicates for treatment 1 and 6 replicates for treatments 2–4. Experimental diets were formulated to contain 4 incremental concentrations of digestible P by means of increasing concentrations of monocalcium phosphate (MCP). In the second experiment, 480-d-old male Ross 308 broilers were divided in groups of 12 birds per pen resulting in 16 replicates for the basal diet and 6 replicates for each test diet. A total of 4 inorganic P sources, MCP, monodicalcium phosphate (MDCP), dicalcium phosphate (DCP) and defluorinated phosphate (DFP) were added to the basal diet to determine the precaecal P digestibility. Three of the 4 inorganic P sources (MCP, MDCP and DCP) represented a mix of batches from different producers. At the end of both experiments, the chyme of the posterior part of the small intestine was collected. Digestibility of P and Ca was determined using titanium dioxide as indigestible marker.

In Experiment 1, a reduction in precaecal digestibility of P was observed above an estimated precaecal digestible dietary P concentration of 4.8 g/kg.

The precaecal P digestibility of the tested inorganic P sources in Experiment 2 was 78.3% for MCP, 59.0% for DCP, 70.7% for MDCP and 31.5% for DFP.

INTRODUCTION

Phosphorus (P) is an essential mineral for all farm animals and plays a key role in cellular metabolism, cellular regulatory mechanisms and in bone mineralisation (Suttle, Furthermore, next to calcium (Ca), it is the most abundant mineral in the body of farm animals including poultry (Georgievskii *et al.*1982) In most diets for poultry, a source of inorganic P, for example, monocalcium phosphate (MCP) or dicalcium phosphate (DCP), is included to fulfil the P requirement of the animals. There is a need to reduce dietary concentrations of total P and inclusion of these phosphate products because of the limited inorganic mineral P reserves in the world and because of the negative environmental impact of P in excreta on the environment (Liu *et al.*2008). This requires adequate insight in the digestibility and relative value of different sources of commonly used phosphate products. In the past, P digestibility of ingredients used in poultry nutrition was generally determined by calculating the difference between P consumed and P excreted in the excreta. This measurement may result in an underestimation of true digestible P, in the case that absorbed but unutilised P is excreted via the urine. Precaecal determination of P digestibility at the distal part of the ileum avoids this possible underestimation of digestible P of ingredients (Rodehutscord *et al.*2012) and presumably results in P digestibility coefficients that are more reliable and consistent. However, precaecal digestible P values may be affected by the level of retainable P in the diet. When the retainable P content reaches or exceeds the requirements of the birds, the absorption of P from the digestive tract may be reduced. Therefore, it is essential to know in which range of dietary P content a constant P digestibility can be expected. Furthermore, Shastak *et al.* (2014)showed that precaecal digestibility of phytate-bound P decreases with increasing concentrations of non-phosphate P in the diet. This implies that the observed digestibility of P from a test ingredient is confounded by a decreasing contribution of P released from phytate. Hence, basal diets used in P digestibility studies should contain no or minimal amounts of phytate P in order to minimise disturbing effects of phytate P on measured P digestibility coefficients of test products. The first aim of this study, comprising two experiments, was to determine in Experiment 1 the relationship between incremental dietary P content and precaecal digestible P (pcdP) and precaecal digestible Ca (pcdCa) content in male broilers. The second aim was to determine the pcdP digestibility from the inorganic P sources MCP, dicalcium phosphate (DCP), monodicalcium phosphate (MDCP) and defluorinated phosphate (DFP) at marginal levels of P supply in male broilers, based on semi-purified diets containing minimal concentrations of phytate P to ensure a minimum level of interference from phytate P in Experiment 2. The third aim of this study was to make a limited comparison between apparent total tract retention (ATTR) of P and pcdP for MCP. The protocol developed by the World Poultry Science Association (WPSA, 2013) on the determination of pcdP digestibility was largely followed in this study.

MATERIALS AND METHODS

Experimental diets

Diets used in Experiments 1 and 2 are presented in Table 1 and Table 2, respectively. In Experiment 1 (conducted in 2014), 4 diets were formulated to contain incremental concentrations of retainable P based on CVB (2011) by replacing diamol with increasing levels of monohydrated MCP (Aliphos, Bulgaria). In Experiment 2 (conducted in 2014), the precaecal P digestibility of the inorganic P sources MCP, DCP, MDCP and DFP was determined (chemical composition of the 4 inorganic P sources is shown in Table 3). Therefore, in addition to a basal diet, 4 diets were prepared including 2 g P/kg diet from either MCP, MDCP, DCP or DFP in a constant Ca:P ratio of 1.40 and using $CaCO_3$ (grinding fineness of $CaCO_3$ less than 250 µm) to include the required amount of Ca. In both experiments, titanium dioxide (TiO_2) was used as marker to determine the digestibility. The MCP was acquired from 4 suppliers for monohydrated MCP and mixed in equal portions (Aliphos, the Netherlands; Aliphos, Bulgaria; Yara, Sweden and Timab, France). The MDCP was acquired from 3 suppliers and mixed in equal portions (Aliphos, the Netherlands, semi-monohydrated; Aliphos, Bulgaria, semi-monohydrated and Timab, France, monohydrated). The DCP was acquired from 3 suppliers and mixed in equal portions (Aliphos, the Netherlands, semihydrated; Aliphos, Bulgaria, semi hydrated and Timab, France, anhydrous). The DFP was acquired from 1 supplier (Phosphorit, Eurochem group, Russia).

Table 1. Composition of ingredients and nutrients in the diets used in Experiment 1 (in g/kg as fed unless otherwise indicated).

Item	1.0 rP (basal)	2.7 rP	4.3 rP	6.0 rP
Ingredients				
Maize starch gelatinised	220.0	220.0	220.0	220.0
Native maize starch	168.3	168.3	168.3	168.3
Soybean meal	110.0	110.0	110.0	110.0
Egg white powder	100.0	100.0	100.0	100.0
Sugar	100.0	100.0	100.0	100.0
Casein	100.0	100.0	100.0	100.0
Oat hulls	70.0	70.0	70.0	70.0
Diamol	70.0	57.9	45.8	33.4
Soya oil	40.0	40.0	40.0	40.0
Premix[1]	5.0	5.0	5.0	5.0
Titanium dioxide	5.0	5.0	5.0	5.0
Limestone[2]	3.3	6.1	8.9	11.7
Monocalcium phosphate[3]	1.7	11.1	20.5	30.0
Sodium chloride	0.2	0.1	0.1	0.1
Potassium carbonate	3.6	3.6	3.6	3.6
Sodium bicarbonate	1.0	1.0	1.0	1.1
Magnesium oxide	1.0	1.0	0.9	0.9
L-arginine	0.9	0.9	0.9	0.9
Nutrients calculated				
Crude protein	228	228	228	228
Crude fibre	25	25	25	25
Crude fat	45	45	45	45
Metabolisable Energy (MJ/kg)	12.74	12.74	12.74	12.74
Nutrients analysed				
Dry matter	896	892	893	896
Ash	88	87	84	83
Ca	2.45	5.01	7.50	10.04
Phosphorus, total	2.26	4.51	6.47	8.60
Ca:P	1.08	1.11	1.16	1.17
Phytate P	0.85	0.85	0.83	0.85
Ti	3.05	3.09	3.03	3.10

Calculated concentrations of retainable P (rP; g/kg product) were based on CVB (2011) table values, derived from faecal digestibility studies.

[1] Premix (provided per kg of diet): 4.1 mg retinyl acetate, 60 µg cholecalciferol, 45.5 mg dl-a-tocopherol, 1.5 mg menadione, 2.0 mg thiamine, 7.5 mg riboflavin, 10.0 mg d-pantothenic acid, 35 mg niacin, 2000 µg biotin, 20 µg cyanocobalamin, 1.0 mg folic acid, 3.5 mg pyridoxine-HCl, 460 mg choline chloride, 80 mg Fe (as $FeSO_4 \cdot H_2O$), 12 mg Cu (as $CuSO_4 \cdot 5H_2O$), 60 mg Zn (as $ZnSO_4 \cdot H_2O$), 85 mg Mn (as MnO), 0.4 mg Co (as $CoSO_4 \cdot 7H_2O$), 0.8 mg I (as KI), 0.15 mg Se (as Na_2SeO_3) and 125 mg antioxidant.

[2] Grinding fineness of limestone was less than 250 µm.

[3] Analysed composition of monocalcium phosphate (g/kg): moisture 17, ash 806, phosphor 208 and calcium 182.

Table 2. Content of ingredients and nutrients in experimental diets used in Experiment 2 (in g/kg as fed unless otherwise indicated).

	Basal	MCP	MDCP	DCP	DFP
Ingredients					
Maize starch, gelatinised	250.0	250.0	250.0	250.0	250.0
Maize starch, native	226.6	226.6	226.6	226.6	226.6
Soybean meal	110.0	110.0	110.0	110.0	110.0
Sucrose	100.0	100.0	100.0	100.0	100.0
Potato protein	87.2	87.2	87.2	87.2	87.2
Egg white powder	65.0	65.0	65.0	65.0	65.0
Oat hulls	60.0	60.0	60.0	60.0	60.0
Soya oil	40.0	40.0	40.0	40.0	40.0
Lignocellulose	20.0	20.0	20.0	20.0	20.0
Diamol	16.1	4.3	4.8	5.7	9.0
Premix[1]	5.0	5.0	5.0	5.0	5.0
Potassium carbonate	5.4	5.4	5.4	5.4	5.4
Sodium bicarbonate	2.9	2.9	2.9	2.9	1.0
Magnesium oxide	1.0	1.0	1.0	1.0	1.0
Potassium chloride	0.7	0.7	0.7	0.7	0.7
DL-methionine	0.9	0.9	0.9	0.9	0.9
L-arginine	1.2	1.2	1.2	1.2	1.2
Titanium dioxide	5.0	5.0	5.0	5.0	5.0
Limestone[2]	3.0	6.0	5.1	2.3	1.3
MCP	0.0	8.8	0.0	0.0	0.0
MDCP	0.0	0.0	9.2	0.0	0.0
DCP	0.0	0.0	0.0	11.1	0.0
DFP	0.0	0.0	0.0	0.0	10.7
Nutrients calculated					
rP	0.98	2.68	2.56	2.54	2.54
Crude protein	181	181	181	181	181
Crude fibre	36	36	36	36	36
Crude fat	47	47	47	47	47
Phytate P	0.58	0.58	0.58	0.58	0.58
Metabolisable Energy (MJ/kg)	13.13	13.13	13.13	13.13	13.13
Nutrients analysed					
Dry matter	881	884	886	880	876
Ash	41	39	39	40	42
Ca	1.83	4.42	4.42	4.36	4.35
Phosphorus, total	1.36	3.39	3.34	3.28	3.24
Ca:P	1.35	1.30	1.32	1.33	1.34
Ti	2.78	2.79	2.80	2.78	2.76

Concentrations of retainable P (rP; g/kg as fed) based on apparent total tract retention experiments were obtained from CVB (2011).

[1]Premix (provided per kg of diet): 4.1 mg retinyl acetate, 60 μg cholecalciferol, 45.5 mg dl-a-tocopherol, 1.5 mg menadione, 2.0 mg thiamine, 7.5 mg riboflavin, 10.0 mg d-pantothenic acid, 35 mg niacin, 2000 μg biotin, 20 μg cyanocobalamin, 1.0 mg folic acid, 3.5 mg pyridoxine-HCl, 460 mg choline chloride, 80 mg Fe (as $FeSO_4 \cdot H_2O$), 12 mg Cu (as $CuSO_4 \cdot 5H_2O$), 60 mg Zn (as $ZnSO_4 \cdot H_2O$), 85 mg Mn (as MnO), 0.4 mg Co (as $CoSO_4 \cdot 7H_2O$), 0.8 mg I (as KI), 0.15 mg Se (as Na_2SeO_3) and 125 mg antioxidant.

[2]Grinding fineness of limestone less than 250 μm.

Table 3. Composition of inorganic phosphorus sources tested in Experiment 2 (in g/kg product).

Item	Moisture (basal)	Ash	Phosphorus	Calcium
MCP[1]	27	804	218	180
MDCP[1]	24	839	207	217
DCP[1]	21	895	184	282
DFP	1	997	190	323

[1]Analysis based on the mix of 4 sources of MCP, 3 sources of MDCP, and 3 sources of DCP, all of these mixtures being mixed in equal portions.

Birds and management

In Experiment 1, a number of 260-d-old male Ross 308 broilers were obtained from a commercial hatchery. The birds were vaccinated against avian infectious bronchitis and Newcastle disease and were group-housed on wood shavings between 0 and 14 d of age. In this period, the birds received a commercial starter diet (nutrient concentrations: 216 g/kg RE, 9.0 g/kg Ca, 5.9 g/kg P and 0.01% added commercial phytase; Natuphos 5000 G, BASF, Germany). From d 14 until the end of the experiment at d 24, the birds received the experimental diets and were divided over pens with plastic slatted floors (Jansen Poultry Equipment, Barneveld, the Netherlands) in groups of 10 birds per pen resulting in 8 replicates with 10 birds per replicate for treatment 1 and 6 replicate pens for treatment 2, 3 and 4. Temperature was set at 34°C at d 1 and gradually reduced thereafter to 21°C at d 21. Lighting schedule was 24 L:0D during d 1 and 2, 18 L:6D between d 3 and 21 and 24 L:0D between d 22 and 24 in order to ensure a steady-state situation with a homogenous distribution of feed intake during the last 3 d of the experiment. Feed and water were provided *ad libitum* during the complete period of the experiment.

Experiment 2 was part of a larger study and was carried out in two rounds with the same number of treatments and replicates per round. In total, 2200 (1100 per round)-d-old Ross 308 broilers were obtained from a commercial hatchery, of which 480 were used for Experiment 2. Birds were vaccinated against avian infectious bronchitis and Newcastle disease and were group-housed on wood shavings during the first 18 (round 1) and 17 (round 2) days. During these periods, the birds received a commercial starter diet (nutrient concentrations: 199 g/kg crude protein, 10.9 g/kg Ca and 7.0 g/kg P, no addition of exogenous phytase).

From 17/18 d of age until the end of the experiment at d 28, birds were housed in pens with plastic slatted floors (Jansen Poultry Equipment, Barneveld, the Netherlands) in groups of 12 birds per pen resulting in a total number of 40 pens of which 16 were used for the basal diet (8 pens per round) and 6 pens for each of the other diets (3 pens per diet per round). During this period, birds received the experimental diets.

Sample collection and measurements

In Experiment 1, total body weight (BW) per pen was determined at d 14 and 24 of age. Cumulative feed intake was recorded as the difference between feed provided during d 14–24 and feed remaining at d 24. At the end of Experiment 1 at d 24, birds were anaesthetised with 1 ml/kg BW of a 5:3 mixture of Sedanum and Ketamine via injection in the breast muscle and subsequently killed with an injection of 0.5 ml of T61 in the wing artery. Then, the birds were dissected and the posterior third of the small intestinal section between Meckel's diverticulum and 2 cm before the junction to the caeca was removed and emptied by flushing the separated gut segment with distilled water. After collection, the samples were pooled per pen and immediately frozen and stored at −20°C pending analysis.

In Experiment 2, total BW per pen was determined at d 18 (round 1) and d 17 (round 2). Cumulative feed intake was recorded during d 18–27 for round 1 and d 17–27 for round 2 in the same way as for Experiment 1. At d 25, 26 and 27 of age, representative samples of excreta from the basal diet and the MCP diet were semi-quantitatively collected per pen, two times a day at 8:00 and 16:00 h, over a period of three full days of 24 h. Collection of excreta was semi-quantitative as excreta contaminated with feathers and feed were discarded. After collection, samples were immediately stored at −30°C. Then samples were pooled per pen and stored again at −30°C pending analysis. At d 27 and 28 of age, birds were anaesthetised and killed, and the contents of the posterior ileum were collected, pooled per pen and frozen as described for Experiment 1.

Chemical analyses and calculations

The MCP product used in Experiment 1 was analysed for P and Ca (ISO 11885, 1998*c*). The test products in Experiment 2 were analysed for P (ISO 6491, 1998*a*) and Ca (ISO 6869, 2000). Furthermore, for the two experiments, identical chemical analyses and calculations were used. Diets were analysed for dry matter (DM) (ISO 6496, 1998*b*), crude ash (ISO 5984, 2002), crude protein (ISO 5983, 1997), Ca and P (ISO 11885, 1998*c*) and Ti (van Bussel *et al.*2010)

Ileal digesta and excreta of birds were analysed for DM (ISO 6496, 1998*b*), crude ash (ISO 5984, 2002), Ca and P (ISO 11885, 1998*c*) and Ti (van Bussel *et al.*2010)

The precaecal digestibility and ATTR of P, Ca, DM, ash and organic matter was calculated as follows:

$$Y(\%) = 100 - [100 \times (Ti_{diet} \times Nutrient_{digesta}) / (Ti_{digesta} \times Nutrient_{diet})]$$

where Ti_{diet} and $Ti_{digesta}$ are the analysed concentrations of Ti in the diet and digesta or excreta (g/kg DM), respectively, and $Nutrient_{diet}$ and $Nutrient_{digesta}$ are the analysed concentrations of nutrients in the diet and digesta or excreta, respectively. The pcdP content (g/kg) was determined as the pcdP digestibility (%) × P content (g/kg)/100.

Statistical analysis

In both experiments, pen was the experimental unit and statistical analyses were carried out using Genstat, 16th edition (Experiment 1) and 17th edition (Experiment 2). In both Experiments 1 and 2, performance characteristics were analysed using Analysis of variance (ANOVA) with phosphate level (Experiment 1) or phosphate product (Experiment 2) as fixed effect, and for Experiment 2, round was included in the model as a random effect. Because in Experiment 1 a significant quadratic effect ($P < 0.001$) in the response of pcdP to dietary P content was observed, a broken stick regression analysis was carried out to estimate the relationship between pcdP and P (g/kg) (Figure) using the R2Lines procedure of Genstat. This broken stick model included an intercept, two slopes and a break point value. The estimated slopes represent the digestibility of P at inclusion levels below and above the breakpoint.

Figure. Relationship between precaecal digestible P concentration in broiler diets (pcdP; g/kg) (treatments means and standard deviation) and dietary P content (g/kg) analysed with a broken stick regression model represented by the lines. Goodness of fit of the model: $R^2 = 0.979$ and SE = 0.221. Parameter estimates: break point value of precaecal digestible P = 4.80 ± 0.168 g/kg and dietary P = 5.81 ± 0.284 g/kg, slope before break point = 0.880 ± 0.0516, slope after break point = 0.201 ± 0.0735.

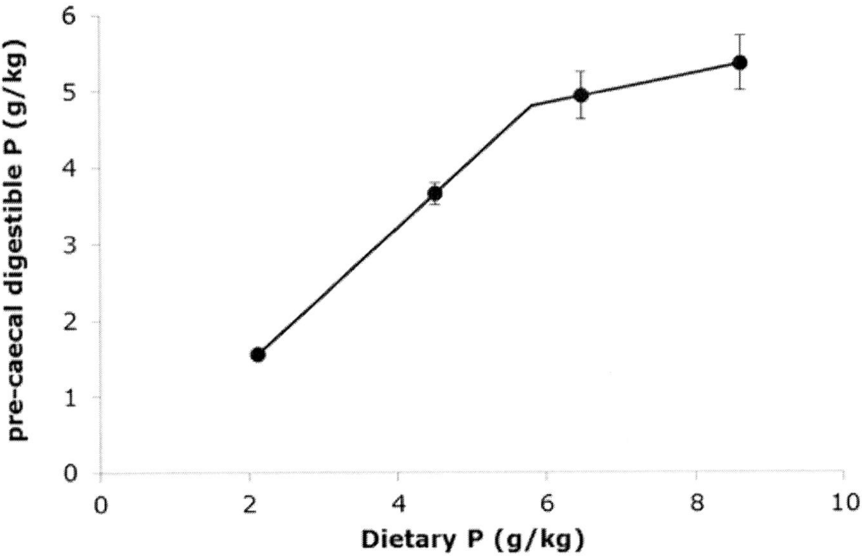

In Experiment 2, digestibility of Ca and P for each test product was determined as described in the WPSA (2013)protocol for P using a multiple linear regression model between pcdP content and added dietary P content including a common intercept value for all products and allowing separate regression coefficients for the products tested using the regression procedure in Genstat. In deviation of the WPSA protocol, the determination of the digestibility of Ca and P was based on the inclusion of one rather than two inclusion levels of each test product after having shown a linear response below a P content of 4.8 ± 0.17 g/kg in Experiment 1.

RESULTS

Experiment 1

One statistically determined outlier value for the lowest dietary P concentration was removed from the data set and 3 outlier values for the highest dietary P concentration were removed from the data set before fitting the broken stick model. The 3 outlier observations for the highest dietary P concentration were considered outlier values as they had (1) organic matter digestibility values that deviated more than 1.9 standard deviations from the mean organic matter digestibility of 94 ± 1.7% and (2) negative pcdCa digestibility values. In the Figure, the broken stick relationship between pcdP content (g/kg) and dietary P (g/kg) as analysed for the results in Experiment 1 shows that 98% of variation in pcdP content could be explained by the broken stick model. The pcdP digestibility of MCP, reflected by the slope, was estimated to be 88.0 ± 5.16% before the break point value at a dietary P concentration of 5.8 ± 0.28 g/kg and 20.1 ± 7.35% after this break point value. The performance characteristics are given in Table 4. The rP concentration had a significant ($P < 0.001$) effect on BW at d 24, average daily gain (ADG) and average daily feed intake (ADFI). Birds receiving the lowest retainable P (rP; g/kg) level diet had lower BW at d 24, lower ADG and lower ADFI compared to birds receiving the other diets with rP concentrations ranging from 2.7 to 6.0. No differences in BW, ADG and ADFI were observed between the diets with rP concentrations ranging from 2.7 to 6.0. The pcdP and pcdCa digestibility values of the diets are shown in Table 5. The rP concentration had a significant ($P < 0.001$) effect on pcdP and pcdCa digestibility. The pcdP digestibility of the 6.0 rP diet was lower compared to the other rP concentrations. Furthermore, the 2.7 rP diet had numerically the highest pcdP digestibility and had a higher pcdP digestibility compared to the 1.0 and the 6.0 rP diets. The pcdCa digestibility of the 1.0 and 2.7 rP diets were higher than the 4.3 and 6.0 rP diets.

Table 4. Performance characteristics of broilers in Experiment 1 from d 14 until d 24 supplied with diets varying in retainable P (rP; g/kg product).

	1.0 rP (basal)	2.7 rP	4.3 rP	6.0 rP	Standard error of the mean (SEM)	*P*-value
BW d 14 (g)	471	470	466	465	4.2	0.476
BW d 24 (g)	1136^{b}	1286^{a}	1269^{a}	1266^{a}	23.7	<0.001
ADG (g/d)	66.5^{b}	81.7^{a}	80.3^{a}	80.0^{a}	2.27	<0.001
ADFI (g/d)	87.0^{b}	109.7^{a}	108.8^{a}	108.7^{a}	2.65	<0.001
Feed:gain (g/g)	1.311	1.345	1.355	1.358	0.0276	0.286
Mortality (%)	0.0	0.0	1.7	0.0	0.79	0.359

[a, b]Means in a row with no common superscript letter differ significantly at $P < 0.05$.

Table 5. Precaecal diet digestibility (%) of P and Ca in Experiment 1 with diets varying in retainable P.

	1.0 rP (basal)	2.7 rP	4.3 rP	6.0 rP	SEM	*P*-value
P	73.5^{b}	81.2^{a}	76.3^{ab}	62.4^{c}	2.09	<0.001
Ca	65.2^{a}	69.3^{a}	48.1^{b}	32.1^{b}	5.96	<0.001

[a, b, c]Means in a row with no common superscript letter differ significantly at $P < 0.05$.

Experiment 2

The performance characteristics of the birds in Experiment 2 as presented in Table 6 were significantly ($P < 0.001$) affected by the dietary treatments for BW at d 27, ADG, ADFI and feed:gain ratio. Birds fed the basal diet had the lowest ADG, ADFI and highest feed:gain ratio compared to the other treatments. Overall, performance results were highest for MCP and MDCP and lowest for DFP.

Table 6. Performance characteristics of broilers in Experiment 2 from d 18 (Round 1)/d 17(Round 2) until d 27 supplied with a low-P basal diet or the same diet supplemented withmonocalcium phosphate (MCP), monodicalcium phosphate (MDCP), dicalcium phosphate(DCP) and defluorinated phosphate (DFP).

	Basal	MCP	MDCP	DCP	DFP	SEM	*P* value
BW d 17/18 (g)	640	637	631	636	645	5.7	0.661
BW d 27 (g)	1208[d]	1452[a]	1414[ab]	1375[bc]	1361[c]	16.1	< 0.001
ADG (g/d)	60.1[c]	85.9[a]	82.7[a]	78.0[b]	75.7[b]	1.31	< 0.001
ADFI (g/d)	107.3[d]	128.7[a]	125.5[ab]	120.7[bc]	121.7[bc]	1.85	< 0.001
Feed:gain (g/g)	1.799[a]	1.500[c]	1.522[c]	1.554[bc]	1.615[b]	0.0277	< 0.001
Mortality (%)	2.6	1.4	0.0	0.0	1.4	1.43	0.467

[a, b, c, d]Means in a row with no common superscript letter differ significantly at $P < 0.05$.

In Table 7, the precaecal digestibility of DM, ash, P and Ca of the experimental diets is presented. Highly significant effects ($P < 0.001$) on precaecal digestibility of ash, Ca and P were observed. The dietary precaecal digestibility of ash, P and Ca was significantly lower for the DFP diet compared to the other diets (except the basal diet). The pcdP digestibility was higher for MCP and MDCP diets compared to the DCP and DFP diets and the pcdP digestibility of the DCP diet was higher than that of the DFP diet. The pcdCa digestibility of the MCP diet was higher than that of the DCP diet.

In Table 8, the pcdP digestibility values for the 4 test products are presented. Results in Table 8 show that pcdP digestibility is highest for MCP followed by MDCP, DCP and DFP. Compared to MCP, pcdP digestibility of the MDCP, DCP and DFP were significantly lower ($P = 0.042$ for MDCP and $P < 0.001$ for DCP and DFP).

For the MCP diet and the basal diet, the ATTR of P and Ca was determined as well. The ATTR of P and Ca for the basal diet was 65.4 ± 0.99 and −1.7 ± 2.62%, respectively, and ATTR of P and Ca for the MCP diet was 71.3 and 57.2%, respectively. For the test product MCP, the ATTR of P was 75.3 ± 1.1.

Table 7. Precaecal diet digestibility (%) of DM, ash, P and Ca of a low-P basal diet or the same diet supplemented with monocalcium phosphate (MCP), monodicalcium phosphate (MDCP), dicalcium phosphate (DCP) and defluorinated phosphate (DFP), determined in broilers at 27 d of age.

	Basal	MCP	MDCP	DCP	DFP	SEM	*P*-value
DM	81.9^{b}	83.7^{a}	83.2^{ab}	84.0^{a}	83.1^{ab}	0.61	0.032
Ash	34.8^{c}	49.1^{a}	47.0^{ab}	47.5^{ab}	35.8^{c}	1.65	<0.001
P	64.0^{bc}	72.6^{a}	68.0^{ab}	61.1^{c}	45.1^{d}	2.37	<0.001
Ca	67.2^{a}	66.0^{a}	64.4^{ab}	59.9^{b}	37.4^{c}	1.91	<0.001

[a, b, c, d]Means in a row with no common superscript letter differ significantly at $P < 0.05$.

Table 8. Precaecal digestibility (%) of P (pcdP%) and Ca (pcdCa%) of monocalcium phosphate (MCP), monodicalcium phosphate (MDCP), dicalcium phosphate (DCP) and defluorinated phosphate (DFP), determined in broilers at 27 d of age.

	MCP	MDCP	DCP	DFP
pcdP%	78.3 ± 2.97	70.7 ± 3.06	59.0 ± 3.12	31.5 ± 3.16
P-value of contrast to MCP	NA	0.042	<0.001	<0.001

DISCUSSION

Experiment 1

Experiment 1 was conducted to estimate the dietary (digestible) P content above which a decrease in precaecal P digestibility is to be taken into account. From the published literature, this was not clear. For example, Rodehutscord *et al.* (2012)observed a linear increase in pcdP with increasing concentrations of dietary P content up to 8.14 g/kg. Results of Experiment 1 show that the marginal efficiency of precaecal absorption of P declined above a certain level of P intake and absorption. The broken line model indicated that this decline started at a pcdP concentration of 4.8 ± 0.17 g/kg feed. Rodehutscord *et al.* (2012)determined the pcdP of MCP (g/kg) in an experiment with 5 incremental concentrations of dietary P, ranging from 2.39 to 8.14 g/kg, by adding MCP to a basal ration. The range in determined pcdP was 0.94–3.0 g/kg. For this range, a linear relationship between pcdP and dietary P (g/kg) was observed, that is, no break point could be established. In contrast, a break point was established in the response of P in the excreta to an incremental dietary P content. This indicates that excretion of excess P in the urine progressively increased at a concentration of dietary P below the level of dietary P at which precaecal absorption of dietary P started to decline. In the study of Rodehutscord *et al.* (2012)a progressive increase in excretion of P was observed above a dietary P concentration of 5.2 g/kg, corresponding with a pcdP concentration of 2.46 g/kg and an ATTR P concentration of 2.71 g/kg. This ATTR P concentration of 2.71 g/kg agreed well with the break point value of ATTR of P of 2.4 g/kg observed in a study from Leske and Coon (2002Leske, K. & Coon, C. (2002) combined with the results of Experiment 1 in the present study indicate that a decline in ATTR may be observed at a lower P intake than a decline in precaecal digestibility, presumably due to a progressive increase in P excretion in the urine. The precaecal digestibility of P seems reliable below a pcdP concentration of at least 3.7 g/kg feed and probably even below 4.8 g/kg, because in the present study, in Experiment 1, the model estimated a relation linear up to 4.8 g pcdP/kg. The range in observed pcdP levels in the experiment was sufficiently large to estimate a break point value. However, because of the limited number of pcdP levels created in Experiment 1, it remains unknown whether or not there truly is a distinct break point value. The determined pcdP digestibility of MCP of 88.0% in the current study agrees well with the ATTR of P values from CVB (2011)for monohydrate MCP of 85% but is higher than the pcdP content observed in Experiment 2 of 78.3%.

Experiment 2

The results of pcdP digestibility for the diets shown in Table 7, combined with the P contents of the same diets shown in Table 2, show that the highest observed pcdP was 2.46 g/kg for the MCP diet, which is well below the break point value of 4.8 ± 0.17 g/kg established in Experiment 1. This indicates that the conditions of Experiment 2 were adequate for testing a reliable pcdP value for the different feed phosphates.

Growth performance of the birds receiving the basal and DFP diets were considerably below the performance of birds of the MCP, MDCP and DCP treatments (Table 6), indicating that retainable P supply for these treatments was below the optimum. Similar negative effects of a low dietary P concentration on performance were observed in other studies as well (Leske and Coon, 2002Leske, K. & Coon, C. (2002) Yan and Waldroup, Mello *et al.*, (2012) A difference in pcdP digestibility for MCP was observed between Experiment 1 (88.0%) and Experiment 2 (78.3%). Different MCP sources were used for Experiment 1 (Aliphos Bulgaria) and Experiment 2 (a mixed product of sources from different origins; Aliphos, the Netherlands; Aliphos, Bulgaria; Yara, Sweden and Timab, France). Probably, differences in chemical composition and production processes between the MCP sources used in Experiments 1 and 2 caused these differences in pcdP digestibility. Another explanation for the difference in pcdP digestibility between Experiments 1 and 2 is the higher Ca:P ratio in Experiment 2 (1.33 average) compared to Experiment 1 (1.16 average), which might have contributed to more phytate–Ca complex formation and consequently a decreased absorption of phytate P (Tamim *et al.*, 20014 Amerah *et al.*, This is supported by results of a study of Rodehutscord *et al.* (2012) and in the present study, diets low in phytate P were used and the P contents in the basal diets were low (<2.40 g/kg). In contrast to the explanation of a negative effect of a high Ca:P ratio, in a study from Van der Klis and Versteegh (1992*a*) for MCP a high ATTR of P of 79% was observed at a Ca:P ratio of 2.27, where MCP was regarded to be the only dietary P source. In the study of Van der Klis and Versteegh (1992*a*)the MCP digestibility was determined at a low retainable P concentration of 2.05 g/kg.

The pcdP digestibility of the anhydrous DCP of 59.0% observed in this study is in between the ATTR values of 76% and 53% for dihydrated DCP and anhydrous DCP, respectively, as measured in the study from Van der Klis and Versteegh (1992*b*)but closer to the value for the anhydrous form. However, the pcdP digestibility value of anhydrous DCP of 59.0% established in this study is substantially higher than the pcdP digestibility of anhydrous DCP of 25.0–30.0% observed in the study of Shastak *et al.* (2012) In the same study of Shastak *et al.* (2012)a pcdP digestibility of 54–67% for monosodium phosphate was found which is low compared to reported monosodium phosphate digestibility (Van der Klis and Versteegh, 1996) and indicates that the differences in pcdP digestibility between DCP observed in this study and observed in the study of Shastak *et al.* (2012)may be attributed to systematic differences in experimental conditions and not to differences in DCP properties per se.

The pcdP digestibility of MDCP of 70.7% observed in this study was substantially lower than the ATTR value of 80.3% observed in a study from Van der Klis and Versteegh (1998*a*Van der Klis, J.D. & Versteegh, H.A.J. (1998a) in which MDCP was regarded as the only P source and tested at a retainable P concentration of 3.13 g/kg with a Ca:P ratio of 1.79. In another study of Van der Klis and Versteegh

(1992*b*)two MDCP sources were tested resulting in precaecal P digestibility values of 72.2% and 81.8%, indicating that substantial differences in MDCP qualities exist. Unfortunately, no information was provided with respect to the degree of hydration of the two MDCP sources and their DM, Ca and P content.

For DFP, no literature data was found to be used for comparison with the observations in the present study. The results of this study indicate a much lower pcdP digestibility compared to the other commonly used phosphate products.

Because of the large differences in observed digestibility of inorganic P sources between studies, it might be argued that in future research, in addition to absolute pcdP digestibility values, more focus should be paid to differences between inorganic phosphate sources within studies and on describing the degree of hydration of the products tested. Compared to other studies, in the present study, a comparison is possible between the major inorganic phosphate sources.

At the basal diet, an unexpected low ATTR of Ca of −1.7 ± 2.62% was observed. This low ATTR of Ca is likely caused by the low P supply in the basal diet and the preferential retention of P in soft tissue (i.e. muscle and organs) rather than in bone as hydroxy apatite. Since the Ca content in soft tissue is very low (Ca:P is 1:35 in broiler meat according to Fuentes Zapata *et al.* (1998)the majority of Ca is excreted. This finding emphasises that total tract digestibility studies cannot be used for the determination of Ca digestibility in feed stuffs, but rather reflect the retention in body tissue.

In this study, P sources from different suppliers were pooled to form MCP, MDCP and DCP mixtures. The determined precaecal P and Ca digestibility of these pooled P mixtures can be considered as average values for these type of P sources being used in Europe but it should be realised that digestibility values of individual MCP, MDCP and DCP products might differ from these average values as a result of differences in chemical composition and production processes (Viljoen, 2001) This becomes also evident from the difference in pcdP digestibility of MCP between Experiments 1 and 2 in the current study.

It is concluded that the response in pcdP digestibility to increments in dietary P concentration is linear over a wider range of dietary P than the response in P retention based on ATTR. A reduction in precaecal absorption of P was observed above an estimated pcdP concentration of 4.8 g/kg. In Experiment 2, in which pdcP digestibility of inorganic P sources were determined at marginal dietary pcdP contents of maximal 2.5 g/kg, the pcdP digestibility of the P sources, representing a mixture of available products in the market, was 78.3% for MCP, 59.0% for DCP, 70.7% for MDCP and 31.5% for DFP.

REFERENCES

1. Amerah, A.M., Plumstead, P.W., Barnard, L.P. & Kumar, A. (2014) Effect of calcium level and phytase addition on ileal phytate degradation and amino acid digestibility of broilers fed corn-based diets. *Poultry Science*, 93: 906–915. doi:10.3382/ps.2013-03465,
2. CVB (Centraal Veevoederbureau) (2011) *Table on Feed Ingredients. Data on the Chemical Composition, Digestibility and Nutritional Value of Feedstuffs*. (Product Board Animal Feed, Den Haag, The Netherlands).
3. Fuentes Zapata, J.F., dos Reis Moreira, R.S., Freire Fuentes, M.D.F., Miranda Sampaio, E. & Morgano, M. (1998)Meat mineral content in broilers fed diets without mineral and vitamin supplements. *Pesquisa Agropecuaria Brasileira*, 33: 1815–1820.
4. Georgievskii, V.I., Annenkov, B.N. & Samokhin, V.I. (1982) *Mineral Nutrition of Animals. Studies in the Agricultural and Food Sciences*. (Butterworth & Co, London). ISBN 978-0-408-10770-7.
5. International Organization for Standardization (1997) *Animal Feeding Stuffs. Determination of Nitrogen Content and Calculation of Crude Protein Content. Kjeldahl Method. ISO 5983*. (Geneva, Switzerland, Int. Organ. Standardization).
6. International Organization for Standardization (1998a) *Animal Feeding Stuffs. Determination of Phosphorus Content – Spectrometric Method. ISO 6491*. (Geneva, Switzerland, Int. Organ. Standardization).
7. International Organization for Standardization (1998b) *Animal Feeding Stuffs. Determination of Moisture and Other Volatile Matter Content. ISO 6496*. (Geneva, Switzerland, Int. Organ. Standardization).
8. International Organization for Standardization (1998c) *Water Quality. Determination of 33 Elements by Inductively Coupled Plasma Atomic Emission Spectroscopy. ISO 11885*. (Geneva, Switzerland, Int. Organ. Standardization).
9. International Organization for Standardization (2000) *Animal Feeding Stuffs. Determination of the Contents of Calcium, Copper, Iron, Magnesium, Manganese, Potassium, Sodium and Zinc – Method Using Atomic Absorption Spectrometry. ISO 6869*. (Geneva, Switzerland, Int. Organ. Standardization).
10. International Organization for Standardization (2002) *Animal Feeding Stuffs. Determination of Crude Ash. ISO 5984*. (Geneva, Switzerland, Int. Organ. Standardization).
11. Leske, K. & Coon, C. (2002) The development of feedstuff retainable phosphorus values for broilers. *Poultry Science*, 81: 1681–1693. doi:10.1093/ps/81.11.1681
12. Liu, Y., Villalba, G., Ayres, R.U. & Schroder, H. (2008) Global phosphorus flows and environmental impacts from a consumption perspective. *Journal of Industrial Ecology*, 12: 229–247. doi:10.1111/j.1530-9290.2008.00025.x
13. Mello, H.H.C., Gomes, P.C., Rostagno, H.S., Albino, L.F.T., da Rocha, T.C., de Almeida, R.L. & Calderano, A.A. (2012)Dietary requirements of available phosphorus in growing broiler chickens at a constant calcium: available phosphorus ratio. *Revista Brasileira de Zootecnia*, 41: 2323–2328. doi:10.1590/S1516-35982012001100004
14. Rodehutscord, M., Dieckmann, A., Witzig, M. & Shastak, Y. (2012) A note on sampling digesta from the ileum of broilers in phosphorus digestibility studies. *Poultry Science*, 91: 965–971. doi:10.3382/ps.2011-01943
15. Shastak, Y., Witzig, M., Hartung, K. & Rodehutscord, M. (2012) Comparison of retention and prececal digestibility measurements in evaluating mineral phosphorus sources in broilers. *Poultry Science*, 91: 2201–2209. doi:10.3382/ps.2011-02063
16. Shastak, Y., Zeller, E., Witzig, M., Schollenberger, M. & Rodehutscord, M. (2014) Effects of the composition of the basal diet on the evaluation of mineral phosphorus

sources and interactions with phytate hydrolysis in broilers.*Poultry Science*, 93: 2548–2559. doi:10.3382/ps.2014-03961

17. Suttle, N.F. (2010) *Mineral Nutrition of Livestock: Fourth Edition*, Wallingford, UK, CABI pp. 1–547.
18. Tamim, N.M., Angel, R. & Christman, M. (2004) Influence of dietary calcium and phytase on phytate phosphorus hydrolysis in broiler chickens. *Poultry Science*, 83: 1358–1367. doi:10.1093/ps/83.8.1358
19. van Bussel, W., Kerkhof, F., van Kessel, T., Lamers, H., Nous, D., Verdonk, H., Verhoeven, B., Boer, N. & Toonen, H.(2010) Accurate determination of titanium as titanium dioxide for limited sample size digestibility studies of feed and food matrices by inductively coupled plasma optical emission spectrometry with real-time simultaneous internal standardization. *Atomic Spectroscopy*, 31: 81–88.
20. Van der Klis, J.D. & Versteegh, H.A.J. (1992a) De beschikbaarheid van fosfor voor slachtkuikens in de voederfosfaten Rükana en Rükana-phos. Verslag No. 130. (Beekbergen, The Netherlands, COVP-DLO).
21. Van der Klis, J.D. & Versteegh, H.A.J. (1992b) De beschikbaarheid van fosfor voor slachtkuikens in plantaardige en dierlijke veevoedergrondstoffen en in voederfosfaten. Verslag No. 132. (Beekbergen, The Netherlands, COVP-DLO).
22. Van der Klis, J.D. & Versteegh, H.A.J. (1996) Phosphorus nutrition of poultry, in: Garnsworthy, P.C., Wiseman, J. &Haresign, W. (Ed) *Recent Advances in Animal Nutrition*, pp. 71–83 (Nottingham, UK, Nottingham Univ. Press).
23. Van der Klis, J.D. & Versteegh, H.A.J. (1998a) The absorbable phosphorus content in mono-dicalcium phosphate (Kynofos 21) in broilers. Report No. 98.051. (Lelystad, The Netherlands, ID-DLO).
24. Van der Klis, J.D. & Versteegh, H.A.J. (1998b) The absorbable phosphorus content in feed phosphates in broilers. Report No. 98.056. (Lelystad, The Netherlands, ID-DLO).
25. Viljoen, J. (2001) Quality of feed phosphate supplements for animal nutrition, in: *South African Animal Science*, 2 (1):pp. 13–19.
26. WPSA (WORLDS'S POULTRY SCIENCE ASSOCIATION, WORKING GROUP 2 - NUTRITION- OF THE EUROPEAN FEDERATION OF BRANCHES) (2013) Determination of phosphorus availability in poultry. *World's Poultry Science Journal*, 69: 687–698. doi:10.1017/S0043933913000688,
27. Yan, F. & Waldroup, P.W. (2006) Nonphytate phosphorus requirement and phosphorus excretion of broiler chicks fed diets composed of normal or high available phosphate corn as influenced by phytase supplementation and vitamin D source. *International Journal of Poultry Science*, 5: 219–228. doi:10.3923/ijps.2006.219.228

CHAPTER 9

Grain Sorghum: A Conundrum for Chicken-Meat Production

Sonia Y. Liu [1,*], Glen Fox [2], Ali Khoddami [3], Karlie A. Neilson [4], Ha H. Truong [1], Amy F. Moss [1] and Peter H. Selle [1]

[1]*Poultry Research Foundation, Faculty of Veterinary Science, The University of Sydney, 425 Werombi Road, Camden, NSW 2570, Australia*

[2]*Queensland Alliance for Agriculture and Food Innovation (QAAFI) Centre for Nutrition and Food Science, The University of Queensland, Toowoomba, QLD 4350, Australia*

[3]*Department of Plant and Food Sciences, Faculty of Agriculture and Environment, The University of Sydney, NSW 2006, Australia*

[4]*Australian Proteome Analysis Facility, Macquarie University, Sydney, NSW 2109, Australia*

ABSTRACT

The inclusion of grain sorghum in diets for broiler chickens is quite common; however, under Australian conditions, the utilisation of starch/energy by birds offered sorghum-based diets appears inadequate. Various factors inherent in sorghum, including kafirin, phenolic compounds and phytate, may limit energy utilisation. The recent quantification of kafirin, the dominant protein fraction in sorghum, has allowed its nutritional significance to be assessed. This is important as indirect evidence suggests that kafirin concentrations in local sorghums are increasing as an unintended consequence of breeding programs. Presently, Australian sorghums do not contain condensed tannin but, from analyses and assessments of other polyphenolic compounds and phenolic acids, "non-tannin" phenols appear to be negative influences. Anecdotally, white sorghums are considered to be superior to red varieties thus the fact that polyphenolic pigments are responsible for the "redness" of sorghum assumes relevance. Inclusions of sulphite reducing agents in broiler diets have generated promising responses but seem dependent on sorghum properties. Preliminary studies have shown the possibilities of using rapid visco-analyser (RVA) starch pasting profiles, promatest protein solubilities and grain textures to indicate sorghum quality and further studies are required to confirm these hypotheses. These assessments may indicate which sorghums will best respond to reducing agents such as sodium metabisulphite. Finally, the usually modest responses of broilers to exogenous feed enzyme inclusions in sorghum-based are considered in this review.

Keywords: enzyme; kafirin; phenolic compounds; phytate; poultry; protein; Rapid Visco-Analysis; sorghum; starch

1. INTRODUCTION

Primarily for the provision of energy density, grain sorghum is quite frequently included in rations for pigs, poultry and feed-lot cattle. However, in a recent meeting held by the Australian Feed Grain Partnership in December 2012, nutritionists from the three industries expressed concern about the efficiency of energy utilisation in animals offered sorghum-based diets. Essentially, the energy provided by sorghum is derived from starch and, indeed, Black, *et al.* [1] had already drawn attention to this shortcoming.

In order to address the conundrum that is quantifying the true feed quality traits of grain sorghum the concept of the "Bermuda triangle" was developed on the premise that three factors inherent in sorghum are responsible for sub-standard starch and energy utilisation. The three factors nominated, or the apices of the Bermuda triangle, were (i) kafirin, the dominant protein fraction in grain sorghum; (ii) phenolic compounds, potentially including condensed tannin and (iii) phytate or *myo*-inositol hexaphosphate (IP_6). The inference is that sorghum starch is not intrinsically problematic but these three starch-extrinsic factors combine to compromise energy/starch utilisation. It has been proposed that the kafirin proportion of total sorghum protein is escalating in Australia which, if valid, would compromise the nutrient quality of sorghum protein [2]. However, this was not then seen as a major issue from the stand-point of starch utilisation. Almost certainly, contemporary Australian sorghum crops no longer contain condensed tannin; nevertheless, the likelihood is that "non-tannin" phenolic compounds may negatively influence energy utilisation. Phytate may limit starch utilisation and/or glucose absorption thereby negatively influencing energy utilisation. Thus the prime objective of this review is to consider the three apices of the sorghum Bermuda triangle in order to address the conundrum that is grain sorghum as a feedstuff for chicken-meat production.

2. BACKGROUND

In Australia, sorghum (*Sorghum bicolor* L.) is a common energy source in diets for pigs, poultry and beef cattle as it can be grown under drier conditions than those suitable for wheat and maize [3]. Grain sorghum may be exported but very little is used for human consumption. In Australia, sorghum is the second feed grain to wheat as the basis of diets for broiler chickens and there has been a considerable research focus on sorghum initiated by the reports of Perez-Maldonado and Rodrigues [4] and Bryden, *et al.* [5].

Almost certainly, sorghum still suffers from the stigma of condensed tannin (CT), a potent polyphenolic anti-nutritive factor associated with "bird-proof" sorghums. In a recent local survey of pig and poultry nutritionists, the majority believed that sorghum contained some CT and a minority thought CT was present in sufficient concentrations to compromise animal performance. However, the Poultry Research Foundation has subjected more than 60 sorghum samples to the quantal Clorox bleach test [6] and none possessed a pigmented testa; this indicates that they do not contain CT. This has been confirmed with vanillin assays to quantify CT concentrations and local sorghums are now "tannin-free" Type I varieties. Thus our contention, contrary to the report by Perez-Maldonado and Rodrigues [4], is that contemporary Australian sorghum crops do not contain condensed tannin.

The perception is that sorghum is routinely inferior to wheat or even "sub-optimal" but our impression is that this is not necessarily valid. Nevertheless, it is conceded that broilers offered wheat-based diets are advantaged by better pellet quality given the lower starch gelatinisation temperature of wheat in comparison to sorghum. Moreover, broiler chickens offered wheat-based diets are far more likely to respond to feed enzymes, especially non starch polysaccharides (NSP)-degrading enzymes, than their counterparts on sorghum.

Nevertheless, Liu, *et al.* [7] reported that sorghum-based diets supported similar weight gain and feed conversion efficiency as maize-based diets, and both outperformed wheat-based diets. Ao and Choct [8] found that sorghum-based diets generated superior weight gains than broiler diets based on wheat and barley to 21 days post-hatch. In another comparison, Selle, *et al.* [9] showed that broilers offered non-supplemented, sorghum-based diets numerically outperformed wheat in growth performance but this was reversed by the addition of an exogenous xylanase; xylanase in sorghum-based diets did not statistically influence apparent metabolisable energy (AME); whereas, in wheat-based diets xylanase significantly improved AME by 1.21 MJ (14.76 *versus* 13.55 MJ/kg). The failure of a "pure" xylanase feed enzyme to elicit responses in sorghum-based diets is not surprising because sorghum is a "non-viscous" grain containing low levels of NSP that have the capacity to increase digesta viscosity in the gut and impede digestion and absorption of nutrients.

The problematic growth performance of broilers offered sorghum-based diets has been extensively reviewed [2,3,5,10,11]. Briefly, it is recognised that kafirin, the dominant fraction in sorghum protein, is poorly digested due to its hydrophobicity [12,13]. Protein–phytate interactions stem from electrostatic interactions between phytate and basic amino acid residues of arginine, histidine and lysine at pH levels below protein isoelectric points [14]. Kafirin has an unfavourable amino acid profile as it contains a paucity of basic amino acids, especially lysine [15] and, therefore, it is less likely to form protein-phytate complexes. It is well accepted that the potential phytate- protein interaction may reduce the solubility of protein and therefore, impede its digestion [16]. This may partially explain the lack of "extra-phosphoric" effects following the inclusion of phytase in sorghum-based broiler diets.

It is well established that condensed tannin, when present in sorghum, has the capacity to depress protein digestion and compromise bird performance [13]. However, condensed tannin, a polyphenolic compound and potent anti-nutritive factor is almost certainly not present in contemporary Australian sorghum crops. However, the possibility remains that "non-tannin" phenolic compounds in sorghum negatively influence digestion of protein and starch and compromise growth performance in broiler chickens [17]. Liu, *et al.* [18] reported white sorghum-based diets supported better amino acid digestibilities than red sorghum and it may be that soluble conjugated and insoluble bound phenolic acids hold the key to the apparent superiority of white sorghum. Ferulic acid is the most abundant phenolic acid in sorghum but it certainly is not unique to sorghum.

Selle, *et al.* [19] reported sorghum contained an average of 2.92 g/kg total phosphorus (P) and the proportion of phytate-bound P of total P was higher in sorghum (82.7%) than barley (67.3%) and wheat (74.9%). Fifteen sorghum varieties contained an average of 2.41 g/kg phytate-P or 8.55 g/kg IP_6 phytate. The primary purpose for including phytase in broiler diets is to liberate phytate-bound P [20]. This "phosphoric effect" of phytase in sorghum-based diet is in evidence as Truong, *et al.* [21] reported that phytase significantly increased apparent disappearance rates of phytate in the proximal jejunum, distal jejunum, proximal ileum and distal ileum in boilers. Also, Liu, *et al.* [7] showed that phytase significantly increased toe ash from 11.68% *versus* 12.29% in broiler chickens offered sorghum-based diets. However, the "extra-phosphoric" responses to phytase sorghum-based diets often appear to be muted. The formation of protein-phytase complexes in the gut at pH less than the isoelectric point of protein well may be fundamental to the anti-nutritive properties of phytate and their attenuation generates the "extra-phosphoric" protein and energy responses to phytase [16]. However, from first principles, phytate may not readily complex kafirin because it contains a paucity of basic amino acids [11]. Moreover, this may be compounded by the fact that kafirin is contained in discrete protein bodies in sorghum endosperm with outer hydrophobic layers of cross-linked β- and γ-kafirin. Thus, these chemical and physical factors may limit interactions or complex formation between phytate and kafirin and the pursuant "extra-phosphoric" phytase responses.

Liu and Selle [22] proposed the provision of either rapidly digestible protein or slowly digestible starch will enhance bird performance and that the protein digestion rate may be more important than starch in this reciprocal or bilateral relationship. Compared to other grains, sorghum has relatively slower starch digestion rate [23]. Strategies that enhance protein digestion in sorghum may result in a better balance between the post-enteral availability of glucose and amino acid in chickens and improve growth performance. Therefore, the utilisation of exogenous enzymes and sulfite reducing agents is discussed in this review. Because of the inconsistent growth performance and nutrient utilisation in sorghum-based broiler diets, it is imperative to develop methods to determine sorghum grain quality. Starch pasting profiles determined by rapid visco-analyser (RVA), Promatest protein solubilities and Symes particle size index (PSI) textures have been assessed in a series of studies and their relevance is discussed in this review.

3. STARCH

Starch is the most abundant component in sorghum with amylopectin the dominant single polymer at around 75% of total starch, with amylose being around 25%. On average, sorghum contains 70% starch; therefore, it is considered to be the principal grain component used for energy. The utilisation of sorghum starch by broiler chickens as an energy source does appear to be sub-standard in relation to maize. As reviewed by Truong, *et al.* [24] the distal ileal digestibility of maize starch is superior to that of sorghum. These researchers reported that, across 11 studies, maize starch had an average ileal digestibility coefficient of 0.950 (range: 0.873 to 0.993).

Alternatively, across seven studies, sorghum starch had an average ileal starch digestibility coefficient of 0.883 (range: 0.846 to 0.921). On the basis of this comparison, sorghum starch digestibility was inferior to maize by 7.05%. However, this inferiority may not be due to the digestibility of starch *per se* but to extrinsic factors exerting negative impacts. Nevertheless, starch digestibility *per se* is influenced by amylose to amylopectin ratios, the two constituents of starch.

Sang, *et al.* [25] reported that *in vitro* starch digestion in waxy sorghum (0% amylose) was higher than heterowaxy sorghum (14% amylose) and normal sorghum (23.7% amylose). However, kafirin protein bodies were observed to be more numerous and closely associated with starch granules in the endosperm of non-waxy sorghums [26]. Moreover, sorghum grains are reported to contain about 6.5% resistant starch, and this may be increased following cooling of cooked sorghum due to starch recrystallisation [27].

The amylose:amylopectin ratio in grains may contribute to variations in starch digestion rates. However, inconsistent findings for the impact of amylose on broiler performance and energy utilisation have been reported. Svihus, *et al.* [28] reported that high amylose contents in cereal grains depress starch digestibility. Ertl and Dale [29] found significant differences in metabolisable energy in different hybrids of maize; however, there was no significant difference of metabolisable energy between normal and waxy maize from the same hybrid. Pirgozliev, *et al.* [30] investigated the performance of broilers offered wheat-based diets supplemented with three levels of dietary maize starch and three amylose:amylopectin ratios. Theses researchers found no significant differences in feed intake, weight gain and feed conversion ratio between dietary treatments. In contrast, Collins, *et al.* [31] reported that waxy-maize-based diets had significantly higher weight gain (2136 g/bird *versus* 1910 g/bird) and feed intake (3882 g/bird *versus* 3366 g/bird) from 18–42 days post-hatch and this could be due to the pellet durability index in diets based on waxy maize which were significantly higher than normal maize in grower (75.4% *versus* 54.6%) and finisher diets (77.9% *versus* 61.3%). Therefore, it remains unclear as to whether or not high amylose content in cereals depresses broiler performance. Based on the discussion above, amylose may retard starch digestion and synchronise starch and protein digestion but, alternatively, high amylose contents may negatively influence pellet quality and depress broiler performance by reducing feed intake. It seems that waxy sorghum may have more rapid protein digestion coupled with more complete starch digestion; therefore, waxy sorghum may perform better in diets for broiler chickens. Unlike maize and wheat, the application of waxy-sorghum in broiler diets is highly limited and it is difficult to clarify its possible impact on growth performance in poultry.

Further to our understanding of starch, amylose and amylopectin until recently has been somewhat limited to measuring the amount of these. A recent reported identified a genetic change in one of the starch synthesis enzymes (namely pullulanase) which resulted in major changes in amylopectin structural properties including digestibility when measure *in vitro* [32]. This suggests variation in starch structure as well as using specific starch types, such as waxy, may contribute to improved digestibility.

4. APICES OF THE SORGHUM BERMUDA TRIANGLE

4.1. Kafirin

Kafirin is the dominant fraction of sorghum protein and Virupaksha and Sastry [33] reported that kafirin constituted 54.1%, glutelin 33.4%, globulin 7.0% and albumin 5.6% of endosperm protein in sorghum. Kafirin is located in discrete protein bodies which are embedded in the glutelin protein matrix of sorghum endosperm where both kafirin and glutelin are intimately associated with starch granules. Subsequently, Taylor, *et al.* [34] reported average concentrations of 54.0 g/kg kafirin and 27.1 g/kg glutelin in 41 grain sorghum samples with an average crude protein content of 111 g/kg. Thus kafirin and glutelin comprised 48.0% and 27.7% of sorghum protein, respectively. However, kafirin, as a proportion of protein, was positively correlated ($r = 0.469$; $p < 0.005$) with sorghum protein concentrations; whereas, the proportion of glutelin was negatively correlated ($r = -0.401$; $p < 0.01$). This may suggest that screening sorghum quality by its protein concentration may result in sorghum grains with high kafirin concentrations.

Selle, *et al.* [35] proposed kafirin concentrations in Australian sorghums have increased in recent decades probably as an unintended consequence of breeding for increased yield and the use of high input nitrogen fertilizers. While this proposal is based on circumstantial evidence, the likely implications are important. In a quest to enhance grain weathering resistance, Henzell [36] stated breeding programs commonly target sorghums with a relatively dense or corneous endosperm. As such, selecting sorghums with hard endosperms will result in higher kafirin contents. Importantly, the textures of Australian sorghums are relatively high by international standards. Our group has determined the Symes PSI texture [37] of 38 sorghum varieties; the mean PSI value was 9.84 ± 1.603 which is in the more rigid end of the "very hard" scale and only two varieties or 5.3% of samples were "softer" in this scale. These findings are in contrast to an extensive global survey recorded by de Alencar Figueiredo, *et al.* [38]. In this survey the mean PSI texture of 117 sorghum samples was 12.7 but 44% of samples had softer textures than "very hard" as opposed to 5.3% in the local sorghum varieties. Moreover, the likelihood is that "soft" sorghums are superior in performance and nutrient utilisation to "hard" sorghums as the basis for broiler diets [39].

It is instructive to compare the amino acid profiles of local sorghum varieties from related reports compiled by Ravindran, *et al.* [40] and subsequently by Bryden, *et al.* [5], where the same analytical methods were adopted. The mean protein content of sorghums in the two studies was very similar (98.7 g/kg *versus* 103.6 g/kg). However, sorghums in the latter study contained 21.3% less lysine (1.89 g/kg *versus* 2.40 g/kg; $p < 0.015$) and 25.7% less arginine (3.32 *versus* 4.47 g/kg; $p < 0.005$). Given that kafirin contains low levels of lysine and arginine, these reductions in concentrations of more than 20% are entirely consistent with the proposal that kafirin constitutes a greater proportion of total sorghum protein in the latter study.

This is a disadvantageous development because kafirin contains low levels of lysine but high levels of leucine. Moreover, kafirin is a poor source of digestible amino acids due to its inherent hydrophobicity and the structure of protein bodies. It is generally accepted disulphide cross-linking in the β- and γ-kafirin fractions, which are located in the periphery of the protein body, impedes the digestion of the central α-kafirin core and this is exacerbated by hydrothermal processes as they induce disulphide cross-linkages [41].

Clearly, kafirin compromises sorghum protein quality [2] but this may not be crucial as kafirin constitutes in the order of 15% of total protein in a sorghum-based broiler diet. However, if kafirin levels in local sorghums are indeed increasing then the real possibility that kafirin compromises starch/energy utilisation is pivotal. The mechanisms by which kafirin compromises starch utilisation have not been clarified but arguably their genesis stems from the close physical proximity of kafirin protein bodies, starch granules and glutelin protein matrices in sorghum endosperm. As discussed by Gidley, *et al.* [42] the quantity of starch in sorghum endosperm exceeds that of kafirin protein, which suggests that the magnitude of biophysical starch-protein interactions is capped. Intuitively, biochemical starch-protein interactions may hold more importance but despite the recognition of the importance [43] they have yet to be precisely defined.

The proposition that kafirin compromises starch/energy utilisation, as enunciated by Taylor [44], is generally accepted although reservations have been expressed by Gidley, *et al.* [42]. In sorghum endosperm the diameter of starch granules (~15 μm) is considerably larger than kafirin protein bodies (1–3 μm) and the endosperm contains considerably higher concentrations of starch than kafirin. Starch granules and kafirin protein bodies are both discrete structures in sorghum endosperm. The concept that hydrothermally induced disulphide cross-linkages in the cysteine-rich periphery of protein bodies impede the digestion of the central α-kafirin core is straightforward because α-kafirin would be "protected" by a layer of relatively insoluble protein. It is more difficult to conceptualise that these reactions are also a direct impediment to starch digestion, although they may physically hinder starch gelatinisation. Intuitively, it seems any interactions between starch and protein that might impede starch gelatinisation/digestion are more likely to occur between starch and the glutelin protein matrix because starch granules are embedded in the protein matrix. Another possible inherent impediment to starch digestion is the endosperm cell walls. Taylor [44] considered it was probable that these cell walls constitute something of a barrier to digestion. The cell walls are rich in insoluble glucuronoarabinoxylan type pentosans and are intimately bound to the glutelin protein matrix, quite possibly via ferulic acid, a phenolic compound that is abundant in sorghum.

De Mesa-Stonestreet, *et al.* [45] discussed that kafirins are located primarily in spherical protein bodies, which are embedded in a glutelin protein matrix, and the integrity of which is dependent on disulphide cross-linkages. This is allied to studies using three-dimensional fluorescence microscopy completed by Ezeogu, *et al.* [46] and their conclusion was that low starch digestibility of cooked sorghum flour was the result of "the more disulfide-bonded protein matrix". Another possibility, as considered by Truong, *et al.* [24], are interactions involving disulphide linkages between β- and γ-kafirin on the periphery of protein bodies and starch-granule associated proteins on the surface of starch granules may have the capacity to impede starch digestion. Collectively, it seems possible that disulphide cross-linkages involving kafirin protein bodies, glutelin protein matrices and starch-granule associated proteins in sorghum endosperm are a set of connections forming a web-like protein network that impedes starch utilisation.

Following the quantification of kafirin concentrations in sorghum, a series of feeding studies completed by Poultry Research Foundation indicated that kafirin has negative impact on energy utilisation in broiler chickens (unpublished data). The methodology involved to quantify kafirin has been fully documented by Truong, *et al.* [47]. In these studies, kafirin significantly depressed ME:GE ratios ($p < 0.04$) and tended to decrease AMEn ($p = 0.074$) in broiler chickens from meta-analyses with non-significant ($p > 0.10$) experimental leverages (Table 1). Thus our tentative conclusion is kafirin content and composition compromises starch/energy utilisation in sorghum-based diets and the real possibility that the chicken-meat industry is being confronted with "high-kafirin" sorghums is a predicament that needs to be considered by sorghum breeding programs. Put simply, the availability of relatively low protein sorghums with lesser kafirin proportions should be substantially beneficial to growth performance and energy utilisation in broiler chickens.

Table 1. The relationship between dietary concentrations of kafirin and energy utilisation (ME:GE ratios and AMEn) in broiler chickens offered sorghum-based diets.

Response	Observations (n =)	Experiment Leverage (p =)	Dietary Kafirin Leverage (p =)	Whole Model
ME:GE	10	0.109	0.034	$R^2 = 0.71$, $p = 0.045$
AMEn	13	0.485	0.013	$R^2 = 0.70$, $p = 0.074$

ME: metabolisable energy; GE: General energy; AMEn: Nitrogen corrected apparent metabolisable energy.

There has been an increasing focus on the inclusion of exogenous proteases in pig and poultry diets. More specifically, Taylor [44] suggested that proteases should be evaluated in sorghum-based broiler diets with the caveat that kafirin and glutelin are not readily hydrolysed by proteolytic enzymes. In this context it is relevant that most or all exogenous proteases are hydrolases and inherently lack the capacity to cleave disulphide cross-linkages which requires reductase activity.

Selle, *et al.* [48] reported that a *Bacillus lichenformis*-derived protease improved N digestibility in conventional, sorghum-based broiler diets that were steam-pelleted at an 80 °C conditioning temperature. At the standard inclusion rate (300 units/gram), this protease significantly increased N digestibility coefficients by 16.5% in the distal jejunum (0.627 *versus* 0.538), by 5.91% in the proximal ileum (0.770 *versus* 0.727) and by 7.09% in the distal ileum (0.770 *versus* 0.719). Subsequently, Liu, *et al.* [49] reported that this protease significantly improved the digestibility of arginine (4.63%), histidine (8.23%), isoleucine (9.87%), leucine (12.7%), phenylalanine (10.5%), threonine (7.36%), valine (10.4%), alanine (12.8%), aspartic acid (9.60%), glutamic acid (9.38%), glycine (9.73%), proline (14.6%), serine (11.5%), and tyrosine (12.1%). Only the responses in lysine and methionine digestibility were not significant and this was probably because the diets contained quite high levels of these two amino acids in "free" or synthetic forms. In addition, protease supplementation significantly ($p < 0.05$ - < 0.005) increased the rates of digestion of 12 of the 16 amino acids assessed. Moreover, as proposed by Liu and Selle [22], it is probable digestion rates of amino acids were more indicative of broiler performance than static amino acid digestibility coefficients.

Instructively, this protease also significantly increased the digestibility of starch in the distal jejunum by 13.6% (0.770 *versus* 0.678; $p < 0.005$) and by 4.80% (0.851 *versus* 0.812; $p < 0.05$) in the proximal ileum and numerically enhanced starch digestibility in the proximal jejunum (15.8%) and the distal ileum (3.15%). Based on the presumption that this enzyme preparation did not contain any starch-degrading side-activities, these findings are consistent with the proposition that kafirin and/or glutelin are interfering with starch digestion when sorghum-based diets are offered to poultry.

4.2. "Non-tannin" Polyphenolic Compounds and Phenolic Acids

Phenolic compounds are a diverse group of phytochemicals ranging from highly-polymerised inert lignins to simple C_7-C_9 phenolic acids [50] and their concentrations in sorghum (170–10,260 mg/100 g) are substantially more than maize (30.9 mg/100 g) and wheat (22–40 mg/100 g) [51]. The polyphenolic compound condensed tannin is an anti-nutritive factor that depresses broiler performance [52] and has been even described as a toxic factor [53]. Sorghum cultivars may be divided into three categories depending on their genotypes and condensed tannin contents [54]. Type I sorghums do not have a pigmented testa and are "tannin-free", Type II sorghums have a pigmented testa layer that contains condensed tannin and Type III ("bird-proof") sorghums contain condensed tannin in both the testa and the peripcarp. Dykes and Rooney [55] reported condensed tannin concentrations of 0.28, 4.48 and 11.95 g/kg in Type I, Type II and Type III sorghums, respectively. As discussed, it is most likely Australian sorghums no longer contain condensed tannin; however, there is the very real possibility non-tannin phenolic compounds negatively influence digestion of starch and protein to compromise growth performance in broiler chickens.

Taylor [44] concluded grain sorghum cultivars contain higher levels of phenolic compounds than other cereals and that red (non-tannin) sorghums are highly pigmented with polyphenols, anthocyanins and anthocyanidins and, importantly, these phenols bind strongly to starch. Also ferulic acid has been shown to influence starch pasting profiles which implies phenolic acids have the capacity to interact with starch [17]. Thompson and Yoon [56] found tannic acid depressed wheat starch digestion *in vitro* and Thompson, *et al.* [57] reported intakes of phenolic compounds by humans were negatively correlated with blood glycaemic indices. Phenolic compounds and phytate are both located in the periphery of sorghum grain (pericarp and aleurone layer) rather than in the central endosperm where starch granules are located. Thus in native grain sorghum it seems unlikely any inter-reactions between phenolics and phytate with starch would take place; however, they may be initiated by processing (e.g., hammer-milling, steam-pelleting) and take place in the avian gut, perhaps particularly in the gizzard.

Phenolic compounds appear to be susceptible to processing and extrusion has been shown to reduce the molecular weight of phenolic polymers in sorghum [58]. More recently, Dlamini, *et al.* [59] reported that extrusion reduced average total phenol contents by 51% (14.92 *versus* 7.37 mg catechin equivalents/g) in the whole grain of five sorghum varieties. The impact of decortication or dehulling was more profound with a reduction of 68% (14.92 *versus* 4.80 mg catechin equivalents/g) due to the location of phenolics in the outer layers of the grain. The effects of extrusion suggest steam-pelleting sorghum-based diets may reduce the size of phenolic polymers, which may influence any subsequent interactions between phenolics and starch in the avian gut.

Zhu [60] reported non-covalent interactions involving starch and phenolic compounds influence physicochemical and nutritional properties of feedstuffs. Earlier, Tomasik and Schilling [61] and Barros, *et al.* [62] opined that phenolics readily form starch complexes and are probably more likely to form starch-phenolic complexes with amylose than amylopectin. Kandil, *et al.* [63] reported that phenolic acids play an important role in the resistance of starch to hydrolysis in a study involving barley, maize, triticale and wheat but not sorghum. Welsch, *et al.* [64] found phenolic compounds were capable of inhibiting Na^+-K^+-ATPase or the "sodium pump" which suggested intestinal uptakes of nutrients including glucose via Na^+-dependent transporter systems could be compromised.

A wide range of polyphenolic compounds and phenolic acids have been quantified in a number of diverse sorghum samples by this institute. As a consequence, our contention is that the anti-nutritive properties of phenolic compounds are not the sole domain of CT and to some extent are shared by the balance of polyphenols and phenolic acids found in grain sorghum. Instructively, Elkin, *et al.* [65] concluded that condensed tannin is only partially responsible for variations in the nutrient quality. Also, phenolic compounds and phytate appear to share several anti-nutritive properties [11]. Provisionally, there was a negative linear relationship ($r = -0.569$; $p = 0.042$) between dietary levels of total phenolic compounds and ME:GE ratios in broilers offered diets based on nine "tannin-free" sorghum varieties across five of feeding studies. This preliminary outcome supports the contention and prompts identification of which polyphenols or phenolic acids are responsible. Initial indications are that the flavan-4ols (polyphenolic compounds) and soluble conjugated and insoluble bound ferulic acid (a phenolic acid) might be involved.

The colour of red sorghums is attributable to polyphenolic pigments (anthocyanins) and axiomatically red sorghums contain more polyphenols than white sorghum varieties. Indeed, Beta and Corke [66] reported that four red sorghums had a greater concentration of total polyphenols than six white sorghums by more than a three-fold factor. In Australia probably all white sorghums are of the Liberty variety. Interestingly, we have found that Liberty contains lower concentrations of phenolic acids than a number of red sorghum varieties albeit in a limited number of samples. Three categories of phenolic acid concentrations were quantified, free and conjugated phenolic acids which are soluble and the insoluble, bound phenolic acids. In two projects, nine red sorghums contained an average of 500 μg/g total phenolic acids and 392 μg/g total ferulic acid; whereas, in contrast, two white Liberty sorghums contained 328 μg/g phenolic acids and 235 μg/g ferulic acid. Thus red sorghums contained 28% more phenolic acids and 40% more ferulic acid than white sorghums in this comparison. Ferulic acid is the dominant phenolic acid in sorghum, representing 78% of total phenolic acids across the 11 varieties; however, ferulic acid is certainly not unique to sorghum amongst grains as it is present in maize and wheat.

Dykes, *et al.* [67] reported a positive correlation between total phenols and flavan-4ols ($r = 0.94$; $p < 0.001$) in non-tannin sorghums. A recent study at this institute involved offering sorghum-based diets (580 g/kg) based on six different varieties to broiler chickens. In this study, flavan-4ols in sorghum *per se* was negatively correlated with ME:GE ratios ($r = -0.919$; $p = 0.010$) which supports the contention that "non-tannin" polyphenolic compounds in sorghum have the capacity to compromise energy utilisation in poultry.

4.3. Phytate

The inclusion of phytase in broiler diets is now a very common practice and the reciprocal impacts of dietary phytate and exogenous phytase on broiler performance have been reviewed extensively [16,20,68,69]. Briefly, to avoid repetition, sorghum invariably contains phytate at relative and absolute concentrations that are often somewhat higher than other cereal grains [19,70] and it is generally accepted that phytate negatively influences protein and energy utilisation in poultry [20,68,71]. The polyanionic IP_6 phytate molecule may form binary or ternary complexes with proteins depending on the isoelectric point of protein and environmental pH [70]. Phytic acid may affect starch digestion by directly binding starch through phosphate linkages or indirectly via starch granule-associated proteins [72]. Also, phytate is a potent chelator of divalent cations including zinc and calcium. However, the fundamental role of phytase is to liberate phytate-bound P and this enhanced P utilisation is of astonishing environmental importance. Nevertheless, the suggestion is that the response of starch and protein digestion to phytase [73] in sorghum-based broiler diets appears to be less robust in comparison to maize- and wheat-based diets [74]. The likelihood is that the "extra-phosphoric" effects of phytase were largely a consequence of phytase attenuating protein-phytate complex formation.

As proposed by Cosgrove [14], electrostatic attractions between positively-charged arginine, histidine, and lysine residues with polyanionic phytate molecules are pivotal to the initiation of binary protein-phytate complex formation. This takes place at pH levels less than the isoelectric point of proteins and the isoelectric point (iP) of kafirin is 5.9 [75] which is considerably higher than pH levels usually found in the crop, proventriculus and gizzard. However, from first principles, it follows that phytate may not readily bind kafirin due to its paucity of basic amino acids. Also, the structure of the discrete, spherical kafirin protein body with its hydrophobic periphery may limit interactions with phytate. Both factors may in turn limit "extra-phosphoric" responses to exogenous phytases in sorghum-based diets simply because kafirin is not readily bound by phytate in binary complexes. It is also follows that if phenolic compounds are limiting performance in sorghum-based broiler diets, then phytase and other standard exogenous enzymes do not have the capacity to counteract this impact.

Both phytate and the various phenolic compounds are located in the outer layers of grain sorghum including the aleurone. Perhaps because of this we are consistently finding that concentrations of phytate and phenolics in sorghums are positively correlated. In this context it is relevant that Selle, *et al.* [11] gave consideration to the analogous anti-nutritive properties of phytate and phenolic compounds. As one related example, Kreydiyyeh, *et al.* [76] found that tea extracts reduced intestinal mucosal Na^+-K^+-ATPase or sodium pump activity and uptakes of glucose and sodium in rats. Subsequently, Kreydiyyeh [77] reported that tannic acid, a constituent of tea extracts, inhibited intestinal sodium pump activity *in vitro* and phenylalanine uptakes in rats. This was attributed to direct inhibition of Na^+-K^+-ATPase activity by tannic acid and dissipation of the Na gradient required for intestinal uptakes of phenylalanine. Alternatively, Liu, *et al.* [78] demonstrated the addition of 1000 FTU/kg phytase to corn-soy diets significantly increased Na^+-K^+-ATPase activity in the duodenum and jejunum of broilers by approximately 20%, and increased glucose concentrations in enterocytes. The implication is phytate depresses sodium pump activity and intestinal uptakes of glucose. Overall, the suggestion is that both phytate and phenolic compounds can impede sodium pump activity along the small intestine thereby compromising intestinal uptakes of glucose and amino acids via Na^+-dependent transport systems.

5. WHITE *VERSUS* RED SORGHUM

White sorghums are considered to be superior to red sorghums by both pig and poultry nutritionists, which was reflected in our survey. One possibility is that higher concentrations of both polyphenolic compounds and phenolic acids in red sorghums are contributing to their apparent inferiority in comparison to white sorghums. While phenolic acids are not related to grain colour their relatively low concentrations in two samples of Liberty may hold relevance. Interestingly, flavan-4ols *per se* are not responsible for "redness" in sorghum but they are precursors of red polyphenolic pigments. Therefore, it follows that there would be less flavan-4ols in white than red sorghums and our analyses have confirmed this to be the case.

The majority of grain sorghums harvested in Australia possess red pericarps while the minority (Liberty) have white pericarps. This reflects a perception that red sorghums varieties are more viable agronomically than white sorghum and another reason is that white sorghums have a grey appearance following exposure to moisture. Perez-Maldonado and Rodrigues [4] offered broiler chickens diets based on 31 different sorghums including three white sorghums (Liberty) harvested in 2004 and 2005. As shown in Table 2, the mean AME of white sorghum was 0.45 MJ higher (15.87 *versus* 15.42 MJ/kg dry matter (DM)) than red sorghum and the AMEn was 0.39 MJ higher (15.33 *versus* 14.94 MJ/kg DM) across both 2004 and 2005 harvest years. A larger number of white sorghum samples would have been preferable; nevertheless, the differences in AME ($p < 0.125$) and AMEn ($p < 0.05$) either approached or were statistically valid. Thus the implication is that the energy/starch component of white sorghum is better utilised than red sorghums.

Selle, *et al.* [48] compared a white Liberty (98 g/kg protein) with a red Buster (121 g/kg protein) sorghum as the dietary basis for broiler chickens. Both sorghums were hammer-milled through a 3.2 mm screen prior to being incorporated into diets that were steam-pelleted at a conditioning temperature of 95 °C. Liberty supported better weight gains by 4.69% (1540 *versus* 1471 g/bird; $p < 0.03$) and feed conversion ratios by 2.88% (1.485 *versus* 1.529; $p < 0.05$) from 7 to 28 days post-hatch. These improvements in growth performance were associated with significantly higher starch digestibility coefficients in four small intestinal segments, higher starch digestion rates, a higher predicted glycaemic response and a greater quantity of potentially digestible starch. Also, from the same

Table 2. Key characteristics of the sorghum varieties assessed by Perez-Maldonado and Rodrigues [4].

Parameter	Red Sorghums (n = 28)	White Sorghums (n = 3)
Starch (g/kg)	637	656
Protein (g/kg)	117	105
Kafirin index [1]	5.48	4.27
Phytate (g/kg)	7.37	6.27
"Phenolic compounds"	6.99	1.53
Apparent ileal starch digestibility	0.881	0.908
Apparent metabolisable energy (AME)	15.42 MJ/kg	15.87 MJ/kg
Range	*14.3–16.2*	*15.6–16.1*
N-corrected AME (AMEn)	14.94 MJ/kg	15.33 MJ/kg
Range	*14.0–15.2*	*15.2–15.5*

[1] Kafirin Index = $leucine_{(g/kg)}$ − (arginine + histidine + lysine) (g/kg).

comparison, Liu, *et al.* [18] reported that white sorghum-based diet had significantly higher ($p < 0.05$) average amino acid digestibility coefficients than red sorghum-based diet in the proximal ileum (0.797 *versus* 0.769) and distal ileum (0.833 *versus* 0.806); white sorghum-based diet also had significantly higher apparent digestibility coefficients of leucine, alanine, glutamic acid and proline in the distal jejunum and 14.1% higher average amino acid digestion rate than red sorghum-based diets (2.47 *versus* 2.17 × 10^{-2} min^{-1}). The more rapid protein digestion rate in white sorghum-based diets may have contributed to the better feed conversion efficiency via synchronizing starch and protein digestion and the availability of glucose and amino acids.

In this comparison, white sorghum contained less phytate (6.74 *versus* 9.22 g/kg) and less total phenolic compounds (2.02 *versus* 4.13 GAE/g) than the red sorghum (Buster). However, as discussed, non-tannin phenolic compounds may negatively influence protein and starch digestion and compromise growth performance in broiler chickens. Phenolic acids are found in the outer layers of the kernel (pericarp, testa, and aleurone), and the bound phenolic acids are associated with the cell walls [55]. The white sorghum contained less free, conjugated and bound phenolic acids (Figure 1) and less flavan-4ols, which may play an important role in mould resistance [55]. In broiler diets based on sorghum- or sorghum-wheat blends, sorghum is an important energy source and phenolic compounds may have impeded starch digestion and glucose absorption in this comparison.

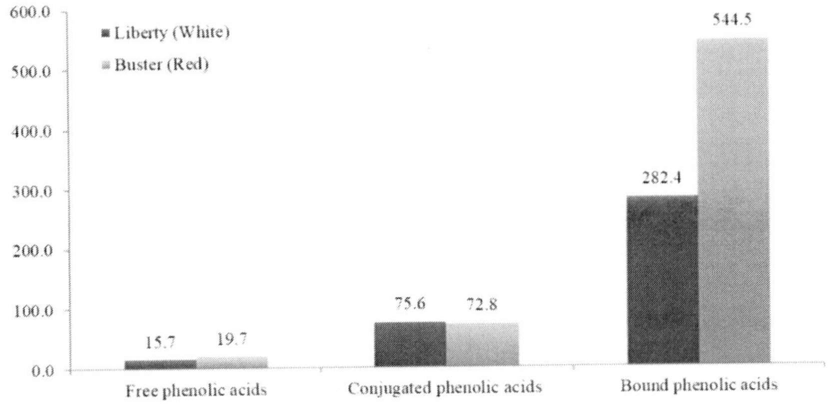

Figure 1. The concentrations of free, conjugated and bound phenolic acids in white and red sorghum.

6. STARCH AND PROTEIN DIGESTIVE DYNAMICS IN SORGHUM-BASED BROILER DIETS

Accumulating evidence suggests that the provision of some slowly digestible starch in broiler diets is beneficial for feed conversion efficiency [79,80,81]. One possibility is that retarded starch digestion rates provide a greater supply of glucose to the lower small intestine which spares amino acids from catabolism for energy to meet the gut mucosa's requirements thereby permitting the entry of amino acids into the systemic circulation to become available for protein deposition [22]. Based on an *in vitro* starch digestion evaluation in various cereal grains, Giuberti, *et al.* [23] reported that sorghum had the least rapidly digestible starch (119 g/kg), potential digestible starch (70.4 g/100 g dry starch) and the highest resistant starch (275 g/kg). The authors also showed that sorghum had a low starch digestion rate (0.018 min^{-1}) and the lowest predicted glycaemic index (Table 3). This is consistent with findings of Weurding, *et al.* [82] where starch digestion rates in the small intestine of broiler chickens differ across feed ingredients and sorghum had the lowest and slowest starch digestion rate (Table 4).

Table 3. Starch fractions and digestion rates of starch in various cereal grains [23].

Item	Rapidly Digested Starch (g/kg DM)	Slowly Digested Starch (g/kg DM)	Resistant Starch (g/kg DM)	Potential Digestible Starch (g/kg DM)	Rate of Starch Digestion (min^{-1})
Maize	147	367	191	95.0	0.017
Barley	151	264	143	88.6	0.024
Wheat	181	379	53	92.8	0.035
Sorghum	119	342	275	70.4	0.018

DM: dry matter.

Table 4. Starch digestion characteristics in the small intestine of broilers fed diets containing different sources of starch [82].

Starch Source	Jejunal Digestibility	Ileal Digestibility	Potential Digestible Starch (%)	Starch Digestion Rate (h^{-1})
Wheat	0.882	0.944	93.9	2.51
Corn	0.888	0.969	96.8	2.55
Sorghum	0.837	0.953	95.4	1.81
Barley	0.898	0.981	98.5	2.51
Peas	0.574	0.804	85.3	1.03
Potato starch	0.198	0.329	38.4	0.55

Liu, *et al.* [83] found starch and nitrogen digestive dynamics were more indicative of feed conversion efficiency and weight gain than their static ileal digestibility coefficients. These findings suggest that the rate of amino acids absorption in the small intestine in relation to glucose absorption may have a tangible bearing on protein deposition and feed conversion efficiency. Similar investigations have been conducted in pigs; van den Borne, *et al.* [84] defined nutrient asynchrony as a partial temporal separation of amino acid and glucose availability and assessed the effects of synchronicity of amino acids and glucose on protein and energy metabolism. They found that the efficiency of protein utilisation was significantly enhanced in so-called synchronous diets (56.7%) in comparison to asynchronous diets (47.1%). It was concluded that the asynchronous availability of glucose and amino acids increased amino acid oxidation and compromised protein utilisation. Because sorghum has relatively slower starch digestion than maize and wheat, it may have the potential to provide a diet with better balanced digestion of starch and protein. However, accelerating protein digestion and amino acid absorption in sorghum-based diets is a challenge which may be met by inclusions of sulfite reducing agents and exogenous proteases.

7. INCLUSIONS OF SULFITE REDUCING AGENTS

The negative impact of "moist-heat" or "wet-cooking" on sorghum protein digestibility is recognised. Wet cooking has been shown to reduce *in vitro* protein digestibility from 78%–100% to 45%–55% [85]. This may be associated with the formation of disulphide cross-linkages in kafirin coupled with its inherent hydrophobicity and secondary protein structural changes [11]. Kafirin is less soluble than zein in maize and gluten in wheat [13]. The observed lower *in vitro* protein digestibility of wet-cooked sorghum in comparison to maize and wheat may be related to the structure of the kafirin protein body. The formation of disulphide cross-linkages in the periphery (β- and γ-kafirin) may generate a relatively insoluble protein barrier which "protects" the central core (α-kafirin) from digestion by limiting the access of enzymes. The inclusion of reducing agents (dithiothreitol and sodium bisulphite) with sorghum during wet-cooking has been shown to significantly improve protein digestion [86]. Interestingly, disulphide-mediated polymerisation of zein formed in cooked maize did not result in reduced protein digestibility [87]; this, coupled with protein body structure, may be attributed to the formation of more disulphide-bounded protein oligomers in sorghum than in maize [88].

Sulphite reducing agents, including sodium sulphite (Na_2SO_3), sodium bisulphite ($NaHSO_3$) and sodium metabisulphite ($Na_2S_2O_5$), have the capacity to cleave disulphide cross-linkages and have been shown to improve *in vitro* starch and protein digestibility in sorghum [86,89,90]. Selle, *et al.* [91] reported that sodium metabisulphite (SMBS) significantly decreased disulphide cross-linkages and increased free sulphydryl groups *in vitro.* Moreover, *in vivo*, SMBS significantly enhanced energy utilisation as it increased AME by up to 0.56 MJ in broilers offered sorghum-based diets containing a range of SMBS inclusion levels. Subsequently, Selle, *et al.* [92] found that graded SMBS inclusions linearly decreased concentrations of disulphide bonds ($r = -0.775$; $p < 0.001$), linearly increased free sulphydryl groups ($r = 0.890$; $p < 0.001$) and protein solubility of the diets ($r = 0.943$; $p < 0.001$). In the same study, the inclusion of 1.50 g/kg SMBS in conventional sorghum-based broiler diets significantly improved AME by 0.36 MJ (13.81 *versus* 13.45 MJ/kg; $p = 0.005$), AMEn by 0.45 MJ (12.30 *versus* 11.85 MJ/kg; $p = 0.001$) and feed conversion ratio (FCR) by 2.47% (1.422 *versus* 1.458; $p = 0.025$).

From the preliminary study [91], it appears broilers chickens were able to tolerate dietary SMBS inclusions approaching 5.00 g/kg before feed intakes are depressed; however, inclusion levels under 5.00 g/kg have the capacity to enhance nutrient utilisation of sorghum-based diets. With the exception of one sorghum (the red sorghum evaluated in the Truong, *et al.* [93] study), responses in energy utilisation (AME, ME:GE ratio and AMEn) to SMBS inclusions up to 5.25 g/kg were all positive across five studies as shown in Table 5. In studies 2 to 5 inclusive (complete sorghum-based diets as opposed to sorghum-cased diets used in study 1), the average SMBS inclusion of 2.84 g/kg increased AME by 0.27 MJ, AMEn by 0.32 MJ, and ME:GE ratios by 1.90%. SMBS also improved FCR in broilers offered complete sorghum-based diets by an average of 1.37% or a median response of 1.47%. It would be advantageous to ascertain if these improvements stem mainly from reductions of disulphide cross-linkages and increased solubility of proteins, perhaps especially kafirin, or from starch depolymerisation [94,95]. Given the importance placed on starch-protein interactions in sorghum specifically [43], and in general [24], this is only complicated by the likelihood that the two impacts are interdependent. There is the real possibility SMBS modifies both starch and protein digestive dynamics to benefit energy utilisation in broiler chickens [22]. It seems likely that SMBS responses with certain grain sorghum varieties will be more pronounced than others. Inclusions of SMBS displays promise in commercial practice as it has positive effects on energy utilisation and efficiency of feed conversion and its dietary inclusion cost is largely negated by its replacement of sodium bicarbonate to maintain dietary sodium at intended levels.

Table 5. Responses to sodium metabisulphite (SMBS) inclusions of up to 5.25 g/kg on energy utilisation parameters and feed conversion ratio (FCR) in five studies conducted by Poultry Research Foundation involving a total of eight sorghum varieties [1].

SMBS Inclusion	Study/ Sorghum	AME Response (MJ)	ME:GE Response (%)	AMEn Response (MJ)	FCR Response (%)
1.25	1/#12 [2]	0.39	1.82	0.29	17.4
2.50	1/#12 [2]	0.17	0.54	0.08	5.67
5.00	1/#12 [2]	0.53	2.58	0.51	20.7
1.50	2/#13	0.36	2.22	0.45	2.47
2.25	2/#13	0.35	2.47	0.38	1.65
3.00	2/#13	0.29	1.73	0.31	2.26
3.75	2/#13	0.36	1.73	0.30	3.02
4.50	2/#13	0.40	2.47	0.51	2.33
5.25	2/#13	0.34	2.35	0.43	2.47

SMBS Inclusion	Study/ Sorghum	AME Response (MJ)	ME:GE Response (%)	AMEn Response (MJ)	FCR Response (%)
1.75	3/FW	−0.42	−1.34	−0.37	3.13
1.75	4/MP	0.40	2.61	0.45	0.39
3.50	4/MP	0.25	1.65	0.38	−1.64
1.75	4/JM	0.41	3.27	0.47	−1.82
3.50	4/JM	0.31	2.32	0.48	−1.63
1.75	5/HP	0.22	1.45	0.22	0.53
3.50	5/HP	0.34	2.46	0.35	8.57
1.75	5/Tiger	0.09	1.25	0.06	−0.60
3.50	5/Tiger	0.29	1.94	0.29	1.47
1.75	5/HFQ	0.06	0.42	0.17	0.13
3.50	5/HFQ	0.50	3.38	0.59	0.54
(studies 2–5)	g/kg				
Mean	2.84	0.27 MJ	1.90%	0.32 MJ	1.37%
Median	3.00	0.34 MJ	2.22%	0.38 MJ	1.47%

[1] Results were aggregated from five studies; [2] Sorghum-casein diets; ME: metabolisable energy; GE: General energy; AMEn: Nitrogen corrected apparent metabolisable energy.

8. RESPONSES OF EXOGENOUS FEED ENZYMES

The inclusion of exogenous enzymes in sorghum-based diets have been reviewed by Selle, *et al.* [11] and Liu, *et al.* [3]. It is generally accepted that phytate negatively influences protein utilisation in poultry [20,68]. This is reflected in the Ravindran, *et al.* [96] study where *Aspergillus niger* phytase (1200 FTU/kg) increased mean apparent ileal digestibility (AID) coefficients of 14 amino acids by 6.46% (0.791 *versus* 0.743) in broilers offered "all-sorghum" diets. However, the low protein content of the sorghum (73.3 g/kg) used in this study probably indicates a relatively small kafirin proportion of total protein. There is the suggestion that the "extra-phosphoric" responses to phytase in sorghum are less pronounced than in other grains [74]. As proposed by Cosgrove [14], at pH levels less than the isoelectric point of protein, electrostatic attractions between positively-charged arginine, histidine, and lysine residues with polyanionic phytate molecules are crucial to binary protein-phytate complex formation. If so, it follows that phytate would not readily bind kafirin due to its paucity of basic amino acids. This may in turn limit "extra-phosphoric" responses to exogenous phytases in sorghum-based diets.

A study completed by the Poultry Research Foundation in 2013 evaluated the effects of an *Escherichia coli*-derived phytase (1000 FTU/kg) on starch digestibility coefficients in four small intestinal sites in broilers offered conventional sorghum-based diets steam-pelleted at 70 °C. Phytase significantly increased average starch digestibility coefficients by 7.27% (0.812 *versus* 0.757; $p < 0.04$) in the four small intestinal sites.

Subsequently, Liu, *et al.* [7] showed that there were numerical improvements in weight gain, feed conversion efficiency and N retention in sorghum-based diets following phytase supplementation (1000 FTU/kg). Truong, *et al.* [93] found that phytase did not significantly influence nutrient utilisation and apparent digestibility of starch and nitrogen along the small intestine in birds offered sorghum-based diets. Truong, *et al.* [21] reported that phytase (1000 FTU/kg) increased apparent disappearance rates of phytate in four small intestinal segments which indicates that phytase has the capacity to degrade phytate and generate "phosphoric responses" in sorghum-based diets. Thus phytase inclusions in sorghum-based diets have the potential to reduce levels of inorganic P source and reduce feed costs but the "extra-phosphoric" responses may be somewhat muted.

Investigations into the inclusion of other exogenous enzymes in sorghum-based diets are limited. Liu, *et al.* [49] reported that a *Bacillus lichenformis*-derived protease significantly increased the digestibility of 13 amino acids in distal ileum, which ranged from a 4.5% increase in histidine (0.820 *versus* 0.785, $p < 0.05$) to a 7.9% increase in proline (0.779 *versus* 0.722, $p < 0.01$); it also significantly increased digestion rates of 12 amino acids from 34.2% for histidine (2.57 *versus* 3.45×10^{-2} min^{-1}, $p < 0.05$) to 55.7% for proline (2.28 *versus* 3.55×10^{-2} min^{-1}, $p < 0.01$). Taylor [44] suggested that kafirin may not be readily degraded due to its poor solubility and disulphide linkages. The positive impact of this protease on amino acid digestibility in sorghum-based diets is encouraging; ideally, it should be confirmed in further studies.

Liu, *et al.* [97] examined an enzyme combination of xylanase, amylase and protease in maize-, wheat- and sorghum- based broiler diets and found that enzyme supplementation did not influence growth performance and nutrient utilization but significantly increased apparent digestibility coefficients of starch in the proximal ileum and apparent digestibility coefficients of nitrogen in the distal jejunum and proximal ileum. Although sorghum contains low concentrations of soluble non-starch polysaccharide (NSP) and is considered to be a "non-viscous" grain, it is often combined with wheat in broiler diets to enhance pellet quality and NSP degrading enzymes should have the potential to improve growth performance and nutrient utilisation in sorghum-wheat blended diets. However, few studies have investigated this issue and future studies are required to determine the effects of exogenous enzymes on sorghum-wheat-blend diets in broiler chickens.

9. INNOVATIVE ASSESSMENTS OF SORGHUM QUALITY

The capacity to assess the quality of sorghum as a feedstuff for broiler chickens is clearly a desirable objective. Therefore, in this section consideration is given to three approaches which, in concert, may enable nutritionists to rank the quality of a sorghum shipment and formulate diets accordingly to maximise broiler performance.

9.1. Promatest Protein Solubility

Odjo, *et al.* [98] has described the Promatest procedure to determine the salt-soluble protein content of maize. Briefly, the solubility of extracted protein is compared to the standard of bovine serum albumin via a colorimetric reaction with coomassie brilliant blue G250 (Merck, Germany). Importantly, Gehring, *et al.* [99] found that Promatest protein solubilities of 12 maize samples were positively correlated with N-corrected AME (r = 0.81) and ileal digestible energy (r = 0.48) to significant ($p < 0.01$) extents in broiler chickens. The determination of salt-soluble protein contents in sorghum cultivars in relation to broiler performance could be considered to be innovative and our preliminary investigations are encouraging. For example, Selle, *et al.* [100] found a significant correlation between protein solubilities of diets based on three sorghum varieties with red, white and yellow pericarps offered to broiler chickens as either unprocessed mash or as re-ground mash following steam-pelleting at a conditioning temperature of 90 °C with N retention, as shown in Figure 2. Overall steam-pelleting reduced protein solubility from 56.3% to 38.3% with a corresponding reduction in N retention from 60.6% to 55.0%. Moreover, Promatest protein solubilities were positively correlated with concentrations of free sulphydryl groups (r = 0.585, $p < 0.001$) and negatively correlated with disulfide cross-linkages (r = −0.518, $p < 0.001$).

The majority broiler diets are steam-pelleted and sorghum is known to be vulnerable to wet-cooking or hydrothermal processing [41]. Hamaker, *et al.* [89] reported that wet-cooking resulted a 30.3% reduction of *in vitro* pepsin digestibility in sorghum in comparison to 13.9% in barley, 10.0% in rice, 5.9% in wheat and 4.9% in maize. Duodu, *et al.* [13] suggested that this could be due to the factors including disulfide and non-disulfide cross-linking and alterations to the secondary structure of sorghum protein. Given this, it appears that Promatest protein solubility may hold special relevance for sorghum in poultry.

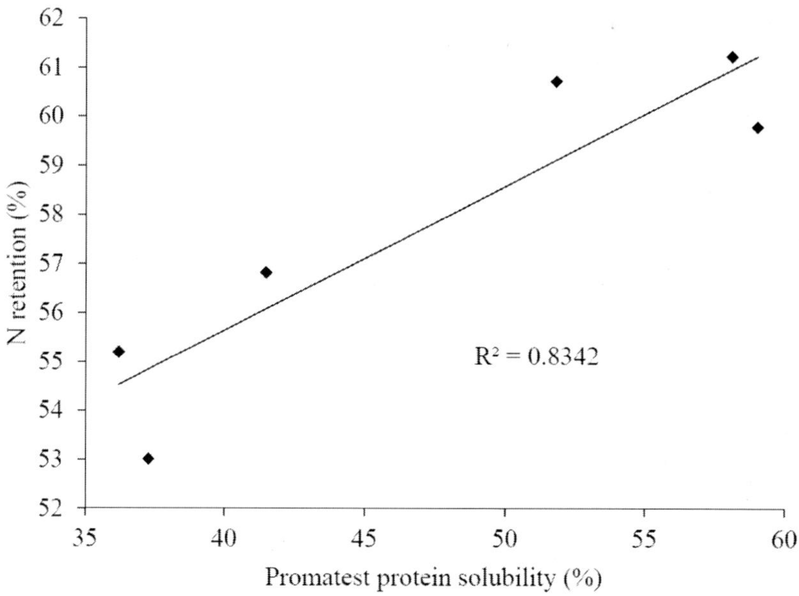

Figure 2. Linear relationship ($r = 0.881$ $p = 0.021$) between protein solubility and N retention in broiler chickens offered sorghum-based diets (adapted from Selle, *et al.* [100]).

Subsequently, in red sorghum-based diets, Selle, *et al.* [48] reported that Promatest protein solubility was positively correlated with concentrations of free sulphydryl groups ($r = 0.720, p < 0.01$). In an investigation of including SMBS into broiler diets, Selle, *et al.* [92] found that graded inclusions of SMBS linearly decreased disulphide bonds ($r = -0.775, p < 0.001$) and increased concentrations of free sulphydryl groups ($r = 0.890, p < 0.001$) and protein solubility ($r = 0.943, p < 0.001$). Also, protein solubility was correlated with the concentrations of both disulphide bonds and free sulphydryl groups in this study. Therefore, it should be possible to apply Promatest protein solubility as an indicator of protein quality in sorghum *per se* and sorghum-based broiler diets. The correlation between protein solubility and N retention is promising; nevertheless, further research is required to determine the relevance of Promatest protein solubility with *in vivo* growth performance and nutrient utilisation in broiler chickens.

9.2. RVA Starch Pasting Profiles

Starch is the major energy source in diets for broiler chickens; therefore, starch pasting profiles of the constituent grain and/or complete diets could be relevant to broiler performance. RVA starch pasting profiles are used extensively for human foods but they have not been widely applied to animal nutrition and their relevance is considered to be somewhat questionable. However, in an as yet unpublished study, sorghum-based diets were offered to broiler chickens and RVA viscosities of the constituent sorghum varieties were found to be correlated with various parameters of growth performance, nutrient utilisation and protein digestibility as documented in Table 6. Here, sorghum-SBM based diets were offered to broiler chickens from 7 to 28 days post-hatch and digesta samples were collected at the end of the study to determine digestibility coefficients in the proximal jejunum, distal jejunum, proximal ileum and distal ileum. Importantly, holding and final RVA viscosities were negatively correlated to feed conversion efficiency to significant extents; thus the implication is as final RVA viscosities of sorghum grains increase FCR is enhanced. This is reflected in the positive correlations between RVA viscosities and parameters of nutrient utilisation and protein digestibility where five observations were significant ($p < 0.05$) and a further seven approached significance ($p < 0.10$).

Table 6. Correlations between rapid visco-analyser (RVA) starch pasting profiles and parameters of growth performance, nutrient utilisation and protein digestibility coefficients in broiler chickens (PJ, proximal jejunum; DJ, distal jejunum; DI, distal ileum).

RVA Viscosity	Category	Parameter	Correlation Coefficient	Significance
Holding	Growth performance	FCR	$r = -0.821$	$p = 0.045$
Final			$r = -0.832$	$p - 0.040$
Peak	Nutrient	AME	$r = 0.789$	$p = 0.062$
Breakdown	utilisation		$r = 0.774$	$p = 0.071$
Peak		ME:GE ratio	$r = 0.800$	$p = 0.056$
Breakdown			$r = 0.767$	$p = 0.075$
Final		N retention	$r = 0.880$	$p = 0.021$
Setback			$r = 0.774$	$p = 0.071$
Peak		AMEn	$r = 0.835$	$p = 0.039$
Breakdown			$r = 0.834$	$p = 0.039$
Final	Protein (N)	PJ	$r = 0.803$	$p = 0.054$
Setback	digestibility		$r = 0.896$	$p = 0.016$
Setback		DJ	$r = 0.798$	$p = 0.057$
Setback		DI	$r = 0.862$	$p = 0.027$

ME: metabolisable energy; GE: General energy; AME: apparent metabolisable energy; FCR: feed conversion ratio.

Proteins with disulfide bonds may restrict starch granule swelling during gelatinization, and Zhang and Hamaker [101] reported that dithiothreitol, a reducing agent, treatment of sorghum flour influenced RVA pasting profiles. Moreover, phenolic compound was also reported to influence starch pasting properties [17]. Instructively, Liu, *et al.* [102] showed that SMBS significantly ($p < 0.001$) reduced all five starch viscosity parameters and graded SMBS supplementation linearly reduced final starch viscosities ($r = -0.986$; $p < 0.001$). The profound impact of SMBS on RVA starch pasting profiles may be attributed directly to the depolymerisation of starch polysaccharides and indirectly to the reduction of disulphide cross-linkages in kafirin, glutelin and possibly starch granule-associated proteins. Therefore, it does appear that RVA starch pasting profiles will provide innovative and valid assessments of sorghum quality.

9.3. Grain Texture: Symes Particle Size Index (PSI)

Symes [37] established the particle size index methodology to determine grain texture or "hardness" in wheat. Assessments of grain texture are not straightforward in practice and we have adopted the Symes PSI method to determine the hardness of sorghum [48,100,103]. Poultry Research Foundation has determined the Symes texture of 38 sorghum varieties in Australia as shown in Figure 3. The mean PSI value was 9.84 ± 1.603 (co-efficient of variation: 16.3%) which is at the harder end of the "very hard" spectrum of 8 to 12 inclusive and only three varieties fell outside this category.

On the basis of this method Australian sorghums are noticeably "harder" with lower PSI values than those from an extensive global survey recorded by de Alencar Figueiredo, *et al.* [38]. In this survey the mean texture of 117 sorghum samples was in the order of 12.7% but 44% of samples had softer textures than "very hard" as opposed to 5.3% (2 ex 38) in Australian sorghums. However, the likelihood is that "soft" sorghums are superior in performance and nutrient utilisation to "hard" sorghums as the basis for broiler diets. For example, Cabrera, *et al.* [39] reported that broilers offered diets based on a soft sorghum with mean particle sizes of 1000, 800 and 600 μm outperformed those offered hard sorghum by 7.14% in weight gain (945 *versus* 882 g/bird) and 4.38% in FCR (1.639 *versus* 1.714) in a 21-day feeding study. Kafirin is associated with hard textures of grain sorghums and in 13 sorghum varieties with quantified kafirin concentrations evaluated by Poultry Research Foundation, there was a relationship ($r =$

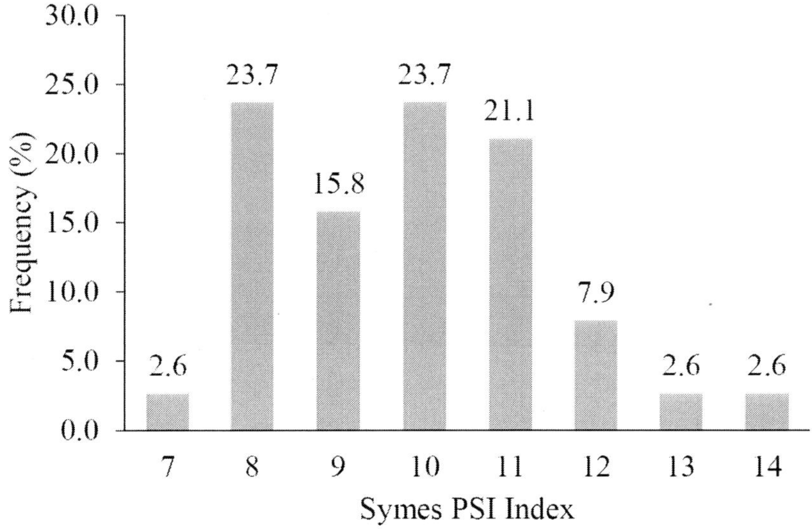

Figure 3. Frequency of Symes particle size index (PSI) texture across 38 Australian sorghums.

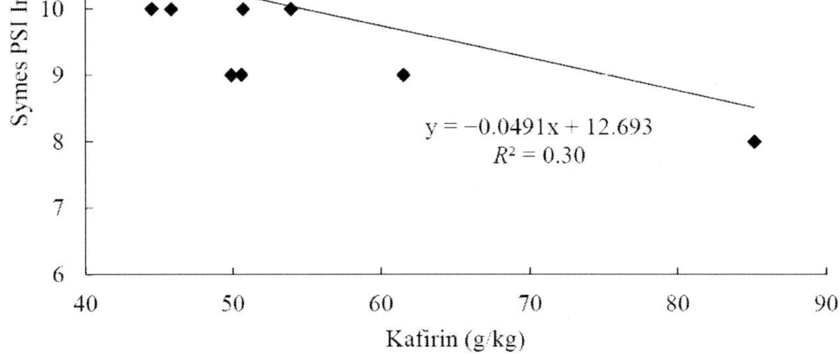

Figure 4. Linear relationship (r = −0.546; p = 0.054) between kafirin concentrations and grain texture (Symes PSI) in 13 Australian sorghums.

−0.546; $p = 0.054$) between kafirin and harder textures that closely approached significance (Figure 4). Chiremba, *et al.* [104] used a tangential abrasive dehulling device was used to determine grain texture and reported that hard textures were positively correlated with concentrations of phenolic acids in maize and sorghum. This is a potentially important outcome but we have not detected similar relationships between total phenolic acids or ferulic acid with Symes PSI texture. This is may be because the limited number of sorghums (13 sorghums) that have been examined, future research is required to clarify the relationship between sorghum grain texture and intrinsic factors, including phenolic acids and kafirin.

10. IMPLICATIONS

Sorghum may be a conundrum as a feedstuff for chicken-meat production because of the relatively inefficient utilisation of dietary energy it provides but the likelihood is that kafirin and non-tannin phenolic compounds may be largely responsible for this situation. Therefore, the real possibility that kafirin levels in local sorghums are increasing as an inadvertent consequence of breeding programs should be addressed. Another viable approach would be the production of more white sorghum varieties with lower concentrations of "non-tannin" phenolic compounds. White sorghums with low protein and kafirin levels and axiomatically low phenolic compound levels should be a highly acceptable feedstuff. The assessment of the nutritive value of a delivery of grain sorghum to a feed-mill is an interesting challenge. However, it seems likely that a combination of Promatest protein solubilities, RVA starch pasting profiles and Symes PSI grain textures would be indicative and certainly merit further investigation as innovative approaches to meet this challenge. Sorghum may be a feed grain with issues for chicken-meat production but this could be, and should be, rectified.

ACKNOWLEDGMENTS

The authors gratefully acknowledge the financial support received from the Chicken-meat Committee of the Rural Industries Research and Development Corporation (RIRDC) for the series of sorghum projects entitled "Steam pelleting temperature of sorghum-based broiler diets" (PRJ-003810), "Evaluation of sodium bisulphite in sorghum-based broiler diets" (PRJ-007639) and "The factors influencing sorghum starch digestibility in broiler chickens" (PRJ-008695).

REFERENCES

1. Black, J.L.; Hughes, R.J.; Nielsen, S.G.; Tredrea, A.M.; MacAlpine, R.; van Barneveld, R.J. The energy value of cereal grains, particularly wheat and sorghum, for poultry. *Proc. Aust. Poult. Sci. Symp.* 2005, *17*, 21–29.
2. Selle, P.H. The protein quality of sorghum. *Proc. Aust. Poult. Sci. Symp.* 2011, *22*, 147–160.
3. Liu, S.Y.; Selle, P.H.; Cowieson, A.J. Strategies to enhance the performance of pigs and poultry on sorghum-based diets. *Anim. Feed Sci. Technol.* 2013, *181*, 1–14.
4. Perez-Maldonado, R.A.; Rodrigues, H.D. *Nutritional Characteristics of Sorghums from Queensland and New South Wales for Chicken Meat Production*; RIRDC Publication No. 09/170; Rural Industries Research and Development Corporation: Barton, Australia, 2009.
5. Bryden, W.L.; Liu, X.; Ravindran, G.; Hew, L.I.; Ravindran, V. *Ileal digestible amino acids values in feedstuffs for poultry*; RIRDC Publication No. 09/71; Rural Industries Research and Development Corporation: Barton, Australia, 2009.
6. Waniska, R.D.; Hugo, L.F.; Rooney, L.W. Practical methods to determine the presence of tannins in sorghum. *J. Appl. Poult. Res.* 1992, *1*, 122–128.
7. Liu, S.Y.; Cadogan, D.J.; Péron, A.; Truong, H.H.; Selle, P.H. Effects of phytase supplementation on growth performance, nutrient utilisation and digestive dynamics of starch and protein in broiler chickens offered maize-, sorghum- and wheat-based diets. *Anim. Feed Sci. Technol.* 2014, *197*, 164–175.
8. Ao, Z.; Choct, M. Effect of early feeding and grain type on growth and performance of broilers. *Proc. Aust. Poult. Sci. Symp.* 2004, *16*, 116–119.
9. Selle, P.H.; Cadogan, D.J.; Ru, Y.J.; Partridge, G.G. Impact of exogenous enzymes in sorghum- or wheat-based broiler diets on nutrient utilization and growth performance. *Int. J. Poult. Sci.* 2010, *9*, 53–58.
10. Bryden, W.L.; Selle, P.H.; Cadogan, D.J.; Liu, X.; Muller, N.D.; Jordan, D.R.; Gidley, M.J.; Hamilton, W.D. *A review of the nutritive value of sorghum for broilers*; RIRDC Publication No. 09/007; Rural Industries Research and Development Corporation: Barton, Australia, 2009.
11. Selle, P.H.; Cadogan, D.J.; Li, X.; Bryden, W.L. Implications of sorghum in broiler chicken nutrition. *Anim. Feed Sci. Technol.* 2010, *156*, 57–74.
12. Belton, P.S.; Delgadillo, I.; Halford, N.G.; Shewry, P.R. Kafirin structure and functionality. *J. Cereal Sci.* 2006, *44*, 272–286.
13. Duodu, K.G.; Taylor, J.R.N.; Belton, P.S.; Hamaker, B.R. Factors affecting sorghum protein digestibility. *J. Cereal Sci.*2003, *38*, 117–131.
14. Cosgrove, D.J. Chemistry and biochemistry of inositol polyphosphates. *Rev. Pure Appl. Chem.* 1966, *16*, 209–224.
15. Mosse, J.; Huet, J.C.; Baudet, J. The amino-acid composition of whole sorghum grain in relation to its nitrogen-content. *Cereal Chem.* 1988, *65*, 271–277.
16. Selle, P.H.; Cowieson, A.J.; Cowieson, N.P.; Ravindran, V. Protein-phytate interactions in pig and poultry nutrition: A reappraisal. *Nutr. Res. Rev.* 2012, *25*, 1–17.
17. Beta, T.; Corke, H. Effect of ferulic acid and catechin on sorghum and maize starch pasting properties. *Cereal Chem.*2004, *81*, 418–422.
18. Liu, S.Y.; Selle, P.H.; Cowieson, A.J. Influence of white- and red-sorghum varieties and hydrothermal component of steam-pelleting on digestibility coefficients of amino acids and kinetics of amino acids, nitrogen and starch digestion in diets for broiler chickens. *Anim. Feed Sci. Technol.* 2013, *186*, 53–63.
19. Selle, P.H.; Walker, A.R.; Bryden, W.L. Total and phytate-phosphorus contents and phytase activity of australian-sourced feed ingredients for pigs and poultry. *Aust. J. Exp. Agric.* 2003, *43*, 475–479.
20. Selle, P.H.; Ravindran, V. Microbial phytase in poultry nutrition. *Anim. Feed Sci. Technol.* 2007, *135*, 1–41.

21. Truong, H.H.; Yu, S.; Peron, A.; Cadogan, D.J.; Khoddami, A.; Roberts, T.H.; Liu, S.Y.; Selle, P.H. Phytase supplementation of maize-, sorghum- and wheat-based broiler diets, with identified starch pasting properties, influences jejunal and ileal digestibilities of phytate (IP6) and sodium. *Anim. Feed Sci. Technol.* 2014, *198*, 248–256
22. Liu, S.Y.; Selle, P.H. A consideration of starch and protein digestive dynamics in chicken-meat production. *Worlds Poult. Sci. J.* 2015, *71*, 297–310.
23. Giuberti, G.; Gallo, A.; Cerioli, C.; Masoero, F. *In vitro* starch digestion and predicted glycemic index of cereal grains commonly utilized in pig nutrition. *Anim. Feed Sci. Technol.* 2012, *174*, 163–173.
24. Truong, H.H.; Liu, S.Y.; Selle, P.H. Starch utilisation in chicken-meat production: The foremost influential factors. *Anim. Prod. Sci.* 2015.
25. Sang, Y.J.; Bean, S.; Seib, P.A.; Pedersen, J.; Shi, Y.C. Structure and functional properties of sorghum starches differing in amylose content. *J. Agric. Food Chem.* 2008, *56*, 6680–6685.
26. Wong, J.H.; Lau, T.; Cai, N.; Singh, J.; Pedersen, J.F.; Vensel, W.H.; Hurkman, W.J.; Wilson, J.D.; Lemaux, P.G.; Buchanan, B.B. Digestibility of protein and starch from sorghum (*Sorghum bicolor*) is linked to biochemical and structural features of grain endosperm. *J. Cereal Sci.* 2009, *49*, 73–82.
27. Taylor, J.R.N.; Emmambux, M.N. Developments in our understanding of sorghum polysaccharides and their health benefits. *Cereal Chem.* 2010, *87*, 263–271.
28. Svihus, B.; Uhlen, A.K.; Harstad, O.M. Effect of starch granule structure, associated components and processing on nutritive value of cereal starch: A review. *Anim. Feed Sci. Technol.* 2005, *122*, 303–320.
29. Ertl, D.; Dale, N. The metabolizable energy of waxy *vs.* Normal corn for poultry. *J. Appl. Poult. Res.* 1997, *6*, 432–435.
30. Pirgozliev, V.R.; Rose, S.P.; Bedford, M.R. The effect of amylose: Amylopectin ratio in dietary starch on growth performance and gut morphology in broiler chickens. *Arch. Fur Geflugelkunde* 2010, *74*, 21–29.
31. Collins, N.E.; Moran, E.T.; Stilborn, H.L. Performance of broilers fed normal and waxy corn diets formulated with chick and rooster derived apparent metabolizable energy values for the grains. *J. Appl. Poult. Res.* 2003, *12*, 196–206.
32. Gilding, E.K.; Frere, C.H.; Cruickshank, A.; Raba, A.K.; Prentis, P.J.; Mudge, A.M.; Mace, E.S.; Jordan, D.R.; Godwin, I.D. Allelic variation at a single gene increases food value in a drought-tolerant staple cereal. *Nat. Commun.* 2013.
33. Virupaksha, T.K.; Sastry, L.V.S. Studies on protein content and amino acid composition of some varieties of grain sorghum. *J. Agric. Food Chem.* 1968, *16*, 199–203.
34. Taylor, J.R.N.; Schussler, L.; Vanderwalt, W.H. Fractionation of proteins from low-tannin sorghum grain. *J. Agric. Food Chem.* 1984, *32*, 149–154.
35. Selle, P.H.; Liu, S.Y.; Cowieson, A.J. Sorghum: An enigmatic grain for chicken-meat production. In *Sorghum: Production, Growth Habits and Health Benefits*; Parra, P.C., Ed.; Nova Publishers Inc.: Hauppauge, NY, USA, 2013; pp. 1–44.
36. Henzell, R.G. Grain sorghum breeding in Australia: Current status and future prospects. In Proceedings of the 2nd Australian Sorghum Conference, Gatton, Australia, 4–6 February 1992; pp. 70–80.
37. Symes, K.J. Inheritance of grain hardness in wheat as measured by particle size index. *Aust. J. Agric. Res.* 1965, *16*, 113–123.
38. De Alencar Figueiredo, L.F.; Davrieux, F.; Fliedel, G.; Rami, J.F.; Chantereau, J.; Deu, M.; Courtois, B.; Mestres, C. Development of nirs equations for food grain quality traits through exploitation of a core collection of cultivated sorghum. *J. Agric. Food Chem.* 2006, *54*, 8501–8509.
39. Cabrera, M.R.; Hancock, J.D.; Bramel-Cox, P.J.; Hines, R.H.; Behnke, K.C. Effects of corn, sorghum genotype, and particle size on milling characteristics and growth performance in broiler chicks. *Poult. Sci.* 1994, *73*, 11.

40. Ravindran, V.; Hew, L.I.; Bryden, W.L. *Digestible amino acids in poultry feedstuffs*; RIRDC Publication No. 98/9; Rural Industries Research and Development Corporation: Barton, Australia, 1998.
41. Selle, P.H.; Gill, R.J.; Downing, J.A. The vulnerability of sorghum to "moist-heat". *Proc. Aust. Poult. Sci. Symp.* 2010, *21*, 68–71.
42. Gidley, M.J.; Flanagan, B.M.; Sharpe, K.; Sopade, P.A. Starch digestion in monogastrics—Mechanisms and opportunities. *Proc. Recent Adv. Anim. Nutr. Aust.* 2011, *18*, 207–213.
43. Rooney, L.W.; Pflugfelder, R.L. Factors affecting starch digestibility with special emphasis on sorghum and corn. *J. Anim. Sci.* 1986, *63*, 1607–1623.
44. Taylor, J.R.N. Non-starch polysaccharides, protein and starch: Form function and feed-highlight on sorghum. *Proc. Aust. Poult. Sci. Symp.* 2005, *17*, 9–16.
45. De Mesa-Stonestreet, N.J.; Alavi, S.; Bean, S.R. Sorghum proteins: The concentration, isolation, modification, and food applications of kafirins. *J. Food Sci.* 2010, *75*, R90–R104.
46. Ezeogu, L.I.; Duodu, K.G.; Taylor, J.R.N. Effects of endosperm texture and cooking conditions on the *in vitro* starch digestibility of sorghum and maize flours. *J. Cereal Sci.* 2005, *42*, 33–44.
47. Truong, H.H.; Neilson, K.A.; McInerney, B.V.; Khoddami, A.; Roberts, T.H.; Liu, S.Y.; Selle, P.H. Performance of broiler chickens offered nutritionally-equivalent diets based on two red grain sorghums with quantified kafirin concentrations as intact pellets or re-ground mash following steam-pelleting at 65 or 97 °C conditioning temperatures. *Anim. Nutr.* 2015. in press.
48. Selle, P.H.; Liu, S.Y.; Cai, J.; Cowieson, A.J. Steam-pelleting temperatures, grain variety, feed form and protease supplementation of mediumly-ground, sorghum-based broiler diets: Influences on growth performance, relative gizzard weights, nutrient utilisation, starch and nitrogen digestibility. *Anim. Prod. Sci.* 2013, *53*, 378–387.
49. Liu, S.Y.; Selle, P.H.; Court, S.G.; Cowieson, A.J. Protease supplementation of sorghum-based broiler diets enhances amino acid digestibility coefficients in four small intestinal sites and accelerates their rates of digestion. *Anim. Feed Sci. Technol.* 2013, *183*, 175–183.
50. Mangan, J.L. Nutritional effects of tannins in animal feeds. *Nutr. Res. Rev.* 1988, *1*, 209–231.
51. Bravo, L. Polyphenols: Chemistry, dietary sources, metabolism, and nutritional significance. *Nutr. Rev.* 1998, *56*, 317–333.
52. Nyachoti, C.M.; Atkinson, J.L.; Leeson, S. Sorghum tannins: A review. *Worlds Poult. Sci. J.* 1997, *53*, 5–21.
53. McClymont, G.L.; Duncan, D.C. Studies on nutrition of poultry. III. Toxicity of grain sorghum for chickens. *Ausi. Vet. J.* 1952, *28*, 229–233.
54. Asquith, T.N.; Izuno, C.C.; Butler, L.G. Characterization of the condensed tannin (proanthocyanidin) from a group II sorghum. *J. Agric. Food Chem.* 1983, *31*, 1299–1303.
55. Dykes, L.; Rooney, L.W. Sorghum and millet phenols and antioxidants. *J. Cereal Sci.* 2006, *44*, 236–251.
56. Thompson, L.U.; Yoon, J.H. Starch digestibility as affected by polyphenols and phytic acid. *J. Food Sci.* 1984, *49*, 1228–1229.
57. Thompson, L.U.; Yoon, J.H.; Jenkins, D.J.A.; Wolever, T.M.S.; Jenkins, A.L. Relationship between polyphenol intake and blood-glucose response of normal and diabetic individuals. *Am. J. Clin. Nutr.* 1984, *39*, 745–751.
58. Awika, J.M.; Dykes, L.; Gu, L.W.; Rooney, L.W.; Prior, R.L. Processing of sorghum (*Sorghum bicolor*) and sorghum products alters procyanidin oligomer and polymer distribution and content. *J. Agric. Food Chem.* 2003, *51*, 5516–5521.
59. Dlamini, N.R.; Taylor, J.R.N.; Rooney, L.W. The effect of sorghum type and processing on the antioxidant properties of African sorghum-based foods. *Food Chem.* 2007, *105*, 1412–1419.

60. Zhu, F. Interactions between starch and phenolic compound. *Trends Food Sci. Technol.* 2015, *43*, 129–143.
61. Tomasik, P.; Schilling, C.H. Complexes of starch with inorganic guests. *Adv. Carbohydr. Chem. Biochem.* 1998, *53*, 263–343.
62. Barros, F.; Awika, J.M.; Rooney, L.W. Interaction of tannins and other sorghum phenolic compounds with starch and effects on *in vitro* starch digestibility. *J. Agric. Food Chem.* 2012, *60*, 11609–11617.
63. Kandil, A.; Li, J.H.; Vasanthan, T.; Bressler, D.C. Phenolic acids in some cereal grains and their inhibitory effect on starch liquefaction and saccharification. *J. Agric. Food Chem.* 2012, *60*, 8444–8449.
64. Welsch, C.A.; Lachance, P.A.; Wasserman, B.P. Dietary phenolic-compounds—Inhibition of Na+-dependent D-glucose uptake in rat intestinal brush-border membrane-vesicles. *J. Nutr.* 1989, *119*, 1698–1704.
65. Elkin, R.G.; Freed, M.B.; Hamaker, B.R.; Zhang, Y.; Parsons, C.M. Condensed tannins are only partially responsible for variations in nutrient digestibilities of sorghum grain cultivars. *J. Agric. Food Chem.* 1996, *44*, 848–853.
66. Beta, T.; Corke, H. Genetic and environmental variation in sorghum starch properties. *J. Cereal Sci.* 2001, *34*, 261–268.
67. Dykes, L.; Rooney, L.W.; Waniska, R.D.; Rooney, W.L. Phenolic compounds and antioxidant activity of sorghum grains of varying genotypes. *J. Agric. Food Chem.* 2005, *53*, 6813–6818.
68. Selle, P.H.; Ravindran, V.; Caldwell, R.A.; Bryden, W.L. Phytate and phytase: Consequences for protein utilisation. *Nutr. Res. Rev.* 2000, *13*, 255–278.
69. Selle, P.H.; Cowieson, A.J.; Ravindran, V. Consequences of calcium interactions with phytate and phytase for poultry and pigs. *Livest. Sci.* 2009, *124*, 126–141.
70. Doherty, C.; Faubion, J.M.; Rooney, L.W. Semiautomated determination of phytate in sorghum and sorghum products. *Cereal Chem.* 1982, *59*, 373–378.
71. Selle, P.H.; Ravindran, V.; Cowieson, A.J.; Bedford, M.R. *Phytate and Phytase*; CABI Publishing: Oxon, UK, 2010; pp. 160–205.
72. Oatway, L.; Vasanthan, T.; Helm, J.H. Phytic acid. *Food Rev. Int.* 2001, *17*, 419–431.
73. Ravindran, V.; Sands, J.S.; Hruby, M.; Partridge, G.G. *Influence of Phyzyme tm xp Phytase on Nutrient Utilization in Broilers Fed Diets Containing Graded Levels of Phytate*; Poultry Research Foundation: Sydney, Australia, 2005; pp. 238–241.
74. Wu, Y.B.; Ravindran, V.; Hendriks, W.H. Influence off exogenous enzyme supplementation on energy utilisation and nutrient digestibility off cereals for broilers. *J. Sci. Food Agric.* 2004, *84*, 1817–1822.
75. Csonka, F.A.; Murphy, J.C.; Jones, D.B. The iso-electric points of various proteins. *J. Am. Chem. Soc.* 1926, *48*, 763–768.
76. Kreydiyyeh, S.I.; Baydoun, E.A.H.; Churukian, Z.M. Tea extract inhibits intestinal-absorption of glucose and sodium in rats. *Comp. Biochem. Physiol. C-Pharmacol. Toxicol. Endocrinol.* 1994, *108*, 359–365.
77. Kreydiyyeh, S.I. Inhibitors in tea of intestinal absorption of phenylalanine in rats. *Comp. Biochem. Physiol. C-Pharmacol. Toxicol. Endocrinol.* 1996, *113*, 67–71.
78. Liu, N.; Ru, Y.J.; Li, F.D.; Cowieson, A.J. Effect of diet containing phytate and phytase on the activity and messenger ribonucleic acid expression of carbohydrase and transporter in chickens. *J. Anim. Sci.* 2008, *86*, 3432–3439.
79. Weurding, R.E.; Enting, H.; Verstegen, M.W.A. The relation between starch digestion rate and amino acid level for broiler chickens. *Poult. Sci.* 2003, *82*, 279–284.
80. Li, T.J.; Dai, Q.Z.; Yin, Y.L.; Zhang, J.; Huang, R.L.; Ruan, Z.; Deng, Z.; Xie, M. Dietary starch sources affect net portal appearance of amino acids and glucose in growing pigs. *Animal* 2008, *2*, 723–729.
81. Enting, H.; Pos, J.; Weurding, R.E.; Veldman, A. Starch digestion rate affects broiler performance. *Proc. Aust. Poult. Sci. Symp.* 2005, *17*, 17–20.
82. Weurding, R.E.; Veldman, A.; Veen, W.A.G.; van der Aar, P.J.; Verstegen, M.W.A. Starch digestion rate in the small intestine of broiler chickens differs among feedstuffs. *J. Nutr.* 2001, *131*, 2329–2335.

83. Liu, S.Y.; Selle, P.H.; Cowieson, A.J. The kinetics of starch and nitrogen digestion regulate growth performance and nutrient utilisation in coarsely-ground, sorghum-based broiler diets. *Anim. Prod. Sci.* 2013, *53*, 1033–1040.
84. Van den Borne, J.; Schrama, J.W.; Heetkamp, M.J.W.; Verstegen, M.W.A.; Gerrits, W.J.J. Synchronising the availability of amino acids and glucose increases protein retention in pigs. *Animal* 2007, *1*, 666–674.
85. Axtell, J.D.; Kirleis, A.W.; Hassen, M.M.; Mason, N.D.; Mertz, E.T.; Munck, L. Digestibility of sorghum proteins. *Proc. Natl. Acad. Sci. USA Biol. Sci.* 1981, *78*, 1333–1335.
86. Choi, S.J.; Woo, H.D.; Ko, S.H.; Moon, T.W. Confocal laser scanning microscopy to investigate the effect of cooking and sodium bisulfate on *in vitro* digestibility of waxy sorghum flour. *Cereal Chem.* 2008, *85*, 65–69.
87. Batterman-Azcona, S.J.; Hamaker, B.R. Changes occurring in protein body structure and alpha-zein during cornflake processing. *Cereal Chem.* 1998, *75*, 217–221.
88. Duodu, K.G.; Nunes, A.; Delgadillo, I.; Parker, M.L.; Mills, E.N.C.; Belton, P.S.; Taylor, J.R.N. Effect of grain structure and cooking on sorghum and maize *in vitro* protein digestibility. *J. Cereal Sci.* 2002, *35*, 161–174.
89. Hamaker, B.R.; Kirleis, A.W.; Butler, L.G.; Axtell, J.D.; Mertz, E.T. Improving the *in vitro* protein digestibility of sorghum with reduing agents. *Proc. Natl. Acad. Sci. USA* 1987, *84*, 626–628.
90. Rom, D.L.; Shull, J.M.; Chandrashekar, A.; Kirleis, A.W. Effects of cooking and treatment with sodium bisulfite on *in vitro* protein digestibility and microstructure of sorghum flour. *Cereal Chem.* 1992, *69*, 178–181.
91. Selle, P.H.; Liu, S.Y.; Cai, J.; Caldwell, R.A.; Cowiesona, A.J. Preliminary assessment of including a reducing agent (sodium metabisulphite) in "all-sorghum" diets for broiler chickens. *Anim. Feed Sci. Technol.* 2013, *186*, 81–90.
92. Selle, P.H.; Liu, S.Y.; Cai, J.; Caldwell, R.A.; Cowieson, A.J. Graded inclusions of sodium metabisulphite in sorghum-based diets: I. Reduction of disulphide cross-linkages *in vitro* and enhancement of energy utilisation and feed conversion efficiency in broiler chickens. *Anim. Feed Sci. Technol.* 2014, *190*, 59–67.
93. Truong, H.H.; Cadogan, D.J.; Liu, S.Y.; Selle, P.H. Addition of sodium metabisulfite and microbial phytase, individually and in combination, to a sorghum-based diet for broiler chickens from 7 to 28 days post-hatch. *Anim. Prod. Sci.* 2015.
94. Paterson, L.; Mitchell, J.R.; Hill, S.E.; Blanshard, J.M.V. Evidence for sulfite induced oxidative reductive depolymerisation of starch polysaccharides. *Carbohydr. Res.* 1996, *292*, 143–151.
95. Paterson, L.A.; Hill, S.E.; Mitchell, J.R.; Blanshard, J.M.V. Sulphite and oxidative-reductive depolymerization reactions. *Food Chem.* 1997, *60*, 143–147.
96. Ravindran, V.; Cabahug, S.; Ravindran, G.; Bryden, W.L. Influence of microbial phytase on apparent ileal amino acid digestibility of feedstuffs for broilers. *Poult. Sci.* 1999, *78*, 699–706.
97. Liu, S.Y.; Cadogan, D.J.; Péron, A.; Truong, H.H.; Selle, P.H. A combination of xylanase, amylase and protease influences growth performance, nutrient utilisation, starch and protein digestive dynamics in broiler chickens offered maize-, sorghum- and wheat-based diets. *Anim. Prod. Sci.* 2014.
98. Odjo, S.; Malumba, P.; Dossou, J.; Janas, S.; Bera, F. Influence of drying and hydrothermal treatment of corn on the denaturation of salt-soluble proteins and color parameters. *J. Food Eng.* 2012, *109*, 561–570.
99. Gehring, C.K.; Bedford, M.R.; Cowieson, A.J.; Dozier, W.A. Effects of corn source on the relationship between *in vitro* assays and ileal nutrient digestibility. *Poult. Sci.* 2012, *91*, 1908–1914.
100. Selle, P.H.; Liu, S.Y.; Cai, J.; Cowieson, A.J. Steam-pelleting and feed form of broiler diets based on three coarsely ground sorghums influences growth performance, nutrient utilisation, starch and nitrogen digestibility. *Anim. Prod. Sci.* 2012, *52*, 842–852.

101. Zhang, G.Y.; Hamaker, B.R. Sorghum (*Sorghum bicolor* L. Moench) flour pasting properties influenced by free fatty acids and protein. *Cereal Chem.* 2005, *82*, 534–540.

102. Liu, S.Y.; Selle, P.H.; Khoddami, A.; Roberts, T.H.; Cowieson, A.J. Graded inclusions of sodium metabisulphite in sorghum-based diets: II. Modification of starch pasting properties *in vitro* and beneficial impacts on starch digestion dynamics in broiler chickens. *Anim. Feed Sci. Technol.* 2014, *190*, 68–78.

103. Selle, P.H.; Liu, S.Y.; Cai, J.; Cowieson, A.J. Steam-pelleting temperatures and grain variety of finely-ground, sorghum-based broiler diets. I. Influence on growth performance, relative gizzard weights, nutrient utilisation, starch and nitrogen digestibility. *Anim. Prod. Sci.* 2014, *54*, 339–346.

104. Chiremba, C.; Taylor, J.R.N.; Rooney, L.W.; Beta, T. Phenolic acid content of sorghum and maize cultivars varying in hardness. *Food Chem.* 2012, *134*, 81–88.

CHAPTER 10

Selenium Requirements and Metabolism in Poultry

Anicke Brandt-Kjelsen[1*], Brit Salbu[1], Anna Haug[2] and Joanna Szpunar[3]

[1]*Department of Environmental Sciences, Norwegian University of Life Sciences (NMBU), Ås, Norway*

[2]*Department of Animal and Aquacultural Sciences, Norwegian University of Life Sciences (NMBU), Ås, Norway*

[3]*Laboratorie de Chimie Analytique Bio-inorganique et Environement, CNRS UMR 5254, Pau, France*

ABSTRACT

As counteract against deficiency in livestock, dietary treatments are supplemented with selenium (Se), usually as the inorganic form sodium selenite (SS). Since Se is considered as toxic as well as an essential element, SS is added to prevent an increase of Se in edible chicken parts. However, in many countries, populations suffer from suboptimal Se intake and even Se deficiency, by increasing the use of organic Se sources such as Se-enriched yeast or wheat in animal feed, there will be a subsequent increase in meat and egg products for human consumption. One could argue that the chickens do not need the extra pool of Se in muscles, as the inorganic form will be sufficient to meet the chickens Se requirements. Since the feed is fortified with selenite, the chickens will always have adequate access to the essential trace element. However, global gene set functional enrichment analysis revealed statistically significant enrichment of a number of biological processes that were dependent on the Se feed sources, such as cell growth, organ development and protein metabolism in favour of organic Se.

Keywords: selenium, requirements, metabolism, feed supplementation, selenoamino acids

1. INTRODUCTION

1.1. The Element Selenium

Selenium (Se) was discovered in 1817 by the Swedish chemist Jöns Jakob Berzelius [1, 2]. The element was named after the moon in Greek; Selene. Selenium is the chemical element number 34, a metalloid with six stable isotopes 76, 77, 78, 79, 80 and 82. The element is in the 16 group in the periodic table together with sulphur (S), oxygen (O) and tellurium (Te). These elements have many common chemical properties. Selenium and S have similar electronegativities, atomic radius and the same oxidation states (-2, 0, +4, +6). Selenium is an S analogue, and substitute S in a series of compounds, such as sulphate (selenate), sulphite (selenite), sulphenic acid (selenic acid). Selenium and S react easily with each other forming selenylsulphide bounds [1]. Despite these similar properties, there are some major differences between S and Se, and substitution of one another results in different chemical properties. One of these differences is illustrated by the acid dissociation constant (pKa) value; H_2Se has a pKa of 3.73 and is a much stronger acid than H_2S with a pKa of 6.96 [1]. Another difference between the two elements is the reduction potential, under standard conditions Se is more readily reduced than S (e.g. SeO_4^{2-} has an $E = 1.15$ volt and SO_4^{2-} has $E = 0.158$ V).

1.2. Seleno Amino Acids

Cysteine (Cys) and methionine (Met) are S amino acids and both have Se analogs, SeCys and SeMet, respectively. As for the acids, the amino acids exhibit different chemical properties depending on the properties of Se and S. While the chemical properties of SeMet and Met are quite similar [3] the pKa (amino group) of SeCys is much lower than for Cys (5.2 vs. 8.3) [4]. At physiological pH the selenol group of SeCys will be in its anionic selenolate form, while the thiols of Cys residues are protonated. These properties make SeCys a more reactive amino acid compared to Cys [1], and give unique Se derived properties to selenoproteins.

1.3. Selenium In The Terrestrial Environment

High Se concentrations are found in sedimentary rocks and shales formed dung the cretaceous period, while lower concentrations are characteristic for igneous (volcanic) rock, sandstone, granite and lime stone [5, 6]. The ubiquitous but uneven Se distribution in bedrock has resulted in Se concentrations in soil ranging from almost zero to 1250 mg Se kg^{-1} in some seleniferous soils in Ireland [7, 8]. High soil Se concentrations can also be a consequence of prolonged deposition from precipitation containing seawater. I Norway the observed Se concentration in humus showed that the Se concentrations in soil decreases with distance from the sea [9]. Through observations in animal production, it is apparent that vast land areas worldwide do not supply sufficient Se for optimal livestock nutrition, and areas with low Se concentrations in soil are much more common than areas with very high Se concentrations in soil [8].

1.3.1. Selenium Transfer From Soil To Plant

The distribution and availability of Se to plants depend on different soil conditions such as pH, organic matter [10], redox conditions, competing ionic species (e.g. sulphate), microbial activity, soil texture, compaction, mineralogy, temperature and moisture [11]. The bioavailability of Se from soil is more important than the total soil Se concentration. Inorganic Se species present in soil are selenide (Se_2^-), elemental Se (Se_0), selenate (SeO_4^{2-}) and selenite (SeO_3^{2-}) [12]. In aerobic soils, with pH around seven, selenate predominates, whereas selenite predominates at lower pH and redox potential. Under strongly reduced soil conditions selenide dominates [13]. Selenate is more mobile and thus available to plants compared to selenite.

1.3.2. Selenium In Plants

The uptake of Se in higher plants depends on plant species and their physiological phase of development. Selenate is more accessible to plants than selenite and organic Se such as SeMet. The uptake of selenate and SeMet follow the active transporters of analog S species and is metabolized through the sulphate assimilation pathway. Selenite, however, is not mediated by membrane sulphite transporters [2]. After selenite has been taken up in the plant it is rapidly transformed to SeMet or SeCys and accumulated in proteins in the roots, while selenate is highly mobile in the xylem and translocated to plant parts above ground [12, 14]. The Se concentrations in plants are related to the general protein content of the plants and their different ability to accumulate Se. Most plants that are used in forage as well as grasses are non-accumulators [2] and, hence, have low Se concentrations. Surveys on Se concentrations in crop have revealed that areas producing crops with Se concentrations too low (<0.1 mg Se kg^{-1}) to meet animal requirements are more common than areas producing toxic levels (>2 mg Se kg^{-1}) of Se in crops, hence supplementation of diets is necessary to meet the animals Se requirements. There has been an increased interest of investigating the beneficial use of medical plants and phytogenetic compunds in poultry diets [15], some of these plants are Se hyperaccumulators [16] and future research will prove if these plants could replace the traditional supplementation used today.

1.3.3. Selenium Requirements In Poultry And Domestic Animals

The toxicity of Se in animal diets was recognized around 1932 [17]. However, a decade later it was obvious that Se deficiency lead to several negative effect. Impaired growth and development, poor feathering, reduced egg production and hatchability, pancreatic degeneration, nutritional muscular dystrophy, and necrotic lesions in liver, muscle and heart are among the effects reported in Se deficient animals, all of which resulted in large economic losses for the industry [18–20]. In Finland, inadequate Se intake caused nutritional disorders in pigs and reduced the profitability of the production. As a result, all commercial animal feeds have been supplemented with selenite since 1969 [8], followed by Denmark in 1975, Norway in 1979 [21] and Sweden in 1980.

In addition, low Se levels reduce the immune function and the animals are more prone to develop diseases such as white muscle disease, cardiac muscle metamorphism, blood capillary disease, cancer, anaemia and liver bleeding [22, 23]. In low Se areas worldwide, livestock producers have adopted methods that ensure adequate Se status in animals. Since selenium was foreseen as a toxic element, the supplementation recommendations were designed to ensure that the Se requirements were met without increasing the Se concentrations in tissues significantly. In that way humans were protected from ingesting "toxic levels" of Se from food products [24]. Sodium selenite (SS) and sodium selenate which do not increase muscle Se levels substantially were therefore approved as supplements to the livestock and poultry industry [24]. Selenite is most often used [24]. The use of Se-enriched yeast (SY) as feed additive is increasing. The European Food Safety Authority has approved a maximum of 0.2 mg Se from different yeast strings in complete Feed [25, 26]. Se-enriched yeast approved as supplement consists of 60–85% SeMet, 2–4% SeCys and <1% selenite or selenate [27]. Based on the legislation from the European Committee the allowed maximum total feed Se concentration is 0.5 mg kg^{-1} DM (dry weight) [26]. These regulations also distinguish between inorganic or organic Se, where the maximum of organic Se is set to 0.2 mg Se kg^{-1}. Numerous reports on the positive effects of organic Se sources in animal feed are available. In addition to reduced frequency of diseases related to Se deficiency, organic Se readily increases tissue Se concentrations [28, 29]. In 2014 DL-Selenomethionine was accepted as an organic Se source after it was showed to be safe for chickens for fattening up to 1.5 mg Se/kg feed, the upper limit of 0.2 mg organic Se/kg feed still stands [30].

2. UNDERSTANDING THE SPECIATION AND METABOLISM OF SELENIUM IN POULTRY

2.1. Bioavailability And Bioaccessibility Of Selenium In A Sample Preparation And Instrument Analysis Perspective

2.1.1. Speciation Analysis Of Selenium

Speciation analysis of Se is applied to identify and quantify Se species present in different matrices. This is of major importance as bioavailability, cross membrane transport and metabolism of Se is highly dependent on the Se species present [31]. While numerous studies on Se species in yeast and Se accumulating plants have been published (e.g. [32–34]), fewer studies have been carried out on Se species in common food. It has therefore been necessary to develop reliable analytical techniques to study the speciation of Se in environmental and biological samples, to understand the biochemical cycle of Se [35, 36].

Selenium is usually present in food and biological samples at low concentrations and can be present as a variety of Se species. To perform speciation analysis of Se in biological matrixes different methodologies are applied. Hyphenated techniques such as liquid chromatography coupled to inductive coupled plasma mass spectrometry (LC-ICP-MS) and liquid chromatography coupled electrospray ionization mass spectrometry (LC-ESI-MS) are two complementary detection methods [37]. The ICP-MS quantifies based on signals from the elements mass/charge ratio; the ionization is virtually species independent and accurate absolute quantification is possible. In ESI-MS the Se species are identified based on molecules mass/charge ratio, where ionization is species and matrix dependent and quantification requires isotopically labelled species. Identification based on retention time matching with well-defined standards by HPLC-ICP MS is tentative, while ESI-MS data alone does not provide sufficient evidence for structural confirmation [33, 38].

With increased knowledge and sensitive instruments, more information on Se species in food will emerge. As many detectors are very sensitive (low detection limits), the challenges lie within sample preparation and chromatographic separations. At present, no standardized approaches to Se speciation analysis exist and different strategies of sample preparation and separation prior to detection are applied.

2.1.2. Sample Preparation And Separation Techniques

Selenium species in food and biological samples can be extracted using different agents. Water extraction has been applied to water-soluble Se

amino acids (MeSeCys and -glutamyl-MeSeCys), driselase is used to release Se bound to cell walls, different proteases for hydrolysis of selenoproteins and seleno-containing proteins [34], and tryptic digestion of water-soluble Se species are applied to identify peptide sequences [39]. Another approach is the use of in vitro digestion steps mimicking the gastro intestinal tract to get information on the bioaccessible fraction of Se from different matricies [40, 41].

Following extraction, different techniques are used to separate Se species. The most used technique is high performance liquid chromatography (HPLC) with reversed phase (RP), ion-paring (IP)-RP, or ion-exchange (IE) columns, sometimes with a pre-concentration step using a size exclusion column (SEC) to collect the low molecular mass fraction [33, 42, 43]. Optimization of the chromatographic separation conditions to obtain narrow and well defined peaks are strived for at all times for best results [37]. Isotope dilution has improved the quality control of analysis and can be used species-specific or species unspecific, and be applied pre or post column to the ICP-MS [44].

During the last years, measurement of Se has focused on a more proteomic approach revealing information on selenoproteins and seleno-containing proteins. Examples are proteins being separated on gel electrophoresis (GE) (sodium dodecyl sulphate polyacrylamide gel electrophoresis, SDS PAGE) for protein identification by matrix-assisted laser desorption/ionization (MALDI)—time of flight (TOF)-MS or ESI-MS or laser ablation ICP-MS to identify the proteins containing Se in a gel [45]. Another way to identify Se species in a SDS PAGE is the use of radioactive labelled ^{75}Se. Neutron activation analysis (NAA) followed by gamma measurements of Se could give information on trace levels of Se in different matrixes. The use of affinity columns separating selenoalbumin (SelA), selenoprotein P (SelP) and glutathione peroxidase (Gpx) has been successfully reported [45].

Isotopic labelling of Se species makes it possible to follow the fate of different species in the environment, either applied in fertilizers to plants and vegetables, in feed or injected to animals. Detailed knowledge on the formation of different Se species has been gained by this approach [46–50], but some questions remain unanswered such as the turnover of Se form plant crops in adequately fed animals, or the uptake and transformation of Se from colon [36]. Speciation analysis of Se is essential to understand the uptake, metabolism, distribution and transformation of Se in biological system.

2.2. Selenium Species Dependent Metabolism Of Selenium

According to [51] all Se-forms (chemically pure forms and different forms in food) are generally well transported over the intestinal membrane (70–95%), but the uptake varies according to the Se source and status of the individual [51]. The same active transporters take up selenate as sulphate, while selenite is taken up by passive transport and does not share the pathway of sulphite. Immediately after selenite has entered the red blood cells, it is reduced to selenide by cellular glutathione [52] or through the actions of thioredoxin reductase [53]. Thereafter, selenide is transported to an organ, usually the liver, to undergo selenosynthesis [52]. Selenate reaches the liver in its intact chemical form where it is reduced to selenide through the same pathways as selenite [54]. In the liver of animals inorganic Se and organic Se is transformed into SeCys.

Organic seleno amino acids also reach the liver in their intact form where the conversion to selenide may follow different pathways. SeCys is transformed to selenide through the -lyase reaction, while SeMet can follow three known reaction pathways;

1. Conversion to SeCys through the intermediate SeCystathionine.
2. Directly by -lyase reaction.
3. SeMet does not enter the selenide pool, but is inserted into proteins as Met by tRNA coding for Met [50, 52, 53]. When SeMet substitutes for Met in proteins the Se concentration increases in the form of Se containing proteins.

Selenide is assumed to be the main precursor of Se metabolism. Selenide has three main conversion routes dependent on the Se status of the organism. Firstly, selenide is transformed to Se-phosphate then to Se-cysteyl tRNA for insertion of SeCys in selenoproteins. Secondly, at lower levels of intake, excess selenide not used in selenoprotein synthesis is also converted into selenosugars for excretion via urine. Thirdly, at high levels of intake, methyltransferases add a methyl group to selenide, leading to a sequential conversion to methylselenol and further to dimethyl selenide (excreted via breath and feces) and trimethyl selenonium (excreted via urine) in addition to excretion of selenosugars [50, 53, 55].

Methylated Se species, MeSeCys, -glut-MeSeCys and selenobetaine follow different pathways compared to the other dietary Se species described. The methylated selenospecies are transformed directly into methylselenol and excretion, or are demethylated to selenide [52].

Bioavailability studies should also address the transformation of Se into biological active metabolites (bioactive Se) [51, 52]. Selenoproteins are considered bioactive as they are essential for animals and humans.

Selenoprotein P (SelP) is the major selenoprotein synthesized in the liver and is released to the blood stream, as a transporter of Se in the body. The liver also releases cellular Gpx to the blood stream whereas the kidneys release extracellular Gpx. According to [53] the uptake of SelP from plasma is by specific receptor-mediated processes of apolipoproteins in brain and testis and megalin in kidneys. The specific uptake mechanisms to other tissues remain unknown.

The use of Se supplements to poultry improves the Se nutritional status by increasing the glutathione peroxidase Gpx activity in plasma. Glutathione peroxidase is the commonly used biomarker for Se status in livestock. When birds are fed with selenium enriched yeast the Gpx3 activity in blood remains higher for a longer period of time after supplementation than if the birds are supplemented with sodium selenite [56]. The Se supplementation and Gpx activity in poultry blood follow a dose response relationship [56–58], that seem to level off at approximately 400 ng/g whole blood [59].

2.3. Se Supplementation Of Life Stock

As mentioned earlier Se-enriched yeast increase chicken muscle Se concentrations significantly compared with sodium selenite [7, 57]. Hence, increasing the use of organic Se sources such as Se-enriched yeast in animal feed will give a subsequent increase in meat, egg and milk products for human consumption.

The aim of the present work was to compare the ability of Se-enriched wheat (SW), Se-enriched yeast (SY) and sodium selenite (SS), as Se supplements in dietary treatments to increase the Se concentration in edible chicken parts for human consumption. To compare chickens fed elevated Se dietary treatments with industrial produced chickens in Norway. The work also included identification and quantification of Se amino acids (SeMet, SeCys) in the edible parts of the chickens and effects on gene expression.

3. ANIMAL EXPERIMENT

The experimental research on animals followed internationally recognized guidelines. All animals were cared for according to laws and regulations controlling Norwegian experiments with live animals according to the Norwegian Animal Research Authority.

Ninety male chicken (*Gallus gallus*) (Ross 308, Samvirkelaget kylling, Norway) were divided into three groups and fed with either inorganic Se as sodium selenite (SS) or organic Se as Se-enriched yeast (SY) or Se-enriched wheat (SW). The calculated Se concentration in the dietary treatments was 0.8 mg Se kg^{-1}. Other main dietary constituents were added in equal amounts. Selenite and Se-enriched yeast were used as control groups to the Se-enriched wheat group, as these two Se supplements have been intensively studied at different concentrations levels [58, 60] among others. As a reference five chickens, 28 days of age, and chicken feed were collected from the chicken industry (Nortura SA chicken slaughterhouse in Rakkestad (Østfold, Norway; Ross 308). The industry use SS as the dietary source of Se, the maximum allowed concentration in the diet is 0.5 mg Se kg^{-1} [27].

The chickens were fed ad-libitum for 33 days at the Animal Production Experimental Centre (SHF) at the Norwegian University of Life Sciences (UMB, Ås, Norway) randomly placed in separate pens. The animals were weighted after 8, 11 and 28 days. After 33 days the chickens were slaughtered, muscles (leg and breast) and liver sampled cut out, vacuum packed and stored frozen (-20°C) prior to analysis. Blood samples were taken by beheading and subsequent collection from the neck in 50 ml tubes (average volume of 30 ml), and stored at -20°C. The tissues were freeze dried prior to total Se measurements and speciation analysis.

3.1 Measurements

3.1.1 Total Se Measurements

Tissue of muscles (breast and leg) and liver from chickens fed SS, SY or SW were freeze dried and homogenized with a mixer mill (Retsch mixer mill MM 200) equipped with zirconium jars. The samples were digested using a microwave assisted (UltraCLAVE, Milestone) nitric acid (5 ml distilled suprapure 14 M HNO_3) decomposition of 0.1 g of dried tissue, or 1 ml (weight) of whole blood at 240°C for 40 min and diluted to 50 ml with 2% ethanol. Total Se was measured using high resolution inductive coupled plasma mass spectrometry (HR-ICP-MS) (Thermo Finnigan Element2, Bremen, Germany) at the National Institute of Occupational Health (STAMI). The instrument was used in high resolution mode. Tellurium was added as internal standard (IS).

The total concentration in the supernatants were digested (1 ml, weight) with sub-distilled nitric acid using the UltraCLAVE and diluted to 50 ml with MQ water. The samples were mixed 1:1 with 4% ethanol using a mixing block connected to the peristaltic pump of the ICP-MS (Perkin Elmer, Elan 6000) and total Se was measured on mass 82. Wheat flour (1567A) and

Bovine liver (1577B) from NIST were used as certified reference material. Tellurium was added as internal standard (IS).

3.1.2 Speciation Analysis

Proteolytic extractions were carried out according to the method from [61] on freeze dried and homogenized samples from three different chickens of each dietary treatment. One ml of 0.1 M Tris-HCl (pH 7.5) was added to 0.1 g meat sample and the suspension mixed with an ultrasonic probe for 1 min. Reduction and carboxymethylation (CAM) of the samples was done by adding 200 µl 0.2 M DTT (in 0.1 M Tris-HCl pH 7.5) and 275 µl 0.5 M IAM. The samples were incubated with careful mixing in the dark at room temperature (20°C). After 2 h 2.5 ml 0.2 M DTT was added and the samples shaken for 1 h in order to destroy excess of IAM. Then, 7 ml of 0.1 M Tris-HCl buffer was added together with protease XIV and lipase. The samples were digested with 30 mg protease and 20 mg lipase over night (×3) at 37 ± 0.5°C in an incubator cupboard on a Roto-Shake (Genie) in the dark. All digested samples were centrifuged at 10,000 × g for 10 min at 4°C and supernatants pooled and freeze dried. The freeze dried supernatants were dissolved in 5 ml of MQ water. One ml was used for total Se measurements and 4 ml for identification and quantification of Se amino acids. Enzymatic extracts were centrifuged at 12,110 g for 5 min (mini Spin, Eppendorf) and filtered through a syringe filter (0.45 µm) before high performance liquid chromatography (HPLC, Perkin Elmer 200 series pump)-ICP-MS (Elan 6000) analysis. The chromatographic separation was done on a reversed phase (RP) Altima C8/Alltech (150 × 4.6 mm, 5 µm) column using an injection volume of 100 µl. Isocratic elution of 2% MeOH and 0.1% HFBA (as an ion pairing agent) with a flow rate of 0.8 or 0.9 ml min^{-1}. Separation of CAM SeCys and CAM SeMet were preformed within 10 min.

The SeCys standard was synthesizes according to the procedure described by [61]. Wheat flour (BC210a) from European Reference Material (ERM) from LGC-standards is verified for SeMet and was added to the speciation analysis for quality assurance.

3.1.3 Amino Acid Measurements

Total (peptide bound and free) amino acids were measured in two breast muscle samples from each treatment (SS, SY and SeW). The measurements were performed on a Biochrom 30 Amino Acid Analyzer (Biochrom Ltd., Cambridge, UK) at Aquaculture Protein Centre (APC) at NMBU, following the procedure described by the Commission Directive 98/64/EC (1998). The samples were oxidized and hydrolyzed prior to HPLC-ultra violet (UV) measurements. The method does not distinguish between salts of amino acids and cannot differentiate between D and L forms of amino acids.

3.1.4 Gene Expression Analysis

Gene expression analyses offer a sensitive and rapid detection of transcriptional changes occurring at the cellular level after different treatments and exposures. Global (un-biased) transcriptional analysis using a oligo-array was applied to see whether SS and SW dietary treatments had different effect on the gene expression profiles of chicken muscle in regard to chicken health. The analysis was performed at Norwegian Institute of Water Research (NIVA).

Isolation of mRNA from fresh frozen muscle samples were performed by TRIzol® extraction, quality controlled by measuring salt and phenol interferences and approximately mRNA concentration by nanodrop (Nanodrop® ND-100 UV-Visible spectrophotometer, NanoDrop Technologies). The samples were diluted to 100 ng μL^{-1} so that RNA integrity could be determined by gel electrophoreses (Bioanalyzer Instrument, Agilent Technologies) samples passing the quality cut-off criteria had RNA Integrity Number (RIN) >9, 260/230 > 1.99 and 260/280 > 2.31.

Microarray analysis was performed following Agilent's protocol "One Color Microarray-Based Gene Expression Analysis" (Quick Amp Labelling, Version 5.7 March 2008) using a 44k Agilent Chicken (V2) array.

3.1.5 Statistics

Statistical analysis was conducted using one way ANOVA with 95% confidence interval and Tukeys simultaneous test to establish statistical difference ($p < 0.05$) among individual treatment means using GraphPad prism version 5.

3.2 Results And Discussion

3.2.1 Feed Se Concentrations

The Se concentration in the reference diet from Nortura SA ($n = 6$) was 0.47 ± 0.052 mg Se kg^{-1}. The Se concentrations in the elevated dietary treatments ($n = 6$) were 0.90 ± 0.030 mg Se kg^{-1} for SS, 0.91 ± 0.040 mg Se kg^{-1} for SY, and 1.0 ± 0.017 mg Se kg^{-1} for SW. The Se concentrations were 0.1–0.2 mg Se kg^{-1} higher than the estimated concentration of 0.8 mg Se kg^{-1}, likely attributed to other constituents in the diet such as fishmeal and soy meal.

3.2.2 Bird Physiological Performance

The mortality was within the expected range (10% in the group fed with SS, 17% in the group fed with SY, and 3% in the group fed with SW) of experiments performed at the Animal Production Experimental Centre. There were no effects of dietary treatment on intake or live weight gain.

3.2.3 Bioavailability Of Se: Total Se Concentrations

The limit of detection was 0.4 μg L^{-1} based on three times the standard deviation of eight blank samples and the mean standard deviations of Se-77, Se-78 and Se-82. The limit of quantification was based on 10 times the standard deviation of the blank samples and was 1.3 μg L^{-1}.

Whole blood (wet weight) Se-concentrations were significantly higher ($p < 0.05$) for all the chickens receiving Se elevated dietary treatments (SS, SY and SW) compared to the reference group (reference SS) (**Figure 1**). The Se concentration in whole blood from the chicken receiving SY and SW were 468 ± 43 μg Se kg^{-1} and 521 ± 43 μg Se kg^{-1}, respectively and significantly higher ($p < 0.05$) than the concentration of 362 ± 32 μg Se kg^{-1} in whole blood of the SS group. There were no differences in blood Se-concentrations between the two groups fed with organic Se. The dose dependent increase of Se in blood (**Figure 1**) with increasing Se in the diet is in line with previous published results in chickens and young turkeys [7, 57, 58]. The results are, however, not in line with results from [58, 59] were no difference in whole blood Se concentrations were observed between SS and SY at the same dietary Se level (in lamb and chickens, respectively). The Se concentration in whole blood (220 ± 27 μg Se kg^{-1}) of the reference group was in accordance with the concentration in whole blood (240 μg L^{-1}) of the SS group of [58] with a dietary treatment of 0.42 mg Se kg^{-1}.

The total Se concentration differed among tissue types; the liver had the highest Se concentrations, followed by the breast muscle and leg muscle (**Figure 2**). The high liver Se concentration is due to the tendency of glandular visceral tissue to have higher Se concentration than skeletal tissue [59], and the fact that the liver handles most of the absorption of nutrients and regulates their release into blood for further distribution or excretion. The Se concentrations in the livers of the reference group were 2.3 ± 0.18 mg Se kg^{-1}. By increasing the Se concentrations in the diets to 0.9–0.10 mg Se kg^{-1} the Se concentrations in livers increased significantly ($p < 0.05$) compared to the reference group (SS: 3.2 ± 0.19 mg Se kg^{-1}, SY: 3.5 ± 0.25 mg Se kg^{-1}, SW: 3.8 ± 0.89 mg Se kg^{-1}) (**Figure 1**). The equal Se concentrations measured in liver of chickens receiving dietary treatments with different Se forms at the same Se concentrations level was in accordance with results obtained by [49].

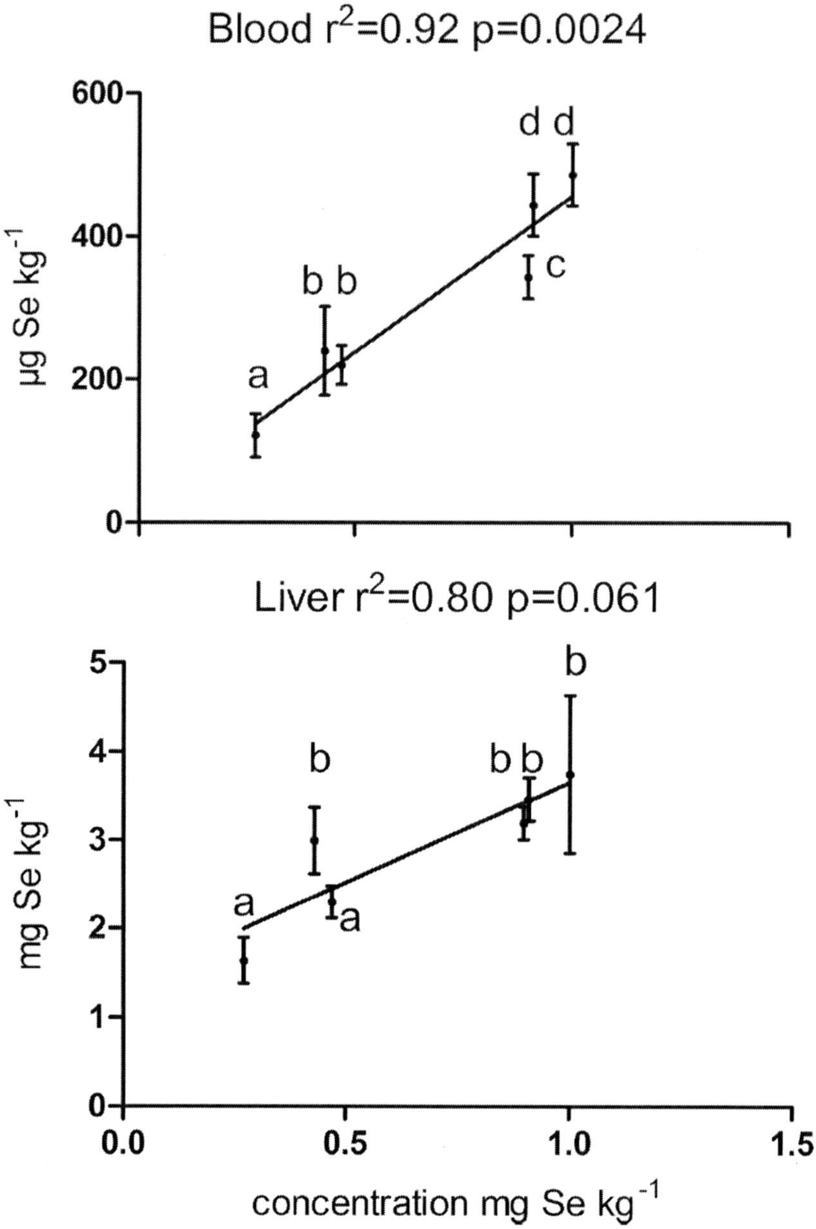

FIGURE 1. Correlation between feed Se concentration and liver and blood Se concentrations.

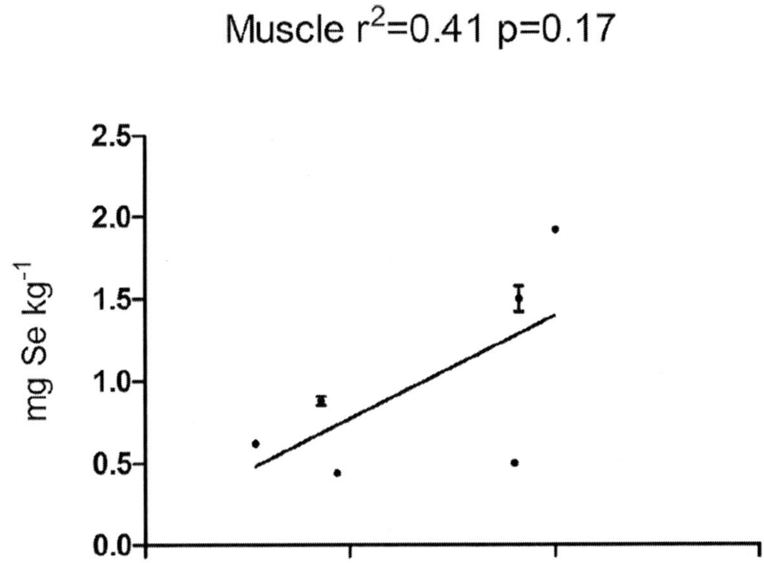
Muscle r²=0.41 p=0.17
mg Se kg⁻¹
2.5
2.0
1.5
1.0
0.5
0.0

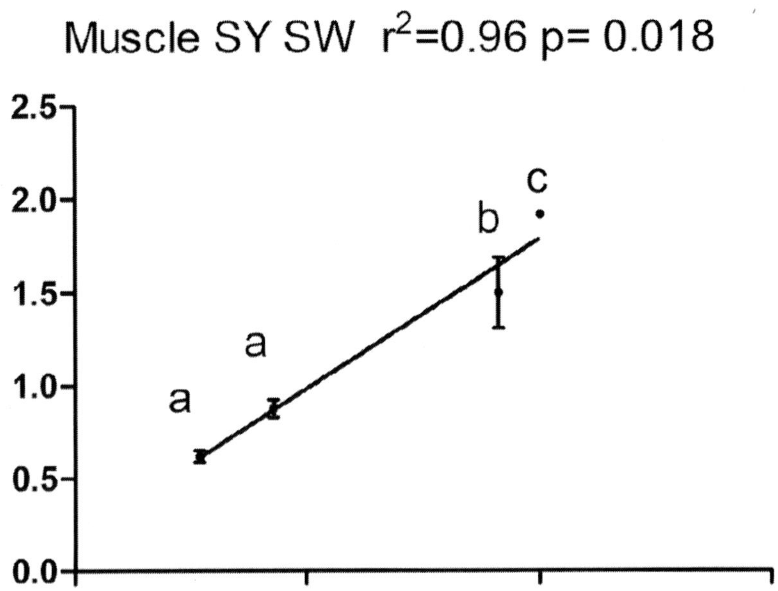
Muscle SY SW r²=0.96 p= 0.018
2.5
2.0
1.5
1.0
0.5
0.0
a
a
b
c

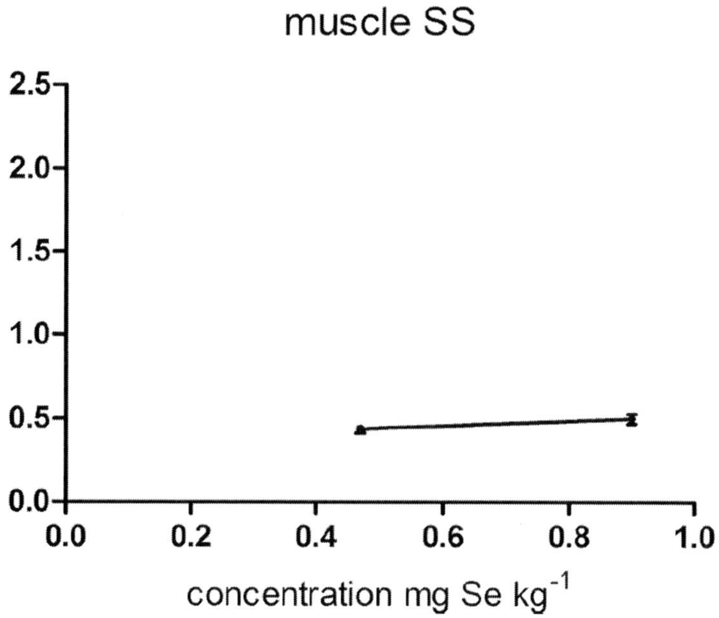

FIGURE 2. Correlation between Se concentration in feed and Se concentration in muscle with regard to Se source added to the feed (SY, Se-enriched yeast; SW, Se-enriched wheat and SS, sodium senenite).

An increase from 0.47 ± 0.052–0.9 ± 0.030 mg Se kg^{-1} of Se as SS in the diets did not result in a significant increase in the breast or leg muscle Se concentrations, 0.44 ± 0.24 and 0.50 ± 0.31 mg Se kg^{-1}, respectively (**Figure 2**). Se-concentration in chickens muscle of chickens fed with organic Se was significantly higher ($p < 0.05$) than in muscles from chickens fed with selenite. The Se concentration in breast muscles were significantly higher ($p < 0.05$) in the group receiving SW compared to SY (1.9 ± 0.020 and 1.5 ± 0.019 mg Se kg^{-1}). No difference between SW and SY diets on the total Se concentrations measured in leg muscle. Most publications on Se in dietary treatments use 0.3 and 0.5 mg Se kg^{-1} (e.g. [57]). The muscle Se concentrations in the chickens receiving SS in the diet was in line with the concentrations reported by [62] after dietary supplement with 0.6 mg Se kg^{-1} as SS and 0.6 mg Se kg^{-1} as SY after 42 days, and in line with the total Se concentrations reported by [63] in breast muscle of Ross 308 chickens.

These data show that Se from SW result in similar Se concentrations in muscle as do SY, and based on these criteria SW could be an Se source for biofortification of Se in chickens, as suggested by [64]. The increased concentration of Se in muscle from organic Se in dietary treatments has been shown by several authors, and is attributed to the unspecific uptake of SeMet in the methionine pathway resulting in an unspecific storage of Se in muscle proteins [64, 66].

A wet weight portion of 170 g of chicken meat from the industrially produced chickens or selenite supplemented groups would give a contribution of 18 ± 1.3 µg Se, which is close to the values given by the Norwegian Food Table of 13 µg (170 g portion). The contribution from 170 g of chicken meat from chickens supplemented with SW or SY would be 67 ± 7 µg Se, and cover the new Se recommendations from the Nordic Nutrition Recommendations [67] of 50 µg Se/day for women and 60 µg Se/day for men.

The bioactive Se is defined as the fraction of Se from feed or food that is converted into biological active selenoproteins [52]. SeCys is incorporated into all selenoproteins and quantification of SeCys will therefore give a measure of the bioactive amounts of Se in different tissues. Hence, the reservoir of SeMet is not related to a greater bioactivity because of the unspecific incorporation into general body proteins were Se is not recognized by the organism for selenoprotein synthesis [52].

The extraction recovery after enzymatic digestion for the determination of Se-species in muscle was between 94 and 109%, whereas the extraction from liver was lower and varied from 77 to 103%, in feed the extraction was between 83 and 104%. The lower extraction efficiencies in some of the liver and feed samples may be related to a high fat content. The problems with the SeMet derivatization is likely the explanation for the low column recovery (~80%) of total Se compared to the extraction efficiency (>95%) in some of the samples, The different chemical structure of the Se binding to —H in the SeCys and $—CH_3$ could give different reaction with IAM. The derivatization of SeCys was fast and complete based on the occurrence of one narrow peak in the chromatogram by RP-ICP-MS and confirmation on ESI-MS of the CAM SeCys standard at mass to charge ratio (m/z) 227 showing the isotopic pattern of Se. ESI spectra of CAM SeMet standard did not give the expected peak at 255 m/z, but rather a peak at 194, which may be explained by the loss of $—CO_2$ (44 m/z) and NH_3 (17 m/z) from CAM SeMet [43]. The RP-chromatograms of the muscle samples showed one distinct peak of SeMet identified by standard addition at 4 min, a smaller peak identified as protein bound SeCys after 2.7 min and four smaller unidentified Se species. In the liver samples CAM SeMet eluted in two peaks (retention time of ca. 5.4 and 5.9 min) and protein bound CAM SeCys at 4 min, as for the muscle samples four smaller unknown peaks were observed in the chromatograms.

3.2.4 Se Amino Acid Concentration In Liver And Muscle

SeMet was the predominant Se amino acid in livers of the chickens receiving dietary treatments with SW or SY, whereas SeCys was the dominating Se amino acid in livers of chickens receiving SS as dietary supplement. These results are not in line with the findings of [57] were SeCys was predominant in liver tissue irrespective of treatment (with feed Se concentrations of 0.3 and 0.5 mg Se kg^{-1}). On the other hand in experiments with lamb [59] and beef cattle [59] receiving SY doses tend to result in SeMet compromising the greater proportion of total Se in the liver tissue which is in accordance with the result in the present work using elevated Se concentrations in the diet (**Figure 3**).

The different Se amino acid portioning may reflect a saturation of SeCys in the liver at higher SW or SY doses and subsequent change from incorporation into selenoproteins to the non-specific incorporation of SeMet into general liver tissue proteins [57].

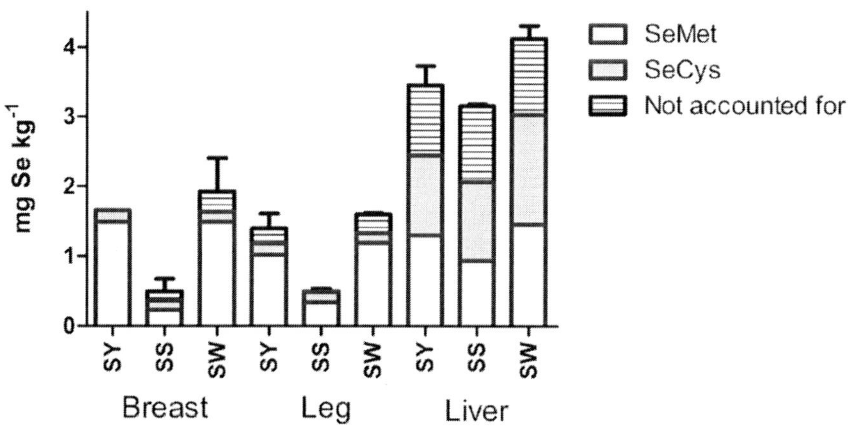

FIGURE 3. Concentration of Se amino acids in muscle and liver with regard to Se source (SY, Se-enriched yeast; SW, Se-enriched wheat and SS, sodium senenite).

The predominance of SeMet in breast tissue of chickens receiving SW or SY in the dietary treatment is in line with the observations of [42, 57].

The concentration of SeCys in breast and leg muscle tissues were irrespective of dietary treatment (0.15 ± 0.02 µg Se g^{-1}; $n = 6$) and in accordance with concentrations reported by [42], where 3 (0.11 ± 0.0, 0.18 ± 0.01 and 0.13 ± 0.08 µg Se g^{-1}) out of 4 (1.11 ± 0.35 µg Se g^{-1}) samples from chicken breast tissue had similar Se concentrations. It seems that the source of Se did not affect the concentration of SeCys in the muscle when there is an excess or adequate concentration of Se in the diet. The low level of SeMet found in tissues of chickens fed with selenite is likely due to other constituents in the diet, e.g. fishmeal and soybean [68]. The Se speciation analysis showed that the SeMet concentration was a factor 5 higher in chickens fed organic Se than in chickens fed with selenite. The present results show that Se from SS was less retained in muscle tissues compared to Se from SY and SW, and the retention observed is attributed to the incorporation as SeCys in proteins as stated by [65]. An increased organic Se source consisting primarily of SeMet would increase the SeMet concentration in the tissues and thereby the retention of Se in muscle tissue.

Even though the concentrations of Se in the liver were the highest, the effect on daily human intake would be of minor significance as the consumption of liver is low compared to the consumption of breast and leg muscle. Therefore focus should be on the increase of Se concentration in chicken breast and leg muscle by feeding organic Se to the chickens (SY and SW). The bioaccessibility of Se from the chicken meat to humans has previously been investigated by an in vitro model [41], showing relatively high accessibility of Se from chicken meat to humans (70–90% depending on the method used). Furthermore, human studies with isotopic labelled Se have demonstrated bioavailability over 90% from different food sources [69]. The Se biofortification of edible chicken meat products will increase the human Se intake and should be considered as a relevant strategy to increase the Se concentration in human plasma.

3.2.5. Amino Acid Composition

The use of organic and inorganic Se in the feed did not affect the overall amino acid distribution in the breast muscle tissues. The composition of amino acids in chicken meat is highly dependent on the composition of the diet. In this experiment the diets were added methionine (Met) ($3\,g\,kg^{-1}$), lysine (Lys) ($4\,g\,kg^{-1}$) and threonine (Thr) ($1\,g\,kg^{-1}$), other important protein sources in the feed was soybean and fish meal. According to [63] a concentration of $3\,g\,Met\,kg^{-1}$ will affect the bioavailability of SeMet and transfer to muscle, as the two amino acids are taken up by the same sodium transport system. This system is specific for neutral amino acids, resulting in inhibition of SeMet by Met as there is only trace concentration of SeMet compared to Met in the diet.

The amino acid distributions were compared to the Danish Food Composition Databank (DFCD) (Ed. 07.01) and What's In The Food You Eat Search Tool from United State Department of Agriculture, the amino acid composition of the breast muscles were comparable to what is reported in the DFCD but not from USA. This is probably because the Danish and

Norwegian diets are wheat based whereas the diets in USA are maize based. The knowledge of the interaction between Se and Met is limited [63] and since all diets had the same Met concentrations it is impossible to extract any information on the influence on SeMet in this experiment.

3.2.6. Gene Expression Of Muscle Samples

Muscular gene expressions in *Gallus gallus* as a result of inorganic versus organic Se supplement (SS and SW) were determined with the use of a 44k Chicken oligonucleotide microarray. The results indicated similar expression of genes coding for selenoproteins, which is in line with the finding of equal SeCys concentrations in the muscles. However, global gene set functional enrichment analysis (Fatiscan) revealed statistically significant enrichment ($p < 0.05$) of a number of biological processes that were dependent on the Se feed sources, such as cell growth, organ development, protein metabolism (**Table 1**). Although Bayers statistics were not able to detect changes in differential expression of single genes, fatiscan analysis based on ranking of statistical values from these gene-centric analysis were able to identify subtle changes affecting a broad set of biological responses. Future effort to elucidate the biological meaning of the findings may potentially provide a mechanistic understanding to characterize molecular signatures associated with organic and inorganic Se supplements.

TABLE 1. Differences in regulations of genes in biological processes in chicken muscle of chickens fed SW ($n = 4$) compared to muscle of chickens fed SW ($n = 4$).

Functional term	Score*
Anatomical structure development	5.67
Organ development	4.34
Anatomical structure morphogenesis	2.79
Protein catabolic process	2.31
Nervous system development	2.18
Nitrogen compound metabolic process	2.18
Skeletal system development	1.96
Cellular nitrogen compound metabolic process	1.96
Regulation of cell size	1.96
Cell growth	1.6
Brain development	1.6

*A term annotation weight can be computed as the number of sequences annotated to that term or as an annotation confluence score. This confluence score (Node Score) takes into account the number of sequences converging at one gene ontology term and penalizes by the distance to the term where each sequence actually was annotated.

[By comparing the fold change of genes involved in Se processes, it could be deduced that the use of Se-enriched wheat positively regulated developmental and metabolic processes in the muscle compared to selenite. The difference may be due to differences in uptake and bioavailability interacting (stimulatory/suppression) with biological processes. Brennan et al. [70] reported similar findings in oviducts of female chickens indicating differences in growth and metabolic patterns being higher when supplemented with SY compared to SS at lower Se concentrations (0.3 mg Se kg^{-1}).

4. CONCLUSION

Organic Se is three times more efficient than inorganic Se in increasing the chicken breast and leg muscle Se concentration, due to the high accumulation of SeMet in muscle proteins. The Se-source neither affects the muscle SeCys concentration nor the amino acid composition in the muscles. The use of SW as dietary supplement in chicken feed is an alternative to the SY based supplements to increase Se and in particular SeMet concentrations in edible chicken parts. The bioactive Se is defined as the fraction of Se from feed or food that is converted into biological active selenoproteins. SeCys is incorporated into all selenoproteins and quantification of SeCys will therefore give a measure of the bioactive amounts of Se in different tissues. A relatively high proportion of the Se from the diets were converted into bioactive SeCys in the liver (35 ± 2.2%), no significant difference between inorganic and organic Se sources. The concentration of SeCys in breast and leg muscle tissues were irrespective of dietary treatment. The results indicated similar expression of genes coding for selenoproteins, which is in line with the finding of equal SeCys concentrations in the muscles.

Due to longer retention time of SeMet in muscle, and possibly other organs, SeMet acts as a reservoir of Se in the body. The SeMet reservoirs could sustain the SeCys status of the chickens over time [57]. To our knowledge there is no known biological functions of SeMet, and the inorganic forms are readily transformed to SeCys through selenide in liver. One could argue that the chickens do not need the extra pool of Se in the form of SeMet in e.g. muscles, as the inorganic form will be sufficient to meet the chickens Se requirements. Since the feed is fortified with selenite the chickens will always have adequate access to the essential trace element. However, global gene set functional enrichment analysis revealed statistically significant enrichment of a number of biological processes that were dependent on the Se feed sources, such as cell growth, organ development and protein metabolism in favour of organic Se.

Selenium from wheat or chicken muscle could be important Se sources to ensure adequate Se intake in humans in Se sub deficient or deficient populations.

5. ACKNOWLEDGEMENTS

The authors would like to acknowledge Dr. Espen Govasmark and Dr. Katarzyna Bierla for their help with speciation analysis and Prof. Knut Erik Tollefsen with statistics and help with gene expression analysis.

REFERENCES

1. Johansson, L., Gafvelin, G., and Arner, E. S. J. (2005). Selenocysteine in proteins properties and biotechnological use. Biochimica et Biophysica Acta-General Subjects 1726, 1–13.
2. Terry, N., Zayed, A. M., de Souza, M. P., and Tarun, A. S. (2000). Selenium in higher plants. Annual Review of Plant Physiology and Plant Molecular Biology 51, 401–432.
3. Chen, G. S., and Metzenbe, R. l., (1974). Isolation and properties of selenomethionineresistant mutants of neurospora-crassa. Genetics 77, 627–638.
4. Huber, R. E., and Criddle, R. S. (1967). Comparison of chemical properties of selenocysteine and selenocystine with their sulfur analogs. Archives of Biochemistry and Biophysics 122, 164–173.
5. Lyons, M. P., Papazyan, T. T., and Surai, P. F. (2007). Selenium in food chain and animal nutrition: lessons from nature. Asian-Australasian Journal of Animal Sciences 20, 1135–1155.
6. Van Metre, D. C., and Callan, R. J. (2001). Selenium and vitamin E. Veterinary Clinics of North America-Food Animal Practice 17, 373–402.
7. Haug, A., Eich-Greatorex, S., Bernhoft, A., Hetland, H., and Sogn, T. (2008). Selenium bioavailability in chicken fed selenium-fertilized wheat. Acta Agriculturae Scandinavica Section A—Animal Science 58, 65–7.
8. Hartikainen, H. (2005). Biogeochemistry of selenium and its impact on food chain quality and human health. Journal of Trace Elements in Medicine and Biology 18, 309–318.
9. Låg, J., and Steinnes, E. (1978). Regional distribution of selenium and arsenic in humus layers of Norwegian forest soils. Geoderma 20, 3–14.
10. Eich-Greatorex, S., Sogn, T. A., Ogaard, A. F., and Aasen, I. (2007). Plant availability of inorganic and organic selenium fertiliser as influenced by soil organic matter content and pH. Nutrient Cycling in Agroecosystems 79, 221–231.
11. Sager, M. (2006). Selenium in agriculture, food, and nutrition. Pure and Applied Chemistry 78, 111–133.
12. Pyrzynska, K. (2002). Determination of selenium species in environmental samples. Microchimica Acta 140, 55–62.
13. Elrashidi, M. A., Adriano, D. C., Workman, S. M., and Lindsay, W. L. (1987). Chemical-equilibria of selenium in soils—a theoretical development. Soil Science 144, 141–152.
14. Govasmark, E., and Salbu, B. (2011). Translocation and re-translocation of selenium taken up from nutrient solution during vegetative growth in spring wheat. Journal of the Science of Food and Agriculture 91, 1367–1372.

15. Manafi, M. (2015) Comparison study of a natural non-antibiotic growth promoter and a commercial probiotic on growth performance, immune response and biochemical parameters of broiler chicks. Journal of Poultry Science, 52, 274–281.
16. Cappa, J. J., Yetter, C., Fakra, S., Cappa, P. J., DeTar, R., Landes, C., Pilon-Smits, E. A. H., and Simmons, M. P. (2015). Evolution of selenium hyperaccummulation in Stanleya (Brassicaceae) as inferred from phylogeny, physiology and X-ray microprobe analysis. New Phytologist, 2, 583–595.
17. Gissel-Nielsen, G., Gupta, U. C., Lamand, M., and Westermarck, T. (1984). Selenium in soils and plants and its importance in livestock and human nutrition. Advances in Agronomy 37, 397–460.
18. Walter, E. D., and Jensen, L. S. (1963). Effectiveness of selenium and noneffectiveness of sulfur amino acids in preventing muscular dystrophy in turkey poult. Journal of Nutrition 80, 327–331.
19. Scott M. L., Olson G., Krook L., and Brown W. R. (1967) Selenium-responsive myopathies of myocardium and of smooth muscle in young poult. Journal of Nutrition 91, 573–583.
20. Cantor and Tarino (1982) Comparative effects of inorganic and organic dietary sources of selenium on selenium levels and selenium-dependent glutathione-peroxidase activity in blood of young turkeys. Journal of Nutrition 112, 2187–2196
21. Karlsen, J. T., Norheim, G., and Froslie, A. (1981). Selenium content of norwegian milk, eggs and meat. Acta Agriculturae Scandinavica 31, 165–170.
22. Surai, P. F. (2002). Selenium in poultry nutrition 2. Reproduction, egg and meat quality and practical applications. Worlds Poultry Science Journal 58, 431–450.
23. Surai, P. F., Karadas, F., Pappas, A. C., Sparks, N. H. C., (2006). Effect of organic selenium in quail diet on its accumulation in tissues and transfer to the progeny. British Poultry Science 47, 65–72.
24. Schrauzer, G. N., Surai, P. F., (2009). Selenium in human and animal nutrition: Resolved and unresolved issues. A partly historical treatise in commemoration of the fiftieth anniversary of the discovery of the biological essentiality of selenium, dedicated to the memory of Klaus Schwarz (1914–1978) on the occasion of the thirtieth anniversary of his death. Critical Reviews in Biotechnology 29, 2–9.
25. European Food and Safety Authority (EFSA), 2011: Scientific Opinion on the safety and efficacy of selenium in the form of organic compounds produced by the selenium-enriched yeast Saccharomyces cerevisiae NCYC R645 (SelenoSource AF 2000) for all species. EFSA Journal 9 (6), 2279.
26. European Food and Safety Authority (EFSA), 2012: Scientific Opinion on safety and efficacy of selenium in the form of organic compounds produced by the selenium-enriched yeast *Saccharomyces cerevisiae* NCYC R646 (Selemax 1000/2000) as feed additive for all species. EFSA Journal 10 (7), 2778.
27. European Food and Safety Authority (EFSA), 2008: Selenium-enriched yeast as source for selenium added for nutritional purposes in foods for particular nutritional uses and foods (including food supplements) for the general population—Scientific Opinion of the Panel on Food Additives, Flavourings, Processing Aids and Materials in Contact with Food. EFSA Journal 766, 1–42.
28. Aspila, P. (1991). Metabolism of selenite, selenomethionine and feed-incorporated selenium in lactating goats and dairy-cows. Journal of Agricultural Science in Finland 63, 9–73.
29. Cozzi, G., Prevedello, P., Stefani, A. L., Piron, A., Contiero, B., Lante, A., Gottardo, F., and Chevaux, E. (2011). Effect of dietary supplementation with different sources of selenium on growth response, selenium blood levels and meat quality of intensively finished Charolais young bulls. Animal 5, 1531–1538.
30. EFSA FEEDAP Panel (EFSA Panel on Additives and Products or Substances used in Animal Feed). (2014). Scientific Opinion on the safety and efficacy of dl-selenomethionine as a feed additive for all animal species. EFSA Journal 2014;12(2):3567, 20 pp. doi:10.2903/j.efsa.2014.3567

31. Pedrero, Z., and Madrid, Y. (2009). Novel approaches for selenium speciation in foodstuffs and biological specimens: a review. Analytica Chimica Acta 634, 135–152.
32. Larsen, E. H., Lobinski, R., Burger-Meÿer, K., Hansen, M., Ruzik, R., Mazurowska, L., Rasmussen, P. H., Sloth, J. J., Scholten, O., and Kik, C. (2006). Uptake and speciation of selenium in garlic cultivated in soil amended with symbiptic fungi (mucorrhiza) and selenate. Analytical Bioanlytical Chemistry 385, 1098–1108.
33. Larsen, E. H., Sloth, S., Hansen, M., and Moesgaard, S. (2003). Selenium speciation and isotope composition in 77Se-enriched yeast using gradient elution HPLC and ICP-MS-dynamic reaction cell-MS. Journal of Analytical Atomic Spectrometry 18, 310–316.
34. Polatajko, A., Sliwka-Kaszynska, M., Dernovics, M., Ruzik, R., Encinar, J. R., and Szpunar, J. (2004). A systematic approach to selenium speciation in, selenized yeast. Journal of Analytical Atomic Spectrometry 19, 114–120.
35. Pyrzynska, K. (1998). Speciation of selenium compounds. Analytical Sciences 14, 479–483.
36. Gammelgaard, B., Jackson, M. I., and Gabel-Jensen, C. (2011). Surveying selenium speciation from soil to cell-forms and transformations. Analytical and Bioanalytical Chemistry 399, 1743–1763.
37. Gammelgaard, B., Gabel-Jensen, C., Sturup, S., and Hansen, H. R. (2008). Complementary use of molecular and element-specific mass spectrometry for identification of selenium compounds related to human selenium metabolism. Analytical and Bioanalytical Chemistry 390, 1691–1706.
38. Dumont, E., Vanhaecke, F., and Cornelis, R. (2006). Selenium speciation from food source to metabolites: a critical review. Analytical and Bioanalytical Chemistry 385, 1304–1323.
39. Encinar, J. R., Ouerdane, L., Buchmann, W., Tortajada, J., Lobinski, R., and Szpunar, J. (2003). Identification of water-soluble selenium-containing proteins in selenized yeast by size-exclusion-reversed-phase HPLC/ICPMS followed by MALDI-TOF and electrospray Q-TOF mass spectrometry. Analytical Chemistry 75, 3765–3774.
40. Moreda-Pineiro, J., Moreda-Pineiro, A., Romaris-Hortas, V., Moscoso-Perez, C., Lopez-Mahia, P., Muniategui-Lorenzo, S., Bermejo-Barrera, P., and Prada-Rodriguez, D. (2011). In-vivo and in-vitro testing to assess the bioaccessibility and the bioavailability of arsenic, selenium and mercury species in food samples. Trac-Trends in Analytical Chemistry 30, 324–345.
41. Brandt-Kjelsen, A., Govasmark, E., Vegarud, G., Haug, A., Szpunar, J., and Salbu, B. (2012). In vitro digestion of selenium from selenium-enriched chicken. Pure and Applied Chemistry 84, 249–258.
42. Bierla, K., Dernovics, M., Vacchina, V., Szpunar, J., Bertin, G., and Lobinski, R. (2008). Determination of selenocysteine and selenomethionine in edible animal tissues by 2D size-exclusion reversed-phase HPLC-ICP MS following carbamidomethylation and proteolytic extraction. Analytical and Bioanalytical Chemistry 390, 1789–1798.
43. Lipiec, E., Siara, G., Bierla, K., Ouerdane, L., and Szpunar, J. (2010). Determination of selenomethionine, selenocysteine, and inorganic selenium in eggs by HPLC-inductively coupled plasma mass spectrometry. Analytical and Bioanalytical Chemistry 397, 731–741.
44. Reyes, L. H., Marchante-Gayon, J. M., Alonso, J. I. G., and Sanz-Medel, A. (2006). Application of isotope dilution analysis for the evaluation of extraction conditions in the determination of total selenium and selenomethionine in yeast-based nutritional supplements. Journal of Agricultural and Food Chemistry 54, 1557–1563,
45. Heras, I. L., Palomo, M., and Madrid, Y. (2011). Selenoproteins: the key factor in selenium essentiality. State of the art analytical techniques for selenoprotein studies. Analytical and Bioanalytical Chemistry 400, 1717–1727.
46. Bugel, S. H., Sandstrom, B., and Larsen, E. H. (2001). Uptake and retention of selenium from shrimps in man. Journal of Trace Elements in Medicine and Biology 14, 198–204.

47. Kapolna, E., Laursen, K. H., Husted, S., and Larsen, E. H. (2012). Bio-fortification and isotopic labelling of Se metabolites in onions and carrots following foliar application of Se and Se-77. Food Chemistry 133, 650–657.
48. Suzuki, K. T. (2005). Metabolomics of selenium: Se metabolites based on speciation studies. Journal of Health Science 51, 107–114.
49. Suzuki, K. T., Doi, C., and Suzuki, N. (2006). Metabolism of Se-76-methylselenocysteine compared with that of Se-77-selenomethionine and Se-82-selenite. Toxicology and Applied Pharmacology 217, 185–195.
50. Tsuji, Y., Suzuki, N., Suzuki, K. T., and Ogra, Y. (2009). Selenium metabolism in rats with long-term ingestion of Se-methylselenocysteine using enriched stable isotopes. Journal of Toxicological Sciences 34, 191–200.
51. Finley, J. W. (2006). Bioavailability of selenium from foods. Nutrition Reviews 64, 146–151.
52. Thiry, C., Ruttens, A., De Temmerman, L., Schneider, Y. J., and Pussemier, L. (2012). Current knowledge in species-related bioavailability of selenium in food. Food Chemistry 130, 767–784.
53. Fairweather-Tait, S. J., Bao, Y. P., Broadley, M. R., Collings, R., Ford, D., Hesketh, J. E., and Hurst, R. (2011). Selenium in human health and disease. Antioxidants & Redox Signaling 14, 1337–1383.
54. Fox, T. E., Atherton, C., Dainty, J. R., Lewis, D. J., Langford, N. J., Baxter, M. J., Crews, H. M., and Fairweather-Tait, S. J. (2005). Uptake of selenium from wheat, garlic, and cod intrinsically labeled with Se-77 and Se-82 stable isotopes. International Journal for Vitamin and Nutrition Research 75, 179–186.
55. Payne, R. L., and Southern, L. L. (2005). Changes in glutathione peroxidase and tissue selenium concentrations of broilers after consuming a diet adequate in selenium. Poultry Science 84, 1268–1276.
56. Juniper, D. T., Phipps, R. H., and Bertin, G. (2011). Effect of dietary supplementation with selenium-enriched yeast or sodium selenite on selenium tissue distribution and meat quality in commercial-line turkeys. Animal 5, 1751–1760.
57. Yoon, I., Werner, T. M., and Butler, J. M. (2007). Effect of source and concentration of selenium on growth performance and selenium retention in broiler chickens. Poultry Science 86, 727–730.
58. Juniper, D. T., Phipps, R. H., Ramos-Morales, E., and Bertin, G. (2009). Effects of dietary supplementation with selenium enriched yeast or sodium selenite on selenium tissue distribution and meat quality in lambs. Animal Feed Science and Technology 149, 228–239.
59. Haug, A., Graham, R. D., Christophersen, O. A., and Lyons, G. H. (2007). How to use the worlds scarce selenium resources efficiently to increase the selenium concentration in food. Microbial Ecology in Health and Disease 19(4), 209–228.
60. Bierla, K. (2008). Analytical approaches to determination of selenium, selenoamino acids and selenium metabolites in biological materials using elemental and molecular mass spectrometry. PhD-thesis.
61. Krstic, B., et al. (2012). Options for the production of selenized chicken meat. Biological Trace Element Research 146(1) 68–72.
62. Skrivan, M., et al. (2008). Effect of dietary selenium on lipid oxidation, selenium and vitamin E content in the meat of broiler chickens. Czech Journal of Animal Science 53(7) 306–311.
63. Govasmark, E., Brandt-Kjelsen, A., Szpunar, J., Bierla, K., Vegarud, G., and Salbu, B. (2010). Bioaccessibility of Se from Se-enriched wheat and chicken meat. Pure and Applied Chemistry 82, 461–471.
64. Combs, G. F. (2001). Selenium in global food systems. British Journal of Nutrition 85, 517–547.
65. Rayman, M. P., Infante, H. G., and Sargent, M. (2008). Food-chain selenium and human health: Spotlight on speciation. British Journal of Nutrition 100, 238–253
66. Nordic Nutrition Recommendations (2012). Integrating nutrition and physical activity. © Nordic Council of Ministers, Copenhagen 2004.

67. Yoshida, S., Haratake, M., Fuchigami, T., and Nakayama, M. (2011). Selenium in seafood materials. Journal of Health Science 57, 215–224.
68. Sloth, J. J., Larsen, E. H., Bugel, S. H., and Moesgaard, S. (2003). Determination of total selenium and Se-77 in isotopically enriched human samples by ICP-dynamic reaction cell-MS. Journal of Analytical Atomic Spectrometry 18, 317–322.
69. Brennan, K. M., Crowdus, C. A., Cantor, A. H., Pescatore, A. J., Barger, J. L., Horgan, K., Xiao, R., Power, R. F., and Dawson, K. A. (2011). Effects of organic and inorganic dietary selenium supplementation on gene expression profiles in oviduct tissue from broiler-breeder hens. Animal Reproduction Science 125, 180–188.

CHAPTER 11

Phosphorus Bioavailability: A Key Aspect for Conserving this Critical Animal Feed Resource with Reference to Broiler Nutrition

Xiuhua Li *, Dagong Zhang, Tsung Y. Yang † and Wayne L. Bryden

Poultry Science Unit, School of Agriculture and Food Sciences, The University of Queensland, Gatton Qld 4343, Australia

ABSTRACT

Phosphorus (P) is an essential element, and the majority of animal feed phosphate is derived from phosphate rock that is a non-renewable resource. Current global P reserves may be depleted in 50–100 years. This poses the challenge of securing future P supply for the global animal feed industries. Currently, nutritionists formulate diets with substantial safety margins to guarantee that animals do not become P deficient. Excessive dietary P concentrations increase, not only the cost of diets, but also P excretion and pollution of the environment. We contend that understanding P bioavailability is central to the sustainable use of this mineral in animal agriculture. Poultry accounts for approximately 50% of animal feed phosphate consumption worldwide and for this reason we use the meat chicken or broiler as a case study to explore the nuances of P bioavailability. We conclude that, to tackle the challenge of dietary P bioavailability, cooperative research on a global scale is needed to standardise measurement procedures in order to produce a robust and reliable database which can be used by nutritionists to formulate diets to meet the bird's P requirements precisely. Achievement of this goal will assist endeavours to sustain the global supply of phosphorus.

Keywords: phosphorus; requirements; available P; bioavailability; poultry; animal nutrition; rock phosphate; global feed supply

1. INTRODUCTION

Phosphorus (P) is an essential element for all forms of life, from single to multicellular organisms. It is required for normal muscle growth and egg formation, is an important component of nucleic acids, the genetic code, and phospholipids, and is also a co-factor of many enzyme systems. Phosphorus plays a vital role in maintaining osmotic and acid-base balance, energy metabolism, amino acid metabolism and protein synthesis. Phosphorus was named "life's bottleneck" by chemist and science writer Isaac Asimov [1], who claims: "Life can multiply until all the phosphorus has gone, and then there is an inexorable halt which nothing can prevent," also writing, "We may be able to substitute nuclear power for coal, and plastic for wood, and yeast for meat, and friendliness for isolation—but for P there is neither substitute nor replacement."

The main ingredients of non-ruminant diets are cereal grains and their by-products. These diets usually require supplementation with inorganic P, because of the low concentrations, and low availability of total P in cereal feed ingredients. Meat and bone meal has relatively high bioavailable P content but is banned from use in animal feed in European countries. This policy increases the demand for inorganic feed phosphates [2]. The majority of feed phosphate is derived from phosphate rock that is a non-renewable resource and becoming increasingly scarce and expensive [3]. Current global P reserves may be depleted in 50–100 years [4,5]. This poses the challenge of securing future P supply for the international and national animal feed industries. Poultry account for approximately 50% of animal feed phosphate consumption worldwide [6]. Feed phosphates make a significant contribution to the P content of any broiler diet; generally providing as much as 60% of the bird's P requirements [7]. There is limited information on biologically determined P values in feed ingredients for poultry [8,9,10]. Therefore, nutritionists have to formulate diets with substantial safety margins to guarantee that birds do not become P deficient. Greater dietary P concentrations increase the cost of poultry production, and P excretion into manure. Excessive amounts of P in manure contribute to the pollution of lakes, streams and wetlands leading to surface water eutrophication. Eutrophication of fresh water globally is accelerated by P accumulation [11,12].

All these concerns have attracted much global attention and stimulated the re-examination of the use of inorganic P in animal diets. This issue is pertinent to all non-ruminant animals that require dietary P supplementation, especially poultry (broilers, layers and breeders) and pigs. In this review, the focus is on meat chickens or broilers as much of the P used in animal nutrition is in poultry diets. In an effort to optimise P use by poultry, the P requirements of the modern broiler have been re-examined [13,14,15,16,17,18] and defining P bioavailability in feedstuffs has commenced [2,19]. Mention of bioavailability, which encompasses digestion, absorption and utilisation of P, evokes much debate and is the thrust of this review.

In many instances, the terms "availability" and "bioavailability" are interchangeable when referring to nutrients. However, to avoid confusion in this review, we use the term bioavailability to encompass the digestion, absorption and utilisation of P, as the term "available P" has a specific connotation in P nutrition as described below. We contend that bioavailability is central to the sustainable use of P in animal agriculture. A thorough understanding of P bioavailability will permit the development of strategies to enhance efficiency of P utilisation, reduce wastage of P resources and reduce environmental pollution by P.

2. PHOSPHORUS METABOLISM

It is important when discussing phosphorus requirements and bioavailability to appreciate the metabolism of phosphorus (Figure 1) and to remember the close association that phosphorus homeostasis has with the metabolism of two other nutrients; calcium (Ca) and vitamin D. The major features of these interactions are briefly reviewed here and the interested reader is referred to detailed reviews on the topic [20,21]. The major store of Ca and P is the skeleton where the cationic and anionic forms, respectively, of these minerals connect to form hydroxyapatite, which confers rigidity on the bone matrix [2]. Bone is continually turned over and to ensure that the biological demands for both Ca and P are met, metabolism of these essential minerals is tightly coordinated.

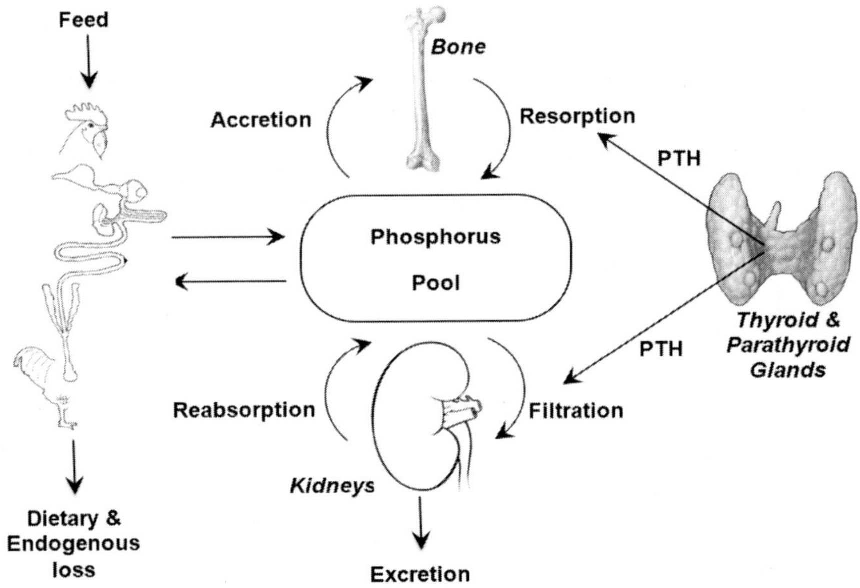

Figure 1. Phosphorus metabolism. Coordination of three organ systems, intestine, kidneys and bone, is central to Ca and P homeostasis. Reduced circulating concentrations of Ca causes increased secretion of parathyroid hormone (PTH) by the parathyroid glands. PTH increases bone resorption and renal Ca reabsorption but increased renal excretion of P. In the kidneys, PTH also stimulates formation of the hormonal form of vitamin D_3 which increases intestinal absorption of Ca and, to a lesser extent, the absorption of P. Elevated blood Ca concentrations as a result of these processes creates a negative feedback loop to ensure homeostasis. If hypercalcaemia occurs, calcitonin is secreted by the thyroid gland to inhibit bone resorption.

Utilisation of Ca and P is modulated by the relative amounts of each in the diet [2]. However, the quantities of each available for their respective metabolic functions is determined by efficiencies of intestinal absorption, glomerular filtration, renal tubular reabsorption, rates of transfer from blood to bone, and intestinal endogenous losses [20]. The efficiencies of these processes in the different tissues are modulated by several hormones, chiefly parathyroid hormone (PTH) and the hormonal form of vitamin D_3 (1,25 dihydroxycholecalciferol; calcitriol) [20,21]. The control of P metabolism differs from that of Ca, with less control in the gut and greater control in the kidneys. Moreover, there is increasing evidence that dietary protein and carbohydrate affect intestinal metabolism of Ca and P [21]. Together, the factors that determine intestinal absorption and kidney excretion along with endocrine regulation of absorption and reabsorption in the kidneys facilitate P homeostasis.

3. PHOSPHORUS REQUIREMENTS

Research on the P requirement of broilers has been the subject of numerous investigations for many decades; however, the minimum requirement for this nutrient has still not been established with any certainty. The published results are variable as different breeds of birds, age of birds, feed ingredients (with or without phytase supplementation), Ca and P sources were used in different studies. In some instances, the differences observed may reflect true differences in requirements, not merely variation resulting from experimental conditions. To further complicate interpretation of results, different criteria, including body weight gain, feed conversion ratio (FCR), tibia breaking strength, tibia ash, or toe ash contents have been used to assess responses to P (Table 1).

Table 1. Summary of available P requirements (g/kg diet) of broilers.

	AP (g/k) (Bird Age, Day)		
Criteria of Assessment	1–21	21–42	Reference
Carcass P analysis	3.7 (day 0–10), 3.0 (day 10–30)	2.4 (day 30–40)	[2]
Bone breaking strength	3.9	3.0	[13]
FCR		1.63	[22]
FCR	3.3–3.9		[23]
Growth and tibia ash	3.2–2.8 (day 18–32)	2.4–1.9 (day 32–42), 1.1 (day 43–49)	[24,25]
Growth, bone ash and processing losses	4.5	4.0	[26]
Weight gain	3.5 (day 0–14)	3.0 (day 15–49)	[15,18]
Weight gain	3.2–3.4		[27]
Weight gain		1.86	[22]
Weight gain	3.2–3.5		[23]
Weight gain	2.8–2.9		[28]
Tibia or toe ash	4.5	3.5	[29]
Tibia ash		3.3	[22]
Tibia ash	3.5–3.9		[23]
Tibia ash	3.7–3.9	3.1 (day 49)	[30]

Evidence in the literature clearly demonstrates that the P requirement of broilers is much lower than NRC [29] recommendations and the values currently used by the industry [2,7,13,15,17,18,22,23,24,25,27,28,31]. The greatest challenge in applying a lower dietary P strategy in practice is to decide what dietary concentrations of P to use in a formulation. This is due to a lack of information on biologically determined P values in feed ingredients for poultry and an agreed P evaluation system [8,9,10].

4. EVALUATION OF PROCEDURES THAT DETERMINE BIOAVAILABLE PHOSPHORUS

There are many approaches for determining P bioavailability of feed and feed ingredients. Some methods are biological and rely on *in vivo* measures including growth, bone strength, bone ash content and blood parameters. Chemical procedures have also been developed to estimate P bioavailability. Most published values using both biological and chemical procedures are qualitative rather than quantitative. Recently, there has been an international move to establish ileal digestible P as the method of choice for determining P bioavailability. An overview of the different approaches for the determination P bioavailability is given below.

4.1. Biologically Determined Values

Initial attempts to quantitate the availability of P for chicks from various phosphate sources used tibia bone ash as the response factor. The method was based on relative percentage of tibia bone ash of chicks fed a test supplement to that of chicks fed a reference standard, thus obtaining relative bioavailability P values [32]. A positive linear relationship was found between tibia bone ash percentage and dietary P intake. A similar relationship was found with toe ash percentages [33,34,35]. Hoshii and Yoshida [36] demonstrated that P bioavailability determined with a slope ratio assay using toe ash was closely associated with the values determined by carcass P retention, suggesting that changes in toe ash concentration reflect changes in net P utilisation.

In the 1980s and '90s, the bioavailability of P was routinely estimated by non-linear bioassays (asymptotic or sigmoidal curves) of body weight and tibia bone or toe ash, using a graded range of P supplements [37,38,39,40,41]. Others [42,43] determined P bioavailability using responses in bone ash content, bone breaking strength, body weight gain or blood inorganic P concentration of birds fed a test P source compared with the response of birds to a standard reference P source. Bioassays based on growth, blood parameters, tibia bone or toe ash percentage and tibia bone breaking strength can only provide a relative biological reference value or ranking [42,43] and do not measure true or apparent P bioavailability *per se*. The values do not define the quantity of bioavailable P and therefore have limited value when formulating diets [44].

Early attempts to determine true or apparent P retention by collecting excreta in balance studies with chicks were unsuccessful [45,46]. Later, Simons and Versteegh [47] were successful using a balance assay based on excreta collection, when birds were fed a basal diet, nearly free of P (5 g/kg Ca), to which test phosphates were added at low concentrations of approximately 2 g/kg AP. At this level, there is negligible urinary P excretion [47].

4.2. Chemically Determined Values

The most common P evaluation system (NRC, [29,48]) is based on the relationship of P with phytate; non-phytate P (NPP) and phytate P (PP). Phytate is the bound form in which P is found in plants; it is discussed in detail elsewhere (see [49,50,51,52,53]). The NPP system has the advantage of being based on the chemical analysis of feed ingredients. Non-phytate P is the P in the feed or ingredient that is not bound as part of phytic acid. It can be determined by subtracting analysed PP from analysed total P [14]. With this approach, it is assumed that NPP, together with P from non-plant sources are completely available to poultry and non-ruminant animals; assuming that PP is not available to non-ruminant animals. If PP in plant feedstuffs is not analysed, it is assumed that NPP is 30% of total plant P and the remaining 70% is PP. Non-phytate P has been referred to as "AP" or used interchangeably with "AP." These assumptions based on simple differentiation from chemical analysis have been adopted in diet formulation worldwide, since publication by the NRC in 1977.

The NPP system has limitations. Chemical analysis determines how much P is present, but does not indicate to what degree, if any, P is utilised by animals and birds. In other words, chemical analysis does not provide information on P bioavailability of the ingredient or diet and thus any changes in PP and P bioavailability of an ingredient will not be reflected in this system. Numerous studies have found that NPP is not 100% bioavailable and up to 80% of PP can be bioavailable to birds [8,54,55]. In fact, no elements are completely absorbed and utilised by poultry [56].

Phytate has been found in the excreta of the chicks fed a phytate-free (casein-gelatin) diet, whereas none was found in the excreta of germ-free chicks fed the same diet. It appears that phytate may be synthesised by the microflora in the avian digestive tract [57]. Therefore, it is possible that the net microbial hydrolysis of phytate occurring during transit of digesta through gastrointestinal tract may be partly offset by microbial synthesis [58]. This makes determination of dietary NPP requirement by total collection or indigestible marker and analysing NPP in excreta unreliable since some NPP may have been incorporated into phytate.

Other chemical or *in vitro* approaches have attempted to relate the solubility of P supplements to P bioavailability determined *in vivo* and have been unsuccessful [59].

4.3. Ileal Digestibility as an Estimate of Phosphorus Bioavailability

In an attempt to overcome the difficulties associated with current approaches to P bioavailability determinations, the Nutrition Working Group of the European Federation of Branches of World's Poultry Science Association (hereafter Rodehutscord and WPSA, [60]) has developed a protocol to determine ileal P digestibility by regression analysis. Two to three diets are formulated for each test P source. A low P basal diet is used, and a minimum of two levels of the P source under test are added to the test diet. At the highest level of inclusion, P supply must not exceed the requirement. With such conditions, the AP of the P source under test can be determined from the slopes of linear regression equations (see [60] for details). The protocol is currently being evaluated in an international collaborative ring test. Collinearity appears to exist between two variables used in the regression analysis and may lead to unreliable and unstable estimates of regression coefficients [61].

Finally, none of the methods outlined above measure P bioavailability *per se* but rather provide an estimate. Accurate measurements of bioavailability would require very expensive and laborious experiments, including the use of radioactive isotopes. What is required is a robust procedure that will provide reliable and repeatable estimates of P bioavailability. It would appear that precaecal P digestibility is the method of choice. A similar situation exists with amino acid availability where precaecal or ileal digestibility is used to estimate availability. Detailed discussions that describe amino acid digestibility assays are available [62].

5. TERMS USED TO DESCRIBE BIOAVAILABLE PHOSPHORUS

Over the years, a number of terms have been suggested to define certain aspects of P utilisation as described by Rodehutscord and WPSA [60] and outlined below. These terms have largely arisen to reflect differing approaches to the determination of bioavailable P.

Digestible P is the portion of dietary total P that is not recovered in faeces. The determination of digestible P requires *in vivo* studies with quantitative determination of dietary P intake and P excretion in faeces. Alternatively, digestible P can be determined by using an indigestible marker. However, as bird excreta consists of faeces and urine, determination of digestible P requires colostomised birds to exclude urinary P content.

Retainable P (RP) indicates the proportion of dietary total P that is retained in the body and can be determined quantitatively following measurement of P intake and P excretion in faeces [60]. It is also called total tract P digestibility and is easy to measure. However, as urine is included in the determination of digestible P, this term is not an appropriate description of the measurement in poultry. Importantly, if a P intake is lower than the requirement, the P content of the urine can be ignored. In these circumstances, the determination of RP should be similar to the determination of digestible P.

Available P (AP) is the quantity of P that is absorbed from the diet by the animal. This definition of AP is different from previous interpretations [29,60] where it has been equated to NPP and RP. In the currently acceptable definition of AP [60], it is determined by including an indigestible marker in the feed and is the proportion of total dietary P that is not recovered in the terminal ileum. It is also called precaecal digestible P or ileal P digestibility [60]. The major advantage of this method is that if P intake exceeds requirements, urinary excretion of P does not confound the results.

There is now general agreement that ileal digestible P is the preferred approach for determining AP as a quantitative estimate of P bioavailability. Rodehutscord and WPSA [60] recommend that retainable P values not be used in the future to avoid confusion. Retainable P reflects both P digestibility and post-absorptive utilisation since bird excreta includes faeces and urine. Dietary supply of P relative to requirement and body status affects total tract P utilisation [63]. In several studies of individual feed ingredients and mixed diets, the RP in total tract is substantially different from AP in the ileum (Table 2). Based on these and other publications that demonstrate a similar trend, both utilisation of P or Ca in feeds and requirement in birds should be expressed on an ileal digestible basis [63]. Furthermore, AP can be used to measure the marginal level of P supplied in animal diets to meet the P requirement of the animal [60].

Table 2. Comparisons of retainable P (RP) with available P (AP) in the ileum of poultry [63].

Ingredient	RP (%)	AP (%)	Reference
Low-phosphorus mixed diet	43	29	[64] [1]
Low-phytate soybean meal	77	94	[65] [2]
Conventional soybean meal	60	94	[65] [2]
Low-phosphorus mixed diet	59	52	[66] [1]
Low-phosphorus mixed diet	45	30	[67] [1]
Canola meal	39	66	[63] [2]

[1] Expressed as apparent digestibility; [2] Expressed as true digestibility derived from regression analysis.

6. PHOSPHORUS SOURCES AND AVAILABILITY

Sources of P in poultry diets include feed ingredients of both plant and animal origin and also inorganic forms of the mineral. These sources differ in P bioavailability.

6.1. Plant Feedstuffs

Plant feedstuffs include cereal grains, oil seeds and the respective by-products. Phytate is the principal storage form of P in many plant tissues. As commercial broiler diets are primarily comprised of feedstuffs of plant origin, most of the P present in these ingredients is in the form of phytate P (PP) [68]. The concentration of PP within the same ingredient can vary considerably. Barrier-Guillot *et al.* [69] found that the concentration of PP ranged from 0.092% to 0.268% dry matter (DM) in wheat and depended on soil fertility, plant variety, and stage of plant maturity at harvest.

The bioavailability of PP for broilers ranges from 0% to 80% (Table 3) [2] depending on diets, bird age and metabolic adaptation. The degree to which PP is utilised by birds will depend, to a large extent, upon the hydrolysis of PP in the gastrointestinal tract. Moreover, the bioavailability of PP in different feed ingredients ranges from 0% to 60% [8,47,67], while the bioavailability of total P within ingredients varies considerably (15% to 75%; Li *et al.*, unpublished data [70]). There are many factors affecting PP utilisation: phytase activity of dietary ingredients, dietary concentrations of fibre, Ca, P and vitamin D_3, age of bird and genotype [58,71].

Table 3. Total P (TP) (g/kg), retainable P (RP) (% of TP), non-phytate P (NPP) (g/kg), phytate P (PP) (g/kg), PP (% of TP) and calculated PP degradability (%) of some of the feedstuffs in three-week-old broilers [2].

Feedstuffs	TP (g/kg)	RP (% TP) *	NPP (g/kg)	PP (g/kg)	PP (% TP)	PP Degradability (%) **
Beans	4.9	52	1.3	3.6	73.5	53
Lupin	3	72	1.5	1.5	50	80
Maize	3	29	0.7	2.3	76.7	16
Peas	4.1	41	1.5	2.6	63.4	23
Rapeseed	10.9	33	3.8	7.1	65.1	10
Rice bran	17.2	16	3.1	14.1	82	2
Soy bean (heat treated)	5.5	54	2	3.5	63.6	49
Soybean meal	7.1	61	2.8	4.3	60.6	61
Sunflower meal	11.9	38	4.2	7.7	64.7	19
Wheat	3.4	48	0.9	2.5	73.5	46
Wheat middlings	10.8	36	2.8	8	74.1	26

* Values were determined in broilers fed diets containing 1.8 g calculated retainable P and 5.0 g Ca per kg diet from 10 days of age onward. P retention was measured in a three-day balance period (from days 21 to 24) in which dietary P intake and excreta P output were measured quantitatively; ** Phytate P degradability was calculated as: 100 × (AP/0.8–NPP)/PP [2].

6.2. Animal Feedstuffs

There are a number of animal-sourced feedstuffs (fish meals, meat meals, meat and bone meals) used in poultry diets mainly as protein supplements. Among these, meat meal is also a potential P source depending on its bone content. Meat meals are prepared by rendering offal from abattoirs. There can be a wide variation between batches depending on the raw materials used and rendering conditions [72]. If the ash content is high, this indicates that it contains a higher amount of bone and is referred to as meat and bone meal. If the ash content is low it is referred to as meat meal.

Meat and bone meal is a variable source of protein for poultry diets, but can be an excellent source of Ca and P and other minerals. Therefore, it is commonly used as a P supplement in poultry diets [7]. However, there are few reports that have measured AP from animal-derived feedstuffs. Limited RP values of animal-derived feedstuffs are shown in Table 4.

Table 4. Total P (TP) (g/kg), retainable P (RP) (% of TP) of animal-derived feedstuffs in three-week-old broilers [2].

Animal Feedstuffs	TP (g/kg)	RP (% TP) *
Bone meal	76	59
Fish meal	22	74
Meat meal	29	65
Meat and bone meal	60	66

* Values were determined in broilers fed diets containing 1.8 g calculated retainable P and 5.0 g Ca per kg diet from 10 days of age onward. P retention was measured in a three-day balance period (from days 21 to 24) in which dietary P intake and excreta P output were measured quantitatively.

6.3. Inorganic Phosphorus Sources

Inorganic P sources are commonly derived from natural rock phosphates. These natural phosphates are not readily utilised by poultry or other animals but must be chemically modified before inclusion in animal feed. Rock phosphates are changed into various orthophosphate forms: monocalcium phosphate, dicalcium phosphate, monodicalcium phosphate, defluorinated rock phosphate and monosodium phosphate [44]. Some commonly used mineral sources and their respective P composition and retention rate are listed in Table 5.

Table 5. Inorganic feed phosphates and their respective total P (TP) (g/kg) and retainable P (RP) (% of TP) [2].

Inorganic P Source	TP (g/kg)	RP (% TP) *	Reference
Calcium sodium phosphate	180	59	[2] *
Dicalcium phosphate	183	83	[74] 1,2
Dicalcium phosphate (anhydrous)	197	55	[2] *
Dicalcium phosphate (hydrous)	181	77	[2] *
Monocalcium phosphate	226	84	[2] *
Monodicalcium phosphate (hydrous)	213	79	[2] *
Monodicalcium phosphate	203	77	[13] 1
Monodicalcium phosphate	200	80	[13] 1
Monodicalcium phosphate	216	81	[13] 1
Monosodium phosphate	224	92	[2] *
Defluorinated phosphate	182	86	[74] 1,2
Defluorinated phosphate	179	76	[74] 1,2

* Values were determined in broilers fed diets containing 1.8 g calculated retainable P and 5.0 g Ca per kg diet from 10 days of age onward. P retention was measured in a three-day balance period (from days 21 to 24) in which dietary P intake and excreta P output were measured quantitatively; 1 Apparent utilisation of P from inorganic sources by broiler as determined under deficiency conditions; 1,2 Retainable P determined through broken line slope response.

Most of these studies determined apparent retention of P from each of the inorganic sources within the deficiency range of P. Leske and Coon [13] showed dramatic reductions in P retention from monocalcium phosphate as dietary P concentration approached the requirement level (98% retention at half of the P requirement to 59% retention at requirement level). Waldroup [73] reported that nearly 50% of excreted P is likely to be of inorganic origin.

Phosphorus must be in the phosphate form to be absorbed by birds. Improper processing and higher temperatures have a detrimental impact on the biological availability of calcium phosphates. As phosphates are heated, pyro- and meta- complexes are formed which greatly reduce the P bioavailability of inorganic sources. The availability of inorganic P sources is substantially affected by hydration, with hydrated phosphate sources generally more biologically available than anhydrous sources of the same type [75]. Feed phosphate with larger particle size has been shown to be more biologically available. Larger particle phosphate is retained longer in the gizzard allowing more complete digestion [76].

7. FACTORS AFFECTING PHOSPHORUS AVAILABILITY

Determination of the bioavailability of P in feedstuffs is a major challenge. There are many factors affecting P bioavailability in birds, including: experimental techniques; the chemical form of P; dietary concentrations of Ca, vitamin D3, energy, protein, fat, and vitamin K; the availability and interaction of other nutrients in the diet, feed processing, and particle size; bird physiological and health factors, such as feed consumption, growth rate, sex and age; management factors, including ambient temperature and lighting program. Some of these factors have been investigated in detail [14] and the major factors are briefly described below.

7.1. Phosphorus and Phytate

High levels of dietary inorganic P reduce PP availability [49,77] due to either the inhibitory effect of P on phytase activity [78], a negative feedback mechanism by excessive amounts of dietary P, or inhibition by products of phytate hydrolysis [49,58].

Phytate is the principal storage form of P in plant feedstuffs and is considered to be an anti-nutritive factor [79]. Phytate carries a strong negative charge and is capable of binding di- and trivalent cations such as calcium, cobalt, copper, iron, magnesium, manganese, nickel and zinc in very stable complexes [58,80], reducing the availability of P as well as these minerals to the animal [2,54,55,81].

Phytate has been shown to inhibit activities of some digestive enzymes such as pepsin, α-amylase and trypsin. Phytate may inhibit proteolysis by changing the protein configuration of digestive enzymes [82]. Inhibition may also result from the chelation of Ca ions which are essential co-factors for the activity of trypsin and α-amylase leading to reduced protein and starch digestion [50,83,84,85,86].

It has also been reported that phytate can reduce fat digestibility by forming insoluble Ca-phytate complexes with fatty acids in the gut lumen [87]. In its chelated form, the phytate molecule is difficult to hydrolyse by phytase. The pH affects the solubility of phytate. Most phytate mineral complexes are soluble at low pHs (less than 3.5) with maximum insolubility occurring between 4 and 7 [88]. Champagne [89] found that Ca-phytate complexes precipitate at pH between 4 and 6, which is the approximate pH of the intestine where the Ca ions should be absorbed. Taylor [90] has suggested that the primary factor affecting PP utilisation is the Ca ion concentration in the small intestine where insoluble Ca-phytate complexes form. A precipitated phytate mineral complex would not be accessible for hydrolysis or absorption in the intestine.

7.2. Calcium

Calcium is an essential element for bone and egg shell formation, blood clotting, muscle contraction and transmission of nerve impulses. Calcium is also an important co-factor for many enzymes and hormones.

Increases in dietary P and Ca concentrations may affect apparent digestion of these nutrients. High levels of dietary Ca and low levels of P had detrimental effects on broiler performance [16,17,91,92]. Elevated dietary Ca levels increase pH in the gut and, as a result, P absorption [93] and retention [91] are decreased. High dietary Ca concentration can significantly increase gastrointestinal pH [94] and decrease not only phytate hydrolysis, but also pepsin activity in the proventriculus/gizzard and reduce apparent ileal protein digestibility [51]. High plasma P levels decrease Ca absorption from the gut [95]. The optimum dietary Ca-to-P ratio is of utmost importance for broiler production. With the Ca: P ratio, there will be a range of values that can meet metabolic demands for any production outcome. When dietary P concentration is high, birds can tolerate a wider range of Ca levels than when dietary P marginally meets the requirement (Li *et al.* unpublished data [96]).

Total Ca values are currently used for feed formulation since the bioavailability of Ca for feed ingredients have not been determined. Bioavailable Ca from plant raw materials is expected to be low. This is due to the high phytate content of these raw materials and the low contribution of Ca to the diet. Also, to some extent, phytate will negatively influence the availability of Ca from this dietary source by binding Ca in phytate-mineral complexes. Nevertheless, there is a need to determine the bioavailability of Ca in feed ingredients to further delineate the interaction of this nutrient with other nutrients and to improve the precision of feed formulation.

7.3. Vitamin D

Vitamin D_3 is essential for absorption of Ca and P utilization as depicted in Figure 1. Several reports have demonstrated that dietary addition of vitamin D_3 can significantly enhance the retention of P in birds [52]. However, excess vitamin $D_3$and its metabolites have not been shown to have a further beneficial effect on bird productivity when birds are already consuming adequate vitamin D_3. Like many other minerals, understanding of the molecular mechanisms of P absorption, especially the role of the hormonal form of vitamin D_3, is still limited.

7.4. Phytase

Numerous studies have demonstrated that dietary addition of phytase increases hydrolysis of PP and improves P utilisation in pigs and poultry [15,16,17,18,53]. Maximum utilisation of phytate by broilers fed diets with supplemental phytase was approximately 50% on average [97] following dietary P supplementation. There are many factors affecting the efficacy of microbial phytase such as phytase source and dose [98], physical factors of feedstuffs, feed particle size, Ca to P ratio [99], Ca bioavailability and inorganic P supplements, vitamin D status [100], bird age [101] and body weight [102], gut pH, digesta retention time, and Ca and P status. Phytase was found to be most efficacious when added to diets containing low AP and adequate Ca concentrations (Li *et al.* unpublished data [96]).

Dietary Ca level affects the efficacy of phytase in broilers [103], but not to the same extent for all phytases [104]. The same source of phytase can produce different responses in different flocks [105,106]. Moreover, the efficacy of feed phytase may be improved by the simultaneous dietary addition of other exogenous enzymes. All these variables make predicting phytase responses very difficult and lead to variation in P availability.

7.5. Birds

The ability of poultry to absorb dietary P is known to be influenced by age and physiological state. Young chicks with a rapidly developing skeletal system are critically sensitive to P deficiency [107] and tend to use P more efficiently than do mature birds [108]. Older birds have a greater ability to use PP, possibly due to more phytase activity in their intestinal tract [56]. A concern often raised by commercial nutritionists is the relevance of digestibility values generated with birds of one age group to birds of another age or other species.

Applegate *et al.* [106] conducted two experiments in which chicks were fed the same diet and found that the apparent PP hydrolysis varied nearly two-fold between the first and second experiment. Similar results have been reported by Nelson and Kirby [109] and Tamim *et al.* [103]. Bird variations make it difficult to predict treatment responses of different flocks. The variation between experiments and between flocks arises from many sources, some of which are summarised in Table 6.

Table 6. Summary of variation in processes associated with feeding of phosphorus to broilers [10].

Source of Variation	Coefficient of Variation (%)
Sampling variation	5 to 10
Analytical variation	5
Mixer variation	5 to 10
Bird utilisation	16
Ingredient variation (corn and soybean meal)	8 to 13

8. CONCLUSIONS

Re-examination of the phosphorus requirement of modern broilers has demonstrated that it is lower than the NRC [29] recommendations and the values currently used by the chicken meat industry. However, there are practical limitations in applying this information to industry as there is limited data on the bioavailability and variability of P in feed ingredients; there are few reports that have measured P bioavailability of feedstuffs for poultry. Furthermore, the different methods used have made comparisons between the studies and application to practice extremely difficult. It is only when commercial poultry nutritionists have this information that they will be in a position to reduce the safety margins for P when formulating diets. To tackle these challenges, cooperative research on a global scale is needed to standardise measurement procedures in order to produce a robust and reliable database which can be used by nutritionists to formulate diets to meet the bird's P requirement precisely; this approach is appropriate as only a couple broiler strains dominate international markets. As a starting point, the universal adoption of ileal P digestibility [60] as the preferred procedure to determine AP would be a major step forward. Moreover, to make sense of the data derived and to allow comparison of the results, it will be necessary to report ingredient composition of experimental diets, dietary concentrations of Ca and non-phytate P, dietary electrolyte balance and mineral content, all of which influence P absorption.

Achievement of these goals will assist endeavours to sustain the global supply of phosphorus.

ACKNOWLEDGMENTS

The Australian Government Rural Industries Research and Development Corporation -Chicken Meat Program, Canberra, ACT, has supported the phosphorus research conducted by the authors.

AUTHOR CONTRIBUTIONS

All the authors contributed equally to the paper.

REFERENCES

1. Asimov, I. *Asimov on Chemistry*; Doubleday: Garden City, NY, USA, 1974.
2. Van der Klis, J.D.; Versteegh, H.A.J. Phosphorus nutrition of poultry. In *Recent Developments in Poultry Nutrition 2*; Garnsworthy, P.C., Wiseman, J., Eds.; Nottingham University Press: Nottingham, UK, 1999; pp. 309–320.
3. Shastak, Y.; Witzig, M.; Hartung, K.; Rodehutscord, M. Comparison of retention and prececal digestibility measurements in evaluating mineral phosphorus sources in broilers. *Poult. Sci.* **2012**, *91*, 2201–2209.
4. Cordel, D.; Drangert, J.; White, S. The story of phosphorus: Global food security and food for thought. *Glob. Environ. Chang.* **2009**, *19*, 292–305.
5. Neset, T.S.; Cordell, D. Global phosphorus scarcity: Identifying synergies for a sustainable future. *J. Sci. Food Agric.***2012**, *92*, 2–6.
6. Devereux, C.; Smart, M.; Kalt, F.P.; Takei, N. Animal feeds: Phosphate supplements. In *CEH Marketing Research Report*; HIS Chemical: Englewood, CO, USA, 1994. [Google Scholar]
7. Waldroup, P.W. Nutritional approaches to reducing phosphorus excretion by poultry. *Poult. Sci.* **1999**, *78*, 683–691.
8. Van der Klis, J.D.; Versteegh, H.A.J. Phosphorus nutrition in broilers. In *Recent Advances in Animal Nutrition*; Garnsworty, P.C., Wiseman, J., Haresign, W., Eds.; Nottingham University press: Nottingham, UK, 1996; pp. 71–83.
9. Cossa, J.; Oloffs, K.; Kluge, H.; Jeroch, H. Investigation into the TP and PP content in different varieties of grain maize. In Proceedings of the 11th European Symposium on Poultry Nutrition; World's Poultry Science Association: Faaberg, Denmark, 1997; pp. 444–446.
10. Applegate, T.J.; Angel, R. Feasibility *versus* Practicality of Phosphorus Reduction in Poultry: Progress and Future Needs. USDA National Agricultural Library: West Lafayette, IN, USA. Available online: http://www.cals.ncsu.edu /waste_mgt/ natlcenter/sanantonio/ApplegateandAngel.pdf
11. Schindler, D.W. Evolution of phosphorus limitation in lakes. *Sciences* **1977**, *195*, 260–262. [Google Scholar] [CrossRef] [PubMed]

12. Sharpley, A.N.; Chapra, S.C.; Wedepohl, R.; Sims, J.T.; Daniel, T.C.; Reddy, K.R. Managing agricultural phosphorus for protection of surface waters: Issues and options. *J. Environ. Qual.* **1994**, *23*, 437–451.
13. Leske, K.; Coon, C.N. The development of feedstuff retainable phosphorus values for broilers. *Poult. Sci.* **2002**, *81*, 1681–1693.
14. Angel, R. Calcium and phosphorus requirements in poultry. In Proceedings of the 1st International Phytase Summit, Washington, DC, USA, 28–30 September 2010; pp. 65–71.
15. Li, X.; Zhang, D.; Huang, K.H.; Bryden, W.L. Available phosphorus requirement of broilers fed sorghum based diet from day 22 to 49. In Proceedings of the 19th European Symposium on Poultry Nutrition, Potsdam, Germany, 26–29 August 2013; p. 200.
16. Li, X.; Zhang, D.; Huang, K.H.; Bryden, W.L. Dietary calcium levels and non-phytate phosphorus requirement of broilers from days 1 to 21. *Aust. Poult. Sci. Symp.* **2013**, *24*, 50.
17. Yang, T.Y.; Setiyawan, E.; Li, X.; Zhang, D.; Bryden, W.L. Dietary calcium levels and non-phytate phosphorus requirement of broilers from days 22–49. In Proceedings of Recent Advances in Animal Nutrition in Australia 2013; University of New England: Armidale, New South Wales, Australia, 2013.
18. Zhang, D.; Li, X.; Huang, K.H.; Bryden, W.L. Available phosphorus requirement of broilers fed sorghum based diet from day 1 to 21. In Proceedings of the 19th European Symposium on Poultry Nutrition, Potsdam, Germany, 26–29 August 2013; p. 196.
19. Li, X.; Zhang, D.; Bryden, W.L. Bioavailability of phosphorus in feedstuffs for broilers. In *Final Report for Rural Industries Research and Development Corporation*; Rural Industries Research and Development Corporation: Kingston, Australia, 2015.
20. Klasing, K.C. *Comparative Avian Nutrition*; CAB International: Wallingford, UK, 1998.
21. Vitti, D.M.S.S., Kebreab, E., Eds.; *Phosphorus and Calcium Utilization and Requirements in Farm Animals*; CAB International: Wallingford, UK, 2010.
22. Yan, F.; Kersey, J.H.; Waldroup, P.W. Phosphorus requirements of broiler chicks three to six weeks of age as influenced by phytase supplementation. *Poult. Sci.* **2001**, *80*, 455–459.
23. Persia, M.E.; Saylor, W.W. Effects of broiler strain, dietary nonphytate phosphorus, and phytase supplementation on chick performance and tibia ash. *J. Appl. Poult. Res.* **2006**, *15*, 72–81.
24. Angel, R.; Applegate, T.J.; Christman, M. Effect of dietary non-phytate phosphorus (nPP) on performance and bone measurements in broilers fed on a four-phase feeding system. *Poult. Sci.* **2000**, *79* (Suppl. 1), 21–22.
25. Angel, R.; Applegate, T.J.; Christman, M.; Mitchell, A.D. Effect of dietary non-phytate phosphorus (nPP) level on broiler performance and bone measurements in the starter and grower phase. *Poult. Sci.* **2000**, *79* (Suppl. 1), 22.
26. Moran, E.T., Jr.; Todd, M.C. Continuous submarginal phosphorus with broilers and the effect of preslaughter transportation: Carcass defects, further-processing yields, and tibia-femur integrity. *Poult. Sci.* **1994**, *73*, 1448–1457.
27. Waldroup, P.W.; Kersey, J.H.; Saleh, E.A.; Fritts, C.A.; Yan, F.; Stilborn, H.L.; Crum, R.C., Jr.; Raboy, V. Non-phytate phosphorus requirement and phosphorus excretion of broilers fed diets composed of normal or high available phosphate corn with and without microbial phytase. *Poult. Sci.* **2000**, *79*, 1451–1459.
28. Yan, F.; Waldroup, P.W. Non-phytate phosphorus requirement and phosphorus excretion of broiler chicks fed diets composed normal or high available phosphate corn as influenced by phytase supplementation and vitamin D source. *Int. J. Poult. Sci.* **2006**, *5*, 219–228.
29. NRC. *Nutrient Requirements of Poultry*, 9th ed.; National Academy Press: Washington, DC, USA, 1994.

30. Yan, F.; Kersey, J.H.; Fritts, C.A.; Waldroup, P.W. Phosphorus requirements of broiler chicks six to nine weeks of age as influenced by phytase supplementation. *Poult. Sci.* **2003**, *83*, 294–300.
31. Dhandu, A.S.; Angel, R. Broiler non-phytin phosphorus requirement in the finisher and withdrawal phases of a commercial four-phase feeding system. *Poult. Sci.* **2003**, *82*, 1257–1265.
32. Gillis, M.B.; Norris, L.C.; Heuser, G.F. Studies on the biological value of inorganic phosphates. *J. Nutr.* **1954**, *52*, 115–125.
33. Grau, C.R.; Zweigart, P.A. Phosphatic clay as a phosphorus source for chicks. *Poult. Sci.* **1953**, *32*, 500–503.
34. Fritz, J.C.; Roberts, T.; Bochne, J.W.; Hove, E.L. Factors affecting the chick's requirement for phosphorus. *Poult. Sci.***1969**, *48*, 307–320.
35. Yoshida, M.; Hoshii, H. Improvement of biological assay to determine available phosphorus with growing chicks. *J. Poult. Sci.* **1977**, *14*, 33–43.
36. Hoshii, H.; Yoshida, M. Comparison of phosphorus availabilities estimated biologically based on toe ash content and carcass phosphorus retention. *J. Poult. Sci.* **1977**, *14*, 279–283.
37. Potchanacom, M.; Potter, L.M. Biological values of phosphorus from various sources for young turkeys. *Poult. Sci.***1987**, *66*, 505–513.
38. Potter, L.M. Bio-availability of phosphorus from various phosphates based on body weight and toe ash measurements. *Poult. Sci.* **1988**, *67*, 96–102.
39. Potter, L.M.; Potchanacom, M.; Ravindran, V.; Kornegay, E.T. Bio-availability of phosphorus in various phosphate sources using body weight and toe ash as response criteria. *Poult. Sci.* **1995**, *74*, 813–820.
40. Ketels, E.; De Groote, G. The relative bio-availability and the ileal digestibility of phosphorus in mineral and animal sources. In Proceedings of the 18th World's Poultry Congress, Nagoya, Japan, 4–9 September 1988; pp. 873–875.
41. De Groote, G.; Ketels, P.; Seynaeve, P. Relative bio-availability of phosphorus in meat meals as influenced by particle size. In Proceedings of the International Poultry Congress, Istanbul, Turkey, 6–10 September 1991; pp. 42–52.
42. De Groote, G. Biological availability of phosphorus in feed phosphates for broilers. In Proceedings of the 4th European Symposium on Poultry Nutrition, Tours, France, 17–20 October 1983; Larbier, M., Ed.; pp. 91–102.
43. Sauveur, B. Bio-availability to poultry of plant-origin phosphorus-Methodological criticisms and results. In Proceedings of the 4th European Symposium on Poultry Nutrition, Tours, France, 17–20 October 1983; Larbier, M., Ed.; pp. 103–113.
44. Payne, S.G. The Phosphorus Availability of Feed Phosphates in Broilers. MS Thesis, University of Stellenbosh, Stellenbosh, South Africa, 2005.
45. Edwards, H.M., Jr.; Gillis, M.B. A chromic oxide balance method for determining phosphate availability. *Poult. Sci.***1959**, *38*, 569–574.
46. Sibbald, I.R. Measurement of mineral bio-availability: Extension of true metabolisable energy methodology. *Poult. Sci.* **1982**, *61*, 485–487.
47. Simons, P.C.M.; Versteegh, H.A.J.; Jongbloed, A.W.; Kemme, P.A.; Slump, P. Improvement of phosphorus availability by microbial phytase in broilers and pigs. *Br. J. Nutr.* **1990**, *64*, 525–540.
48. NRC. *Nutrient Requirements of Poultry*, 7th ed.; National Academy Press: Washington, DC, USA, 1977.
49. Ankra-Badu, G.A.; Aggrey, S.E.; Pesti, G.M.; Bakalli, R.I.; Edwards, H.M., Jr. Modelling of parameters affecting phytate phosphorus bioavailability in growing birds. *Poult. Sci.* **2004**, *83*, 1083–1088.
50. Thompson, L.U.; Yoon, H. Starch digestibility as affected by poly phenols and phytic acid. *J. Food Sci.* **1984**, *49*, 1228–1229.
51. Walk, C.L.; Bedford, M.R.; McElroy, A.P. Influence of limestone and phytase on broiler performance, gastrointestinal pH, and apparent ileal nutrient digestibility. *Poult. Sci.* **2011**, *91*, 1371–1378.

52. Ravindran, V.; Bryden, W.L.; Kornegay, E.T. Phytates: Occurrence, bioavailability and implications in poultry nutrition. *Poult. Avian Biol. Rev.* **1995**, *6*, 125–143. [Google Scholar]
53. Selle, P.H.; Ravindran, V. Microbial phytase in poultry nutrition. *Anim. Feed Sci. Technol.* **2007**, *135*, 1–41.
54. Angel, R.; Saylor, W.W.; Mitchell, A.D.; Powers, W.; Applegate, T.J. Effect of dietary phosphorus, phytase, and 25-hydroxycholecalciferol on broiler chicken bone mineralization, litter phosphorus, and processing yields. *Poult. Sci.***2006**, *85*, 1200–1211.
55. Rodehutscord, M.; Dieckmann, A.; Witzig, M.; Shastak, Y. A note on sampling digesta from the ileum of broilers in phosphorus digestibility studies. *Poult. Sci.* **2012**, *91*, 965–971.
56. Axe, D. Phosphorus value in ingredient sources examined. *Feedstuffs* **1998**, *70*, 17–28.
57. Savage, J.E.; Yohe, J.M.; Pickett, E.E.; O'Dell, B.L. Zinc metabolism in the growing chick. Tissue concentration and effect of phytate on absorption. *Poult. Sci.* **1964**, *43*, 420–426.
58. Wise, A. Dietary factors determining the biological activities of phytase. *Nutr. Abs. Rev.* **1983**, *53*, 791–806.
59. Shastak, Y.; Rodehtscord, M. Determination and estimation of phosphorus availability in growing poultry and their historical development. *World's Poult. Sci. J.* **2013**, *69*, 569–586.
60. Rodehutscord, M.; WPSA (Working Group No. 2: Nutrition of the European Federation of Branches of World Poultry Science Association). Determination of phosphorus availability in poultry. *World's Poult. Sci. J.* **2013**, *69*, 687–698.
61. Allison, P. When Can You Safely Ignore Multicollinearity? 2012. Available online: http://statisticalhorizons.com/multicollinearity
62. Bryden, W.L.; Li, X. Amino acid digestibility and poultry feed formulation: Expression, limitations and application. *Rev. Bras. Zootec.* **2010**, *39* (Suppl.), 279–287.
63. Adeola, O.; Applegate, T.J. Phosphorus and calcium equivalency of phytase. In Proceedings of the 1st International Phytase Summit, Washington, DC, USA, 28–30 September 2010; pp. 45–64.
64. Adedokun, S.A.; Sands, J.S.; Adeola, O. Determining the equivalent phosphorus released by an *Escherichia coli* derived phytase in broiler chicks. *Can. J. Anim. Sci.* **2004**, *84*, 437–444.
65. Dilger, R.N.; Adeola, O. Estimation of true phosphorus digestibility and endogenous phosphorus loss in growing chicks fed conventional and low-phytate soybean meals. *Poult. Sci.* **2006**, *85*, 661–668.
66. Jendza, J.A.; Dilger, R.N.; Sands, J.S.; Adeola, O. Efficacy and equivalency of an Escherichia coli-derived phytase for replacing inorganic phosphorus in the diets of broiler chickens and young pigs. *J. Anim. Sci.* **2006**, *84*, 3364–3374.
67. Nyannor, E.K.D.; Adeola, O. Corn expressing an *Escherichia coli*-derived phytase gene: Comparative evaluation study in broiler chicks. *Poult. Sci.* **2008**, *87*, 2015–2022.
68. Nelson, T.S. The hydrolysis of phytate phosphorus by chicks and laying hens. *Poult. Sci.* **1976**, *55*, 2262–2264.
69. Barrier-Guillot, B.; Casado, P.; Maupetit, P.; Jondreville, C.; Gatel, F. Wheat phosphorus availability: 1-*In vitro* study; Factors affecting endogenous phytasic activity and phytic phosphorus content. *J. Sci. Food Agric.* **1996**, *70*, 62–68.
70. Li, X.; Zhang, D.; Bryden, W.L. University of Queensland: Gatton, Australia Unpublished Work. , 2015. University of Queensland: Gatton, Australia Unpublished Work. , 2015.
71. De Groote, G.; Huyghebaert, G. The bio-availability of phosphorus from feed phosphates for broilers as influenced by bio-assay method, dietary Ca-level and feed form. *Anim. Feed Sci. Technol.* **1997**, *69*, 329–340.

72. Skurray, G.R. The nutritional evaluation of meat meals for poultry. *World's Poult. Sci. J.* **1974**, *30*, 129–136.
73. Waldroup, P.W. Phosphorus, phytase, and the environment—A retrospective look. *Proc. MD Nutr. Conf. Feed Manuf.* **2002**, *49*, 195–202.
74. Coon, C.N.; Seo, S.; Manangi, M.K. The determination of retainable phosphorus, relative biological availability, and relative biological value of phosphorus sources for broiler. *Poult. Sci.* **2007**, *86*, 857–868.
75. Gillis, M.B.; Edwards, H.M., Jr.; Young, R.J. Studies on the availability of calcium orthophosphates for chickens and turkeys. *J. Nutr.* **1962**, *78*, 155–162.
76. Lima, F.R.; Mendonca, C.X., Jr.; Alvarez, J.C.; Garzillo, J.M.F.; Ghion, E.; Leal, P.M. Biological evaluations of commercial dicalcium phosphates as sources of available phosphorus for broiler chicks. *Poult. Sci.* **1997**, *76*, 1707–1713.
77. Sanders, A.M.; Edwards, H.M., Jr.; Roland, G.N. Calcium and phosphorus requirements of the chick as determined response surface analysis. *Br. J. Nutr.* **1992**, *67*, 21.
78. Ravindran, V.; Cabahug, S.; Ravindran, G.; Selle, P.H.; Bryden, W.L. Response of broiler chickens to microbial phytase supplementation as influenced by dietary phytic acid and non-phytate phosphorus levels. II. Effects on apparent metabolisable energy, nutrient digestibility and nutrient retention. *Br. Poult. Sci.* **2000**, *41*, 193–200.
79. Bryden, W.L.; Selle, P.H.; Ravindran, V.; Acamovic, T. Phytate: An anti-nutrient factor in animal diets. In *Poisonous Plants: Global Research and Solutions*; Panter, K.E., Wierenga, T.L., Pfister, J.A., Eds.; CABI Publishing: Wallingford, UK, 2007; pp. 279–284.
80. Maenz, D.D.; Engle-Schaan, C.M.; Newkirk, R.W.; Classen, H.L. The effect of minerals and mineral chelators on the formation of phytase-resistant and phytase susceptible forms of phytic acid in solution and in a slurry of canola meal. *Anim. Feed Sci. Technol.* **1999**, *81*, 177–192.
81. Pallauf, J.; Rimbach, G.; Pippig, S.; Schindler, B.; Most, E. Effect of phytase supplementation to a phytate-rich diet based on wheat, barley and soya on the bioavailability of dietary phosphorus, calcium, magnesium, zinc and protein in piglets. *Agribiol. Res.* **1994**, *47*, 39–48.
82. Singh, M.; Krikorian, A.D. Inhibition of trypsin activity *in vitro* by phytase. *J. Agric. Food Chem.* **1982**, *30*, 799–800.
83. Deshpande, S.S.; Cheryan, M. Effects of phytate, divalent cations and their implications on alpha-amylase activity. *J. Food Sci.* **1984**, *49*, 516–519.
84. Liener, I.E. Anti-nutritional factors in legume seeds: State of art. In *Recent Advances of Research on Anti-Nutritional Factors in Legume Seeds*; Huismen, J., Van Der Poel, T.F.B., Liener, I.E., Eds.; Pudoc: Wageningen, The Netherlands, 1989; pp. 6–13.
85. Knuckles, B.E.; Betschart, A.A. Effect of phytate and other myo-inositol phosphate esters on alpha-amylase digestion of starch. *J. Food Sci.* **1987**, *52*, 719–721.
86. Caldwell, R.A. Effect of calcium and phytic acid on the activation of trypsinogen and the stability of trypsin. *J. Agric. Food Chem.* **1992**, *40*, 43–46.
87. Leeson, S.; Summers, J.D.; Caston, L. Response of brown egg strain layers to dietary calcium or phosphorus. *Poult. Sci.* **1993**, *72*, 1510–1514.
88. Selle, P.H.; Ravindran, V.; Caldwell, R.A.; Bryden, W.L. Phytate and phytase: Consequences for protein utilisation. *Nutr. Res. Rev.* **2000**, *13*, 255–278.
89. Champagne, E.T. Effects of pH on mineral-phytate, protein-mineral-phytate and mineral-fiber interactions. Possible consequences of atrophic gastritis on mineral bioavailability from high fiber foods. *J. Am. Coll. Nutr.* **1988**, *38*, 199–215.
90. Taylor, T.G. Dietary phosphorus and egg shell thickness in the domestic fowl. *Br. Poult. Sci.* **1965**, *6*, 79–87.

91. Keshavarz, K. The effect of dietary levels of calcium and phosphorus on performance and retention of these nutrients by laying hens. *Poult. Sci.* **1986**, *65*, 114–121.
92. Keshavarz, K. Phosphorus requirement of laying hens with and without phytase. In Proceedings of the Cornell Nutrition Conference for Feed Manufacturers, Syracuse, NY, USA, 19–21 October 1999; pp. 81–92.
93. Hurwitz, S.; Bar, A. Absorption of calcium and phosphorus along the intestinal tract of the laying fowl as influenced by dietary calcium and eggshell formation. *J. Nutr.* **1965**, *86*, 433–438.
94. Guinotte, F.; Gautron, J.; Nys, Y.; Soumarmon, A. Calcium solubilisation and retention in the gastrointestinal tract in chicks (*Gallus domesticus*) as a function of gastric acid secretion inhibition and of calcium carbonate particle size. *Br. J. Nutr.* **1995**, *73*, 125–139.
95. Keshavarz, K.; Austic, R.E. Effects of dietary minerals on acid–base balance and eggshell quality in chickens. *J. Nutr.* **1990**, *120*, 1360–1369.
96. Li, X.; Zhang, D.; Bryden, W.L.; Huang, K.H. University of Queensland: Gatton, Australia Unpublished Work. , 2013. University of Queensland: Gatton, Australia Unpublished Work. , 2013.
97. Selle, P.H.; Ravindran, V.; Bryden, W.L.; Scott, T. Influence of dietary phytate and exogenous phytase on amino acid digestibility in poultry: A review. *J. Poult. Sci.* **2006**, *43*, 89–103.
98. Shirley, R.B.; Edwards, H.M., Jr. Graded levels of phytase past industry standards improves broiler performance. *Poult. Sci.* **2003**, *82*, 671–680.
99. Manangi, M.K.; Coon, C. Phytate phosphorus hydrolysis in broilers in response to dietary phytase, calcium, and phosphorus concentrations. *Poult. Sci.* **2008**, *87*, 1577–1586.
100. Mohammed, A.; Gibney, M.J.; Taylor, T.G. The effects of dietary levels of inorganic phosphorus, calcium and cholecalciferol on the digestion of phytate phosphorus by the chick. *Br. Poult. Sci.* **1991**, *66*, 251–259.
101. Kornegay, E.T. Nutritional, environmental and economic considerations for using phytase in pig and poultry diets. In *Nutrient Management of Food Animals to Enhance and Protect the Environment*; Kornegay, E.T., Ed.; Lewis Publishers: New York, NY, USA; CRC Press: Boca Raton, FL, USA, 1996; pp. 277–302.
102. Punna, S.; Roland, D.A., Sr. Influence of supplemental microbial phytase on first cycle laying hens fed phosphorus-deficient diets from day one of age. *Poult. Sci.* **1999**, *78*, 1407–1411.
103. Tamim, N.M.; Angel, R.; Christman, M. Influence of dietary calcium and phytase on phytate phosphorus hydrolysis in broiler chickens. *Poult. Sci.* **2004**, *83*, 1358–1367.
104. Augspurger, N.R.; Baker, D.H. Phytase improves dietary calcium utilisation in chicks, and oyster shell, carbonate, citrate, and citrate-malate forms of calcium are equally bioavailable. *Nutr. Res.* **2004**, *24*, 293–301.
105. Angel, R.; Dhandu, A.S.; Applegate, T.J. Phosphorus requirements of broilers and the impact of exogenous phytases. In Proceedings of the Arkansas Poultry Nutrition Conference; University of Arkansas: Fayetteville, AR, USA, 2002; pp. 1–22.
106. Applegate, T.J.; Angel, R.; Classen, H.L. Effect of dietary calcium, 25-hydroxycholecalciferol, or bird strain on small intestinal phytase activity in broiler chickens. *Poult. Sci.* **2003**, *82*, 1140–1148.
107. Nelson, T.S.; Peeler, H.T. The availability of phosphates from single and combined phosphates to chicks. *Poult. Sci.* **1961**, *40*, 1321–1327.
108. Waldroup, P.W. Bioassays remain necessary to estimate phosphorus, calcium bioavailability. *Feedstuffs* **1996**, *68*, 13–23.
109. Nelson, T.S.; Kirby, L.K. The calcium binding properties of natural phytate in chick diets. *Nutr. Rep. Int.* **1987**, *35*, 949–956.

CHAPTER 12

The Impacts of Hammer-Mill Screen Size and Grain Particle Size on The Performance of Broiler Chickens Offered Diets Based on Two Red Sorghum Varieties

P. H. Selle[a], H. H. Truong[a,b], A. Khoddami[c], A. F. Moss[a], T. H. Roberts[c] and S. Y. Liu[a]

[a]*Poultry Research Foundation, Faculty of Veterinary Science, The University of Sydney, Camden, Australia;*

[b]*Poultry CRC, University of New England, Armidale, Australia;*

[c]*Department of Plant and Food Sciences, Faculty of Agriculture and Environment, The University of Sydney, Sydney, Australia*

ABSTRACT

1. The two red grain sorghums were extensively characterised. Kafirin, polyphenolic compounds, free, conjugated and bound phenolic acids, phytate concentrations and starch pasting profiles were determined.

2. The experiment consisted of a 2 × 4 factorial array of dietary treatments comprising two red sorghum varieties (Tiger and Block I) ground through 4 hammer-mill screen sizes (2.0, 3.2, 4.8 6.0 mm) prior to incorporation into nutritionally equivalent diets. Eight steam-pelleted dietary treatments were each offered to 7 replicates (6 male Ross 308 birds per cage) from 7 to 28 d post-hatch.
3. Effects of dietary treatments on growth performance, relative gizzard and pancreas weights, nutrient utilisation, apparent starch and protein (N) digestibility coefficients and disappearance rates from 4 small intestinal segments were determined.
4. The 2.0-mm hammer-mill screen generated an average geometric mean particle size of 794 μm and the 6.0-mm screen a mean particle size of 1405 μm. However, hammer-mill screen size did not influence weight gain or FCR. The 6.0-mm screen size generated significantly higher starch and protein (N) digestibility coefficients in the distal jejunum and distal ileum than the 2.0-mm hammermill screen.
5. Tiger sorghum was superior to Block I sorghum, as significant advantages were observed for feed conversion ratios (3.25%), AME (0.37 MJ), ME:GE ratios (4.15%), AMEn (0.53 MJ), distal ileal starch digestibility coefficients (2.46%) and protein (N) digestibility coefficients in the distal jejunum (4.66%), proximal ileum (1.96%) and distal ileum (2.16%). The inferior Block I sorghum contained more kafirin (67.1 versus 51.3 g/kg), phytate (9.79 versus 8.40 g/kg), total phenolic compounds (4.68 versus 4.12 mg GAE/g), flavan-4-ols (7.98 versus 5.04 ABS/ml/g), total phenolic acids (554 versus 402 μg/g) and total ferulic acid (375 versus 281 μg/g) in comparison to Tiger sorghum.

INTRODUCTION

This study was part of a project designed to investigate the digestibility and utilisation of sorghum starch by broiler chickens. Reported digestibilities of sorghum starch in broiler chickens are inferior to those of maize (Truong et al., 2015a). The effects of grain particle size on broiler growth performance in steam-pelleted diets are less pronounced than in mash diets (Amerah et al., 2007). However, Liu et al. (2013) suggested that particle size reductions of grain sorghum may enhance nutrient digestion, energy utilisation and growth performance in broiler chickens. This suggestion was basedo n a series of three separate, but very similar, feeding studies (Selle et al., 2012, 2013, 2014) in which identical diets were offered to broiler chickens that were based on the same white sorghum (Liberty) variety. This sorghum was ground through different hammer-mill screen sizes of 2.0, 3.2 and 6.0 mm in the three studies prior to being incorporated into complete steam-pelleted diets. Collectively, the outcomes indicated a quadratic relationship between hammer-mill screen size and feed conversion ratios (FCR) and for this particular grain sorghum a hammer-mill screen size in the order of 3.75 mm was optimal for FCR.

In practice, sorghums with red pericarps are far more common than white sorghums. Therefore, the primary objective of the present study was to confirm the effects of particle size on growth performance and nutrient utilisation and to determine optimal hammer-mill screen sizes for two red sorghum varieties, namely "Block I" and "Tiger". The secondary objective was to compare the performance of broiler chickens offered diets based on these two sorghum varieties as they had been extensively characterised. The relevant data include concentrations of kafirin, polyphenols, phenolic acids, phytate and RVA starch pasting profiles of the two sorghum varieties.

MATERIALS AND METHODS

Block I sorghum was harvested in the Murrumbidgee Irrigation Area (NSW, Australia) in 2012 and Tiger was harvested in the same region the following year. Both sorghums were ground through 2.0-, 3.2-, 4.8- and 6.0-mm hammer-mill screen sizes prior to incorporation into nutritionally equivalent, steam-pelleted diets. It was anticipated that the optimal screen size would lie within this range.

The two sorghum varieties were very extensively characterised as is evident in Tables 1 and 2. This included quantification of kafirin, phytate, total phenolic compounds, various polyphenols, free, conjugated and bound phenolic acids, RVA starch pasting profiles and particle size index (PSI) grain textures (Symes, 1965). The analysis method to quantify kafirin was adapted from procedures developed by Wallace et al. (1990) and Hamaker et al. (1995) and has been described in Truong et al. (2015b). The complex analytical methods to quantify a range of phenolic compounds have been described in detail in Khoddami et al. (2013, 2015). The Clorox bleach test (Waniska et al., 1992) did not detect a pigmented testa in both grain sorghums which indicates that they did not contain condensed tannin. Phytate (IP6) and phytate-P concentrations were determined by a HPLC procedure and total phosphorus (P) and other minerals were determined by inductively coupled plasma mass spectrometry (ICP-MS).

Table 1. NIR AusScan results, concentrations of protein, amino acids, kafirin, minerals and phytate, grain texture and RVA starch pasting profiles of Block I and Tiger sorghum varieties.

Item	Block I	Tiger	Item	Block I	Tiger
NIR AusScan					
AME (MJ/kg as-fed)	13.9	14.6	Kafirin (g/kg)	67.1	51.3
NIR Protein (g/kg)	142	102	Minerals (mg/kg)		
Total starch (g/kg)	756	830	Calcium (Ca)	186	115
Crude fibre	37	25	Phosphorus (P)	3644	3037
Acid detergent fibre	102	76	Copper (Cu)	5.76	3.61
Neutral detergent fibre	140	93	Iron (Fe)	46.3	33.4
Total soluble NSP	< 0.1	< 0.1	Potassium (K)	3748	3371
Total insoluble NSP	30	30	Magnesium (Mg)	1710	1418
Insoluble arabinoxylans	36	36	Manganese (Mn)	28.4	20.6
Hydration capacity (%)	9.9	9.9	Sodium (Na)	5.27	45.65
Amino acid profile			Strontium (Sr)	2.38	0.89
Crude protein	137.1	99.9	Zinc (Zn)	30.6	15.1
Arginine	4.5	3.4			
Histidine	2.9	2.2	Phytate (g/kg)	9.79	8.40
Isoleucine	5.1	3.8	Phytate-P (g/kg)	2.76	2.37
Leucine	17.7	13.0	Total *P* (g/kg)	3.64	3.04
Lysine	2.7	2.2	Proportion (%)	75.8	78.0
Methionine	2.0	1.4	Grain texture		
Phenylalanine	6.8	5.1	Symes PSI (%)	11	9
Threonine	4.2	3.2			
Valine	6.6	5.0	RVA starch pasting profile		
Alanine	12.0	8.8	Peak (cP)	2392	4771
Aspartic acid	8.5	6.4	Holding	2091	2904
Glutamic acid	28.3	20.6	Breakdown	300	1867
Glycine	4.0	3.1	Final	4592	5746
Proline	11.2	7.9	Setback	2501	2846
Serine	6.1	4.5	Peak time (minutes)	5.63	5.13
Tyrosine	2.3	1.9	Pasting temperature (°C)	79.9	75.1

Table 2. Concentrations of total phenolic compounds, polyphenols, free, conjugated, bound and total phenolic acids and ferulic acid in Block I and Tiger sorghum.

Polyphenols	Block I	Tiger	Phenolic acids (µg/g)	Block I	Tiger
Total phenolics (m GAE/g)	4.68	4.12	Free phenolic acids	35.0	28.9
Total flavonoids (µg/g)	927.6	1036.6	Free ferulic acid	2.1	1.9
Anthocyanin (ABS/ml/g)	6.53	10.86	Conjugated phenolic acids	119.8	83.9
Flavan-4-ols (ABS/ml/g)	7.98	5.04	Conjugated ferulic acid	38.4	33.2
Luteolinidin (µg/g)	16.07	7.83	Bound phenolic acids	399.1	288.9
Apigeninidin (µg/g)	14.75	7.25	Bound ferulic acid	334.8	246.3
5-Methoxy luteolinidin (µg/g)	7.69	6.40	Total phenolic acids	553.8	401.8
7-Methoxy apigeninidin (µg/g)	25.29	11.58	Total ferulic acid	375.3	281.2

The starch pasting profiles were determined using a Rapid-Visco-Analyser as outlined by Hernandez et al. (2008). A 28-g mixture of sorghum (or diet) and water (15:85 w/w) was prepared and held at 50°C temperature for 1 min and then heated from 50°C to 95°C. After holding the hot paste at 95°C for 2.5 min, the slurry was again cooled to 50°C, and then held at that temperature for 2 min with a total time interval of 13 min.

The geometric mean particle size of both sorghums following hammer-milling were determined as shown in Table 3. The mean particle size ranged from 794 µm with the 2.0-mm screen to 1405 µm with the 6.0-mm hammermill screen.

Table 3. Geometric mean particle sizes of two sorghum varieties, Tiger and Block I, following grinding through four hammer-mill screen sizes.

Sorghum variety	Hammer-mill screen size (mm)	Geometric mean particle size (μm)	Standard deviation
Tiger	2.0	783	2.19
	3.2	1055	2.13
	4.8	1354	2.17
	6.0	1402	2.12
Block I	2.0	805	2.10
	3.2	1173	2.02
	4.8	1370	2.08
	6.0	1408	2.19

On the basis of the Table 1 data, two nutritionally equivalent diets containing 620 g/kg of either Tiger or Block I sorghums were formulated on the basis of digestible amino acid concentrations with energy densities of 12.95 MJ/kg (Table 4). The diets contained 20.0 g/kg Celite® as an acid-insoluble-ash (AIA) inert dietary marker. Diets were steam pelleted at a conditioning temperature of 84°C and crumbled after both sorghum grains were ground through 4 hammer-mill screen sizes. Feather-sexed, male broiler chicks (Ross 308) were housed in an environmentally controlled facility and were initially offered a proprietary starter ration. Birds had unlimited access to feed and water under a "23-h on-1-h off" lighting regimen. The birds were individually identified (wing bands) and weighed at d 7 and distributed amongst cages so that mean body weights in each cage and their variations were nearly identical. The 8 dietary treatments were offered to 7 replicates (6 birds per replicate cage), or a total of 336 birds, from 7 to 28 d posthatch. Body weights were determined on d 7 and 28 and feed intakes recorded over the entire period to calculate FCR with adjustments made from the weight of any dead or culled birds, which were monitored on a daily basis. Total excreta were collected from 23 to 26 d post-hatch from each cage to determine parameters of nutrient utilisation, which included apparent metabolisable energy (AME), metabolisable energy to gross energy ratios (ME:GE), nitrogen (N) retention and N-corrected AME (AMEn). On d 28, the birds were killed (intravenous injection of Na pentobarbitone) and digesta samples were collected in their entirety from 4 segments of the small intestine to determine nutrient (starch and crude protein) digestibility coefficients and nutrient disappearance rates (g/bird/d). The 4 small intestinal segments included the proximal jejunum, distal jejunum, proximal ileum and distal ileum.

Table 4. Dietary composition and nutrient specifications of sorghum-based diets based on Tiger and block I varieties

Item (g/kg)	Tiger	Block I
Sorghum	620.0	620.0
Soybean meal	226.4	224.6
Canola meal	75.0	75.0
Sunflower oil	25.0	28.0
Dicalcium phosphate	17.2	16.5
Limestone	6.2	7.4
Lysine HCl	3.1	2.5
Methionine	2.5	2.3
Threonine	1.0	0.7
Arginine	1.0	0.5
Sodium chloride	0.8	1.0
Vitamin-mineral premix[a]	2.0	2.0
Sodium bicarbonate	4.8	4.5
Celite	15.0	15.0
Nutrient specifications		
ME (MJ/kg)	12.95	12.95
Protein	190.9	216.6
Starch	387.8	378.4
Fat	51.2	54.2
Calcium	7.5	7.5
Total phosphorus	6.6	7.5
Available phosphorus	3.8	3.8
Lysine[b]	10.8	10.8
Methionine[b]	5.5	5.4
Threonine[b]	6.9	7.0
Isoleucine[b]	7.3	7.4
Tryptophan[b]	2.0	2.0
Cystine[b]	2.5	2.6
Valine[b]	8.3	8.3
Arginine[b]	11.7	11.7
Histidine[b]	4.2	4.2
Leucine[b]	15.1	15.3
Phenylalanine[b]	7.9	8.1
Sodium	1.8	1.8
Potassium	7.5	7.5
Chloride	2.2	2.2

[a]Vitamin-trace mineral premix supplied per tonne of feed; [million international units, MIU] retinol 12, cholecalciferol 5, [g] tocopherol 50, menadione 3, thiamine 3, riboflavin 9, pyridoxine 5, cobalamin 0.025, niacin 50, pantothenate 18, folate 2, biotin 0.2, copper 20, iron 40 manganese 110, cobalt 0.25, iodine 1, molybdenum 2, zinc 90, selenium 0.3.

[b]Digestible amino acids.

The total excreta collection method over a 72-h period was used to determine AME on a dry matter basis, ME:GE ratios, N retention and AMEn. Total excreta were quantitatively collected from each cage and feed intakes recorded for the 72-h collection period. Excreta were dried in a forcedair oven at 80°C for 24 h and the GE of excreta and diets were determined using an adiabatic bomb calorimeter. The AME values of the diets on a dry matter basis were calculated from the following equation:

$AME_{diet(}MJ/kg\ DM) = (feed\ intake \times GE_{diet)} - (excreta\ output \times GE_{excreta})$ feed intake

ME:GE Ratios were calculated by dividing AME by the GE of the appropriate diets. N contents of diets and excreta were determined using a nitrogen determinator (Leco Corporation, St Joseph, MI) and N retentions calculated from the following equation:

$N\ retention\ (\%) = (feed\ intake \times N_{diet}) - (excreta\ output \times N_{excreta}) \times 100$ $(feed\ intake \times N_{diet)}$

N-corrected AME (AMEn MJ/kg DM) values were calculated by correcting N retention to zero using the factor of 36.54 kJ/g N retained in the body (Hill and Anderson, 1958).

Acid insoluble ash (Celite) was included in diets at 2% as an inert marker to determine starch and N digestibility. The small intestines were removed from killed birds and samples of digesta were gently expressed from the proximal jejunum, distal jejunum, proximal ileum and distal ileum in their entirety and pooled for each cage. Proximal jejuna samples were taken from the end of the duodenal loop to the midpoint with Meckel's diverticulum and distal jejuna samples from the midpoint to the diverticulum. Proximal ileal samples were taken from Meckel's diverticulum to the midpoint with the ileocaecal junction and distal ileal samples were taken from below this midpoint. The digesta samples were freeze-dried to determine apparent digestibilities of starch and crude protein (N) using acid insoluble ash (AIA) as the inert dietary marker. Starch concentrations in diets and digesta were determined by a procedure based on dimethyl sulphoxide, α-amylase and amyloglucosidase as described by Mahasukhonthachat et al. (2010). N concentrations were determined as already stated and AIA concentrations were determined by the method of Siriwan et al. (1993). The apparent digestibility coefficients for starch and protein (N) at up to 4 small intestinal sites were calculated from the following equation:

$Apparent\ digestibility\ coefficient = (nutrient/AIA)_{diet} - (nutrient/AIA)_{digesta}$ $(nutrient/AIA)_{diet}$

Starch and protein (N) disappearance rates (g/bird/d) were deduced from feed intakes over the final phase of the feeding period from the following equation:

Nutrient disappearance rate$_{(g/bird/d)}$= feed intake$_{(g/bird/d)}$ × dietary nutrient $_{(g/kg)}$

× nutrient digestibility$_{(apparent\ digestibility\ coefficient)}$

Ratios of starch to protein disappearance rates in the intestinal segments were calculated as this effectively cancels the potential confounding influence of feed intake.

Experimental data were analysed using the IBM® SPSS® Statistics 20 program (IBM Corporation. Somers, NY). The experimental units were cage means and statistical procedures included univariate analyses of variance using the general linear models procedure, Pearson correlations and single and multiple linear regressions and quadratic regressions. A probability level of less than 5% was considered to be statistically significant. The feeding studies complied with specific guidelines approved by the Animal Ethics Committee of Sydney University.

RESULTS

The effects of sorghum variety and hammer-mill screen size on 7 to 28 d post-hatch growth performance of broilers are shown in Table 5; there was a mortality/cull rate of 6.25% but it was unrelated to treatment. Hammer-mill screen sizes did not significantly influence growth performance; however, the interaction between sorghum variety and hammermil screen size closely approached significance (P = 0.056) for weight gain. This was because weight gains for Block I sorghum remained relatively constant in diets with different particle sizes; whereas, weight gains with Tiger sorghumbased diets appeared to increase with larger hammer-mill screen sizes. When Tiger sorghum was ground through a 6.0-mm hammer-mill screen (particle size: 1402 μm) as opposed to a 2.0-mm screen (particle size: 783 μm), there was a 4.81% increase in weight gain (1592 versus 1519 g/ bird; P < 0.02) which was significant on the basis of a pairwise comparison. Indeed, when Tiger sorghum-based diets are considered in isolation, there were significant positive correlations between hammer-mill screen sizes (r = 0.971; P = 0.029), as shown in Figure 1, and mean particles sizes (r = 0.951; P = 0.049) with weight gains.

Table 5. Effects of grain sorghum variety and hammer-mill screen sizes on growth performance of broilers from 7 to 28 days post-hatch.

Treatment		Growth performance parameter			
Sorghum variety	Screen size (mm)	Weight gain (g/bird)	Feed intake (g/bird)	FCR (g/g)	Mortality rates (%)
Tiger	2.0	1519	2334	1.535	7.14
	3.2	1527	2440	1.600	9.52
	4.8	1579	2461	1.560	11.91
	6.0	1592	2385	1.500	4.76
Block I	2.0	1560	2452	1.575	2.38
	3.2	1567	2562	1.634	2.38
	4.8	1519	2426	1.597	7.14
	6.0	1567	2499	1.597	4.76
SEM		21.452	53.273	0.0338	4.1810
Main effect: sorghum variety					
Tiger		1554	2405a	1.548a	8.33
Block I		1553	2485b	1.600b	4.17
Screen size					
2.0		1539	2393	1.555	4.76
3.2		1547	2501	1.617	5.95
4.8		1549	2443	1.579	9.53
6.0		1580	2442	1.547	4.76
Significance (P =)					
Sorghum variety (SV)		0.948	0.040	0.029	0.165
Screen size (SS)		0.260	0.263	0.150	0.630
Interaction SS x SV		0.056	0.386	0.744	0.856

ab Means within columns not sharing common suffixes are significantly different ($P < 0.05$).

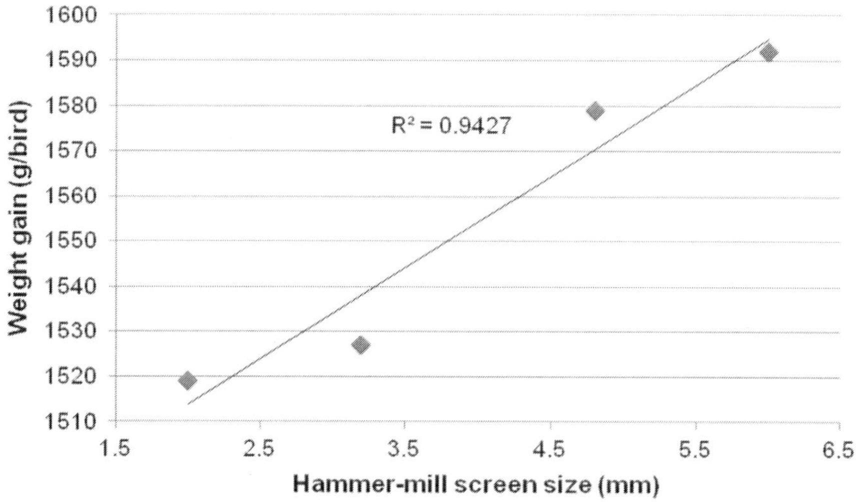

Figure 1. Linear relationship (r = 0.971; P = 0.029) between hammer-mill screen-size and weight gain of broiler chickens offered Tiger sorghum-based diets.

The effects of dietary treatments on relative gizzard and pancreas weights and gizzard pH are shown in Table 6. There were no significant treatment effects on pancreas weights and gizzard pH. However, Block I sorghum supported 7.4% heavier gizzard weights (20.86 versus 19.43 g/ kg; $P < 0.001$) than Tiger sorghum-based diets. Sorghumbased diets ground through a 2.0-mm hammer-mill screen generated significantly 4.94% lighter gizzards (19.43 versus 20.44 g/kg; $P < 0.025$) than the average of the three larger screens.

Table 6. Effects of grain sorghum variety and hammer-mill screen sizes on relative gizzard and pancreas weights and gizzard pH in broilers at 28 days posthatch.

Treatment				
Sorghum variety	Screen size (mm)	Gizzard weight (g/kg)	Pancreas Weight (g/kg)	Gizzard pH
Tiger	2.0	18.14	2.43	3.27
	3.2	20.03	2.52	2.79
	4.8	19.77	2.51	2.94
	6.0	19.79	2.47	2.81
Block I	2.0	20.37	2.58	2.90
	3.2	20.85	2.55	2.95
	4.8	21.08	2.47	2.91
	6.0	21.12	2.48	2.99
SEM		0.4443	0.0978	0.4577
Main effect: sorghum variety				
Tiger		19.43a	2.48	2.95
Block I		20.86b	2.52	2.94
Screen size				
2.0		19.25a	2.50	3.08
3.2		20.44b	2.53	2.87
4.8		20.43b	2.49	2.93
6.0		20.46b	2.47	2.90
Significance (P =)				
Sorghum variety (SV)		< 0.001	0.569	0.881
Screen size (SS)		0.021	0.943	0.473
Interaction SS x SV		0.465	0.796	0.210

ab Means within columns not sharing common suffixes are significantly different ($P < 0.05$).

The nutrient utilisation data arising from this experiment are shown in Table 7. Tiger sorghum was superior to Block I sorghum-based diets by 0.37 MJ in AME (12.89 versus 12.52 MJ/kg; P < 0.01), by 0.53 MJ in N-corrected AME (12.02 versus 11.49 MJ/kg; P < 0.001) and by 4.15% in ME:GE ratios (0.753 versus 0.723; P < 0.001). Hammer-mill screen sizes did not influence any nutrient utilisation parameters and N retention was not influenced by treatments.

The effects of grain sorghum variety and hammer-mill screen sizes on starch digestibility at 4 small intestinal sites are shown in Table 8. Tiger sorghum-based diets supported numerically higher starch digestibilities in the three anterior small intestinal segments including a 3.97% increase in thep roximal jejunum. However, the 2.46% advantage (0.918 versus 0.896; $P < 0.005$) of Tiger over Block I sorghumbased diets was significant in the distal ileum. Overall, larger hammer-mill screen sizes appear to be associated with enhanced starch digestibility. The effect of hammer-mill screen sizes was significant ($P < 0.01$) in the distal jejunum and distal ileum. In the distal jejunum there was a significant difference between 2.0 and 6.0 mm screen sizes of 8.85% (0.849 versus 0.780); similarly, in the distal ileum there was a significant difference of 4.04% (0.926 versus 0.890).

Table 7. Effects of grain sorghum variety and hammer-mill screen sizes on nutrient utilisation of broilers.

Treatment		Nutrient utilisation parameter			
Sorghum variety	Screen size (mm)	AME (MJ/kg DM)	ME:GE (MJ/kg)	N retention (%)	AMEn (MJ/kg DM)
Tiger	2.0	12.65	0.737	63.88	11.82
	3.2	12.78	0.749	65.99	11.91
	4.8	13.17	0.770	67.40	12.24
	6.0	12.97	0.757	65.70	12.10
Block I	2.0	12.46	0.719	63.98	11.48
	3.2	12.47	0.722	66.47	11.36
	4.8	12.44	0.716	63.89	11.44
	6.0	12.72	0.734	65.39	11.69
SEM		0.1852	0.0120	1.4985	0.1526
Main effect: sorghum variety					
Tiger		12.89b	0.753b	65.74	12.02b
Block I		12.52a	0.723a	64.93	11.49a
Screen size					
2.0		12.56	0.728	63.93	11.65
3.2		12.63	0.735	66.23	11.64
4.8		12.81	0.743	65.65	11.84
6.0		12.85	.0745	65.54	11.90
Significance (P =)					
Sorghum variety (SV)		0.007	< 0.001	0.449	< 0.001
Screen size (SS)		0.350	0.370	0.464	0.229
Interaction SS x SV		0.438	0.360	0.524	0.449

ab Means within columns not sharing common suffixes are significantly different ($P < 0.05$)

Table 8. Effects of grain sorghum variety and hammer-mill screen sizes on apparent starch digestibility coefficients at four small intestinal sites in broilers at 28 days post-hatch.

Treatment		Small intestinal site			
Sorghum variety	Screen size (mm)	Proximal jejunum	Distal jejunum	Proximal ileum	Distal ileum
Tiger	2.0	0.703	0.793	0.877	0.907
	3.2	0.740	0.830	0.910	0.916
	4.8	0.735	0.795	0.900	0.917
	6.0	0.757	0.860	0.903	0.933
Block I	2.0	0.697	0.768	0.864	0.873
	3.2	0.686	0.803	0.886	0.889
	4.8	0.700	0.810	0.882	0.901
	6.0	0.741	0.838	0.893	0.920
SEM		0.0293	0.0207	0.0120	0.0120
Main effect: sorghum variety					
Tiger		0.734	0.819	0.897	0.918b
Block I		0.706	0.805	0.881	0.896a
Screen size					
2.0		0.700	0.780a	0.871	0.890a
3.2		0.713	0.817ab	0.898	0.903a
4.8		0.717	0.803a	0.891	0.909ab
6.0		0.749	0.849b	0.898	0.926b
Significance (P =)					
Sorghum variety (SV)		0.204	0.287	0.064	0.004
Screen size (SS)		0.416	0.008	0.079	0.009
Interaction SS x SV		0.868	0.662	0.936	0.719

ab Means within columns not sharing common suffixes are significantly different ($P < 0.05$).

The effects of dietary treatments on starch disappearance rates (g/bird/d) at 4 small intestinal sites are shown in Table 9. Birds offered diets based on Tiger sorghum had 14.7% higher starch disappearance rates (33.51 versus 29.21 g/bird/d; $P < 0.001$) than their Block I counterparts at the proximal jejunum. Tiger sorghum-based diets held a similar advantage of 15.9% (37.44 versus 33.29 g/bird/d; $P < 0.001$) at the distal jejunum. However, hammer-mill screen sizes had a significant effect ($P < 0.005$) at this site where the 6.0-mm screen supported higher starch disappearance rates than the three smaller screens by up to 13.7% (38.07 versus an average of 33.47

g/bird/d). There were significant treatment interactions between sorghum variety and hammer-mill screen size in the proximal ($P < 0.03$) and distal ileum ($P < 0.025$). Overall, increasing hammer-mill screen sizes supported more rapid starch disappearance rates; however, significant increases were only observed with the 6.0-mm screen with Block I sorghum-based diets. In contrast, with Tiger sorghum-based diets, significant increases were observed with the 3.2-mm screen at the proximal ileum and with the 4.8-mm screen at the distal ileum.

Table 9. The effect of grain sorghum variety and hammer-mill screen size on apparent starch disappearance rates (g/bird/day) at four small intestinal sites in broilers at 28 days post-hatch.

Treatment		Small intestinal site			
Sorghum variety	Screen size (mm)	Proximal jejunum	Distal jejunum	Proximal ileum	Distal ileum
Tiger	2.0	30.80	34.74	38.49bc	39.75b
	3.2	34.16	38.37	42.07de	42.35bc
	4.8	34.28	37.22	42.18e	42.98c
	6.0	34.79	39.43	41.42cde	42.79c
Block I	2.0	29.10	32.19	36.17ab	36.52a
	3.2	27.99	32.72	36.09ab	36.20a
	4.8	27.27	31.52	34.31b	35.02a
	6.0	32.47	36.71	39.11bcd	40.29bc
SEM		1.4603	1.1412	1.0788	0.9644
Main effect: sorghum variety					
Tiger		33.51b	37.44b	41.04	41.97
Block I		29.21a	33.29a	36.42	37.01
Screen size					
2.0		29.95	33.47a	37.33	38.13
3.2		31.07	35.55a	39.08	39.27
4.8		30.77	34.37a	38.24	39.00
6.0		33.63	38.07b	40.27	41.54
Significance (P =)					
Sorghum variety (SV)		< 0.001	< 0.001	< 0.001	< 0.001
Screen size (SS)		0.083	0.001	0.057	0.007
Interaction SS x SV		0.184	0.325	0.028	0.023

abcde Means within columns not sharing common suffixes are significantly different ($P < 0.05$).

The effects of sorghum variety and hammer-mill screen sizes on apparent protein (N) digestibility at 4 small intestinal sites are shown in Table 10 where there were not any significant treatment interactions. Tiger sorghum-based diets supported higher N digestibilities in the distal jejunum by 4.66% (0.786 versus 0.751;$P< 0.015$), in the proximal ileumby 1.96% (0.781 versus 0.766; $P < 0.04$) and by 2.16% (0.804 versus 0.787; $P < 0.01$) in the distal ileum. The 6.0-mm screen supported a significantly higher N digestibility by 4.23% (0.788 versus 0.756) than the 2.0 mm screen in the proximal ileum and by 3.32% (0.809 versus 0.783) in the distal ileum.

The effects of dietary treatments on apparent protein (N) disappearance rates are shown in Table 11, where, again, there were no significant interactions. Tiger sorghum-based diets supported more rapid protein disappearance rates (g/ bird/d) in the distal jejunum by 4.53% (18.89 versus 18.07; $P < 0.015$), proximal ileum by 1.95% (18.79 versus 18.43; $P < 0.04$) and distal ileum by 2.17% (19.33 versus 18.92; $P < 0.01$). The 6.0-mm screen supported significantly faster protein disappearance rates by 4.18%

(18.94 versus 18.18) than the 2.0-mm screen in the proximal ileum and by 3.24% (19.44 versus 18.83) in the distal ileum.

Table 10. Effects of grain sorghum variety and hammer-mill screen sizes on apparent protein (N) digestibility coefficients at four small intestinal sites in broilers at 28 days post-hatch.

Treatment		Small intestinal site			
Sorghum variety	Screen size (mm)	Proximal jejunum	Distal jejunum	Proximal ileum	Distal ileum
Tiger	2.0	0.476	0.783	0.771	0.799
	3.2	0.550	0.796	0.791	0.811
	4.8	0.546	0.763	0.760	0.785
	6.0	0.561	0.800	0.804	0.820
Block I	2.0	0.532	0.698	0.742	0.767
	3.2	0.490	0.779	0.779	0.796
	4.8	0.501	0.762	0.772	0.788
	6.0	0.523	0.766	0.772	0.797
SEM		0.0316	0.0169	0.0099	0.0081
Main effect: sorghum variety					
Tiger		0.533	0.786b	0.781b	0.804b
Block I		0.512	0.751a	0.766a	0.787a
Screen size					
2.0		0.504	0.740	0.756a	0.783a
3.2		0.520	0.788	0.785bc	0.804b
4.8		0.524	0.763	0.766ab	0.787a
6.0		0.542	0.783	0.788c	0.809b
Significance (P =)					
Sorghum variety (SV)		0.351	0.012	0.036	0.005
Screen size (SS)		0.714	0.055	0.005	0.005
Interaction SS x SV		0.285	0.136	0.110	0.180

abc Means within columns not sharing common suffixes are significantly different ($P < 0.05$).

Table 11. Effects of grain sorghum variety and hammer-mill screen sizes on apparent protein (N) disappearance rates (g/bird/day) at four small intestinal sites in broilers at 28 days post-hatch.

Treatment		Small intestinal site			
Sorghum variety	Screen size (mm)	Proximal jejunum	Distal jejunum	Proximal ileum	Distal ileum
Tiger	2.0	11.26	18.83	18.53	19.22
	3.2	13.26	19.15	19.03	19.49
	4.8	13.27	18.35	18.27	18.88
	6.0	13.39	19.23	19.32	19.71
Block I	2.0	14.04	16.78	17.83	18.43
	3.2	14.31	18.74	18.74	19.15
	4.8	13.61	18.33	18.57	18.94
	6.0	14.09	18.43	18.56	19.17
SEM		0.8691	0.4463	0.2375	0.1938
Main effect: sorghum variety					
Tiger		12.79	18.89b	18.79b	19.33b
Block I		14.01	18.07a	18.43a	18.92a
Screen size					
2.0		12.65	17.80	18.18ab	18.83a
3.2		13.78	18.95	18.89bc	19.32b
4.8		13.44	18.34	18.42ab	18.91a
6.0		13.74	18.83	18.94c	19.44b
Significance (P =)					
Sorghum variety (SV)		0.053	0.012	0.036	0.005
Screen size (SS)		0.537	0.055	0.005	0.005
Interaction SS x SV		0.516	0.136	0.110	0.180

abc Means within columns not sharing common suffixes are significantly different ($P < 0.05$).

Finally, the apparent starch:protein (N) disappearance rate ratios are shown in Table 12. In the proximal jejunum, the ratio of 2.66 for Tiger sorghum-based diets was significantly ($P < 0.001$) wider than 2.10 for diets based on Block I. There were significant treatment interactions in the three posterior small intestinal segments, especially in both segments of the ileum ($P < 0.001$). For Tiger sorghum- based diets, the narrowest ratio was observed with the 2.0-mm screen and the widest with the 4.8-mm screen. For Block I sorghum-based diets, the narrowest ratio was observed with the 4.8-mm screen and the widest with the 6.0-mm screen.

Table 12. The effect of grain sorghum variety and hammer-mill screen size on apparent starch:protein (N) disappearance rate ratios at four small intestinal sites in broilers at 28 days post-hatch.

Treatment		Small intestinal site			
Sorghum variety	Screen size (mm)	Proximal jejunum	Distal jejunum	Proximal ileum	Distal ileum
Tiger	2.0	2.76	1.84ab	2.07bcd	2.07ab
	3.2	2.58	2.00b	2.21de	2.18 cd
	4.8	2.65	2.03b	2.31e	2.28d
	6.0	2.63	2.05b	2.14 cd	2.17 cd
Block I	2.0	2.11	1.97b	2.03bc	1.98ab
	3.2	1.97	1.75a	1.93ab	1.89a
	4.8	2.01	1.72a	1.84a	1.85a
	6.0	2.32	2.00b	2.11 cd	2.10bc
SEM		0.0934	0.0756	0.0507	0.0493
Main effect: sorghum variety					
Tiger		2.66b	1.98	2.19	2.17
Block I		2.10a	1.86	1.98	1.96
Screen size					
2.0		2.43	1.91	2.05	2.03
3.2		2.28	1.88	2.07	2.03
4.8		2.33	1.87	2.08	2.06
6.0		2.48	2.02	2.13	2.14
Significance ($P =$)					
Sorghum variety (SV)		< 0.001	0.025	< 0.001	< 0.001
Screen size (SS)		0.133	0.174	0.508	0.107
Interaction SS x SV		0.231	0.021	< 0.001	<0.001

abcde Means within columns not sharing common suffixes are significantly different ($P < 0.05$).

DISCUSSION

The effects of hammer-mill screen size, and resultant geometric mean sorghum particle size, were not expected as it was anticipated that there would be quadratic responses and a hammer-mill screen size of less than 6.0 mm would generate the best broiler performance. This assumption was based on three experiments previously completed at this institution, which involved a white sorghum variety as mentioned in the Introduction. This was not the case as hammer-mill screen size did not influence weight gain or FCR to significant extents. Indeed, with Tiger sorghumbased diets, increasing hammer-mill screen size was positively correlated with 7 to 28 d weight gain as illustrated in Figure 1. Moreover, there was a linear correlation ($r = 0.720$; $P = 0.044$) between increasing hammer-mill screen sizes and distal ileal starch digestibility coefficients.

The overall distal ileal starch digestibility coefficient of 0.907 in sorghum-based diets recorded in this experiment is inferior to that of reported values for maize-based diets (Truong et al., 2015a). Moreover, starch digestibility coefficients significantly increased by 4.04% (0.926 versus 0.890) when hammer-mill screen sizes of 2.0 mm and 6.0 mm are compared in favour of the larger screen size. The 6.0-mm hammer-mill screen generated similar geometric mean particle sizes of 1402 and 1408 μm for Tiger and Block I sorghums, respectively, which were less than expected.

The influence of grain particle size on broiler performance has been competently reviewed by Amerah et al. (2007). In this review, the authors suggested that grain particle size should range from 1100 to 1300 μm for 7 to 21 d-old broilers and from 1300 to 1500 μm for older birds. These suggestions are very consistent with the grain particle sizes generated by the larger hammer-mill screen sizes in the present study. Nir et al. (1990) reported that the transition from "fine" to "coarse" sorghum particle sizes (unspecified) significantly increased weight gain by 4.95% and feed intake by 5.45%. Moreover, Nir et al. (1994) found that the performance of 7 to 21 d-old broilers offered maize-, sorghum- and wheat-based diets with a mean grain particle size of 1130–1230 μm was superior to their counterparts offered diets with grain particle size of 570–670 μm. These findings are consistent with the outcomes of the present eeding study. The transition from sorghum ground through a 2.0-mm hammer-mill screen to three larger apertures resulted in a significant increase in relative gizzard weights from 19.25 g/kg to an average of 20.44 g/kg. However, the difference of 6.29% in relative gizzard weights (19.25 versus 20.46 g/kg) between the 2.0 and 6.0-mm hammer-mill screens appears subtle. Nevertheless, when Tiger sorghum-based diets are considered in isolation, there are positive correlations between relative gizzard weights and starch digestibility coefficients ($r = 0.413$; $P = 0.029$) and starch disappearance rates ($r = 0.523$; $P = 0.004$) in the proximal jejunum. The second linear relationship is illustrated in Figure 2. One possible explanation is that as a result of gastroduodenal refluxes, digestac ontaining pancreatic amylase is being recycled back into the gizzard and starch digestibility was enhanced as a result of reverse peristalsis (Liu et al., 2015a). It is established thatw hole-grain feeding generates heavier relative gizzard weights as reviewed by Singh et al. (2014) and Liu et al. (2015a). Heavier relative gizzard weights are presumably indicative of more functional gizzards but they may not bea precise assessment. Interestingly, Nir et al. (1994) found that the transition from fine-to-medium grain particle sizes in broiler diets significantly increased empty gizzard weights but gizzard contents were disproportionately increased to a large extent.

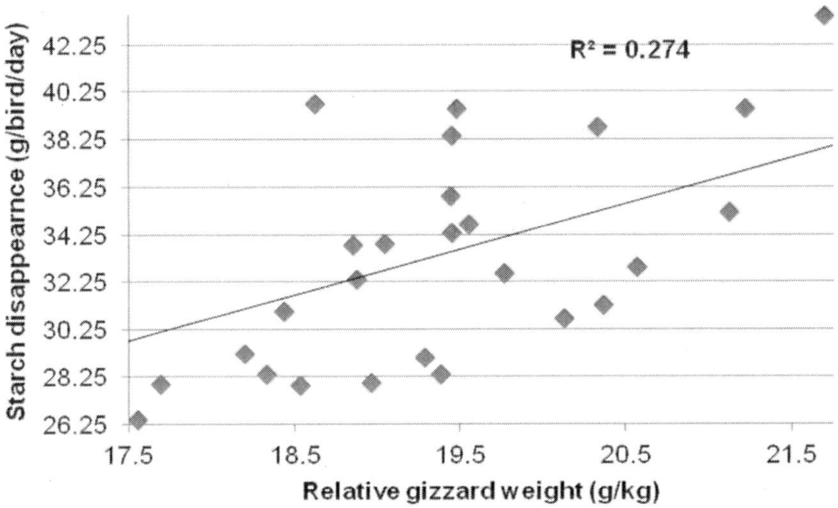

Figure 2. Linear relationship (r = 0.523; P = 0.004) between relative gizzard weights and starch disappearance rates at the proximal jejunum in broiler chickens offered Tiger sorghum-based diets.

Relative gizzard and pancreas weights were positively correlated (r = 0.278; P = 0.038) in the present study. However, with Tiger sorghum-based diets, gizzard weights were correlated with pancreas weights (0.407; P = 0.032), gizzard pH (r = −0.404; P = 0.033) and mean particle size (r = 0.472; P = 0.011). These relationships were not significant with Block I sorghum-based diets. Relative gizzard weights, or the "power of the gizzard", is pivotal to the practice of whole-grain feeding but even in the context of standard diets in which the entire grain component has been ground, the influence of the gizzard and its impact on gut function and reverse peristalsis appears to hold importance.

Liu et al. (2015b) suggested that the triad of kafirin, nontannin phenolic compounds and phytate in grain sorghum negatively influence the utilisation of starch/energy in broilers offered sorghum-based diets. The rationale for this suggestion has been considered in detail elsewhere (Khoddami et al., 2015; Liu et al., 2015b; Truong et al., 2015b). Tiger was clearly superior to Block I sorghum in the present study as significant advantages for feed conversion ratios, energy utilisation, starch and protein (N) digestibility were observed. Therefore, it is noteworthy that Block I contained more kafirin, polyphenols, phenolic acids and phytate than Tiger sorghum. Kafirin is the dominant protein fraction in grain sorghum (Selle et al., 2010) and kafirin has often been implicated as having a deleterious effect on sorghum starch utilisation (Black et al., 2005) However, kafirin may be increasing as a proportion of sorghum protein in Australia as an inadvertent consequence of breeding programs (Selle, 2011). Sorghum contains far more phenolic compounds that other feed grains (Bravo, 1998) and nontannin phenolic compounds may also have deleterious effects on sorghum starch utilisation (Khoddami et al., 2015). Sorghum contains at least as much phytate as other feed grains (Selle et al., 2003) and the anti-nutritive properties of this phosphorus-containing polyanionic molecule in poultry have been documented (Selle and Ravindran, 2007). Finally, it may be instructive that Tiger sorghum had a lower peak time and pasting temperature and higher RVA starch viscosities than Block I sorghum. This was particularly evident for peak RVA viscosity, which was higher for Tiger sorghum by a factor of 1.99 (4771 versus 2392 cP).

In conclusion, Tiger sorghum was superior to Block I sorghum as a feed grain for poultry as it supported an improvement of 3.25% in FCR (1.548 versus 1.600) in broiler chickens. The possibility exists that RVA starch pasting profiles may be predictive of the quality of sorghum as a feed grain in chicken-meat production. However, the effects of hammer-mill screen size, and geometric mean sorghum particle size of grain sorghum on broiler performance were not as expected. The numerically best weight gains and FCR were associated with a hammer-mill screen size of 6.0 mm and a geometric mean particle size in the order of 1400 μm; whereas, it was anticipated that smaller sizes would have been advantageous.

ACKNOWLEDGEMENTS

We would particularly like to thank Dr Karlie Nielsen for quantifying kafirin in sorghum while at the Australian Proteome Analysis Facility (Macquarie University) and Dr David Cadogan (Feedworks) for formulating the diets. We would like to acknowledge the Chicken-meat Committee of the Rural Industries Research and Development Corporation (RIRDC) for their ongoing encouragement and financial support and the Poultry CRC for their support of Ms Ha Truong's PhD candidature.

FUNDING

This work was supported by the RIRDC chicken meat committee [Grant Number PRJ-008695].

ORCID

A. Khoddami http://orcid.org/0000-0002-1774-0000

REFERENCES

1. Amerah, A.M., Ravindran, V., Lentle, R.G. & Thomas, D.G. (2007) Feed particle size: implications on the digestion and performance of poultry. *Worlds Poultry Science Journal*, 63: 439–455. doi:10.1017/S0043933907001560
2. Black, J.L., Hughes, R.J., Nielsen, S.G., Tredrea, A.M., Macalpine, R. & Van Barneveld, R.J. (2005) The energy value of cereal grains, particularly wheat and sorghum, for poultry. *Proceedings of Australian Poultry Science Symposium*, 17:21–29.
3. Bravo, L. (1998) Polyphenols: chemistry, dietary sources, metabolism, and nutritional significance. *Nutrition Reviews*, 56: 317–333. doi:10.1111/j.1753-4887.1998.tb01670.x
4. Hamaker, B.R., Mohamed, A.A., Habben, J.E., Huang, C.P. & Larkins, B.A. (1995) Efficient procedure for extracting maize and sorghum kernel proteins reveals higher prolamin contents than the conventional method. *Cereal Chemistry*, 72: 583–588.
5. Hernandez, J.R., Capareda, S.C., Hays, D.B., Portillo, O.R. & Rooney, W.L. (2008) Effect of grain sorghum protein digestibility on starch gelatinization and enzymatic conversion to glucose. Presented at the 2008 ASABE Annual International Meeting, Providence RI. American Society of Agricultural and Biological Engineers St Joseph, MI. USA.
6. Hill, F.W. & Anderson, D.L. (1958) Comparison of metabolizable energy and productive energy determinations with growing chicks. *Journal of Nutrition*, 64: 587–603.
7. Khoddami, A., Truong, H.H., Liu, S.Y., Roberts, T.H. & Selle, P.H. (2015) Concentrations of specific phenolic compounds in six red sorghums influence nutrient utilisation in broiler chickens. *Animal Feed Science and Technology*, 210: 190–199. doi:10.1016/j.anifeedsci.2015.09.029
8. Khoddami, A., Wilkes, M.A. & Roberts, T.H. (2013) Techniques for analysis of plant phenolic compounds. *Molecules*, 18: 2328–2375. doi:10.3390/molecules18022328
9. Liu, S.Y., Fox, G., Khoddami, A., Neilson, K.A., Truong, H.H., Moss, A.F. & Selle, P.H. (2015b) Grain sorghum: a conundrum for chicken-meat production. *Agriculture*, 5: 1224–1251. doi:10.3390/agriculture5041224
10. Liu, S.Y., Selle, P.H. & Cowieson, A.J. (2013) Strategies to enhance the performance of pigs and poultry on sorghum-based diets. *Animal Feed Science and Technology*, 181: 1–14. doi:10.1016/j.anifeedsci.2013.01.008
11. Liu, S.Y., Truong, H.H. & Selle, P.H. (2015a) Whole-grain feeding for chicken-meat production: possible mechanisms driving enhanced energy utilisation and feed conversion. *Animal Production Science*, 55: 559–572. doi:10.1071/AN13417
12. Mahasukhonthachat, K., Sopade, P.A. & Gidley, M.J. (2010) Kinetics of starch digestion and functional properties of twin-screw extruded sorghum. *Journal of Cereal Science*, 51: 392–401. doi:10.1016/j.jcs.2010.02.008
13. Nir, I., Hillel, R., Shefet, G. & Nitsan, Z. (1994) Effect of grain particle-size on performance .2. grain texture interactions. *Poultry Science*, 73: 781–791. doi:10.3382/ps.0730781
14. Nir, I., Melcion, J.P. & Picard, M. (1990) Effect of particle-size of sorghum grains on feed-intake and performance of young broilers. *Poultry Science*, 69: 2177–2184. doi:10.3382/ps.0692177

15. Selle, P.H. (2011) The protein quality of sorghum. *Proceedings of Australian Poultry Science Symposium*, 22: 147–160.
16. Selle, P.H., Cadogan, D.J., Li, X. & Bryden, W.L. (2010) Implications of sorghum in broiler chicken nutrition. *Animal Feed Science and Technology*, 156: 57–74. doi:10.1016/j.anifeedsci.2010.01.004
17. Selle, P.H., Liu, S.Y., Cai, J. & Cowieson, A.J. (2012) Steam-pelleting and feed form of broiler diets based on three coarsely ground sorghums influences growth performance, nutrient utilisation, starch and nitrogen digestibility. *Animal Production Science*, 52: 842–852. doi:10.1071/AN12026
18. Selle, P.H., Liu, S.Y., Cai, J. & Cowieson, A.J. (2013) Steam-pelleting temperatures, grain variety, feed form and protease supplementation of mediumly-ground, sorghum-based broiler diets: influences on growth performance, relative gizzard weights, nutrient utilisation, starch and nitrogen digestibility. *Animal Production Science*, 53: 378–387. doi:10.1071/AN12363
19. Selle, P.H., Liu, S.Y., Cai, J. & Cowieson, A.J. (2014) Steam-pelleting temperatures and grain variety of finely-ground, sorghum-based broiler diets. I. Influence on growth performance, relative gizzard weights, nutrient utilisation, starch and nitrogen digestibility. *Animal Production Science*, 54: 339–346. doi:10.1071/AN13080
20. Selle, P.H. & Ravindran, V. (2007) Microbial phytase in poultry nutrition. *Animal Feed Science and Technology*, 135: 1–41. doi:10.1016/j.anifeedsci.2006.06.010
21. Selle, P.H., Walker, A.R. & Bryden, W.L. (2003) Total and phytate-phosphorus contents and phytase activity of Australian-sourced feed ingredients for pigs and poultry. *Australian Journal of Experimental Agriculture*, 43: 475–479. doi:10.1071/EA02155
22. Singh, Y., Amerah, A.M. & Ravindran, V. (2014) Whole grain feeding: methodologies and effects on performance, digestive tract development and nutrient utilisation of poultry. *Animal Feed Science and Technology*, 190: 1–18. doi:10.1016/j.anifeedsci.2014.01.010
23. Siriwan, P., Bryden, W.L., Mollah, Y. & Annison, E.F. (1993) Measurement of endogenous amino-acid losses in poultry. *British Poultry Science*, 34: 939–949. doi:10.1080/00071669308417654
24. Symes, K.J. (1965) The inheritance of grain hardness in wheat as measured by the particle size index. *Australian Journal of Agricultural Research*, 16: 113–123. doi:10.1071/AR9650113
25. Truong, H.H., Liu, S.Y. & Selle, P.H. (2015a) Starch utilisation in chicken-meat production: the foremost influential factors. *Animal Production Science*, http://dx.doi.org/10.1071/AN15056
26. Truong, H.H., Neilson, K.A., Mcinerney, B.V., Khoddami, A., Roberts, T.H., Liu, S.Y. & Se lle, P.H. (2015b) Performance of broiler chickens offered nutritionally-equivalent diets based on two red grain sorghums with quantified kafirin concentrations as intact pellets or re-ground mash following steam-pelleting at 65 or 97°C conditioning temperatures. *Animal Nutrition*, 1: 220–228. doi:10.1016/j.aninu.2015.08.002
27. Wallace, J.C., Lopes, M.A., Paiva, E. & Larkins, B.A. (1990) New methods for extraction and quantitation of zeins reveal a high content of gamma-zein in modified opaque-2 maize. *Plant Physiology*, 92: 191–196. doi:10.1104/pp.92.1.191
28. Waniska, R.D., Hugo, L.F. & Rooney, L.W. (1992) Practical methods to determine the presence of tannins in sorghum. *Journal of Applied Poultry Research*, 1: 122–128. doi:10.1093/japr/1.1.122

CHAPTER 13

Technology and Poultry Welfare

Neila Ben Sassi [1], Xavier Averós [1] and Inma Estevez [1,2],*

[1] *Neiker Tecnalia, Arkaute Agrifood Campus, 01080 Vitoria-Gasteiz, Spain;*

nbsassi@ikt.es (N.B.S.); xaveros@neiker.net (X.A.)

[2] *Ikerbasque, Basque Foundation for Science, Maria Diaz de Haro 3, 48013 Bilbao, Spain*

ABSTRACT

Consideration of animal welfare is essential to address the consumers' demands and for the long term sustainability of commercial poultry. However, assessing welfare in large poultry flocks, to be able to detect potential welfare risks and to control or minimize its impact is difficult. Current developments in technology and mathematical modelling open new possibilities for real-time automatic monitoring of animal welfare and health. New technological innovations potentially adaptable to commercial poultry are appearing, although their practical implementation is still being defined. In this paper, we review the latest technological developments with potential to be applied to poultry welfare, especially for broiler chickens and laying hens. Some of the examples that are presented and discussed include the following: sensors for farm environmental monitoring, movement, or physiological parameters; imaging technologies such as optical flow to detect gait problems and feather pecking; infrared technologies to evaluate birds' thermoregulatory features and metabolism changes, that may be indicative of welfare, health and management problems. All these technologies have the potential to be implemented at the commercial level to improve birds' welfare and to optimize flock management, therefore, improving the efficiency of the system in terms of use of resources and, thus, long term sustainability.

Keywords: technology; animal welfare; broiler chickens; laying hens; sensors; images; app; modelling

1. INTRODUCTION

Public concern regarding the conditions in which producing animals are maintained has led to the need for developing methods to verify minimum animal welfare standards. As defined by the World Organization for Animal Health [1], "An animal is in a good state of welfare if (as indicated by scientific evidence) it is healthy, comfortable, well nourished, safe, able to express innate behaviour, and if it is not suffering from unpleasant states such as pain, fear and distress". However, to prove and verify animal welfare requirements in practice is not simple. In intensive poultry production a large number of factors, such as stocking density, environmental deterioration, unsuitable social environments, thermal stress, or difficulties in accessing essential resources can be major sources of stress that can lead to welfare deterioration and reduced performance [2,3,4,5,6,7]. Many of these factors can be controlled through well-established management practices to provide birds with an optimal environment. However, the sharp control of the temperature and relative humidity required to minimize the occurrence of welfare problems in poultry [8,9] might not be easy to achieve under high density or if the available farm equipment is inadequate. In addition, unforeseen situations or potential interactions among factors may be difficult to predict and control, thus potentially impacting on welfare. Welfare assessment serves to verify that the conditions to satisfy welfare standards during production are indeed met.

First attempts to assess animal welfare were resource-based, because assessing minimum resource requirements is generally easier than to evaluate the impact of the production conditions on animals [10]. Resource-based assessment intends to warrant the provision of the necessary environmental conditions for an optimal animal welfare. However, to verify that such conditions did not compromise the welfare of animals it was essential to develop methods based on the impact over the animals themselves. Animal-based assessment methods that were later developed, such as the Welfare Quality (WQ) poultry assessment protocol provides a thorough assessment of the impact of the actual rearing conditions on poultry welfare [11]. At a commercial level, however, the implementation of the full protocol is difficult and time consuming, thus, a simplified and effective protocol was developed for broilers [12]. More recently, a different welfare assessment approach based on the transect method was proposed for commercial broilers and turkeys [13,14]. Initial results suggest that this method is simple and practical for on-farm application and seems to have a good-inter-observer reliability for the detection of major welfare issues in meat poultry.

Despite the important efforts to simplify the available welfare assessment protocols for poultry, their implementation within the European Union (EU) legal framework could impose increased biosecurity challenges and production costs, which may hinder economic profit for farmers [15]. Available technological innovations currently tested for welfare assessment could greatly help to encompass a better environmental control and improve welfare while minimizing costs [15,16,17]. In addition, the reduction in health and welfare problems would lead to a more efficient and sustainable production in the long term.

Precision livestock farming (PLF), defined as the management of livestock production using the principles and technologies of process engineering [18], is based on automatic data acquisition, access, and processing [16]. Data from diverse sources are collected through smart sensors and compiled to a central database, where they will be later analysed to create an automatic management system based on real-time monitoring to control animal performance, health, and welfare [15]. According to Mollo et al. [16], PLF must be able to automatically manage commercial poultry farm equipment (including feeders, fans, heating systems, and sprinklers) based on the collected information. Different studies on broiler chickens and laying hens have shown the importance of technology and PLF to study birds' behaviour and welfare [15,17,19,20]. Although most technologies are still in the experimental phase, some are already available and can be introduced on commercial poultry farms [13,14,21] with good results.

As welfare depends on both management practices and the use of adequate equipment, different technological advances are emerging to improve both. For example, keel bone breakage risk in laying hens is higher in barns and aviaries as compared to cages (Siegford et al. [22]). However, new devices can be installed to detect poor management practices or to identify any behavioural or health issue occurring and, therefore, contribute to improved farm design and equipment use.

This paper reviews available technologies that have the potential to be implemented for a better control of the environment to improve poultry welfare, or to be applied for an automatic welfare assessment. The practical applications and the potential impact of such technologies are discussed.

2. SENSORS

In the last few years' tremendous advances were achieved in sensing technology in terms of diversity, accuracy, and affordability. Sensors, especially wireless sensors, have a wide range of applications in civil and environmental engineering emergency management and agriculture [23,24]. Their application to farming is more recent with first applications aiming to reduce management costs and improve animal health [24]. As sensing technology has become progressively more affordable, and in many cases less complex, research interest into potential applications to assess, control, and improve animal welfare is expanding and it is expected to increase with time. Table 1 summarizes the most relevant sensors tested for application in poultry with potential to benefit welfare.

Table 1. Main sensor technologies and potential applications to improve poultry welfare.

Sensor Type		Reference	Applications
Air quality		[27]	Indoor climatic conditions' assessment
		[28]	Broilers' final weight prediction
Sound	Broiler incubation	[44]	Monitoring hatching windows for better productivity
	Broilers	[35]	Feed intake measurements
		[36]	Growth prediction
		[38]	Thermal comfort estimation within farms
	Laying Hens	[39]	Stress detection induced by environmental temperature variation and fear
		[34]	Determination of feather pecking conditions
Locomotion		[52]	Assessing locomotion deficiency in broilers
		[53]	Use of Geographic Information Systems (GIS) to evaluate space use and different behaviours in laying hens
		[55]	Study of hens' jumps between perches and its impact on bone breakage occurrence
		[56]	Study of hens' use of pop holes and its effect on keel fracture incidence
Health status		[57]	Detection of avian influenza by the measure of broilers' temperature variations
		[58]	Detection of avian influenza by the measure of broilers' activity

2.1. Environmental Sensors

Environmental conditions, in particular inadequate temperature, relative humidity, and the length of exposure have a major impact on broiler welfare, mortality, and performance [8,25]. Exposure to elevated levels of noxious gases like carbon dioxide and ammonia is also known to reduce growth, feed conversion, and immunological response [26]. Even a two weeks' exposure to high carbon dioxide concentrations in one day old chicks is sufficient to increase the incidence of late mortalities and alter heart characteristics [27]. Thus, any efforts to better monitor and control environmental conditions will have a direct impact on bird welfare.

Although real-time multi-sensor monitoring and control of the environmental conditions (besides temperature) is not commonly applied in commercial poultry farms, current advances in sensing technology, with higher capabilities at affordable prices, will permit the development of systems for a precise control of the production environment. Some examples of current developments include multi-sensing systems to monitor environmental temperature, differential indoor atmospheric pressure, and air velocity in broiler flocks [28] to automatically assess the adequacy of the ventilation system design and functioning, which is highly relevant to provide a comfortable environment to poultry. Using sensors to simultaneously collect temperature, relative humidity, carbon dioxide, and ammonia concentrations, Jackman et al. [29] developed a good prediction model to calculate final mean bird weight in broiler flocks. The model showed excellent, house specific prediction ability ($r^2 = 0.89$) between the predicted and observed bird weight based on the conditions of the rearing environment. The development of continuous real-time environmental monitoring combined with advanced modelling tools could be used to provide a warning system to potential deviations from targeted weight gains which may also be a good indication of health or welfare risks; thus, having real potential to assure the optimal and sustained environmental conditions.

2.2. Acoustic Sensors

Bioacoustics studies the characteristics and the biological significance of sounds emitted by living organisms [30]. Birds in particular, rely on acoustic communication for their social interactions and for alarm signalling [17]. Some forms of acoustic signalling can also be considered as reliable stress indicators [31] and, thus, is an interesting approach when looking for reliable welfare assessment indicators. Acoustic studies can range in complexity from the simple establishment of differences on the frequency of emitted vocalizations to intricate analyses on sound physical properties. The later used to be laborious and complex [32], but current available bioacoustics software, like Raven software (Cornell Lab of Ornithology, Ithaca, NY, USA), has made this type of analysis somewhat easier, thus, becoming a practical tool for behavioural and welfare studies.

Using relatively simple acoustic parameters such as vocalization frequency, Koene et al. [33] and Zimmerman et al. [34] were able to detect episodes of food deprivation in broilers and in laying hens, while Bright [35] showed higher rates of squawks and total vocalizations in laying hen flocks with feather pecking problems. Based on complex sound analyses and algorithmic procedures, Aydin et al. [36] were able to distinguish sound signals corresponding to pecking (characterized by a sudden increase in amplitude follow by a sudden decrease) from all other signals in the range of 1000 Hz to 5000 kHz (using a 6th order Butterworth filter). Based on these analyses, together with the recording of the feed uptake (recorded with the traditional feed weighing system), they developed a model to predict feed intake in broilers, which was highly correlated with pecking sounds. A later study, used peak frequency vocalizations emitted by broiler flocks (analysed using Adobe® Audition™ CS6) to predict growth, as they found that these vocalizations were inversely proportional to bird age and weight [37]. Based on these finding, the authors suggested that the automatic analysis of peak frequency vocalizations would allow the development of prediction tools and, therefore, would permit health and welfare assessment at the farm level, with potential to be used as an early warning system.

Sound analyses have been proven to be powerful to determine the adequacy of the thermal environment. Thus, Moura et al. [38] estimated thermal comfort and chick performance based on the analyses of amplitude vocalizations and the noise frequency spectrum (using Cool Edit® and Audacity®) of broiler chicks placed under varying environmental temperatures, while collecting their behavioural response in parallel. They showed that when temperature decreased, the amplitude and frequency of the vocalizations increased as birds grouped together to reduce heat loss, while during thermal comfort the amplitude and frequency of vocalizations stabilized. Pereira et al. [39] identified thermal stress conditions based on broiler vocalizations and verified the existence of different vocalization patterns in heat stressed birds. In this case, the study was based on four vocalization acoustic parameters: energy, bandwidth, and first and second formant (using Praat® and Matlab® software). Thermal stress together with fear induced by routine management practices were the main sources of stress considered by Lee et al. [40], who tested their effects on vocalization patterns of laying hens. They developed an automatic online-monitoring prototype that used bird sounds to notify producers of a stressful situation. The system was developed with support vector machine techniques that were able to classify the sound emitted by laying hens (using Praat 5.3.52 and Weka 3.6 (http://www.cs.waikato.ac.nz/ml)) into categories such as physical stress, thermal stress, and mental stress due to fear. Results were validated with real sound records, showing 96.2% accuracy in detecting stress episodes.

Sound analysis has also been used during incubation as a tool to reduce the hatching window (time interval between the first and the last hatching egg), as it is considered a key factor directly related to broiler welfare and performance [41]. If hatching occurs early, it increases the probability of dehydration and early mortality, but late hatching also increases the risk of poor hatchability, pipped eggs, live-embryo unhatched eggs, and reduced chick quality [42]. Differences in hatching times may also affect feeding behaviour in broilers [43] and pronounced fearfulness in early male hatches [44]. Given the long term consequences of the hatching window on broiler welfare and performance, it is essential to monitor the final phase of the incubation process to minimize the risk of early and late hatches. Exadaktylos et al. [45] conducted a sound analysis within the 2000–4000 Hz region (using a 10th order Butterworth filter) to identify the moment at which embryos reached or passed the internal pipping stage, according to the peak frequency of the sounds, to then apply an adjusted temperature profile and narrow the hatching window. The developed algorithm based on sound analyses detected 93%–98% of the chicks reaching the internal pipping state. Although this was an experimental study, it would not be surprising to see further acoustic analysis applications during incubation to benefit bird health and welfare.

In summary, these studies show a full range of potential applications of sound analysis to optimize the conditions of the rearing environment and to detect behavioural problems such as feather pecking, hunger episodes, or thermal stress. Because sound technology has been around for a number of years, and some parameters are simple to assess, it has real potential for practical commercial implementation to improve health and welfare in poultry.

2.3. Movement Sensors

Freedom of movement is an intrinsic component of animal welfare [46], thus, to assure optimal welfare animals should be able to move freely. However, rearing conditions may hinder movement in poultry due to high density, housing space availability, and design or health condition, among other factors [47,48,49,50]. Thus, movement (or lack thereof) is a direct indicator of the welfare status in poultry. Movement sensors have been used to study different aspects of movement in broilers and in laying hens. Piezoelectric crystal sensors were used to determine locomotion deficiencies, one of the major indicators of broiler welfare [51], by examining the peak vertical force on both feet during walking sequences [52]. With this equipment, the authors were able to detect an asymmetry of the peak force in each foot that led to uneven walking in male broilers, which is a first approach towards real time broiler gait assessment.

A combination of Geographic Information System (GIS) with wireless sensor attached to the birds' body (Mica2 Dot radio mobile) was used by Daigle et al. [53] to study in detail the relationship between movement and behaviour in laying hens. A series of nodes installed through the hens' environment (Mica2 Dot stationary radio nodes) acted as beacons for the detection of the sensors, and a base station collected the data from the stationary nodes. In addition, video recordings allowed the association of the observations with the output of the sensors producing spatial explicit data which permitted analysis of the spatiotemporal variation in individual hens under experimental conditions. The data were used to map the spatial configuration of hens' home range distribution to finally investigate their association with welfare indicators such as health status, expression of natural behaviours, and their emotional state. The results obtained by Daigle et al. [53] showed large inter-individual variability in time spent performing specific behaviours, home range size, and home range overlap with conspecifics. According to the authors, this variability could explain individual hen's condition, and could be indicative of illness, injury, or changes in social dynamics. On the basis of these results, Daigle et al. [53] suggested that a better understanding of confined animals' behaviour and space use, by using GIS technology, may help to advance in-housing design and management practices to improve the welfare of laying hens. However, the implementation of such a technology on the commercial scale would not be cost effective as the authors were actually interested in studying the behaviour of individual non-caged hens.

In alternative housing systems, movement of laying hens across perches and other housing equipment may increase the risk of bone breakage which is a source of poor welfare [54]. Banerjee et al. [54] used a 3-axis ADL335 accelerometer (Analog Devices, Norwood, MA, USA) placed in laying hens and similar communication equipment to the one used by Daigle et al. [53] to calculate landing forces, height of the jumps, and the initiation and landing times when jumping from perches at variable heights to a lower landing surface in an experimental set up. The accelerometers continuously collected the acceleration data in three axes as birds jumped. A pressure mapping system and a video camera were installed as a validation method to measure the landing forces and to record the jumps. Banerjee et al. [55] reported an average landing force of 15.85 and 20.8 KJ when jumping from heights of 41 to 61 cm, respectively. This technological approach capable of assessing the landing forces on laying hens differing in age, size, and plumage integrity may enable the improvement of housing and perching design [55], which is important to resolve some persistent issues on aviary systems that are relevant for the welfare of laying hens.

Richards et al. [56] studied the effect of pop hole use on the percentage of keel fractures in free range laying hens tagged with radio frequency identification (RFID) transponders. Flocks of different ages were monitored during two laying periods and keel fracture incidence was evaluated through regular palpations. It was shown that the average percentage of keel fractures increased with hens' age but higher keel scores were registered when pop holes were less utilized. The authors concluded that fractured hens had a lower ability to use the pop holes to access the outdoor range which reduced the welfare advantages provided by free range housing systems. Siegford et al. [22] reviewed and discussed different technologies used for assessing hens' activity and location and their impact on welfare condition depending on the farm's housing system.

2.4. Sensors for Health Status Detection

Wireless systems equipped with body temperature sensors and accelerometers have been used under experimental conditions to detect chickens infected with highly pathogenic avian influenza six hours before death [57]. The same team later developed more sensitive equipment based on wireless 3-axis accelerometers and a radial lead thermistor that sent the data on activity and temperature to wireless sensor nodes to detect signs of avian influenza. The authors proved the ability of the method to detect abnormal states caused by the disease twice as early as with body temperature sensors alone with a reported detection ratio of 100% [58]. Even though this sensing equipment can prevent economic losses and welfare issues due to disease spread, it would be unpractical and too expensive to fit all individuals with surveillance equipment in a typically large poultry flock. However, sensors could be fit to a subpopulation of sentinel birds, which may be as effective for prevention or as an early detection strategy at least in high risk-areas. In addition, as variation in temperature and reduced activity are common general symptoms for many diseases, this basic equipment could be used as a warning for other health risks as well.

3. IMAGE TECHNOLOGY

3.1. Image Analysis

Skeletal disorders and contact dermatitis are major broiler welfare issues [51] that are still a matter of concern from the welfare and the economic stand point. Good farm management and a good use of novel technology may be of great relevance in the upcoming years to minimize such problems. For instance, in a study conducted in broilers [59], the "Eyenamic Software" [60] was used to calculate birds' activity level by processing calibrated recorded video images. Then they calculated differences in pixel intensity values in comparison to the previous image to calculate an activity index. This system was used to assess the relationship between automatic gaits with gait score obtained by human experts, to later develop an automatic activity index tool capable of detecting leg problems. In this study, video images were processed with the Eyenamic software to calculate an activity index of six birds, which were given a gait score by experts according to Kestin et al. [61]. The method was sensitive to detect severely affected birds with gait scores 4 and 5, but not for moderately affected birds. The authors indicated that if further validation can be obtained, this automatic activity monitoring tool has the potential be used to detect high gait scores (4 and 5) in commercial farms.

An image analysis prototype was evaluated for its adequacy as a tool for automated footpad dermatitis scoring as compared to the traditional human scoring at the slaughter plant [62]. Experimental birds were assessed and selected for each of the five categories of footpad dermatitis according to the Welfare Quality protocol for poultry [11] in semi-commercial conditions. Those birds were transported to the slaughter plant five days later where their feet were first assessed with the automated image-analysis assessment prototype system (Meyn Food Processing technology B.V.) and then assessed by the same expert that performed the first evaluation at the farm. Agreement between both assessment methods was initially poor because the automated system does not consider the depth of the lesion. However, the results improved considerably when considering only those birds for which the automatic system produced a dermatitis scored for both footpads yielding stronger correlations with the expert scoring ($r = 0.68$ and $r = 0.74$ for farm and slaughter plant, respectively) [62]. Even though the system still requires adjustments, the automatic image analysis offers a great potential for automatic footpad dermatitis assessment that could be implemented in a relatively short time period at slaughter plants.

3.2. Optical Flow

A particular type of image analysis is the optical flow analysis (OF) that has been used in many applications including traffic flows [63], movement of glaciers [64], cell and sperm motility [65], and, lately, in the analysis of movement in confined broilers [66,67]. One of the main advantages of OF is that it allows the automatic and continuous assessment of moving images containing hundreds of individuals [68] and, thus, could be a practical approach for the assessment of movement related welfare issues in commercial poultry.

OF detects the brightness change rate in pixels of a moving image and has specific statistical properties such as the mean flow rate, variance, skew, and kurtosis that can be used to detect its association with variations in gait score, pododermatitis, hock burn, and mortalities in broiler flocks [66]. In a study conducted in 24 commercial flocks, Dawkins et al. [66] detected a negative correlation between mean flow and flock mortality in 30-day-old broilers, while both the skew and kurtosis were positively correlated with the incidence of mortalities, culls, hock burns, and mean gait score. However, in a later study conducted with the aim of finding a more direct relationship between OF, behaviour, and welfare, Dawkins et al. [67] were only able to find significant positive correlations for skew and kurtosis with the number of birds walking for at least 10 s, concluding that OF is probably more sensitive to flock uniformity or lack of it.

A combination of OF and Bayesian regression was used by Roberts et al. [69] to predict health and welfare on a continuous basis. Mean, variance, skewness, and kurtosis were estimated daily using an OF algorithm. Gait score was assessed on day 28 in 60 birds/flock. In addition, daily mortality, culls, weights, growth rate, water and food consumption, and total incidence of mortality, culls, pododermatitis, and hock burns were calculated for the rearing period, and were included in the regression model. The model successfully predicted total flock mortality at 15 days of age, gait score became significant from day 13 and was capable of predicting the occurrence of hock burns at one or two days of age. Thus, Roberts et al. [69] showed the powerful predictive power of OF combined with the Bayesian regression model. Recent work indicates that OF technology can even be useful to detect *Campylobacter* infected flocks [68], which has a strong welfare impact as strong inflammatory conditions can lead to diarrhoea, poor litter quality, and deteriorated walking ability in affected birds. In their study, based in 31 commercial flocks, Colles et al. [70] showed that flocks likely to become positive for *Campylobacter* were identified in the first seven to 10 days of life and were characterized by having a lower mean flow rate and consistently higher kurtosis in comparison to non-infected flocks.

Although most OF studies focused in broilers, OF was also used to predict plumage damage in laying hens [71] and to identify the management and/or environmental risk factors involved in plumage deterioration. OF data of 18 commercial laying hen flocks were collected by video recordings at different ages and processed with OF algorithms. A hidden Markov chain was used to identify disturbance periods in the OF dataset.

To validate the method, an expert observer visualized the video recordings and scored a variety of disturbances like birds running or pecking each other. Measures of disturbances were combined with management, environmental, production, and feather damage (scored according to Bright el al. [72]) data for each farm to improve the predictive power of the model. Feather scores in later weeks were predicted using Gaussian linear and nonlinear regression models. The model showed improved prediction of feather damage and a good identification of high prevalent damaged flocks during following weeks.

Considering the positive results obtained with OF analysis, the method seems to be a sensitive tool for the assessment of the health and welfare status in commercial broiler flocks as well as in an experimental setting for laying hens. If positive results continue to be supported by research, this technology may have a major impact on poultry management as it benefits the animals, producers, and consumers by reducing economic losses and improving food safety. Besides, this methodology is non-invasive and it is relatively easy to apply in large flocks. It is probably just a matter of time before OF technology is applied to commercial laying hens or other poultry species.

3.3. Infrared Thermal Imaging

Preventing heat stress is crucial to poultry welfare as it may impact behaviour, immunity, and physiological processes and can cause major mortalities [73,74]. Infrared thermal imaging (IRTI) technology creates infrared images showing the body's superficial temperature distribution from the infrared radiation emitted by objects that is converted into electrical signals. In the thermal image, each colour expresses a specific temperature range related to the defined scale [75], thus it is a practical, non-invasive tool to study welfare aspects related to thermoregulation.

Yahav et al. [76] used IRTI to determine optimal air velocity (AV) for broilers' thermoregulation, while maintaining adequate temperature and relative humidity. Body weight, feed intake, and faecal excretions were collected to estimate the energetic demands for body maintenance, while body heat loss was calculated by radiation and convection using IRTI. With this methodology, the authors showed that 2.0 m/s was the optimal air velocity, allowing the birds to control body temperature with no detrimental effects on performance. It has also been shown that it is possible to monitor changes in the metabolism of broilers associated to thermal variation by analysing body surface temperature through IRTI. Ferreira et al. [77] indicated that IRTI was sensitive enough to identify a reduction in metabolic heat production in birds fed with an oil supplemented diet, which was suggested as a nutrition alternative to minimize heat stress. Work by Giloh et al. [78] corroborated the reliability of using facial surface temperature (measured with IRTI) as an indicator of heat stress by correlating it with changes in body core temperature, corticosterone, thyroid hormones, and arginine vasotocin that are indicative of increased stress levels. Giloh et al. [78] proved the existence of a strong correlation between facial surface temperature and core body temperature and indicated its usefulness as a determining factor to support decisions in a climate-controlled environment farm. In fact, the authors indicated that under experimental conditions, facial surface temperature recorded by IRTI was more informative than ambient temperature regarding the birds' status and, thus, has great potential to be applied at a commercial level to monitor thermal stress levels.

IRTI has also been used in other production phases such as incubation and pre-slaughter. Studies conducted by Shinder et al. [79] using IRTI during the final phase of incubation showed that body weight and body temperature were significantly higher in chicks that were exposed to short periods of cold stress during incubation (days 18 and 19) and had 13% to 18% lower incidence of ascites in comparison to control birds. On the other hand, Naas et al. [80] used IRTI technology to estimate heat exchange between broilers and their environment at pre-slaughter. These estimations were later used to develop a mathematical model to predict broiler surface temperature as a function of air temperature in order to evaluate the effects of pre-slaughter handling and environmental conditions on welfare and mortality. Naas et al. [80] showed that featherless body areas reacted rapidly to environmental changes, affecting homeostasis and increasing deaths on arrival at the slaughterhouse. This non-invasive method can, therefore, help to improve flock management during the pre-slaughter phase, as it permits assessing the birds' welfare conditions and facilitates taking remedial actions to guarantee welfare and to reduce mortalities at the end of the rearing period.

In laying hens, IRTI has been experimentally tested as an assessment method for bumblefoot and plumage condition. Wilcox et al. [81] used IRTI to diagnose subclinical bumblefoot, finding a high correlation between thermal images and the visual score. This correlation was 86.7% in hens classified as clinical, but only 26.7% in hens classified as mildly clinical at day seven post-inoculation with *Staphylococcus aureus*. The authors suggested that IRTI was more sensitive than the visual scoring to detect subclinical cases of bumblefoot, which would facilitate early detection of the inflammation and would reduce associated pain. The method is clearly efficient to detect bumblefoot, even though it is invasive, and collecting feet infrared images requires holding the hens several times during the experiments which is a source of stress and it is detrimental to their welfare [82]. A potential alternative would be assessing a small, representative bird sample to predict the incidence of bumblefoot in the flock in order to take any preventative or remediating step.

Zhao et al. [83] used IRTI to determine its potential to assess feather coverage. In their work they considered three feather coverage categories: excellent feather (EF), fair feather (FF) and no feather (NF) in six different body areas (head, dorsal neck, front neck and crop, back, breast, and belly). For all body areas, the EF surface determined by IRTI was positively correlated with the feather scoring, while FF and NF areas were negatively correlated with feather scoring. The IRTI method also confirmed that feather coverage deteriorated in older hens, which lead to a higher feed to egg conversion rate because of higher sensible heat loss.

3.4. Kinematic Analyses

Kinematics is a branch of classical mechanics that describes the geometry of motion without consideration of the masses and the forces that may have caused the motion [84]. One of the main advantages of the kinematic technology is that it offers the possibility to perform a three-dimensional (3D) evaluation of the motion in a rapid and non-invasive way. In broilers, kinematic analysis was used to identify gait abnormalities [85]. Spherical retro-reflective markers, infrared cameras, and 3D kinematic data processing software were used to compare the gait characteristics of broiler chickens with their ancestral line, the red jungle fowl, and to find a link between broiler gait parameters to better define lameness scores. Caplen et al. [85] found that, while jungle fowl increased their velocity by taking strides of comparably longer duration and length, lame broilers took shorter strides and reduced stride duration to accelerate. They also found that the larger pectoral muscle mass of lame broilers displaces their centre of mass, and this requires their feet to be positioned further forward under the body for support. These findings explained the existing differences between both genetic lines and the consequent impact on broilers' welfare. Kinematic analysis has been also used to study spacing behaviour in laying hens in order to calculate minimum space requirements [86].

4. MOBILE APPS FOR WELFARE ASSESSMENT

A novel approach has been recently proposed for the welfare assessment of commercial broiler and turkey flocks that is based on line transects

[13,14]. The transect method assesses the frequency of broilers or turkeys showing signs of impaired welfare by noting their incidence while walking along predefined paths or transects that are established among drinkers and feeder lines.

The i-Watchbroiler (https://play.google.com/store/apps/details?id =com.daia.iwatchbroiler) and the i-Watchturkey (https://play. google.com / store/apps/ details?id= com.daia.iwatchturkey) software applications for mobile devices based on this methodology, allow assessors to easily record the frequency of birds showing any of the defined welfare incidences by pressing on the touch screen menu (Figure 1a,b), which is standardized by the expected number of birds within each transect.

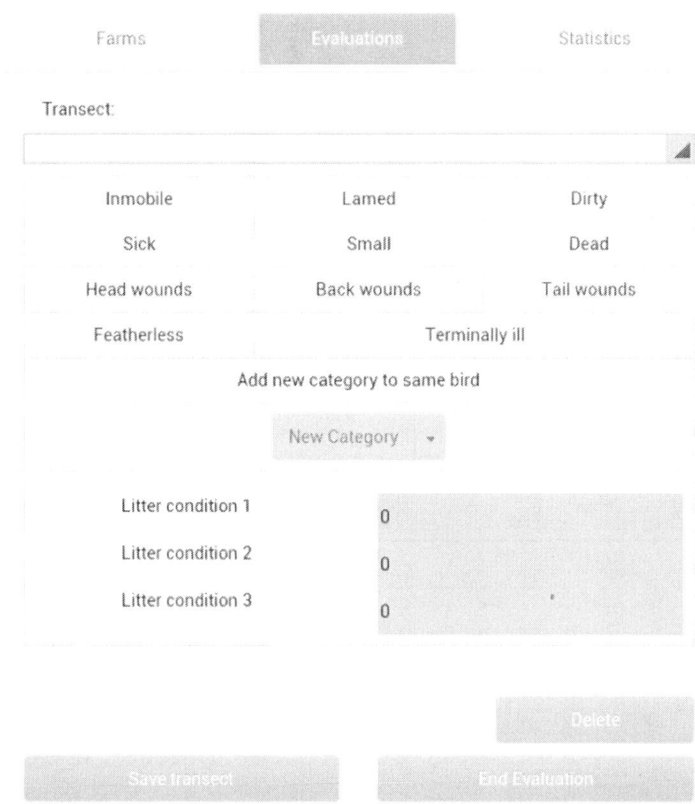

(a)

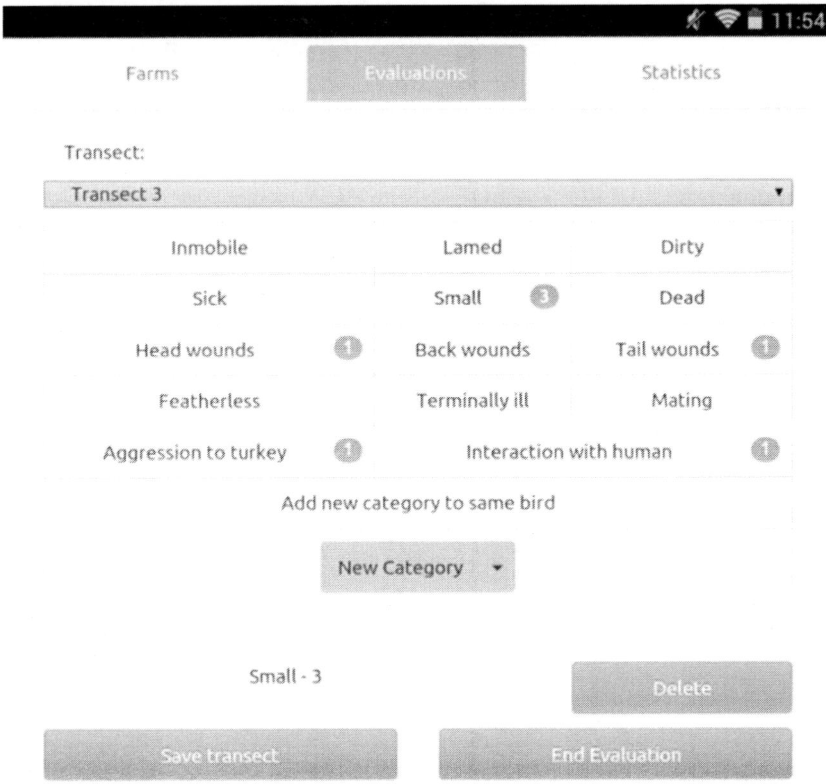

(**b**)

Figure 1. (**a**) The i-Watchbroiler mobile app menu screen. Major welfare indicators assessed include: lame, immobile, sick, small, dirty, terminally ill, featherless, and wounded birds; (**b**) The i-WatchTurkey mobile app menu screen, including specific welfare indicators for turkey assessment.

Both apps permit the inclusion of main features of the housing conditions, flock specific characteristics, and age at assessment so that data files, when exported, contain complete information relative to independent variables and the corresponding assessment results for further statistical analysis. These apps include basic statistics tools so that they are able to provide the mean incidence of each welfare indicator immediately after flock assessment and will calculate potential deviations from previously collected data. The results obtained by Marchewka et al. [13,14] showed a good inter-observer reliability of the methodology for both broilers and turkeys and the method was validated for turkeys [14].

5. MATHEMATICAL MODELLING

Modelling approaches can be used to enhance the application of technology in commercial poultry farms. Indeed, large sets of data produced by sensors and video recordings can be analysed by complex modelling or artificial intelligence algorithms to generate predictions or risk assessment models. Modelling techniques are essential to interpret data from real time monitoring devices in order to develop control systems or to establish risk alerts.

5.1. Environmental Conditions

A good monitoring of temperature, humidity, ventilation and lighting within poultry houses is essential to guarantee optimal rearing conditions and environmental standards for good welfare. Computational fluid dynamics (CFD) is a branch of fluid mechanics that provides a cost effective means of simulating real flows by using governing equations [87]. CFD techniques are based on the resolution of a set of partial differential continuity equations (conservation of mass, conservation of energy, and conservation of momentum) [88]. CFD models have been used to evaluate ventilation efficiency on broiler thermal stress and mortality [87]. Predictions were validated with real environmental data collected with a multisensory system composed of 24 air velocity and temperature sensors and two differential pressure sensors (as explained in earlier sections). According to Bustamante et al. [89], this technique not only permits the improvement of thermal comfort and overall welfare of broilers, but also reduces the electric energy consumption.

CFD has been recently used by Rojano et al. [90] to model and predict climate and air quality parameters (temperature, absolute humidity, and CO_2) in naturally ventilated broiler houses by investigating sensible and latent heat, as well as mass transport and heat transfer emitted by animals, litter, and heaters. The analysis was carried out at two distinct time periods: at the beginning of the growing cycle to assess the influence of heaters and with the birds maintained at low ventilation rates, and at the end of production to assess the influence of stoking density under high ventilation rates and heaters turned off. In the latter case, the effect of sensible and latent heat was examined by simulating three different animal densities in the model. To validate the CFD model, spatial variation of temperature and humidity were collected every 10 min during the production cycle using calibrated sensors as well as indoor and outdoor CO_2 concentrations using photo-acoustic infrared spectrometry. Rojano et al. [88] indicated that for both phases, the predictions of temperature, absolute humidity and CO_2 were, in general, in agreement with experimental data, but recommended taking into consideration the effect of animals, litter, and by-products generated by the heaters to improve CFD model accuracy. CFD modelling not only appears to provide solutions for optimal poultry farm design, but also may enhance future applications in creating a real time automated system capable of controlling house conditions to avoid mortalities due to thermal stress, thus improving animal welfare.

5.2. Spatial Distribution and Activity Modelling

According to the World Organization for Animal Health (OIE) [91], "Changes in the spatial distribution of birds may indicate thermal discomfort or the existence of areas of wet litter or uneven provision of light, food or water". Birds' spatial distribution may evidence problems occurring in a poultry house, thus, recent studies have focused on technological approaches that consider birds' spatial distribution in the context of PLF technologies.

Kashiha et al. [21] used the eYeNamic systems [92], which is an image pre-processing tool, to calculate the number of object pixels in ratio to the background from images captured every five minutes by three cameras placed above a commercial flock. From the pixel ratio, a zone occupation density (ZOD) was calculated (60 ZOD per camera as every image captured was divided into 60 zones), together with the mean occupation rate for the flock; this information was used to calculate an activity index. The authors manipulated lighting periods experimentally in order to design a mathematical model based on the variation of the activity index capable of predicting the response during the next light period. When the measurements deviate from the predicted response calculated by the model it indicates that an event might have been occurring in the house (malfunctioning of feeders, drinkers, heating, ventilation, or a visiting human). The model was validated using the farmer's logbook, where all problems occurring within the house were registered. The comparison of predicted and measured distributions showed that the method could report successfully 95.24% of events in real time during a complete growing period while generating no false alarms. This fully automated technology has been already introduced at a commercial level allows for the identification of problems in broiler flocks and helps farmers to conduct real time monitoring of their animals more efficiently.

Lately, Youssef et al. [93] aimed to predict the behavioural response (activity levels) of broilers under different micro-environmental conditions by introducing a model-based predictive controller (MPC). The dynamic MPC should be able to predict the system output, broiler activity in this case, in response to changes in the control variable (inlet temperature and ventilation rate). In this study, 45 seven-day old broiler chicks were housed in a test chamber where 30 temperature and air velocity sensors as well as CCD cameras were installed to measure temperature inlet, air velocity, and chickens' activity. During the experiment, combinations of ventilation rate and inlet temperature increases were applied. The airflow pattern was estimated to investigate the spatial temperature distribution in relation to the local velocity distribution in the test chamber. This estimation was later compared with the bird's zonal occupation and activity level. A dynamic activity index was calculated on the basis of the variations in the pixel intensity between consecutive frames. Finally, a dynamic modelling of the activity index was calculated to describe the static and dynamic responses of the chicken's activity index in response to variations in air temperature and ventilation rate. With this system, Youssef et al. [91] were able to detect that non-homogeneous airflow patterns in the test chamber resulted in a heterogeneous spatial distribution of the chickens, with those undergoing heat stress tending to occupy high air velocity areas, and vice versa [93]. Even though this technology is still at an experimental stage, it might be useful to correct environmental parameters according to real time behavioural bird response.

5.3. Precision Feeding

The detrimental effects of elevated stocking densities, suboptimal environmental conditions, or inadequate lighting regimes reflect on broilers' feed intake [94]. In addition, the use of different feeding strategies can help birds to better cope with different sources of environmental stress, and to prevent the onset of skeletal disorders [6]. Consequently, the control of feed intake through the use of precision feeding tools would be of great interest to improve flock management and bird welfare.

Gates and Xin [95] developed two algorithms to determine the feeding behaviour of broiler chickens and laying hens, with the aim of assessing the impact of environmental stressors. The algorithm was used to predict feeding patterns such as the number of meals, time at feeder, and meal size, as well as to discriminate between feeding bouts and stereotyped pecking. While laying hens were subject to heat stress, feeding behaviour of broilers was assessed when presented with a specialized sesame diet. The study was validated using video observations of the birds' behaviour. Both algorithms showed robustness in providing parameters like meal size, time at the feeders, and were able to discriminate between eating at the feeder versus stereotyped pecking, all of which were in agreement with the video recorded observations.

5.4. Monitoring Performance, Stress and Health Status

It is well known that the welfare status of laying hens has a direct impact on egg quality [96,97,98]; therefore, a drop in egg production or quality may be indicative of ongoing welfare problems.

Mertens et al. [99] used statistical process control, a technique that permits the formulation of quality limits based on natural process variability, to develop an intelligent control chart to monitor hens' variation on egg weight. Large scale experimental flock data were first used to construct and train the model in order to detect the natural increment in egg weight with increasing age. In a second phase, average egg weight was daily registered and all occurring events (technical failure, mortality, treatments) were recorded in a log file for the validation process. In order to investigate the ability of the control chart to detect drops in egg weight, two main stressors were tested: heat stress and red mite infestation. Results showed that the model was able to detect egg weight losses caused by heat stress, red mite infestation, and other management problems even though some false alarms were registered. The authors could detect abnormalities within two days after the onset of the tested challenges. A more sophisticated algorithm was later developed by Mertens et al. [100] to control daily egg production. The system was able to detect feed intake decline resulting in reduced egg production. Similarly, an error in feed formulation produced an alarm soon after feed administration.

Transmission colour value (TCV) of the egg shell measured by visible-near infrared transmission spectroscopy was used by Mertens et al. [101] to monitor flock stress and health in laying hens. TCV was calculated as the ratio between the transmission at 643 nm (maximum absorbance of the pigmentation molecule protoporphyrin IX) and the transmission at 610 nm (a reference wavelength). In addition, the algorithm based on Mertens et al. [99,100] was used to construct a control chart to monitor the course of TCV and to investigate if changes could relate to stressful events. This technique was successful at detecting a significant variation on eggshell pigmentation due to heat stress, infectious bronchitis, and after an abrupt transition to phase two feeding that caused a decline in feed intake. Variation in TCV values warned about the occurrence of a problem four days earlier than the consequent drop of the average egg weight. The authors concluded that tracking daily variations in eggshell colour might be useful as a relevant stress and health status indicator.

Another technique that may have future applications is Support Vector Machines (SVM), a type of machine learning algorithm, used by Hepworth et al. [102] to identify risk factors for hock burn incidence. SVM are a set of supervised learning algorithms which perform classification by finding the hyperplane that maximizes the margin between two classes of variables [103], and it is used in epidemiology for classification, diagnosis, and risk factor identification. Hepworth et al. [102] recorded data relative to farm management conditions (stocking density, number and age of parent flocks, sex, and rearing system) together with daily water consumption, average weekly weight, mortality, and slaughterhouse outcomes (rejections, downgrades, and hock burns). Test and training data were performed by repeating random division of the collected data in two halves. After ten repetitions, the hierarchical structure was retained in each half of the data. Then, SVM classifiers with linear kernels were built and compared to manually build logistic regression models in order to test SVM classifiers' reliability on predicting hock burn prevalence. As indicated by Hepworth et al. [102], this technique has an enormous potential to improve poultry health and welfare as it has proved robustness for a broad range of complex data sets. Furthermore, SVM does not rely on restrictive assumptions about the distribution and independence of data, in contrast to logistic regression modelling.

6. DISCUSSION

A wide range of technical developments, complex data processing, and modelling tools have emerged in the past few years with the potential to assess, control, and improve poultry welfare. Although some technologies are still in a developmental phase, others have already been implemented under commercial conditions. In fact, many of the technologies here presented could be integrated in farm management processes to enhance poultry welfare and farm efficiency while facilitating the decision making process during the growing cycle. As one of the fastest-growing production species, with very similar management strategies around the world and with a high level of integration, poultry production and especially broilers' offers the ideal conditions for the application of the latest technological developments. Indeed, as the production cycle of broiler chickens is short (40 to 45 days), large data sets containing a substantial variety of information are relatively easy to acquire, which facilitates testing and implementing such technologies and a continuous improvement of welfare condition in next flocks. The objective of PLF technologies is to address and prevent major poultry welfare issues while providing farmers with better and faster management solutions that would result in higher efficiency and economic profit. This review highlighted the most important technological advances that have the potential to be applied to improve the welfare of broilers and laying hens as well as of other poultry species.

It is clear that environmental conditions and noxious gas concentrations have a major impact on the birds' welfare, health, and performance [8,26,27]. Thus, a better and faster control of the environmental conditions along the production process would permit the improvement of birds' health and welfare as well as productive efficiency. Environmental sensors permit real time monitoring of the production conditions in a relatively simple and efficient manner at an affordable cost. However, large environmental data sets are not helpful unless data are processed adequately in order to extract relevant, meaningful information for the end users. Customized algorithms and other complex mathematical techniques allow processing of the collected information to detect variations and their potential consequences, leading to the development of quite precise alert or risk assessment systems. Alerts allow the end user to easily detect when a threshold has been reached, thus facilitating the application of control measures to resolve the problem or to minimize its impact. Hence, automated data collection, processing, and interpretation would permit farmers to fulfil the PLF challenge of improving animal welfare, health, and environmental sustainability [104] as higher performance will be obtained from a set amount of resources.

Regarding the value of technological advances to address specific poultry welfare issues, research has shown that piezoelectric sensors [52] and kinematic technology [85] can be useful to investigate locomotion characteristics and gait abnormalities in broilers, while wireless acceleration sensors can be used to determine the effect of height on the incidence of bone breakage in laying hens [55]. Even though these examples of technological developments are still at an experimental phase and would need further research for commercial implementation, these approaches can be helpful to understand bird locomotion characteristics and to detect locomotion abnormalities, at least in experimental studies.

Analysing and processing data derived from different imaging technologies appears to be suitable to assess gait [59], walking ability [66], and footpad dermatitis [62] in broilers, or to detect bumblefoot incidence in laying hens [81]. The Dawkins et al. [66] study conducted on commercial broiler flocks showed a great potential for a fast detection of abnormal walking behaviour and for a consequent implementation of mitigating strategies. Likewise, SVM classifiers' modelling techniques have been refined to identify risk factors causing hock burn in broiler flocks [102]. At the health level, initial studies in optical flow and wireless sensors indicate that such technology can also be applied to detect infectious diseases that have a major economic and social cost such as *Campylobacter* [70] and avian influenza [58] before the appearance of the first signs of the disease. Considering welfare issues that are more specific to laying hens, a focus on feather pecking was undertaken using optical flow [71], image radio telemetry imaging [83], and sound sensing [35] technologies. A future implementation of these technologies at a commercial level could be efficient to prevent the development of health issues that have major implications for the welfare and performance of meat and egg producing poultry. Production diseases are estimated to cause a 10%–15% reduction in performance in poultry farming [105]. Imaging and sensing technologies able to detect changes in birds' behaviour, health, and welfare would be of great help to minimize economic losses due to this cause.

Providing birds with the possibility to express their basic behavioural repertoire is an important welfare aspect. Spatial requirements for basic behaviours and use of space in laying hens have been addressed using kinematic analysis [86] and geographic information systems sensors [53]. These studies, undertaken at the experimental level, aimed to better understand hens' behaviour and use of space in order to adjust building design and to enrich hens' environment according to scientific knowledge. Continuous real time automatic monitoring of flocks' spatial distribution index [21] and activity [93] should allow a good control of the flocks' behavioural and welfare state, permitting the detection of deviations from the "normal flock behaviour" in a timely manner in order to prevent, or at least minimize, major behavioural and welfare problems that should lead to better farm management, production efficiency, product quality, energy consumption, and, therefore, should improve long term sustainability.

Sound sensors, that have been around for a number of years can be used for a wide range of applications such as to estimate feed intake [36] and to predict growth [36] in broilers, or to detect environmental conditions leading to heat stress in broilers [38] and laying hens [40], with the possibility of notifying farmers regarding such conditions. On the other hand, infrared image technology appears to be suitable to provide modelling tools to calculate optimal air velocity rate for good thermoregulation and growth rate [76], as well as for establishing incubation programs that may improve the ability of broilers to cope with heat stress and reduce the incidence of other health issues later in life [79].

Given the advances in technology and their application to the field of animal health and welfare, it seems that, in the near future, the existence of plug-in equipment containing a full range of environmental and sound sensors, image processing, and analytic capabilities might be a reality, allowing a precise automatic welfare assessment and intelligent management at a commercial level. However, different technologies are still facing major limitations for their implementation at a commercial scale. Indeed, data collection and processing refinement is still needed, robust equipment must be developed to resist the harsh farm conditions and must be cost-efficient. Major technical and software advances have yet to take place in order to develop plug and play systems that provide reliable results. Although some technologies are already being used under commercial conditions, flock welfare assessment is generally carried out by applying existing welfare protocols. In order to facilitate the practical application of welfare protocols in meat poultry, mobile apps based on the transect methodology [13,14] have been recently developed. This non-invasive technique may provide a good depiction of the welfare condition within meat poultry in a simple and affordable manner. This method of assessment, if applied on a regular basis by producers, should be able to provide an early warning of the main broiler welfare and health issues and, thus, would also permit reducing economic losses and improve sustainability of the production.

It is crucial from a welfare perspective to validate such technologies before their implementation and to be mindful of their added value to poultry welfare [106]. Indeed, it is stressed that the main goal of using technology in poultry production is not only to facilitate improvement in farmers' lives and enhance production but also to develop our capacity of understanding birds' behaviour in commercial conditions and to improve their quality of life. A better quality of life will also mean that birds will grow healthier and more efficiently, thus, improving welfare is a direct road towards sustainable poultry production systems.

7. CONCLUSIONS

Despite of all the technological developments achieved in the last ten years, the challenges for full achievement of PLF goals are still important. Further research is needed to improve the data processing and modelling accuracy and to integrate all the suitable sensing, image processing, and data analysis in a plug-in system that is reliable, simple to understand, and economically viable for widespread use. Once available, PLF technologies will certainly provide added value to farmers, especially in regard to improved welfare and reduced environmental impact and long term system sustainability. Indeed, one of the objectives of installing technological devices in a poultry farm is an efficient and early detection tool of potential abnormal situations. Yet, an initial economic investment would be required to acquire and install such devices. Some of the technologies already used at the commercial stage are resilient and affordable. However, demonstrating and verifying the economical, welfare and environmental advantages of the technology in the medium and long term are critical [107]. Proving the coherence of advanced technology and modelling tools to reduce costs by improving welfare should be a clear convincing argument to facilitate the implementation of these technologies at a commercial level.

ACKNOWLEDGMENTS

To the Spanish Ministry of Economy and Competitivity, General Directorate for Science and Technology, National Research Program "Retos de la Sociedad" Project #AGL2013-49173-C2-1-R P. I. Inma Estevez. This project supported N. Ben Sassi with a Ph.D. assistantship.

REFERENCES

1. OIE (Office International des Epizooties). *Terrestrial Animal Health Code*; OIE: Paris, France, 2011.
2. Muiruri, H.K.; Harrison, P.C. Effect of peripheral foot cooling on metabolic rate and thermoreregulation of fed and fasted chicken hens in a hot environment. *Poult. Sci.* **1991**, *70*, 74–79.
3. Appleby, M.C.; Smith, S.F.; Hughes, B.O. Nesting, dust bathing and perching by laying hens in cages: Effects of design on behaviour and welfare. *Br. Poult. Sci.* **1993**, *34*, 835–847.
4. Baxter, M.R. The welfare problems of laying hens in battery cages. *Vet. Rec.* **1994**, *134*, 614–619.
5. Balnave, D.; Muheereza, S. Improving eggshell quality at high temperatures with dietary sodium bicarbonate. *Poult. Sci.* **1997**, *76*, 588–593.
6. Meluzzi, A.; Sirri, F. Welfare of Broiler Chicken. *Ital. J. Anim. Sci.* **2009**, *8*, 161–173.
7. Tactacan, G.B.; Guenter, W.; Lewis, N.J.; Rodriguez-Lecompte, J.C.; House, J.D. Performance and welfare of laying hens in conventional and enriched cages. *Poult. Sci.* **2009**, *88*, 698–707.
8. Dawkins, M.S.; Donnelly, C.; Jones, T. Chicken welfare is influenced more by housing conditions than by stocking density. *Nature* **2004**, *427*, 342–344.
9. Appleby, M.C.; Mench, J.A.; Hughes, B.O. *Poultry Behaviour and Welfare*; CABI Publishing: Wallingford, UK, 2004.
10. Mench, J.A. Assessing animal welfare at the farm and group level: A United States Perspectives. *Anim. Welf.* **2003**, *12*, 493–503.
11. Welfare Quality. *Assessment Protocol for Poultry, Broiler and Laying Hens*; Welfare Quality: Lelystad, The Netherlands, 2009.
12. De Jong, I.C.; Hindle, V.A.; Butterworth, A.; Engel, B.; Ferrari, P.; Gunnink, H.; Perez Moya, T.; Tuyttens, F.A.M.; van Reenen, C.G. Simplifying the Welfare

Quality assessment protocol for broiler chicken welfare. *Animal* **2015**, *10*, 117–127.

13. Marchewka, J.; Watanabe, T.T.N.; Ferrante, V.; Estevez, I. Welfare assessment in broiler farms: Transect walks versus individual scoring. *Poult. Sci.* **2013**, *92*, 2588–2599.
14. Marchewka, J.; Estevez, I.; Vezzoli, G.; Ferrante, V.; Makagon, M.M. The transect method: A novel approach to on-farm welfare assessment of commercial turkeys. *Poult. Sci.* **2015**, *94*, 7–16.
15. Berckmans, D. Precision livestock farming technologies for welfare management in intensive livestock management. *Rev. Sci. Tech.* **2014**, *33*, 189–196.
16. Mollo, M.N.; Vendrametto, O.; Okano, M.T. Precision livestock tools to improve products and processes in Broiler production: A review. *Braz. J. Poult. Sci.* **2009**, *11*, 211–218.
17. Corkery, G.; Ward, S.; Kenny, C.; Hemmingway, P. Incorporating smart sensing technologies into the poultry industry. *J. World Poult. Res.* **2013**, *3*, 106–128.
18. Wathes, C.M.; Kristensen, H.H.; Aerts, J.M.; Berkmans, D. Is livestock precision farming is an engineer's daydream or a nightmare, an animal's friend or foe, and a farmer's panacea or pitfall. *Comput. Electron. Agric.* **2008**, *64*, 2–10.
19. Moura, D.J.; Naas, I.A.; Pereira, D.F.; Silva, R.B.T.R.; Carmago, G.A. Animal welfare concepts and strategy for poultry production: A review. *Braz. J. Poult. Sci.* **2006**, *8*, 137–148.
20. Robins, A.; Phillips, C.J.C. International approaches to the welfare of meat chickens. *World Poult. Sci. J.* **2011**, *67*, 351–369.
21. Kashiha, M.; Pluk, A.; Bahr, C.; Vranken, E.; Berckmans, D. Development of an early warning system for a broiler house using computer vision. *Biosyst. Eng.* **2013**, *116*, 36–45.
22. Siegford, J.M.; Berezowski, J.; Biswas, S.K.; Daigle, C.L.; Gebhardt-Henrich, S.G.; Hernandez, C.E.; Thurner, S.; Toscano, M.J. Assessing activity and location of individual laying hens in large group using modern technology. *Animals* **2016**, *6*, 10.
23. Meyer, M. Opportunities of Wireless Sensors and Controls for Building Operation. *Energy Eng.* **2005**, *102*, 27–48.
24. Ruiz-Garcia, L.; Lunadei, L.; Barreiro, P.; Robla, J.I. A review of Wireless sensor technologies and applications in agriculture and food industry: State of the art and current trends. *Sensors* **2009**, *9*, 4728–4750.
25. Jones, T.A.; Donnelly, C.; Dawkins, M.S. Environmental and management factors affecting the welfare of chickens on commercial farms in the United Kingdom and Denmark stocked at different densities. *Poult. Sci.* **2005**, *84*, 1155–1165.
26. Wang, Y.M.; Meng, Q.P.; Guo, Y.M.; Wang, Y.Z.; Wang, Z.L.; Shan, T.Z. Effect of atmospheric ammonia on growth performance and immunological response of Broiler chickens. *Asian J. Anim. Vet. Adv.* **2010**, *9*, 2802–2806.
27. Olanrewaju, H.A.; Dozier, W.A.; Purswell, J.L.; Branton, S.L.; Miles, D.M.; Lott, B.D.; Pescatore, A.J.; Thaxton, J.P. Growth performance and physiological variables for broiler chickens subjected to short-term elevated carbon dioxide concentrations. *Poult. Sci.* **2008**, *7*, 738–742.
28. Bustamante, E.; Guijaro, E.; García-Diego, F.J.; Balach, S.; Hospitaler, A.; Torres, A.G. Multisensor systems for isotemporal measurements to assess indoor climatic conditions in poultry farms. *Sensors* **2012**, *12*, 5752–5774.
29. Jackman, P.; Ward, S.; Brennan, L.; Corkery, G.; McCarthy, U. Application of wireless technologies to forward predict crop yields in the poultry production chain. *Agric. Eng. Int.* **2015**, *17*, 287–295.
30. Tefera, M. Acoustic signals in domestic chickens (*Gallus gallus*): A tool for teaching veterinary ethology and implications for language learning. *Ethiop. Vet. J.* **2012**, *16*, 77–84.
31. SCAHAW. *The Welfare of Chickens Kept for Meat Production (Broilers)*; Report of the Scientific Committee in Animal Health and Animal Welfare; European

Commission, Health and Consumer Protection Directorate General: Brussels, Belgium, 2000.
32. Marx, G.; Leppelt, J.; Ellendorff, F. Vocalisation in chicks (*Gallus gallus* dom.) during stepwise social isolation. *Appl. Anim. Behav. Sci.* **2001**, *75*, 61–74.
33. Koene, P.; Ruiten, S.; van Bokkers, E.A.M. The effect of increasing broiler behaviour possibilities by giving extra furniture and a slimmer body: The effects of perches and feed restriction. In Proceedings of the 33rd International Congress of the International Society for Applied Ethology, Lillehammer, Norway, 17–21 August 1999; Boe, K.E., Bakken, M., Braastad, B.O., Eds.; p. 136.
34. Zimmerman, P.H.; Koene, P.; Van Hooff, J.A. The vocal expression of feeding motivation and frustration in the domestic layinh hens *Gallus gallus domesticus*. *Appl. Anim. Behav. Sci.* **2000**, *69*, 265–273.
35. Bright, A. Vocalisation and acoustic parameters of flock noise from feather pecking and non-feather pecking laying flocks. *Poult. Sci.* **2008**, *49*, 241–249.
36. Aydin, A.; Bahr, C.; Viazzi, S.; Exadaktylos, V.; Buyse, J.; Berckmans, D. A novel method to automatically measure the feed intake of broiler chickens by sound technology. *Comput. Electron. Agric.* **2014**, *101*, 17–23.
37. Fontana, I.; Tullo, E.; Butterworth, A.; Guarino, M. An innovative approach to predict the growth in intensive poultry farming. *Comput. Electron. Agric.* **2015**, *119*, 178–183.
38. Moura, D.J.; Nääs, I.A.; Alves, E.C.S.; Carvalho, T.M.R.; Vale, M.M.; Lima, K.A.O. Noise analysis to evaluate chick thermal comfort. *Sci. Agric.* **2008**, *65*, 438–443.
39. Pereira, E.M.; Naas, I.; Garcia, R. Identification of acoustic parameters for broiler welfare estimate. *Eng. Agric.* **2014**, *34*, 413–421.
40. Lee, J.; Byeongjoon, N.; Jang, S.; Park, D.; Chung, Y.; Chang, A.-H. Stress detection and classification of laying hens by sound analysis. *Asian-Australas J. Anim. Sci.* **2015**, *28*, 592–598.
41. Van de Ven, L.J.F.; van Wagenberg, V.; Debonne, M.; Decuypere, E.; Kemp, B. Hatching system and time effects on broiler physiology and posthatch growth. *Poult. Sci.* **2011**, *90*, 1267–1275.
42. The Poultry Site. Available online: http://www.thepoultrysite. com/articles /1117/hatchery-management-the-hatch-window/
43. Nielsen, B.L.; Juul-Madsen, H.R.; Steenfeldt, S.; Kjaer, J.B.; Sørensen, P. Feeding activity in groups of newly hatched broiler chicks: Effects of strain and hatching time. *Poult. Sci.* **2010**, *89*, 1336–1344.
44. Løtvedt, P. Effects of Hatching Time on Behavior and Weight Development of Chickens. *PLoS ONE* **2014**, *9*.
45. Exadaktylos, V.; Silva, M.; Berckmans, D. Real-time analysis of chicken embryo sounds to monitor different incubation stages. *Comput. Electron. Agric.* **2011**, *75*, 321–326.
46. Brambell Committee. *Report of the Technical Committee to Enquire into the Welfare of Animals Kept under Intensive Livestock Husbandry Systems*; Command Paper 2836; Her Majesty's Stationery Office: London, UK, 1965.
47. Newberry, R.C.; Hall, J.W. Use of pen space by broiler chickens: Effects of age and pen size. *Appl. Anim. Behav. Sci.***1990**, *25*, 135–136.
48. Cornetto, T.L.; Estévez, I. Behavior of the domestic fowl in presence of vertical panels. *Poult. Sci.* **2001**, *80*, 1455–1462.
49. Leone, E.H.; Estévez, I. Use of space in the domestic fowl: Separating the effects of enclosure size, group size, and density. *Anim. Behav.* **2008**, *76*, 1673–1682.
50. Naas, I.; Paz, I.; Baracho, M.; Mezenes, A.; Bueno, L.; Almeida, I.; Moura, D.J. Impact of lameness on broiler well-being. *J. Appl. Poult. Res.* **2009**, *18*, 432–439.
51. De Jong, I.; Berg, C.; Butterworth, A.; Estevez, I. *Scientific Report Updating the EFSA Opinion on the Welfare of Broilers and Broiler Breeders*; External Scientific Report; EFSA: Parma, Italy, 2012; p. 116.

52. Naas, I.; Paz, I.; Baracho, M.; Gomes de Mezenes, A.; Oliveira de Lima, K.; Freitas Bueno, L.; Neto, M.; Ciaco de Carvalho, V.; Almeida, I.; Luiz de Souza, A. Assessing locomotion deficiency in broiler chicken. *Sci. Agric.* **2010**, *67*, 129–135.
53. Daigle, C.L.; Banerjee, D.; Montgomery, R.A.; Biswas, S.; Siegford, J. Moving GIS Research Indoors: Spatiotemporal analysis of agricultural animals. *PLoS ONE* **2014**, *9*, e104002.
54. Nasr, M.; Nicol, C.; Murrell, J. Do laying hens with keel bone fractures experience pain? *PLoS ONE* **2012**, *7*, e42420.
55. Banerjee, D.; Daigle, C.L.; Dong, B.; Wurtz, K.; Newberry, R.C.; Siegford, J.M.; Biswas, S. Detection of jumping and landing force in laying hens using wireless wearable sensors. *Poult. Sci.* **2014**, *93*, 2724–2733.
56. Richards, G.J.; Wilkins, L.J.; Knowles, T.G.; Booth, F.; Toscano, M.J.; Nicol, C.J.; Brown, S.N. Pop hole use by hens with different keel fracture status monitored throughout the laying period. *Vet. Rec.* **2012**, *170*, 494.
57. Okada, H.; Itoh, T.; Suzuki, K.; Tsukamoto, K. Wireless sensor system for detection of avian influenza outbreak farms at an early stage. In Proceedings of the 2009 IEEE Sensors Conference, University of Warwick, Warwick, UK, 25–28 October 2009.
58. Okada, H.; Suzuki, K.; Tsukamoto, K.; Itoh, T. Applicability of Wireless activity sensor network to avian influenza monitoring system in poultry farms. *J. Sens. Technol.* **2014**, *4*, 18–23.
59. Aydin, A.; Cangar, O.; Eren Ozcan, S.; Bahr, C.; Berckmans, D. Application of a fully automatic analysis tool to assess the activity of broiler chickens with different gait scores. *Comput. Electron. Agric.* **2010**, *73*, 194–199.
60. Leroy, T.; Vranken, E.; Van Brecht, A.; Struelens, E.; Sonck, B.; Berckmans, D. A computer vision method for on-line behavioural quantification of individually caged poultry. *Trans. ASABE* **2006**, *49*, 795–802.
61. Kestin, S.C.; Kowles, T.G.; Tinch, A.E.; Gregory, N.G. Prevalence of leg weakness in broiler chicken and its relationship with genotype. *Vet. Rec.* **1992**, *131*, 190–194.
62. Vanderhasselt, R.F.; Sprenger, M.; Duchateau, L.; Tuyttens, F.A. Automated assessment of footpad dermatitis in broiler chickens at the slaughter-line: Evaluation and correspondence with human expert scores. *Poult. Sci.* **2013**, *92*, 12–18.
63. Bellomo, N.; Bianca, C.; Delitatla, M. Complexity analysis and mathematical tolos towards modelling of living systems. *Phys. Life Rev.* **2009**, *6*, 144–175.
64. Giles, A.B.; Massom, R.A.; Warner, R.C. A method for sub-pixel scale feature-tracking using Radarset images applied to the Mertz Glacier Tongue, East Antarctica. *Remote Sens. Environ.* **2009**, *113*, 1691–1699.
65. Cheng, J.R.; Koh, E.G.I.; Ahmed, S.; Rajapakse, J. Tracking of cell morphology and motion. *Lect. Notes Bioinform.* **2009**, *5780*, 36–45.
66. Dawkins, M.S.; Cain, R.; Roberts, S.J. Optical flow, flock behaviour and chicken welfare. *Anim. Behav. Sci.* **2012**, *84*, 219–223.
67. Dawkins, M.S.; Cain, R.; Merelie, K.; Roberts, S.J. In search of the behavioural correlates of optical flow patterns in the automated assessment of broiler chicken welfare. *Appl. Anim. Behav. Sci.* **2013**, *145*, 44–50.
68. Sonka, M.; Hlavac, V.; Boyle, R. *Image Processing Analysis and Machine Vision*, 3rd ed.; PWS Publishing: London, UK, 1999.
69. Roberts, S.J.; Cain, R.; Dawkins, M.S. Prediction of welfare outcomes for broiler chickens using Bayesian regression on continuous optical flow data. *J. R. Soc. Interface* **2012**, *9*, 3436–3443.
70. Colles, F.; Cain, R.; Nickson, T.; Smith, A.; Roberts, S.; Maiden, M.; Lunn, D.; Dawkins, M.S. Monitoring chicken flock behaviour provides early warning of infection by human pathogen *Campylobaster. Proc. R. Soc. B* **2016**, *283*.
71. Lee, H.-J.; Roberts, S.J.; Drake, K.A.; Dawkins, M.S. Prediction of feather damage in laying hens using optical flows and Markov models. *J. R. Soc. Interface* **2010**, *8*, 489–499.

72. Bright, A.; Jones, T.A.; Dawkins, M. A non-intrusive method of assessing plumage conditions in commercial flocks of laying hens. *Anim. Welf.* **2006**, *15*, 113–118.
73. Estevez, I.; Tablante, N.L.; Pettit-Riley, R.; Carr, L. Use of cool perches by broiler chickens. *Poult. Sci.* **2002**, *81*, 62–69.
74. Lara, L.; Rostagno, M. Impact of heat stress on poultry production. *Animals* **2013**, *3*, 356–369.
75. Naas, I.A.; Garcia, R.G.; Caldara, F.R. Infrared thermal image for assessing animal health and welfare. *J. Anim. Behav. Biometeorol.* **2014**, *2*, 66–72.
76. Yahav, S.; Straschnow, A.; Luger, D.; Shinder, D.; Tanny, J.; Cohen, S. Ventilation, sensible heat loss, broiler energy, and water balance under harsh environmental conditions. *Poult. Sci.* **2004**, *83*, 253–258.
77. Ferreira, V.; Francisco, N.S.; Belloni, M.; Aguirre, G.M.Z.; Caldara, F.R.; Naas, I.A.; Garcia, R.G.; Almeida Paz, I.C.L.; Polycarpo, G.V. Infrared thermography applied to the evaluation of metabolic heat loss of chicks fed with different energy density. *Braz. J. Poult. Sci.* **2010**, *13*, 113–118.
78. Giloh, M.; Shinder, D.; Yahav, S. Skin surface temperature of broiler chickens is correlated to body core temperature and is indicative of their thermoregulatory status. *Poult. Sci.* **2012**, *91*, 175–188.
79. Shinder, D.; Rusal, M.; Giloh, M.; Yahav, S. Effect of repetitive acute cold exposures during the last phase of broiler embryogenesis on cold resistance through life span. *Poult. Sci.* **2009**, *88*, 636–646.
80. Naas, I.A.; Bites Romanini, C.E.; Neves, D.P.; Rodrigues do Nasciemiento, G.; Vercellino, R.A. Broiler Surface temperature distribution of 42 day old chickens. *Sci. Agric.* **2010**, *67*, 497–502.
81. Wilcox, C.S.; Patterson, J.; Cheng, H.W. Use of thermography to screen for subclinical bumblefoot in poultry. *Poult. Sci.* **2009**, *88*, 1176–1180.
82. Lay, D.C.; Fulton, R.M.; Hester, P.Y.; Karcher, D.M.; Kjaer, J.B.; Mench, J.A.; Mullens, B.A.; Newberry, R.C.; Nicol, C.J.; O'Sullivan, N.P.; et al. Hen welfare in different housing systems. *Poult. Sci.* **2011**, *90*, 278–294.
83. Zhao, Y.; Xin, H.; Dong, B. Use of infrared thermography to assess laying hen feather coverage. *Poult. Sci.* **2013**, *92*, 295–302.
84. Beggs, J.S. *Kinematics*, 2nd ed.; Hemisphere Publishing Corporation: Washington, DC, USA, 1983.
85. Caplen, G.; Hothersall, B.; Murrell, J.C.; Nicol, C.; Waterman-Pearson, A.E.; Weeks, C.A.; Colborne, G.R. Kinematic analysis quantifies gait abnormalities associated with lameness in broiler chickens and identifies evolutionary gait differences. *PLoS ONE* **2012**, *7*.
86. Mench, J.A.; Blatchford, R.A. Determination of space use by laying hens using kinematic analysis. *Poult. Sci.* **2014**, *93*, 794–798.
87. Chung, T.J. *Computational Fluid Dynamic*, 2nd ed.; Cambridge University Press: Cambridge, UK, 2010.
88. Patankar, S.V. *Numerical Heat Transfer and Fluid Flow*; Hemisphere Publishing Corporation: Washington, DC, USA, 1980.
89. Bustamante, E.; García-Diego, F.J.; Calvet, S.; Estellés, F.; Beltrán, P.; Hospitaler, A.; Torres, A.G. Exploring ventilation efficiency in poultry buildings: The validation of computational fluid dynamics (CFD) in a cross-mechanically ventilated broiler farm. *Energies* **2013**, *6*, 2605–2623.
90. Rojano, F.; Bournet, P.E.; Hassouna, M.; Robin, P.; Kacira, M.; Choi, C.Y. Modelling heat and mass transfer of a broiler house using computational fluid dynamic. *Biosyst. Eng.* **2015**, *136*, 25–38.
91. OIE (Office International des Epizooties). *Terrestrial Animal Health Code*; OIE: Paris, France, 2015.
92. Costa, A.; Borgonovo, F.; Leroy, T.; Berckmans, D.; Guarino, M. Dust concentration variation in relation to animal activity in a pig barn. *Biosyst. Eng.* **2009**, *104*, 118–124.

93. Youssef, A.; Exadactylos, V.; Berckmans, D. Towards real time control of chicken activity in a ventilated chamber. *Biosyst. Eng.* **2015**, *135*, 31–43.
94. Ferket, P.R.; Gernat, A.G. Factors that affect feed intake for meat birds: A review. *Int. J. Poult. Sci.* **2006**, *5*, 905–911.
95. Gates, R.; Xin, H. Extracting poultry behaviour from time-series weigh scale records. *Comput. Electron. Agric.* **2008**, *62*, 8–14.
96. Lin, H.; Mertens, K.; Kemps, B.; Govaerts, T.; De Ketelaere, B.; De Baerdemaeker, J.; Decuypere, E.; Buyse, J. New approach of testing the effect of heat stress on eggshell quality: Mechanical and material properties of eggshell and membrane. *Br. Poult. Sci.* **2004**, *45*, 476–482.
97. Ebeid, T.A.; Suzuki, T.; Sugiyama, T. High temperature influences eggshell quality and calbindin-D28k localization of eggshell gland and all intestinal segments of laying hens. *Poult. Sci.* **2012**, *91*, 2282–2287.
98. Barbosa Filho, J.; Silva, M.A.N.; Silva, I.J.O.; Coelho, A.A.D. Egg quality in layers housed in different production systems and submitted to two environmental conditions. *Braz. J. Poult. Sci.* **2006**, *8*, 23–28.
99. Mertens, K.; Vaesen, I.; Loffel, J.; Ostyn, B.; Kemps, B.; Kamers, B.; Bamelis, F.; Zoons, J.; Darius, P.; Decuypere, E. Data-based designs of an intelligent control chart for the daily monitoring of the average egg weight. *Comput. Elecron. Agric.* **2008**, *61*, 222–232.
100. Mertens, K.; Vaesen, I.; Loffel, J.; Kemps, B.; Kamers, C.; Zoons, J.; Darius, P.; Decuypere, E.; De Baerdemaeker, J.; De Ketelaere, B. An intelligent control chart for monitoring of autocorrelated egg production process data based on a synergic control strategy. *Comput. Elecron. Agric.* **2009**, *69*, 100–111.
101. Mertens, K.; Vaesen, I.; Loffel, J.; Kemps, B.; Kamers, B.; Perianu, C.; Zoons, J.; Darius, P.; Decuypere, E.; De Baerdemaeker, J.; et al. The transmission colour value: A novel egg quality measure for recording Shell colour used for monitoring the stress and health status of a Brown layer flock. *Poult. Sci* **2010**, *89*, 609–617.
102. Hepworth, P.J.; Nefedov, A.V.; Muchnick, I.B.; Morgan, K.L. Broiler chickens can benefit from machine learning: Support vector machine analysis of observational epidemiological data. *J. R. Soc. Interface* **2012**, *9*, 1934–1942.
103. Vapnik, V. *The Nature of Statistical Learning Theory*, 2nd ed.; Jordan, M., Lauritzen, S.L., Lawless, J.F., Nair, V., Eds.; Springer: New York, NY, USA, 1995; pp. 154–196.
104. Banhazi, T.M.; Lehr, H.; Black, J.L.; Crabtree, H.; Schofield, P.; Tscharke, M.; Berckmans, D. Precision livestock farming: An international review of scientific and commercial aspects. *Int. J. Agric. Biol. Eng.* **2012**, *5*, 1–9.
105. ProHealth Project. Available online: http://www.fp7-prohealth.eu/ (accessed on 15 June 2016).
106. Daigle, C.L. Incorporating the philosophy of technology into animal welfare assessment. *J. Agric. Environ. Ethics* **2014**, *27*, 633–647.
107. Banhazi, T.M.; Lehr, H.; Black, J.L.; Crabtree, H.; Schofield, P.; Tscharke, M.; Berckmans, D. Precision livestock farming: Scientific concept and commercial reality. In Proceedings of the XVth International Congress on Animal Hygiene: Animal Hygiene and Sustainable Livestock Production, Vienna, Austria, 3–7 July 2011.

CHAPTER 14

Ranging Behaviour of Commercial Free-Range Laying Hens

Leonard Ikenna Chielo *, Tom Pike and Jonathan Cooper

Animal Behaviour, Cognition and Welfare Research Group, University of Lincoln, Lincoln LN6 7TS, UK

ABSTRACT

In this study, the range use and behaviour of laying hens in commercial free-range flocks was explored. Six flocks were each visited on four separate days and data collected from their outdoor area (divided into zones based on distance from shed and available resources). These were: apron (0–10 m from shed normally without cover or other enrichments); enriched belt (10–50 m from shed where resources such as manmade cover, saplings and dust baths were provided); and outer range (beyond 50 m from shed with no cover and mainly grass pasture). Data collection consisted of counting the number of hens in each zone and recording behaviour, feather condition and nearest neighbour distance (NND) of 20 birds per zone on each visit day. In addition, we used techniques derived from ecological surveys to establish four transects perpendicular to the shed, running through the apron, enriched belt and outer range. Number of hens in each 10 m × 10 m quadrat was recorded four times per day as was the temperature and relative humidity of the outer range. On average, 12.5% of hens were found outside. Of these, 5.4% were found in the apron; 4.3% in the enriched zone; and 2.8% were in the outer range. This pattern was supported by data from quadrats, where the density of hens sharply dropped with increasing distance from shed. Consequently, NND was greatest in the outer range, least in the apron and intermediate in the enriched belt. Hens sampled in outer range and enriched belts had better feather condition than those from the apron. Standing, ground pecking, walking and foraging were the most commonly recorded activities with standing and pecking most likely to occur in the apron, and walking and foraging more common in the outer range. Use of the outer range declined with lower temperatures and increasing relative humidity, though use of apron and enriched belt was not affected by variation in these measures. These data support previous findings that outer range areas tend to be under-utilized in commercial free-range flocks and suggest positive relationships between range use, feather condition and increased behavioural opportunities and decline in the use of range in cold and/or damp conditions.

Keywords: ranging behaviour; free-range laying hens; feather condition; enrichment; ecological survey

1. INTRODUCTION

Free-range egg production has become popular due to consumer interest in welfare friendly products and the banning of conventional wire cages across the European Union (EU) in January 2012. As a consequence, free-range production approaches 50% of the fresh egg market in the UK [1]. EU Council directive [2] requires that stocking density must not exceed 2500 hens per hectare, which is equivalent to four square metres per bird. A number of quality assurance schemes include further requirements, for example The Royal Society for Prevention of Cruelty to Animals (RSPCA)'s Assured, British Egg Industry Council's Lion Brand and Noble Food's Happy Egg brand stipulate flock sizes of no more than 16,000 birds, and where flocks exceed 6000 hens, the flock should consist of colonies or sub-flocks of no more than 4000 birds. To meet these requirements, commercial free-range systems typically consist of large flocks of up to 16,000 birds, housed in a large permanently built shed in a large field (about six hectares for a 16,000 bird flock).

Several studies have reported limited outdoor use in free-range laying hens [3,4,5,6,7] and this pattern in the use of range may be associated with a number of welfare problems, e.g., feather pecking, cannibalism and parasitic fouling of pasture in the extensively used areas [8,9,10]. The number of hens found outdoors has been reported to be inversely related to flock size [4,9,11,12,13] with a smaller fraction of the population using the range in larger flocks. Ranging patterns of hens have also been found to be influenced by strain differences [14], season and/or weather conditions [3,4,15], early outdoor rearing experience [3], age of flock [11,14], pop-hole availability [3,16], light intensity in the shed [3] and presence of keel bone fractures [15]. Overall, tree cover and artificial shelters have been utilized to attract hens into the range [3,5,17]. These resources are thought to also provide additional behavioural opportunities to the hens, though behaviour has tended not to be studied in detail, except for direct and indirect assessment of feather pecking.

Hens are thought to accrue a number of welfare benefits when they use the range [7,18]. Savory [19] reported a link between tree cover availability, use of range and injurious feather pecking. Increased use of the range has also been associated with lower prevalence of injurious feather pecking in free-range laying flocks in a number of studies [10,11,17,20,21,22]. Nicol *et al.* [17] reported a beneficial effect of increased use of range with hens showing a nine-fold reduction in feather pecking activities when more than 20% of hens used the range on sunny days, whilst Bright *et al.* [10] found that feather damage correlated negatively with percentage of canopy cover in end-of-lay hens. They suggested that providing 5% cover within 20–25 m distance from the laying hen house is beneficial to the improvement of feather condition and that injurious feather pecking was reduced when a higher proportion of hens use the range.

This study further explored the ranging behaviour of free-range laying hens. Whilst previous studies have tended to focus on flock level measures of condition and use of outdoor areas, this study aimed to provide a more detailed assessment of dispersal and behavioural patterns. The outdoor area was divided into 3 zones based on proximity to shed and available resources. These were: apron (0–10 m from shed normally without cover or other enrichments); enriched belt (10–50 m from shed where resources such as cover, trees, bushes or saplings and dust-baths were provided); and outer range (beyond 50 m from shed with no cover and mainly grass pasture). The feather condition, NND and behaviour of the hens in different outdoor areas were sampled to determine the impact of location on these parameters. In addition, this study used line transects (a technique derived from ecological census) as a further means of measuring hen numbers and dispersal patterns in the three outdoor zones. We predicted that there would be a decline in the number of hens per unit area with distance from shed and that increasing flock size would reduce range use as found in previous work. This study aimed to provide evidence on how these two factors interacted and the differences in behaviour, dispersal patterns and feather condition between zones.

2. MATERIALS AND METHODS

2.1. Animals and Management

This study was approved by the University of Lincoln's College of Science Ethics Committee and was carried out using six flocks of commercial laying hens at four farms in Lincolnshire. All six flocks supplied the Happy Egg Company established by Noble Foods UK Ltd. (Oxford, UK). The study population consisted of medium hybrid lines commonly used in egg production: three flocks of Hyline and three flocks of Lohmann Brown hens, with population sizes ranging from 3900 to 23,548 birds and aged between 27 and 55 weeks of age.

Continuous lighting was provided in the sheds for a minimum of eight hours each day and the hens had daytime access to the outdoor range at twenty weeks of age through the pop holes measuring 45 cm × 2 m, located on the two sides of the sheds. There was at least one pop hole per six hundred hens usually opened at 9:00 a.m. each morning and closed at dusk (4:00 p.m. to 6:00 p.m. during the study). The hens were provided with range enrichments and trees to comply with specification of Happy Egg Company. In addition to complying with the requirements of the EU Council Directive [2], the hens in this system had access to additional resources (activity kits) outside the shed and were the same on each farm, comprising of one set of mini shelter, dust bath and a perch per 4000 hens in the outdoor area. Trees were planted between the distance of 10 m and 50 m from the sheds, with the majority of manmade structures located 15–35 m away from the shed. These resources were thought to encourage hens to utilize the outdoor area by providing shelter and increasing behavioural opportunities, though at the time of study the trees were saplings, so canopy cover was limited.

The study was completed between November 2011 and February 2012. Each flock was visited on four different occasions (*i.e.*, each flock was visited four times before another flock was visited) for data collection giving a total of twenty-four sample visits for the study. At least a 48-hour gap was allowed between farm visits to comply with Noble Food's bio-security requirements.

2.2. Sampling Areas

On first arrival, each flock was surveyed for key environmental features including location of shed within field, field boundaries, location and number of pop-holes and the distribution of outdoor resources. Bamboo poles (1 m in height) were then used to divide the outdoor area into zones and to produce 4 lines of transects running perpendicular to the shed. Pilot studies for previous studies [23] had indicated that whilst these poles provide a short term point of interest to hens when initially placed in the ground, the hens rapidly habituated to their presence and the poles had no influence on hens' location after half an hour. Poles were placed in pairs, every 10 m from the shed for 110 m to indicate the quadrats. This had the effect of firstly providing sighting lines parallel to the shed to allow the distance from the shed to be estimated. They also produced the line transect from the edge of shed to 110 m that was made up of eleven 10 m × 10 m quadrats. Where sheds were located centrally in fields and had over 100 m on either side, then two transects were arranged on either side of the shed. Where the sheds were on the edge of the field, with all or most useable area on one side, all four transects were arranged on that side.

Outdoor areas of commercial free-range systems often have features that can be used to differentiate discrete zones located at specific distances from the shed and with different enrichment resources. In this study we defined three discrete zones whose features were common across all the six flocks. These were the apron, enriched and outer range zones. The apron zone was defined as the area between 0 m to 10 m from the shed. There were no additional enrichment resources in this area and no tree cover. Ground vegetation was sparse, with soil, slats, concrete or pebbles covering most of the area. This constituted on average 4.1% of the available outdoor area. The enriched belt covered the area between 10–50 m from the shed and manmade enrichments were located in this area with natural cover in the form of plantations of tree saplings. Ground vegetation varied from low grass pasture, with patches of taller perennial plants such as nettle and patches of bare earth, particularly where the hens had formed scrapes or dust baths. This constituted on the average 21.2% of the available outdoor area. The outer range was defined as the outdoor area 50 m and beyond from the shed and this was the largest part of the outdoor area spanning from the end of the enriched zone to the field boundaries and covered on average 74.7% of available outdoor area. This area mainly consisted of grass pasture, which tended to have low sward during the study period. The boundary for all flocks was 2 m tall electrified wire fence to prevent hens leaving field and deter ground predators from entering the enclosure.

2.3. Data Collection

A general head count was conducted around 12:00 p.m. to determine the total number of hens outdoors in each flock. This involved a brisk walk of the flocks to count the number of hens in apron, enriched and outer range zones, which were combined to provide the estimate for the entire flock. These were used to calculate the percentage of the flock in each area based on farm records of flock size at time of survey. Furthermore, the population density could be calculated in terms of hens per square metre and range area per hen in apron, enriched belt and outer range.

Hen numbers and distribution across zones was also recorded using ecological census techniques utilized by Cooper and Hodges [23]. This was conducted by counting the hens in each quadrat of each transect four times during each visit. Head counts were carried out at 10:00 a.m., 11:00 a.m., 1:00 p.m. and 2:00 p.m. and during the counts, the potential observer influence associated with head counts was minimized by maintaining a distance of 15 meters from the hens. In addition to heads counts, twenty hens were sampled in each zone for NND, feather condition and behaviour. Where there were less than twenty hens in a quadrat, all the hens were sampled. The hen closest to the observer and every second hen was sampled and their immediate activity was categorized using an ethogram of 17 mutually exclusive behaviours ([24], Table 1).

Table 1. Ethogram of behaviours under observation (Adapted from Buijs, [24]).

Activity	Description
Standing	Not moving on two feet, body not touching the floor
Sitting	Body and both hocks touching the floor underneath or directly on either side of the bird
Lying	Lying on its side, with both feet on the same side of the bird
Walking	Slow locomotion, the first foot is put down on the floor before the second one is lifted (without pecking or scratching)
Pecking	Pecking on the ground or objects
Foraging	Walking with pecking and scratching
Running	Rapid locomotion, the second foot is lifted before the first is set down
Preening	Moving the beak over the feathers
Feeding	Pecking at the feed in the feeder, or between such pecks
Drinking	Pecking at the drinker, followed by tilting of the head
Shaking	Rapid whole body movement mostly associated with ruffling of the feathers or shaking dust from the plumage
Aggression	Fights including pecking at another chicken
Dust bathing	Foot scratching and bill-raking the litter or lose soil, followed by vertical wing shaking, head rubbing, bill-raking and/or scratching with one leg whilst lying
Stretching	Elongation of the leg not associated with walking
Comfort behaviour	Includes wing flapping, body shaking, feather ruffling and tail wagging but not preening
Head flick	Rapid head movements in horizontal plane
Scratching	Stepping backwards whilst raking the feet across the floor

A visual assessment of plumage condition of four different body parts (neck, chest, back and sides) was carried out using a six point scoring scale [25]. In this method, values from 0 (best feathers) to 5 (worst feathers) were assigned to each body parts (see Table 2 for scale values and descriptions) with respect to the degree of damage or no damage to the feathers. The feather condition of the hens was effectively scored from a distance of 5 m to minimize disturbance to the focal animal and flock in general. Bright *et al.* [26] reported a strong positive correlation between feather condition scores recorded from distance and scores from close inspection following capture, which suggests that distance feather scoring techniques are reliable and practical on a commercial scale. Feather scoring and behavioural observations were carried out after the head counts.

Table 2. Description of feather scoring method used to evaluate feather condition of the hens (adapted from Bilcik and Keeling, [25]).

Score	Body Feathers	Flight Feathers
0	Intact feathers	Intact feathers
1	Some feathers scruffy up to 3 missing feathers	Few feathers separated but not broken or missing
2	More damaged feathers, greater than 3 feathers missing	A lot of feathers separated and/or a few broken or missing
3	Bald patch <5 cm diameter or <50% of area.	All feather separated, a lot of broken or missing feathers
4	Bald patch >5 cm diameter or greater than 50% of area	Most of feathers missing or broken
5	Completely denuded area	Almost all feathers missing

NND was measured using 1 m poles to mark the locations of the focal bird and its nearest neighbour and the distance measured by running a portable 25 m tape between the poles. The approximate location of chickens and nearest neighbour was noted and measured using measuring tape with the help of a field assistant. This approach was effective at estimating distances to nearest 0.2 m over distances of up to 2 m, but accuracy declined above this distance so distances above 2 m where estimated to the nearest 1 m.

2.4. Weather Measurements

The temperature and relative humidity of sites were measured using a simple indoor/outdoor thermo-hygrometer. Upon arrival, the thermo-hygrometer was positioned in the open outdoor area mid-way between shed and end of range at 2 m above the ground level. The temperature and relative humidity was recorded 4 times per day on each visit at 10:00 a.m., 11:00 a.m., 1:00 p.m. and 2:00 p.m.

2.5. Statistical Analysis

All the data collected was analysed using IBM SPSS 20.0 statistical software (IBM Corporation, Armonk, NY, USA). Temperature and relative humidity data met requirements of parametric statistics, so the effects of time of day were assessed using general linear model analysis of variance (GLM ANOVA) with time of day as fixed factor and flock as a random factor. The average measure of temperature and relative humidity for each visit day was then used to investigate the effect of these measures on total number of hens out of shed. We had planned to investigate the effects of flock size, age and strain as factors in the analysis, however, as it was not possible to balance these between flocks (See Table 3), we instead treated flock identity as a random factor across all analysis.

Table 3. Flocks and their characteristics including strain, number of hens at time of study and age in weeks. Means for % of flock out of shed, hens per quadrat and feather score are reported for each flock.

Farm	Strain of Hens	Size at Time of Study	Age (Weeks)	% of Flock Out of Shed	Hens per Quad	Feather Score
1	Hyline	3900	55	35.1 ± 3.8^{a}	9.8 ± 0.5^{a}	0.14 ± 0.01^{a}
2	Hyline	7300	48	20.1 ± 2.4^{b}	8.1 ± 0.5^{b}	0.21 ± 0.01^{b}
2	Hyline	15,573	27	6.3 ± 0.8^{c}	5.3 ± 0.4^{d}	0.24 ± 0.01^{c}
3	Lohmann Brown	15,470	49	$4.6 \pm 1.1^{c,d}$	6.6 ± 0.4^{c}	0.24 ± 0.01^{c}
4	Lohmann Brown	15,797	51	8.8 ± 0.7^{c}	6.7 ± 0.4^{c}	0.25 ± 0.01^{c}
4 *	Lohmann Brown	23,548	52	3.0 ± 0.5^{d}	5.1 ± 0.4^{d}	0.23 ± 0.01^{c}

[a,b,c,d] Means within a column with different superscripts differ significantly ($p < 0.05$); * Single shed housing two flocks of approximately 12,000 birds each.

GLM analysis was performed on the quadrat distribution data to determine if the fixed factors (zone and time of the day) had influence on the distribution of the hens. Flock identity was treated as a random factor whereas age, strain, flock size, temperature, and relative humidity were fitted as covariates. As there were a number of potential explanatory variables used in the analysis, a step-wise model simplification process was carried out. The residual of the model used in the analysis was found to be normally distributed using a histogram and therefore no data transformation was required before analysis.

NND data was also analysed using GLM approach. A separate model was developed using a similar step-wise simplification procedure as the hen distribution to determine if the location of hens influenced their distances away from the nearest hens. In this model, zone was fitted as a fixed factor and flock was treated as a random factor.

Feather scores from four key body parts (head, neck, chest and back) were explored for descriptive statistics and a GLM analysis was carried out to determine if there was a relationship between feather condition of the hens and the outdoor zone they were found. A fitted feather score model was achieved using a similar approach as in the above analysis.

Behavioural data were explored for descriptive statistics to determine the relative abundance of each of the behaviours recorded. The results of the descriptive analysis revealed that some of the behaviours were not recorded at all or rare and for this reason, only the major behaviours were analysed further using GLM. A step-wise model simplification process was also carried out as in the models above to determine the influence of the zones on the behaviours.

The means and standard error of means for the estimates of distribution, feather scores, NND, behavioural occurrence, temperature and relative humidity are presented in the result section of this article. Further *post hoc* tests were carried out on all the significant variables and interactions in each model using Bonferroni correction factor to determine the pairs that were significantly different from each other.

3. RESULTS

3.1. Weather

Average temperature sampled on visits was 8.56 ± 0.42 °C (range from 0.00 °C to 17.5 °C), and the relative humidity was on average 72.6 ± 1.5% (range from 42.0% to 99.0%). The temperature and relatively humidity varied across the study period, broadly in line with seasonal variations and consequently there was significant variation in both temperature (F_5 = 4.95, $p < 0.001$) and relative humidity (F_5 = 3.38, p = 0.008) between flocks. There was no effect of time of day on either temperature or relative humidity.

3.2. Number of Hens Outdoors and Their Distribution

The mean number of hens out of shed was 1142 ± 91 which represented an average 12.5% of the flock. An average head count of 530 ± 37 was found for the apron area near the shed (5.4% of flock), whereas 401 ± 51 hens were recorded in the enriched belt (4.3%), and 211 ± 44 hens (or 2.8% of flocks) were recorded in the outer range. As the majority of the outdoor area was the outer range, with apron and enriched belts only covering about a quarter of available outdoor area, these resulted in considerable variation in stocking density between the three areas. There were on average 0.31 hens/m^2 on the apron (equivalent to 3.23 m^2 per hen), compared with 0.048 hens/m^2 in the enriched belt (or 20.8 m^2 per hen) and 0.0086 hens/m^2 of outer range giving on average 116 m^2 per hen. There was a significant difference between flocks in percentage of hens on the range ($F_5 = 20.1$, $p < 0.001$), which varied from 35.1% in the smallest flock to 3.0% in the largest flock (Table 3).

3.3. Number of Hens in the Quadrats

The distribution of hens across quadrats was influenced by the time of day ($F_3 = 63.97$, $p < 0.05$) with the greatest number of hens outside at 10:00 a.m. (Figure 1). Outdoor use decreased significantly through the day but showed similar number of hens between 1:00 p.m. and 2:00 p.m. There were more hens in the apron area least in the outer range and intermediate in the enriched zone at all the time periods (Table 4: $F_6 = 30.59$, $p < 0.05$).

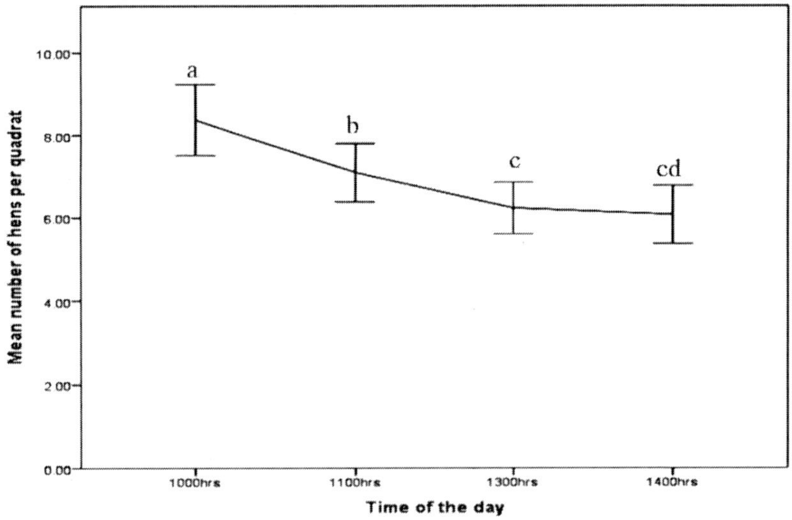

Figure 1. The means of number of hens (±SE) recorded at different times of the day. [a,b,c,d] Means of times with different superscripts are significantly different ($p < 0.05$)

Table 4. Mean numbers of hens in the apron, enriched belt and outer range areas for all the time periods.

	Zones				
Time of the Day	Apron	Enriched	Range	F-Value	SEM
10:00 a.m.	41.49 [a]	9.97 [b]	1.79 [c]	834.76	0.36
11:00 a.m.	29.98 [a]	9.31 [b]	1.79 [c]	460.67	0.35
1:00 p.m.	24.35 [a]	8.27 [b]	1.83 [c]	340.77	0.33
2:00 p.m.	27.67 [a]	8.31 [b]	0.97 [c]	402.67	0.35

[a,b,c] Means within rows with different superscripts are significantly different ($p < 0.05$).

The counts of hens in the quadrats that made up the line transects were consistent with the findings of the overall head counts, with numbers falling with distance from shed ($F_2 = 352$, $p < 0.001$; Figure 2). Quadrat 1, which included the apron, had an average of 30.87 ± 0.44 hens, quadrats 2 to 5 in the enriched belt had on average 8.97 ± 0.22 hens, whilst quadrats 6–11 in the outer range had a mean of 1.60 ± 0.18 hens per 100 m^2 quadrant. These counts equated to about 3.24 m^2 per hen in the apron quadrats, 11.1 m^2 per hen in the quadrats from the enriched belt and 62.6 m^2 per hen in the outer range quadrats. There was also an effect of flock ($F_5 = 29.1$, $p < 0.001$), with smaller flocks having higher numbers of hens per quadrat than larger flocks (Table 3).

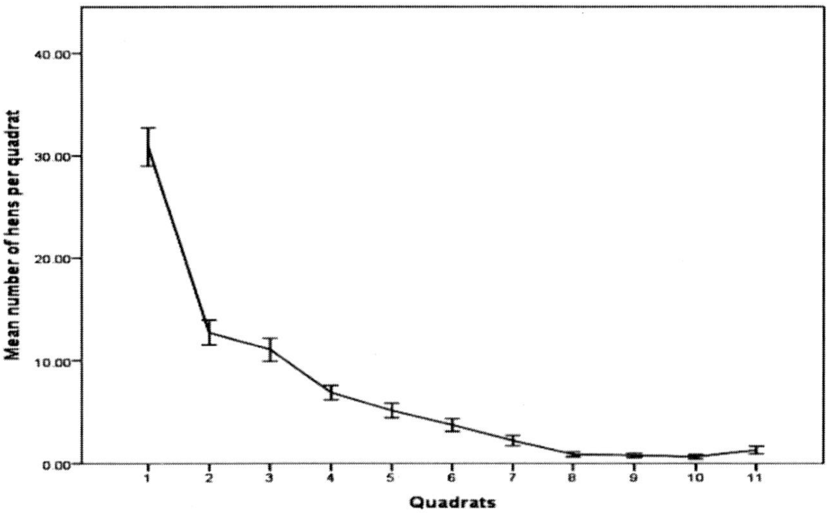

Figure 2. The mean (±SE) number of hens per 100 m^2 quadrat from first quadrat on apron, through enriched belt (2–5) and outer range (6–11) showing sharp decline in numbers with distance from shed.

There was an effect of temperature on the number of hens per quadrat (F_1 = 3.02, $p < 0.001$) with more hens as temperature rose. There was however, no overall effect of relative humidity (F_1 = 3.02, $p > 0.05$) on outdoor use.

3.4. Nearest Neighbour Distance

The NND of hens outdoors and was found to differ between the zones (F_2 = 435, $p < 0.001$). NND was found to be greatest in the outer range (5.67 ± 0.15), least in the apron (1.62 ± 0.05) and intermediate in the enriched belt (2.40 ± 0.08) so distance between hens increased with increasing distance from the shed (Table 5). There was no effect of flock on the NND (F_5 = 2.69, $p > 0.05$).

Table 5. NND and feather condition for hens in apron, enriched belt and outer range.

	Zones				
	Apron	Enriched	Range	F-Value	SEM
NND	1.62^{a}	2.40^{b}	5.67^{c}	435	0.14
Feather condition scores					
Neck	1.064^{a}	0.610^{b}	0.154^{c}	538.38	0.028
Chest	0.296^{a}	0.211^{b}	0.070^{c}	78.27	0.018
Side	0.006^{a}	$0.002^{a,b}$	0.000^{b}	3.52	0.012
Back	0.213^{a}	0.014^{b}	0.007^{b}	192.83	0.002
Mean feather scores	0.395	0.214	0.058		

^{a,b,c} Means within rows with different superscripts are significantly different ($p < 0.05$).

3.5. Feather Condition of the Hens

The results showed that feather loss was worst in the neck area (0.609 ± 0.013), followed by chest (0.192 ± 0.008) and back (0.078 ± 0.005), with the side having the best condition scores (0.003 ± 0.001) (F_3 = 1461, $p < 0.001$). *Post hoc* tests revealed that the hens in the range zone had the best overall feather condition whereas apron had poorer feather condition (see Table 5). Feather condition also differed between flocks (F_5 = 12.9, $p < 0.001$; Table 3) with the smallest flock having the best feather condition, followed by second smallest flock and no difference between the four larger flocks.

3.6. Behaviour of the Hens

Standing (24.8% of samples), pecking (19.8%), walking (26.6%) and foraging (20.6%) were the most recorded behaviours, representing over 90% of the overall activity of hens that were sampled. Of the remaining behaviours only preening (3.4%), sitting (2.6%) and ground scratching were found in more than 1% of samples. Running (0.4%), dust-bathing (0.2%), perching (0.1%) and shaking (0.1% of samples) were rarely sampled. The remaining activities in the ethogram including aggression and comfort behaviours such as wing flapping were not recorded.

To provide a measure of the effect of location on the relative abundance of each activity we compared the incidence of the most common activities in each zone. The results of the relative occurrence of these behaviours in different zones are presented in Table 6. The results showed that standing behaviour differed between the three zones ($F_2 = 52.9, p < 0.001$) and was recorded most in the apron zone. Pecking behaviour was less common in the outer range than in the other zones ($F_2 = 24.3, p < 0.001$). Walking ($F_2 = 14.4, p < 0.001$) and foraging ($F_2 = 53.6, p < 0.001$) activities occurred most in the outer range zone with mean difference between the zones being significant. There was no effect of flock on behavioural measures.

Table 6. The mean occurrence of the four most recorded behaviours (from each daily sample of 20).

Behavioural States	Apron	Enriched	Range	F-Value	SEM
Standing	16.6 ^{a}	8.9 ^{b}	4.2 ^{c}	52.87	0.61
Pecking	9.5 ^{a}	10.9 ^{a}	3.4 ^{b}	24.27	0.57
Walking	9.2 ^{a}	8.3 ^{a}	13.2 ^{b}	14.40	0.49
Foraging	1.7 ^{a}	7.1 ^{b}	15.8 ^{c}	53.63	0.68

^{a,b,c} Means within rows with different superscripts are significantly different.

4. DISCUSSION

The findings of the present study were broadly in line with other research in commercial free-range layer systems regarding range use, in particular that population density declines with distance from shed (e.g., [5,16,17]) and that a lower proportion of the flock are found outdoors as flock size increased (e.g., [4,12,13]). This study also found evidence for a positive relationship between good range use and feather condition (e.g., [17,21,22]) and reduced general range use in colder (e.g., [3,4,15]). The novel sampling methods and additional measures used in this study provided more detailed data on the use of range as well as the relationship between range use, behaviour and feather condition.

The total head count showed that 12.5% of the hens in the sampled flocks were found outdoors at noon. In this study, we did not directly assess the effect of flock size in our analysis due to confounding effects of other flock variables such as age and strain. Nevertheless the variation in both total percentage of the hens counted out of shed and the number of hens counted using quadrat approach suggests that as flock size increased, the use of ranging areas decreased and this is consistent with the findings of a number of other studies [4,9,11,12,13]. It may have been more demanding for hens in larger flocks to come out of the shed because of the number of hens they will encounter before accessing the outdoor area. In addition, hens in larger flocks require bigger sheds and this will mean that they will walk longer distance to access the range.

Most of the hens found outdoors were recorded in the apron zone and the preference for a closer proximity could be an anti-predator strategy as suggested by Nagle and Glatz [5]. They reported more hens in the range (including the farther areas) when shelterbelts were provided, and suggested these provided shelter from predation [6,27] though shelter from wind, rain or direct sunshine would also be a benefit. Hegelund *et al.* [4] also observed that artificial cover attracted more hens into the range and away from the immediate strips of the shed. The work of Zeltner and Hirt [28] supported the findings of this study and showed that hens were less likely to use the range when no overhead cover was provided. The trees planted in the outdoor range of the flocks used for this study were mainly saplings, which did not provide high levels of cover for the hens. Previous work [23] suggests that although saplings can increase range use, full or nearly full canopy cover has greatest impact on the range use.

It is worth noting that in this study, there were differences between the hen densities derived from head counts and those from the quadrats for the outer range and enriched belt, but not the apron. The head counts of hens out of shed gave an estimate of 0.31 hens/m^2 of which is equivalent to 31.0 hens per 100 m^2 of apron, and similar to the average of 30.9 calculated from quadrat counts. In contrast, twice as many hens were recorded in the enriched belt (9.0 compared with 4.8 based on data from entire enriched belt) and outer range (1.6 hens compared with 0.9 across entire outer range) using the quadrat method compared to general head counts. This suggests that the area covered by the line transects in the enriched belt and outer range were more attractive than other areas of enriched belt and outer range zones. This may simply be a matter of ease of access, in that transects were in line with the sides of sheds where most pop holes would be located or because of the movement of the observer along the transect lines during observations which has the potential to attract more hens to this area compared to the less frequently walked areas of the range.

Alternatively, the higher hen density recorded in the quadrat counts may be related to the specific enrichments provided in these locations e.g., the young tree plantations found across the entire enriched belt, the sampled areas were also the locations of the activity kits provided as part of the Happy Egg Companies flock requirements. These normally consisted of covered dust baths, perch and mini-shelters with one set of kits per 4000 hens, and may have provided additional shelter or behavioural opportunities to attract birds compared to young trees alone. A more detailed study of the impact of additional manmade enrichments and their locations on hen movement would be able to differentiate between these two explanations.

The hens used the range most in the morning but the number of hens outdoors dropped as the day progressed. The interaction between zone and time of the day showed that the hens used the apron most at all times. This study supports other findings [4,5,15] where free-range hens were reported to use the outdoor run most in the morning and evening. In the present study, the hens were observed between 10:00 a.m. to 2:00 p.m. and because of this, the use of range beyond the time was not assessed. The differences in atmospheric conditions across the day may explain why the hens prefer to range more in the morning (e.g., Nicol *et al.* [17]), however in this study, although temperature did affect numbers hens sampled, these factors did not consistently vary with time across the sampling days. Alternatively, the higher numbers at 10:00 a.m. may reflect high numbers of hens using range resources soon after pop-hole opening at 9:00 a.m.

There was a relationship with general distribution of hens and of the two weather variables measured; higher temperature had largely positive effect on the numbers counted in quadrats, whereas negative impact of relative humidity that would be predicted from other studies was found in this study. Hegelund *et al.* [4] reported a parabolic relationship between temperature and number of hens in the range and they showed that range use increased up to a maximum temperature of 17 °C then a corresponding decrease in range use at temperatures greater than 17 °C. The maximum recorded temperature for this study was 17.5 °C and appeared to encourage more hens to range. Gilani *et al.* [3] reported a similar range use pattern and evidence from their work found that more hens ranged away from the sheds when the relative humidity level was low, on cooler days and with low rainfall. Hegelund *et al.* [4] also reported that wet and rainy conditions had negative influence on the use of range. These findings point to the potential importance of shelter to protect from adverse weather.

The results suggested that location of the hens affected their behaviour. The most commonly recorded activities were standing, pecking, walking and foraging. Comparison of the number of hens engaged in each of these behaviour indicated that standing and pecking were most often observed in the hens found in the apron area, and that walking and foraging were commonly observed in the outer range. The latter may be a reflection of the availability of the large, open space of the outer range and the distance hens would travel to search for resources in the outer range. In contrast the apron provided relatively little additional environmental resources but was close to the indoor environment so required little movement. It is also noteworthy that no aggressive behaviour was recorded in any of the hens sampled, suggesting this behaviour was rare outside of the shed environment. This study did not sample behaviour inside the shed, which may have provided greater potential to detect incidences of feather pecking and aggression due to higher stocking density and different resource availability.

The results indicated that the NND increased as you go away from the shed. The hens showed the least NND in the apron, greatest in the outer range zone and intermediate in the enriched belt. Based on the results of the total and quadrat head counts, apron zone had the greatest number of hens and covered less than a quarter of the available outdoor area, thereby making it the most densely populated outdoor zone. As the number of hens dropped from the shed, there was more usable space available to individual hens (decreased stocking density), which in turn resulted in the increasing NND recorded in this study. The hens in the outer range may have moved to this area to avoid aggression or competition for the resources in the zones near the shed and may have preferred to maintain greater distances from other hens to avoid such aversive experiences.

The results of feather score analysis showed that the sides had the most intact feathers and neck had the most damaged feathers. There was an effect of flock on feather condition with the two smallest flocks having significantly better feather scores than the four larger flocks. Before drawing too strong a conclusion from this, it is worth noting that only six flocks were included in this study, that they represented a relatively restricted age range (five flocks between 48 and 55 weeks of age and one flock of 27 weeks), and that the feather condition of the hens was generally very good. Normally feather condition would be expected to decline with age, and a larger number of flocks covering a greater age range would be expected to find a significant age effect. Nevertheless, the better feather condition associated with smaller flocks in this study that made more use of the outside environment is noteworthy.

Hens in the outer range zone had the best feathers (in all body parts) whereas apron recorded the worst feathers. The feather condition of hens in the enriched zone was better than those found in the apron but worse than their counterparts in the outer range zone. Although we found some variation in plumage, overall feather condition was excellent and many hens in outer range, enriched and apron zones showed no evidence of feather damage. Bilcik and Keeling [25] suggested that ease of feather removal and the ease of access to different parts of the body may have resulted in specific feathers being attractive targets for pecking in laying hens; they have been reported to prefer shorter semi-plumes than longer ones [29]. Wing feathers have stronger shaft and are longer than the neck feathers, which may have resulted in less damages recorded in the wing feathers. High concentration of birds and greater pecking activities in the apron may have resulted in the poor feather quality of hens in this area. Range zone offered more foraging opportunities and hens in this zone foraged more and had better chances of avoiding competitive locations. Huber-Eicher and Wechsler [30] reported an inverse relationship between feather pecking and foraging behaviour in laying hens and in the present study, more hens in the range zone performed foraging behaviour at greater NND, which may have resulted in less feather damage recorded for the hens found in this area.

5. CONCLUSIONS

These data supports previous studies that reported few hens in the outer range of free-range flocks. Hens that range further from shed were, however, more likely to be engaged in walking and foraging, compared to more sedate birds in the apron and enriched area, and generally had better feather condition. These findings suggest hens that make use of outer range have better welfare than those that remain close to the shed and the direction of any causal relationships warrants further investigation. For example, better feathered hens may be more likely to use range because they find it less aversive than other hens, or their feather condition may be improved because of spending more time out on range. Similarly, hens may travel to outer range because they are actively seeking opportunities to engage in activities such as foraging, or they may tend to naturally disperse further from sheds, and exhibit more foraging in response to the cues they encounter once they reach the outer range.

ACKNOWLEDGMENTS

The authors wish to thank Noble Foods UK Ltd. for part sponsorship of this project and allowing us access to their farms for data collection.

REFERENCES

1. Department for Environment, Food and Rural Affairs. United Kingdom Egg Statistics, Quarter 1, 2015. Available online: https://www.gov.uk/ government/uploads /system/uploads/attachment_data/file/425838/eggs-statsnotice-07may15.pdf (accessed on 18 June 2015).
2. Council Directive 1999/74/EC. Laying down minimum standards for the protection of laying hens. *Official J. Eur. Commun.* 1999, *L 203*, 53–57.
3. Gilani, A.M.; Knowles, T.G.; Nicol, C.J. Factors affecting ranging behaviour in young and adult laying hens. *Brit. Poult. Sci.* 2014, *55*, 127–135.
4. Hegelund, L.; Sørensen, J.; Kjær, J.; Kristensen, I. Use of the range area in organic egg production systems: Effect of climatic factors, flock size, age and artificial cover. *Brit. Poult. Sci.* 2005, *46*, 1–8.
5. Nagle, T.A.D.; Glatz, P.C. Free-range hens use the range more when the outdoor environment is enriched. *Asian-Australas. J. Anim. Sci.* 2012, *4*, 584–591.
6. Nicol, C.J.; Caplen, G.; Edgar, J.; Browne, W.J. Associations between welfare indicators and environmental choice in laying hens. *Anim. Behav.* 2009, *78*, 413–424.
7. Pettersson, I.C.; Freire, R.; Nicol, C.J. Factors affecting ranging behavior in commercial free-range hens. *World Poult. Sci. J.* 2016, *72*, 137–150.
8. Appleby, M.C.; Hughes, B.O.; Elson, H.A. Welfare of laying hens in cages and alternative systems-environmental, physical and behavioral aspects. *World Poult. Sci. J.* 1991, *47*, 109–128.
9. Bubier, N.E.; Bradshaw, R.H. Movement of flocks of laying hens in and out of the hen house in four free-range systems. *Brit. Poult. Sci.* 1998, *39*, S5–S6.
10. Bright, A.; Brass, D.; Clachan, J.; Drake, K.A.; Joret, A.D. Canopy cover is correlated with reduced injurious feather pecking in commercial flocks of free-range laying hens. *Anim. Welf.* 2011, *20*, 329–338.
11. Bestman, M.W.P.; Wagenaar, J.P. Farm level factors associated with feather pecking in organic laying hens. *Livest. Prod. Sci.* 2003, *76*, 559–563.
12. Kijlstra, A.; Traag, W.A.; Hoogenboom, L.A.P. Effect of flock size on dioxin levels in eggs from chickens kept outside. *Poult. Sci.* 2007, *86*, 2042–2048.
13. Whay, H.R.; Main, D.C.J.; Green, L.E.; Heaven, G.; Howell, H.; Morgan, M.; Pearson, A.; Webster, A.J.F. Assessment of the behaviour and welfare of laying hens on free-range units. *Vet. Rec.* 2007, *161*, 119–128.
14. Icken, W.; Thurner, S.; Cavero, D.; Schmutz, M.; Wendl, G.; Preisinger, R. Analysis of the free-range behaviour of laying hens and the genetic and phenotypic relationships with laying performance. *Brit. Poult. Sci.* 2008, *49*, 533–541.
15. Richards, G.J.; Wilkins, L.J.; Knowles, T.G.; Booth, F.; Toscano, M.J.; Nicol, C.J.; Brown, S.N. Continuous monitoring of pop hole usage by commercially housed free-range hens throughout the production cycle. *Vet. Rec.* 2011, *169*, 338.
16. Keeling, L.J.; Hughes, B.O.; Duncan, P. Performance of free-range laying hens in a polythene house and their behaviour on range. *Farm Build. Prog.* 1988, *94*, 21–28.
17. Nicol, C.J.; Pötzsch, C.; Lewis, K.; Green, L.E. Matched concurrent case-control study of risk factors for feather pecking in hens on free-range commercial farms in the UK. *Brit. Poul. Sci.* 2003, *44*, 515–523.
18. Knierim, U. Animal welfare aspects of outdoor runs for laying hens: A review. *Wag. J. L. Sci.* 2006, *54*, 133–145.
19. Savory, C.J. Feather pecking and cannibalism. *World Poult. Sci. J.* 1995, *51*, 215–219.
20. Green, L.E.; Lewis, K.; Kimpton, A.; Nicol, C.J. Cross-sectional study of the prevalence of feather pecking in laying hens in alternative systems and its associations with management and disease. *Vet. Rec.* 2000, *147*, 233–238.
21. Horton, L. *A Study into the Effect of Tree Cover on the Range on the Welfare of Free-Range Layer Hens by Observing Animal Behaviour*; Royal Veterinary College, University of London: London, UK, 2006.

22. Lambton, S.L.; Knowles, T.G.; Yorke, C.; Nicol, C.J. The risk factors affecting the development of gentle and severe feather pecking in loose housed laying hens. *Appl. Anim. Behav. Sci.* 2010, *123*, 32–42.
23. Cooper, J.; Hodges, H. Effect of tree cover on ranging behaviour of free-range hens. In Proceedings of the 44th Congress of the International Society for Applied Ethology, Uppsala, Sweden, 4–7 August 2010.
24. Buijs, S. Using Spatial Distribution and Behaviour to Determine Optimal Space Allowances for Poultry and Rabbits. Ph.D. Thesis, Faculty of Veterinary Medicine and Animal Science, Swedish University of Agricultural Sciences, Uppsala, Sweden, 2011.
25. Bilcik, B.; Keeling, I.J. Changes in feather condition in relation to feather pecking and aggressive behaviour in laying hens. *Brit. Poult. Sci.* 1999, *40*, 444–451.
26. Bright, A.; Jones, T.A.; Dawkins, M.S. A non-intrusive method of assessing plumage condition in commercial flocks of laying hens. *Anim. Welf.* 2006, *15*, 113–118.
27. Browne, W.J.; Caplen, G.; Edgar, J.; Wilson, L.R.; Nicol, C.J. Consistency, transitivity and inter-relationships between measures of choice in environmental preference tests with chickens. *Behav. Proc.* 2010, *83*, 72–78.
28. Zeltner, E.; Hirt, H. Effect of artificial structuring on the use of laying hen runs in a free-range system. *Brit. Poult. Sci.*2003, *44*, 533–537.
29. McKeegan, D.E.F.; Savory, C.J. Feather eating in layer pullets and its possible role in the aetiology of feather pecking damage. *Appl. Anim. Behav. Sci.* 1999, *65*, 73–85
30. Huber-Eicher, B.; Wechsler, B. Feather pecking in domestic chicks: Its relation to dustbathing and foraging. *Anim. Behav.* 1997, *54*, 757–768.

CHAPTER 15

Effects of Dark Brooders on Behavior and Fearfulness in Layers

Anja B. Riber [1],* and Diego A. Guzman [2]

[1] *Department of Animal Science, Faculty of Science and Technology, Aarhus University, Blichers Allé 20,*

[2] *Instituto de Investigaciones Biológicas y Tecnológicas (CONICET-UNC) and Instituto de Ciencia y Tecnología de los Alimentos, Facultad de Ciencias Exactas, Físicas y Naturales, Universidad Nacional de Córdoba, Av. Vélez Sarsfield 1611 (5000), Córdoba X5016, Argentina*

ABSTRACT

Chicks require heat to maintain body temperature during the first weeks after hatch. This may be provided by dark brooders; *i.e.*, horizontal heating elements equipped with curtains. The objective was to test effects of rearing layer chicks with dark brooders on time budget and fearfulness. Behavioral observations were performed during the first six weeks of age. Three different fear tests were conducted when the birds were age 3–6, 14–15 and 26–28 weeks. During the first four days, brooder chicks rested more than control chicks whereas they spent less time drinking, feather pecking and on locomotion ($p \leq 0.009$). On days 16, 23, 30 and 42, brooder chicks spent less time on feather pecking, locomotion and fleeing ($p \leq 0.01$) whereas foraging and dust bathing occurred more often on day 42 ($p \leq 0.032$). Brooder birds had shorter durations of tonic immobility at all ages ($p = 0.0032$), moved closer to the novel object at age 15 weeks ($p < 0.0001$), and had shorter latencies to initiate locomotion in the open-field test at age 28 weeks ($p < 0.0001$). Results support the suggestion that dark brooders can be a successful method of reducing or preventing fear and feather pecking in layers.

Keywords: activity; behavior; brooders; fear; laying hen; novel object; open-field; poultry; tonic immobility; welfare

1. INTRODUCTION

Chicks require extra heat to maintain body temperature during the first weeks after hatch [1]. In many countries, the extra heat for chicks reared under intensive conditions is provided by heating the entire rearing house. An alternative is to simulate the brooding behavior of a mother hen by using dark brooders. Dark brooders are horizontal heating elements equipped with curtains so that dark and warm areas are created for the chicks to creep under when they are in need of extra heat. The ambient temperature outside the brooders may then be kept at around 18 °C–24 °C [2,3], potentially reducing the energy consumption required to heat the poultry house during rearing of chicks.

Already at a young age, chicks manage to control their body temperature by behavioral thermoregulation. Brooding by a mother hen is usually initiated by chicks approaching the hen, sometimes pecking her breast feathers, and then starting to nuzzle under her [4]. When reared without a broody hen, 1- to 3-day old chicks are capable of choosing the correct arm in a maze or pecking a key to receive warmed air or radiation from a heat lamp [5,6,7]. The brooding behavior of the entire group is not completely synchronized as some chicks enter and leave the brooding hen before others [4,8]. In the red junglefowl, termination of brooding is initiated by the brooded chicks themselves 80% of the time [4]. The remaining 20% of the brooding bouts are ended by the broody hen leaving the brooding posture with chicks still under her [4]. The number of chicks under the broody hen, when she leaves the brooding position, is higher in cold compared to warm conditions [4]. Thus, in one out of five times, the broody hen plays an active role in initiating activity bouts in her chicks affecting more chicks the colder the ambient temperature is. She further encourages the chicks to start feeding by adding gestural and vocal food calling elements to her pecking behavior [9,10,11]. It has been found that brooded layer chicks indeed do perform more feeding and foraging activities than non-brooded chicks [11,12,13].

Rearing chicks with a broody hen during the first weeks of life also has an effect on their emotional and social reactivity later in life. Laying hens reared with a broody hen have been found to be less neophobic at 22 weeks of age than laying hens reared without broody hens as they will go nearer a novel object in their home pens [14]. Likewise, it has been shown that brooded chicks are less fearful than non-brooded chicks in a fear test performed in their home pen where the response to the sudden standing up of a human observer was recorded at age 1–8 weeks [15]. When placed in a very novel situation (*i.e.*, fear tests outside the home pens), 4- to 6-week old brooded layer chicks showed less fear-related behavior than non-brooded chicks [13,16], but later in life this difference disappeared [14,17]. Brooded layer chicks have been shown to be less aggressive than non-brooded chicks [18], and this difference persisted throughout the rearing and laying periods [17]. At 28–29 weeks of age, laying hens that were brooded by a mother hen during rear were more likely to seek proximity to a conspecific than laying hens that were non-brooded during rear [14].

In some aspects, the dark brooders resemble a hen in the brooding posture. Although the brooders, unlike the broody hen, do not interact actively with the chicks, rearing with dark brooders increases the synchrony of activity within the group compared to groups of non-brooded chicks [8]. Previous experiments have shown that rearing layer chicks with dark brooders has a reducing effect on feather pecking behavior during rear and early-lay [3,19], a positive effect on plumage condition during the laying period [2], and a reducing effect on cannibalistic incidences during rear and early lay [3]. In addition to having a preventive effect on feather pecking and cannibalism, the brooders may affect other types of behavior of the birds. For example, caretakers and researchers testing brooders, both in commercial settings and in experimental set-ups, have reported anecdotally that chicks, pullets and adult hens reared with brooders seem to be calmer and less flighty than those reared without brooders [2,20].

Until now, no studies have investigated the effects of dark brooders on time budget and fearfulness in poultry. Some breeders have expressed reluctance against using dark brooders, as they fear an increased number of non- or slow-starter chicks, *i.e.*, chicks that never learn to use brooders efficiently and therefore either do not enter under the brooders when in need of extra heat or do not leave the brooder when in need of food or water. The main objective of the present study was to test short- and long-term effects of rearing layer chicks with dark brooders on time budget and fearfulness. We also aimed to examine whether time budget and fearfulness were affected (a) by the size allowance of the brooder allocated per chick and (b) by raising the brooders for short periods during the first four critical days where chicks have to learn to find food and water. The latter was thought to simulate the active role played by the broody hen in initiating activity bouts. We hypothesized that brooder chicks would be less fearful in the three different fear tests performed (tonic immobility test, open-field test and novel object test) and that the effect would be long-lasting. Also, we predicted that brooder chicks would feather peck less, perform fewer aggressive interactions and show less fleeing than non-brooded chicks. We expected to find more drinking, feeding and foraging during the first four days in the brooder treatments where the brooders were raised for short periods compared to the brooder treatments where the brooders were maintained at the same height during the entire brooding period.

2. EXPERIMENTAL SECTION

2.1. Animals and Housing Conditions

Isa Warren layers (n = 2254) were obtained from a commercial hatchery (TopÆg Aps, Viborg, Denmark) as newly hatched non-beak-trimmed female chicks and transported to the experimental poultry facility at AU Foulum (Tjele, Denmark). Immediately upon arrival, the chicks were randomly assigned to one of 22 pens located in two different experimental rooms, B and C.

2.1.1. Brooder and Control Groups

In Room B, brooders (one per pen) were used as the primary heat source for the chicks. Upon arrival, chicks were placed under the brooders. The temperature under the brooders was kept constant at 34 °C during the first 3 days and then lowered half a degree each day until 20 °C were reached on day 28. Also, the ambient temperature in Room B was kept at 24 °C on Days 0 and 1, 22 °C on Days 3 and 4, and then kept constant at 20 °C. In room C, no brooders were used, and the ambient temperature in the room was kept at the same temperature schedule as used for the temperature under the brooders. Sixteen groups were housed in Room B and six in Room C. The conditions in the rooms were otherwise exactly the same. The heat sources used for keeping the schedules for the ambient temperature in Rooms B and C were water heated vertical radiators mounted on the outer walls, which is the traditional way to heat poultry houses in Denmark.

2.1.2. The Four Brooder Treatments

Pens in Room B were randomly assigned to one of four treatments that differed in the size of the brooder placed in each pen (Large or Small) and in the management of the height of the brooders (raised for 10 min every 4 hours on Days 1–4 or maintained at the same height during brooding; Movable or Fixed, respectively). Hence four types of brooders/management combinations were tested: LM (Large and Movable), LF (Large and Fixed), SM (Small and Movable) and SF (Small and Fixed). The large brooders measured 120 cm × 60 cm corresponding to an average available area under the brooder of 72 cm^2 per chick whereas the small brooders measured 90 cm × 60 cm (54 cm^2/chick).

The regular height positioning of the brooders was adjusted only once at 29 days of age to compensate for the growth of the chicks (shifted from 16 cm at the beginning of the experiment to 25 cm on Day 29). All brooders were permanently raised to 2.0 m on Day 41.

2.1.3. Available Resources and Management

Each pen (2 m × 4 m) housed a group of 100–103 chicks, resulting in a stocking density similar to that used for organic pullets and around 2/3 of that used for barn pullets under commercial conditions. Pens were provided with seven automatic water nipples and two feeders (251 cm feeder space, totally). In addition, chicks were fed on paper during the first 4 days. Seventy centimeter high white sheets of hard plastic were attached to the bottom sides of the pens to avoid potential visual contact between individuals from neighboring pens. Bedding was provided in the form of an approximately 5 cm deep layer of wood shavings, and birds were fed a commercial diet (DLG Group, Denmark) according to their rearing stage that was provided *ad libitum*. To enable data collection on time budget from video recordings during the first weeks, where chicks are quite small, we tried to enhance the image quality of the chicks by halving the available floor area using wooden partitions during the first 18 days of the experiment. At Day 18, each pen was provided with two perches (each 2 m long, at 1 m height).

In order to minimize potential undesired minimal differences between rooms/pens influencing the data collection, all groups were reallocated among the 22 available pens in both rooms after completion of the first round of all the fear tests at Day 44 (3 days after the brooders had been permanently raised). This was done in a balanced way so that one group from each of the four brooder treatments and two control groups were housed in Room C and the remaining groups in room B. The allocation of pens within rooms was random, except that a group could not be allocated to its original pen. Time spent daily by caretakers and observers in the two rooms from arrival of the chicks up to the day of reallocation (Day 44) of the groups in the two rooms was approximately the same. Both caretakers and observers noted how much time they spent in Room B and then spent the same amount of time in Room C. This was done in order to avoid differences between treatments in amount of human contact, which potentially could have influenced the fear testing.

At Days 93–94, each pen was provided with six nest boxes (40 cm × 30 cm × 34 cm; W × D × H) and additional perches arranged in a ladder layout providing easy access to the nest boxes. When pullets were 113–114 days old, all groups were reduced to 50 individuals per pen to maintain the stocking density similar to that used for organic hens and around 2/3 of that used for barn hens under commercial conditions during the laying period. Birds removed were mainly those that had been tested in fear tests during 3–6 and 14–15 weeks of age (described below). On Days 118–119 the bedding was removed and replaced by a new layer of approximately 5 cm wood shavings. A computerized system allowed control of light and ventilation, which followed commercial standard practices.

2.1.4. Ethical Note

All procedures involving animals were in accordance with the Danish Ministry of Justice Law No. 382 (10 June 1987) and Acts 333 (19 May 1990), 726 (9 September 1993) and 1016 (12 December 2001). This study is part of a larger project that also evaluates the effects of brooders on welfare and production parameters.

2.2. Data Collection

2.2.1. Behavioral Time Budget

Immediately upon arrival and housing, 10 chicks per group were randomly selected and marked on their back with non-toxic, green spray paint (suitable for animals). To ensure further identification, chicks were remarked on Days 7, 14, 21 and 28. Marked chicks were used as focal animals [21] for time budget determination. Originally, the intention was to collect data on activity from video recordings, but even with the temporarily reduced floor area it was impossible to identify the focal animals on the video recordings. Therefore, we decided to do live observations. The live observations were conducted when the chicks were 1, 2, 3, 4, 9, 16, 23, 30, and 42 days of age using instantaneous scan sampling. The observer slowly walked from pen to pen. When in front of a pen (visible to the chicks) the observer stood still for 15 s before recording the behavior of the marked chicks. In order to minimize disturbances from the schedule of raising brooders in the movable treatments, two different behavioral time budget sampling protocols were used depending on the age of the chicks. The first protocol was used for sampling on Days 1, 2, 3 and 4 of age (Phase I) and consisted of three sets of 12 samples each at regular intervals in the periods between 0612 and 0952 h, 1152 and 1532 h and 1812 and 2152 h, respectively (photophase 2–24 h). The second protocol was used on Days 9, 16, 23, 30 and 42 of age (Phase II) and consisted of 27 samples at regular intervals during the period of time between 0612 and 1702 h (photophase 6–18 h). The experimental environment (scheduled changes in the light regime and raising of the brooders), our operational capabilities and Daigle and Siegford's [22] recommendations regarding time interval between scans were taken into consideration when determining each of these two protocols.

Within each time-point, the number of marked chicks per group performing any of the following 11 behavioral categories was recorded: drinking, feeding, locomotion, foraging, resting, aggression, fleeing, comfort behavior (defined as the chicks stretching its legs and/or wings or wing flapping), dust bathing, exploration and feather pecking. Fleeing was defined as the chick running from frightening stimuli, typically the arrival of the observer, and included possible "freezing" of the chick at the end of the flight, *i.e.*, standing erect and still for a moment until other types of behavior were resumed. Within the brooder groups, resting was split into the following three subcategories: resting outside, under, or upon the brooder. On Day 42, this subdivision was not applicable as the brooders were no longer available for the chicks. Data on time budget were expressed as daily proportions of marked chicks per pen engaged in each of the behavioral categories registered.

2.2.2. Novel Object Test

A Novel Object (NO) test was performed in the home pens at three different developmental stages; Days 43 (chicks), 107 (pullets) and 189 (adult hens) of age. Different objects may elicit different levels of fear, and presenting an object multiple times may result in habituation to the object [23]. Therefore, three different objects were presented in a balanced way between treatments and ages so that one replicate of each of the brooder treatments and two of the control replicates were presented with the same novel object per round of testing. The novel objects used were a red soft drink can, a blue folder and a yellow bucket. To fit this balanced schedule, only 18 pens were tested (six control groups and three groups from each of the four brooder treatments; the same groups were tested across the different test rounds). Each test lasted 5 minutes and began when the object was placed on the floor within an arm's reach from the door, thus avoiding that a person had to enter the pen. Feed and water were located at least 1 m from the NO.

Data on the birds' proximity to the novel object were collected from digital video recordings using instantaneous scan sampling at 15, 30, 60, 120, 180, 240 and 300 s after introduction of the novel object. On the obtained pen images, three similar sized zones were drawn on the screen (using reference marks placed in the pens). These zones were delimited in relation to the distance to the novel object: nearest, middle and most distant from the object. For each time-point sampled, the number of birds in each area and the number of birds touching the novel object were counted. Then, those counts were transformed into percentages to enable comparisons between different rounds. Lastly, according to

Hocking *et al.*, [23], for each time-point sampled a Mean Score Proximity was calculated using the following calculation:

$$Mean\ Score\ Proximity = \frac{1 \times \%Touching + 2 \times \%Nearest + 3 \times \%Middle + 4 \times \%Distant}{100}$$

The position scores ranged from one for the percentage of birds found in the closest proximity to four for the percentage of birds found in the distant area. Thus, a low Mean Score Proximity indicates low avoidance of the novel object.

2.2.3. Open-Field Test

Individual open-field (OF) responses were tested at three different developmental stages (chicks 19–23; pullets 98–106; and adult hens 192–199 days old, from here on referred to as Days 21, 101, and 196) as a measure of fear and exploration [24,25]. Ten naive birds (neither previously tested in OF nor in the tonic immobility test) randomly chosen from each pen were tested at each age. Eight unpainted plywood OF arenas measuring 1.22 m × 1.22 m × 0.81 m (width × depth × height) were located on concrete floor (six in room B and two in room C). The OF arena was divided into 25 squares (24.4 × 24.4 cm) using white markings (chalk). A wire mesh lid was used to prevent potential escapes on Days 101 and 196. Briefly, to begin a test, a bird was caught in her home pen and individually placed in the center of the empty OF arena. Her behavior was video recorded for 10 min by using a closed-circuit television system with a video camera suspended above the OF arena. This arrangement made certain that the experimenter was completely hidden from the bird's view during testing. Upon completion of the test, birds were marked (using plastic leg bands) before being returned to their respective pens to prevent birds from being reevaluated on following tests or rounds. All birds within a pen were tested before continuing to a new pen. The order of pens tested was random, decided by drawing a number from a pot. This procedure was repeated for each round of testing. From the first 520 s of the video recordings of the test, the following behavioral categories were registered using focal sampling: latency to initiate locomotion and number of lines crossed in the square gridline floor. The number of droppings was counted when 10 min had passed and the OF arena was cleaned after each individual.

2.2.4. Tonic Immobility Test

Tonic Immobility (TI) responses were tested at three different developmental stages (chicks 25–27; pullets 95–98; and adult hens 179–183 days old, from here on referred to as Days 26, 96 and 181) as a measure of fearfulness [25]. Ten naive birds (neither previously tested in

OF nor in TI) randomly chosen from each pen were tested at each age. Briefly, the bird to be tested was carefully lifted from its home pen and placed on its back in a wooden cradle situated in the same room. The observer held the bird softly but firm with one hand over the chest and the other hand over the neck/head for 15 s. After releasing the bird, the observer slowly took one step backwards and avoided eye contact with the bird. If the bird remained motionless for 10 s after release, it was deemed to have entered TI. If it moved within 10 s, the procedure was repeated and the number of induction attempts recorded. If TI could not be induced by the third attempt, the bird was deemed to be non-susceptible and a TI duration of 0 s was given. A test ceiling of 10 min was set, and any bird that failed to right itself before this time elapsed was given a maximum score of 600 s, and the test was interrupted. The observer wore similar clothes with a similar color as the caretakers.

The order of pens tested was random, decided by drawing a number from a pot. To prevent birds from being re-tested on following tests or rounds, upon completion of testing time, birds were marked (using plastic leg bands) before being returned to their respective pens. For each TI test performed, the following parameters were registered: the number of inductions, the latency to first head movement (scanning movement), and the duration of TI. During the first round of fear testing, the order of the tests were OF, TI, and NO. During the second and third rounds the order of the tests was TI, OF and NO.

2.3. Statistical Analysis

Similar statistical procedures were performed to evaluate the effect of treatments on the expression of each of the behavioral categories registered. Because two different protocols were used for Phase I and II, data from each Phase were analyzed independently. Briefly, daily proportions of marked chicks per pen engaged in each of the behavioral categories were calculated for each day of data collection per treatment and pen. Differences in expression of each of the behavioral categories were determined by a mixed model ANOVA with treatment (LM, LF, SM, SF and C) as fixed effect and days of age (Phase I: 1, 2, 3 and 4; Phase II: 9, 16, 23, 30 and 42) as a repeated measure. Pen was included as a random effect. Due to low occurrences, *i.e.*, large numbers of zeros, the behavioral categories fleeing, aggression, dust bathing, and exploration during Phase I and aggression during Phase II could not be statistically analyzed.

Differences on Mean Score Proximity to the NO were determined by a mixed model ANOVA with treatment (LM, LF, SM, SF and C) as fixed effect and sampling time after introduction of the NO (15, 30, 60, 120, 180, 240 and 300 s) and age at testing (Days 43, 107 and 189) as repeated measures. Pen was included as a random effect. Differences on variables registered during OF and TI tests were determined by a mixed model ANOVA with treatment (LM, LF, SM, SF and C) as fixed effect and age at testing (OF: Days 21, 101, and 196; TI: Days 26, 96, and 181) as repeated measures. Pen was included as a random effect.

In all models used, heteroscedasticity was corrected by specifying the variance structure of the errors whenever necessary. Stepwise reduction of the models was conducted and non-significant interactions were removed from the model. Models were fit using the *nlme* R library through a user-friendly interface implemented in InfoStat [26]. Whenever significant main effects were detected, a *post-hoc* DGC test (hierarchical multiple comparison procedure) was performed [27]. One of the four SM groups had to be removed from all statistical analyses since we discovered the presence of a rooster in the group when the birds were 85 days old; statistical analysis of the NO test was conducted without change as this SM group was not initially selected for testing.

3. RESULTS

The results on the time budget for the different treatments are summarized in Table 1 for Phase I and in Table 2 for Phase II. In general, the time budgets for the brooder treatments differed more from the time budget of C than from one another.

During Phase I, there were significant interactions between treatment and age for five behavioral categories. The highest percentages of chicks feeding during Days 1 and 2 were found in C while the lowest percentages were found in SM and SF on Days 2 and 4, respectively ($F_{12,48} = 3.78$; $p < 0.001$). Additionally, the highest percentage of feeding chicks on Day 4 was found in SM. Fewer chicks rested in C on all days recorded while the highest percentage of chicks resting was found in SM on Day 2 ($F_{12,48} = 3.21$; $p = 0.002$). The highest percentage of chicks performing comfort behavior was found in C on all days recorded ($F_{12,48} = 2.32$; $p = 0.019$). No differences between days or treatments were found in the percentage of chicks resting under the brooder, except on Day 2 where a higher percentage of chicks was resting under the brooder in SM ($F_{9,33} = 2.88$; $p = 0.012$). More variation was found in the percentage of chicks resting outside the brooders with a higher percentage for all treatments on Day 1 compared to Days 3 and 4 ($F_{9,33} = 3.14$; $p = 0.007$). Additionally, there was a treatment effect for the behavioral categories drinking, locomotion and feather pecking: the highest percentages of chicks drinking ($F_{4,16} = 4.83$; $p = 0.009$), performing locomotion ($F_{4,16} = 21.09$; $p < 0.001$) and feather pecking

($F_{4,16}$ = 11.27; $p < 0.001$) were found in C regardless of the age of the chicks. There was an increase in percentage of chicks foraging with increasing age ($F_{3,48}$ = 29.4; $p < 0.0001$; Means ± SE: 1.9 ± 0.2, 2.5 ± 0.2, 3.5 ± 0.3, and 6.9 ± 0.5 on Days 1, 2, 3 and 4, respectively). In Phase I, the behavioral categories aggression, fleeing, dust bathing, and exploration constituted less than 1% of the time budget in all treatments.

During Phase II, there were significant interactions between treatment and age for 11 behavioral categories. A general pattern of decreasing percentages of resting chicks with age was found for all treatments, but the decrease from day to day varied between treatments ($F_{16,64}$ = 3.76; $p < 0.001$). A higher percentage of chicks was resting under the brooders in the treatments with small brooders (SF and SM) on Day 9 ($F_{9,33}$ = 2.18; p = 0.049). No differences were found between brooder treatments in percentage of chicks resting outside the brooders during any of the days in Phase II ($F_{9,33}$ = 1.05; p = 0.42), but the percentages increased with age ($F_{3,33}$ = 32.53; $p < 0.001$). Likewise, the percentages of chicks resting on the brooders increased with age but faster in the movable brooder treatments (SM and LM; $F_{9,33}$ = 5.77; $p < 0.001$). More chicks were found feeding on the last day (Day 42) compared to the first day (Day 9) of observation in Phase II, and a higher percentage of chicks from the brooder treatments compared to C were feeding on Day 42 ($F_{16,64}$ = 3.88; $p < 0.001$). A similar pattern was found for percentage of chicks foraging ($F_{16,64}$ = 3.13; $p < 0.001$).

Table 1. Proportion of layer chicks (%) performing various behavioral categories when 1, 2, 3, and 4 days old. Gray shaded areas indicate no available brooders.

Behavioral measurements (%) [1]	SF[2] (Day)				SM[2] (Day)				LF[2] (Day)				LM[2] (Day)				C[2] (Day)			
	1	2	3	4	1	2	3	4	1	2	3	4	1	2	3	4	1	2	3	4
Resting	82.0 ± 1.6^{b}	81.8 ± 1.6^{b}	79.3 ± 2.3^{b}	79.0 ± 2.6^{b}	84.0 ± 1.8^{b}	88.1 ± 1.9^{a}	80.2 ± 2.7^{b}	76.1 ± 3.0^{b}	83.4 ± 1.6^{b}	83.4 ± 1.6^{b}	76.2 ± 2.3^{b}	73.9 ± 2.8^{b}	84.0 ± 1.6^{b}	83.3 ± 1.6^{b}	78.3 ± 2.3^{b}	75.0 ± 2.6^{b}	71.3 ± 1.3^{c}	64.6 ± 1.3^{d}	65.5 ± 1.9^{d}	64.6 ± 2.3^{d}
Resting under the brooder	73.7 ± 0.9^{b}	77.0 ± 1.2^{b}	75.9 ± 2.5^{b}	76.6 ± 2.0^{b}	75.4 ± 0.9^{b}	84.3 ± 1.0^{a}	76.4 ± 3.7^{b}	73.7 ± 5.0^{b}	74.2 ± 0.5^{b}	78.9 ± 0.9^{b}	73.9 ± 0.8^{b}	71.6 ± 2.7^{b}	77.4 ± 2.1^{b}	77.6 ± 1.0^{b}	74.5 ± 1.9^{b}	70.6 ± 2.3^{b}				
Resting outside the brooder	8.3 ± 1.6^{a}	4.7 ± 0.7^{b}	3.3 ± 0.8^{b}	2.4 ± 0.3^{c}	8.6 ± 1.5^{a}	3.8 ± 0.6^{b}	3.8 ± 0.4^{b}	2.4 ± 0.1^{c}	9.1 ± 1.0^{a}	4.4 ± 0.4^{b}	2.3 ± 0.5^{c}	2.3 ± 0.2^{c}	6.8 ± 0.6^{a}	5.6 ± 0.9^{a}	3.6 ± 0.6^{b}	4.5 ± 0.3^{b}				
Feeding	5.6 ± 0.3^{c}	6.0 ± 1.0^{c}	7.0 ± 1.6^{c}	4.0 ± 0.7^{d}	8.2 ± 0.4^{b}	3.7 ± 0.3^{d}	5.1 ± 0.6^{c}	8.5 ± 2.6^{b}	5.5 ± 0.5^{c}	6.0 ± 0.5^{c}	9.1 ± 0.6^{b}	6.4 ± 1.1^{c}	6.5 ± 0.9^{c}	6.0 ± 0.5^{c}	7.6 ± 1.7^{c}	6.5 ± 0.9^{c}	8.7 ± 0.5^{b}	12.7 ± 1.1^{a}	10.4 ± 1.4^{b}	6.4 ± 0.8^{c}
Comfort	0.5 ± 0.2^{c}	0.0 ± 0.2^{c}	1.1 ± 0.2^{b}	1.4 ± 0.2^{b}	0.2 ± 0.1^{c}	0.0 ± 0.1^{c}	0.6 ± 0.1^{c}	1.4 ± 0.1^{a}	0.3 ± 0.2^{c}	0.1 ± 0.2^{c}	1.5 ± 0.2^{b}	1.9 ± 0.2^{b}	0.4 ± 0.3^{c}	0.1 ± 0.3^{c}	1.1 ± 0.3^{b}	1.5 ± 0.3^{b}	1.3 ± 0.5^{a}	2.2 ± 0.5^{a}	4.3 ± 0.5^{a}	5.0 ± 0.5^{a}
Drinking	3.1 ± 0.3^{b}				3.0 ± 0.4^{b}				3.0 ± 0.4^{b}				3.5 ± 0.4^{b}				4.6 ± 0.3^{a}			
Locomotion	6.1 ± 0.5^{b}				5.5 ± 0.6^{b}				5.5 ± 0.3^{b}				4.8 ± 0.3^{b}				9.7 ± 0.4^{a}			
Feather pecking	0.1 ± 0.4^{b}				0.1 ± 0.4^{b}				0.1 ± 0.4^{b}				0.2 ± 0.1^{b}				0.9 ± 0.1^{a}			

[1] The values represent means ± SE; (%) = Daily average percentages of focal chicks performing a particular behavior (calculated from 36 observations per group per day);
[2] Treatments = SF: small fixed brooder; SM: small movable brooder; LF: large fixed brooder; LM: Large movable brooder; C: Control; [a–d] Values within a behavioral category with no common letters differ significantly ($p < 0.05$).

Table 2. Proportion of layers (%) performing various behavioral categories when 9, 16, 23, 30, and 42 days old. Gray shaded areas indicate no available brooders.

Behavioral measurements (%)[1]	SF[2] (Day)					SM[2] (Day)					LF[2] (Day)				
	9	16	23	30	42	9	16	23	30	42	9	16	23	30	42
Resting	69.6 ± 1.2 a	55.9 ± 2.3 b	49.6 ± 2.3 b	48.9 ± 2.3 b	36.1 ± 1.3 c	73.9 ± 2.8 a	58.4 ± 2.2 c	50.7 ± 2.9 c	43.7 ± 1.5 d	38.6 ± 1.4 e	64.8 ± 1.7 b	53.2 ± 1.2 c	46.7 ± 3.9 d	44.8 ± 2.9 d	37.4 ± 0.9 e
Resting under the brooder	61.5 ± 2.0 a	41.3 ± 2.0 c	35.6 ± 2.7 d	33.3 ± 1.9 d		65.6 ± 2.3 a	42.2 ± 2.3 c	37.4 ± 3.2 d	28.5 ± 2.2 d		56.3 ± 2.0 b	43.4 ± 2.0 c	37.7 ± 2.7 d	32.0 ± 1.9 d	
Resting outside the brooder	8.1 ± 0.1 a	12.5 ± 1.6 a	14.0 ± 1.6 a	15.5 ± 1.4 a		7.4 ± 1.2 a	16.1 ± 1.8 a	13.3 ± 1.8 a	15.1 ± 1.6 a		8.6 ± 1.0 a	12.5 ± 1.6 a	12.7 ± 1.6 a	16.0 ± 1.4 a	
Resting on the brooder	0.7 ± 0.4 c	10.3 ± 1.9 b	12.6 ± 0.8 b	15.2 ± 1.2 a		0.1 ± 0.1 c	14.7 ± 2.3 a	12.3 ± 0.1 b	16.3 ± 2.2 a		2.1 ± 1.5 c	9.9 ± 0.8 b	12.8 ± 0.6 a	14.8 ± 1.7 a	
Feeding	5.8 ± 1.0 c	9.1 ± 0.8 b	11.9 ± 0.9 a	11.8 ± 1.4 a	12.2 ± 0.5 a	5.0 ± 0.6 c	11.3 ± 1.0 a	13.6 ± 0.5 a	11.9 ± 1.9 a	14.9 ± 2.3 a	8.0 ± 1.5 b	9.5 ± 0.8 b	13.2 ± 1.0 a	15.4 ± 0.9 a	13.6 ± 2.0 a
Foraging	10.4 ± 1.3 c	11.2 ± 1.1 c	12.3 ± 0.7 a	12.9 ± 1.0 c	18.6 ± 1.0 a	6.6 ± 0.2 e	10.3 ± 0.6 c	12.6 ± 0.8 c	12.2 ± 0.9 c	15.9 ± 1.3 b	8.3 ± 1.0 d	14.4 ± 1.2 a	12.9 ± 1.0 c	13.1 ± 0.8 c	14.6 ± 0.9 b
Comfort	1.8 ± 0.3 c	2.9 ± 0.7 b	4.7 ± 0.1 b	4.4 ± 0.7 b	7.4 ± 0.9 a	1.5 ± 0.2 c	2.2 ± 0.7 c	3.8 ± 1.5 b	6.0 ± 0.9 b	5.9 ± 1.5 b	1.5 ± 0.5 c	2.7 ± 1.0 a	3.0 ± 0.3 a	5.2 ± 0.5 a	7.6 ± 0.8 a
Drinking	2.7 ± 0.8 b	3.1 ± 0.3 b	3.6 ± 0.8 b	4.2 ± 0.8 b	5.8 ± 0.3 a	1.9 ± 0.9 b	2.4 ± 0.4 b	4.6 ± 0.3 b	5.9 ± 0.3 a	3.5 ± 0.7 b	2.8 ± 0.4 b	3.1 ± 0.6 b	4.1 ± 0.5 b	3.6 ± 0.3 b	4.2 ± 0.4 b
Locomotion	5.7 ± 0.6 c	13.8 ± 1.5 d	14.3 ± 2.0 d	10.1 ± 1.0 d	14.9 ± 1.1 d	11.0 ± 1.5 d	12.2 ± 1.7 d	11.0 ± 0.8 d	11.3 ± 0.8 d	15.6 ± 1.2 d	12.9 ± 2.8 d	13.7 ± 0.3 d	14.2 ± 2.1 d	12.8 ± 0.8 d	17.9 ± 0.9 c
Feather pecking	0.4 ± 0.2 b	1.6 ± 0.2 b	1.2 ± 0.6 b	1.5 ± 0.6 b	1.0 ± 0.5 b	0.1 ± 0.1 c	1.3 ± 0.6 b	0.9 ± 0.4 b	0.4 ± 0.2 b	1.4 ± 0.6 b	0.0 ± 0.1 c	0.7 ± 0.3 b	1.1 ± 0.3 b	0.9 ± 0.7 b	1.4 ± 0.8 b
Fleeing	0.1 ± 0.2 c	0.7 ± 0.2 c	0.3 ± 0.2 c	0.2 ± 0.2 c	0.3 ± 0.2 c	0.1 ± 0.2 c	0.8 ± 0.2 c	0.0 ± 0.2 c	0.3 ± 0.2 c	0.6 ± 0.2 c	0.2 ± 0.3 c	0.5 ± 0.3 c	0.3 ± 0.3 c	0.6 ± 0.3 c	0.5 ± 0.3 c
Exploration	0.5 ± 0.2 b	1.5 ± 0.4 b	1.9 ± 0.5 b	5.1 ± 0.5 a	2.5 ± 0.9 b	0.4 ± 0.2 b	0.9 ± 0.5 b	2.5 ± 0.6 b	7.4 ± 1.6 a	3.5 ± 0.7 a	1.0 ± 0.4 b	1.7 ± 0.4 b	3.9 ± 1.3 a	3.1 ± 1.1 a	2.1 ± 0.7 b
Dustbathing	0.0 ± 0.2 b	0.1 ± 0.2 b	0.1 ± 0.2 b	0.7 ± 0.2 a	0.9 ± 0.2 a	0.1 ± 0.2 a	0.1 ± 0.2 a	0.3 ± 0.2 a	0.9 ± 0.2 a	0.1 ± 0.2 a	0.4 ± 0.2 b	0.5 ± 0.2 a	0.5 ± 0.2 a	0.3 ± 0.2 b	0.6 ± 0.2 a

Behavioral measurements (%)[1]	LM[2] (Day)					C[2] (Day)				
	9	16	23	30	42	9	16	23	30	42
Resting	64.9 ± 2.0 b	58.1 ± 1.4 b	53.2 ± 2.2 b	45.9 ± 2.1 d	40.4 ± 1.7 d	62.9 ± 1.5 b	46.7 ± 1.7 d	42.0 ± 1.9 d	41.9 ± 2.0 d	38.1 ± 1.2 e
Resting under the brooder	56.3 ± 2.0 b	41.4 ± 2.0 c	37.7 ± 2.7 d	32.0 ± 1.9 d						
Resting outside the brooder	8.6 ± 1.0 a	16.7 ± 1.6 a	15.5 ± 1.6 a	13.9 ± 1.4 a						
Resting on the brooder	3.8 ± 2.6 c	17.3 ± 0.8 a	10.2 ± 0.9 b	12.9 ± 2.2 b						
Feeding	8.5 ± 0.6 c	10.2 ± 1.2 a	14.3 ± 1.1 a	12.2 ± 1.0 a	12.1 ± 0.8 a	6.3 ± 0.8 c	11.7 ± 1.2 a	14.3 ± 1.9 a	10.8 ± 0.9 a	9.2 ± 0.6 b
Foraging	9.8 ± 1.2 c	11.8 ± 0.8 c	10.0 ± 0.9 c	16.0 ± 2.2 b	14.9 ± 0.7 b	8.6 ± 0.5 d	10.6 ± 0.8 c	10.9 ± 0.5 c	11.4 ± 1.2 c	9.7 ± 1.0 c
Comfort	1.8 ± 0.3 c	2.3 ± 0.2 c	3.1 ± 0.7 b	4.4 ± 0.9 b	7.0 ± 1.6 a	3.4 ± 0.6 b	3.6 ± 0.7 b	3.8 ± 0.4 b	4.4 ± 0.6 b	3.8 ± 0.5 b
Drinking	2.9 ± 0.5 b	2.6 ± 0.3 b	3.5 ± 0.5 b	4.1 ± 0.8 b	4.1 ± 0.5 b	2.8 ± 0.4 b	3.3 ± 0.9 b	3.3 ± 0.3 b	2.8 ± 0.5 b	3.7 ± 0.5 b
Locomotion	12.8 ± 1.5 d	12.7 ± 0.3 d	12.1 ± 1.0 d	11.2 ± 0.9 d	15.2 ± 1.4 d	13.2 ± 1.1 d	18.7 ± 1.2 c	19.1 ± 1.2 c	23.2 ± 1.3 b	27.9 ± 0.9 a
Feather pecking	0.0 ± 0.1 c	0.6 ± 0.2 b	0.7 ± 0.4 b	0.7 ± 0.2 b	1.0 ± 0.3 b	1.2 ± 0.1 b	2.6 ± 0.6 a	3.2 ± 0.8 a	2.5 ± 0.7 a	2.5 ± 0.3 a
Fleeing	0.4 ± 0.2 c	0.8 ± 0.2 c	0.0 ± 0.2 c	0.3 ± 0.2 c	0.4 ± 0.2 c	0.8 ± 0.4 c	1.6 ± 0.4 b	2.1 ± 0.4 b	1.7 ± 0.4 b	3.5 ± 0.4 a
Exploration	1.0 ± 0.3 b	0.6 ± 0.1 c	2.3 ± 0.6 b	4.5 ± 1.0 a	3.9 ± 0.8 a	0.6 ± 0.3 b	0.8 ± 0.3 b	0.7 ± 0.3 b	0.8 ± 0.2 b	1.3 ± 0.4 b
Dustbathing	0.6 ± 0.2 a	0.2 ± 0.2 b	0.6 ± 0.2 a	0.7 ± 0.2 a	1.0 ± 0.2 a	0.1 ± 0.1 b	0.4 ± 0.1 b	0.3 ± 0.1 b	0.3 ± 0.1 b	0.4 ± 0.1 b

[1] The values represent means ± SE; (%) = Daily average percentages of focal chicks performing a particular behavior (calculated from 27 observations per group per day); [2] Treatments = SF: small fixed brooder; SM: small movable brooder; LF: large fixed brooder; LM: Large movable brooder; C: Control. a–d Values within a behavioral category with no common letters differ significantly ($p < 0.05$).

In the brooder treatments there was an increase in percentage of chicks performing comfort behavior with increasing age whereas no difference was found between ages in C ($F_{16,64} = 3.21$; $p < 0.001$). A higher percentage of chicks from the brooder treatments (except SM) compared to C performed comfort behavior on Day 42 ($F_{16,64} = 3.21$; $p < 0.001$). No differences were found between treatments in percentage of chicks drinking during any of the days in Phase II, except on Days 30 and 42 where SM and SF, respectively, performed more drinking behavior ($F_{16,64} = 5.64$; $p < 0.001$). The percentage of chicks engaged in locomotion remained stable (with minor exceptions) with age in the brooder treatments and with no difference between treatments whereas a steady increase with age occurred in C, resulting in almost twice as many chicks engaged in locomotion in C compared to any of the brooder treatments at 42 days of age ($F_{16,64} = 3.52$; $p < 0.001$). The percentage of chicks engaged in feather pecking was higher in C ($F_{4,16} = 9.52$; $p < 0.001$). Fleeing was continuously rare in the brooder treatments whereas a steady increase with age was found in C ($F_{16,64} = 2.28$; $p = 0.01$). Exploratory behavior was generally low in C but varied more with age in the brooder treatments ($F_{16,64} = 5.11$; $p < 0.001$). Finally, dust bathing occurred at low frequencies in all treatments but was rarest in C ($F_{16,64} = 1.94$; $p = 0.032$). Aggression constituted less than 1% of the time budget in all treatments.

In the NO test, an effect of the interaction between treatment and age was found ($F_{8,260} = 9.60$; $p < 0.0001$); the highest values of Mean Score Proximity to the novel object were obtained by LM on Day 43, by C on day 107, and by SM and C on day 189 (Figure 1).

Regarding the OF tests performed, there was an interaction between treatment and age in latency to initiate locomotion ($F_{8,447} = 5.76$; $p < 0.0001$); the highest latencies were obtained by SM and C on Day 21, by SF, SM, LF and C on Days 101, and by C on Day 196 (Figure 2). An interaction between treatment and age was also found in number of lines crossed ($F_{8,447} = 2.42$; $p = 0.015$). During the OF test on Day 21, only SF differed from the other treatments having a higher number of line crossings. At 101 days of age, more variation was found with a lower number of line crossings by birds from SF, SM and C. Finally, at 196 days of age, no differences between treatments were found (Figure 3). Fewer droppings were produced by LM on Day 196, but otherwise no difference was found between treatments (Figure 4; treatment × age; $F_{8,447} = 3.47$; $p = 0.0007$).

In the TI test, there was an interaction between treatment and age on number of inductions ($F_{8,603} = 2.43$; $p = 0.0138$); fewer inductions were necessary to induce TI in C on Days 96 and 181 than in the brooder treatments, whereas no difference was found on Day 26 (Table 3). For latency to first head movement and duration of TI, there was a treatment effect where the highest values for both variables were obtained by C regardless of the age by which the birds were tested (latency: $F_{4,16} = 7.58$; $p = 0.0013$; duration: $F_{4,16} = 6.25$; $p = 0.0032$).

Table 3. Tonic immobility responses of layers when 26, 96 and 181 days old.

Behavioral measurements [1]	SF[2] (Days)			SM[2] (Days)			LF[2] (Days)			LM[2] (Days)			C[2] (Days)		
	26	96	181	26	96	181	26	96	181	26	96	181	26	96	181
Inductions (*n*)	2.4 ± 0.2 ^{a}	2.6 ± 0.2 ^{a}	2.3 ± 0.2 ^{a}	2.6 ± 0.2 ^{a}	2.6 ± 0.2 ^{a}	1.5 ± 0.2 ^{b}	2.7 ± 0.2 ^{a}	2.5 ± 0.2 ^{a}	2.1 ± 0.2 ^{a}	2.6 ± 0.2 ^{a}	2.5 ± 0.2 ^{a}	2.1 ± 0.2 ^{a}	2.1 ± 0.2 ^{a}	1.8 ± 0.1 ^{b}	1.9 ± 0.1 ^{b}
Latency to first head movement (s)	69.1 ± 10.0 ^{a}			57.0 ± 10.9 ^{a}			49.3 ± 6.4 ^{a}			80.7 ± 11.4 ^{a}			116.0 ± 10.8 ^{b}		
TI duration (s)	105.3 ± 13.2 ^{a}			93.2 ± 12.5 ^{a}			76.6 ± 10.1 ^{a}			112.6 ± 13.0 ^{a}			154.9 ± 12.5 ^{b}		

[1] The values represent means ± SE; (*n*) = number; (s) = seconds; [2] Treatments = SF: small fixed brooder; SM: small movable brooder; LF: large fixed brooder; LM: Large movable brooder; C: Control; $^{a-b}$ Values within rows with no common letters differ significantly ($p < 0.05$).

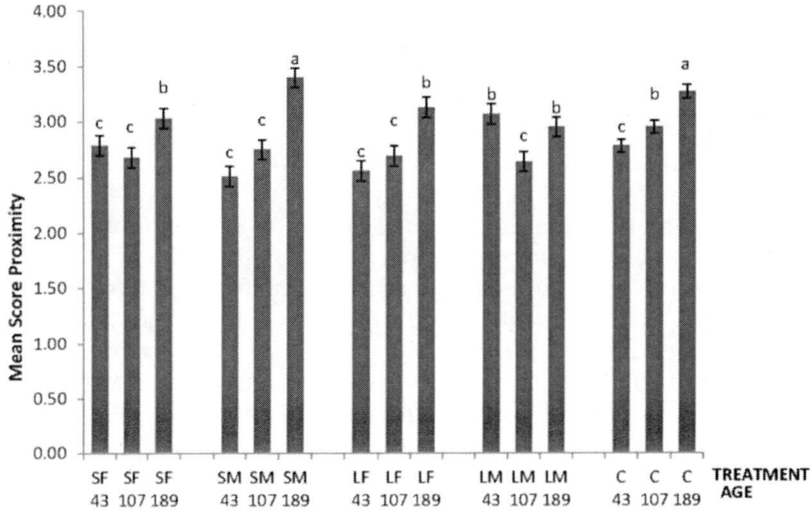

Figure 1. Novel Object test response depicted as Mean Score Proximity of 43, 107 and 189 days old layers from the control and four brooder treatments. Bars with no common letters differ significantly (mixed model ANOVA: $p < 0.05$). Treatments = SF: small fixed brooder; SM: small movable brooder; LF: large fixed brooder; LM: Large movable brooder; C: Control.

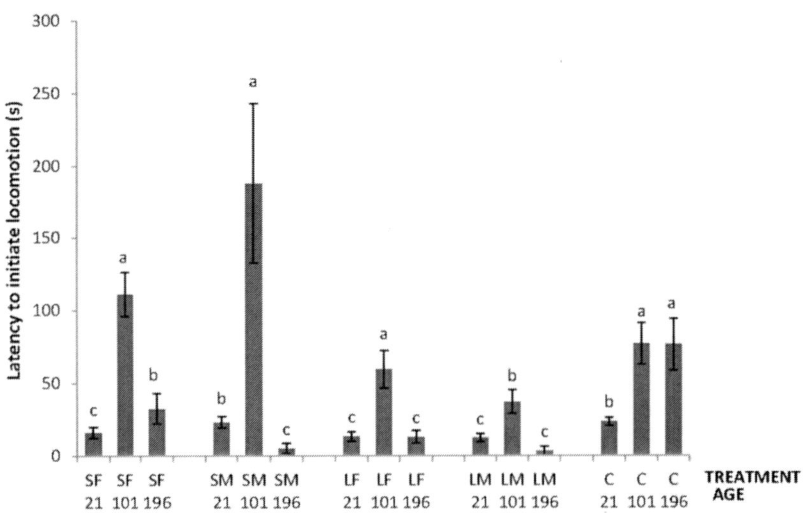

Figure 2. Latency to initiate locomotion in a 520 s open-field test by 21, 101 and 196 days old layers from the control and four brooder treatments. Bars with no common letters differ significantly (mixed model ANOVA: $p < 0.05$). Treatments = SF: small fixed brooder; SM: small movable brooder; LF: large fixed brooder; LM: Large movable brooder; C: Control.

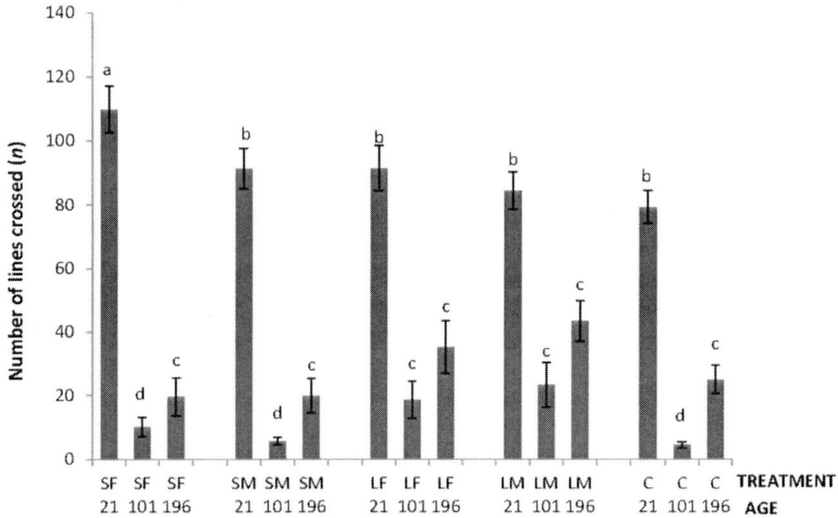

Figure 3. Number of lines crossed in a square gridline floor in a 520 s open-field test by 21, 101 and 196 days old layers from the control and four brooder treatments. Bars with no common letters differ significantly (mixed model ANOVA: $p < 0.05$). Treatments = SF: small fixed brooder; SM: small movable brooder; LF: large fixed brooder; LM: Large movable brooder; C: Control.

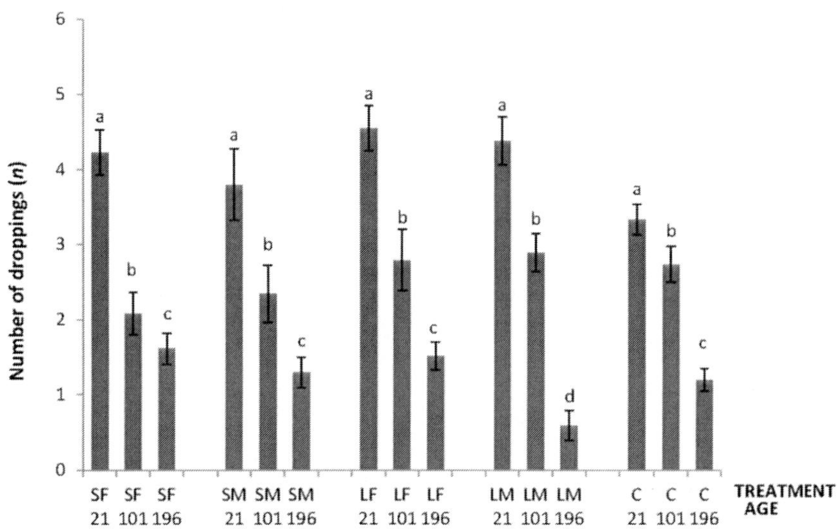

Figure 4. Number of droppings made in 600 s during an OF test by 21, 101, and 196 days old layers from the control and four brooder treatments. Bars with no common letters differ significantly (mixed model ANOVA: $p < 0.05$). Treatments = SF: small fixed brooder; SM: small movable brooder; LF: large fixed brooder; LM: Large movable brooder; C: Control.

4. DISCUSSION

Chicks reared with brooders quickly learned to use the brooders for thermoregulation and to leave the brooders for performance of active, including maintenance, behavior. The main objective of the present study was to test short- and long-term effects of rearing layer chicks with dark brooders on time budget and fearfulness, and it is clear from the results that rearing layer chicks with brooders affects their time budget and fearfulness when compared with chicks reared without brooders. We also aimed to examine whether time budget and fearfulness were affected by the type of brooder, and we found minor differences between brooder treatments although results were not as clear and consistent as when comparing birds reared either with or without brooders.

4.1. Effects of Dark Brooders on Time Budget: Control vs. Brooder Chicks

We predicted that brooder chicks would feather peck less than control chicks, as previous studies have shown that rearing layer chicks with dark brooders has reducing effects on feather pecking behavior during rear and early-lay [3,19]. Riber [28] hypothesized that the positive effects on development of feather pecking were due to the brooders having a separating effect on active and inactive chicks during the sensitive period for learning about food and appropriate dust bathing material. The sensitive period for recognition of food and dust bathing material is imbedded in the first 10 days of life [29,30,31,32]. During this period, molting is very limited, thus the main source of feathers is the plumage of pen-mates. Consequently, avoiding mixing chicks engaged in foraging or dust bathing activities with inactive chicks lying or sitting on the ground will result in fewer ground pecks misdirected to feathers of conspecifics, and as a result, development of feather pecking will be impeded. Indeed, we found that brooder chicks performed less feather pecking behavior during the first 1–4 days of age (Phase I) and that this reduction in feather pecking behavior persisted throughout Phase II (Days 9, 16, 23, 30 and 42 of age).

We also hypothesized that brooder chicks would show less fleeing than non-brooded chicks. During Phase I, observations of fleeing were too infrequent to be analyzed statistically, but in Phase II, the results on fleeing supported our hypothesis: Brooder chicks showed less fleeing than the control chicks, revealing a lower underlying fearfulness [33]. One may argue that the chicks under the brooders could not perceive the arrival of the observer and, therefore, they did not react to that with fleeing, thereby skewing the results. However, the brooders were removed prior to the last day of observation in Phase II (Day 42) without causing an increase in fleeing on Day 42. The lower underlying fearfulness in brooder chicks was also supported by the results from the fear tests (discussed in 4.3.).

Finally, we predicted that brooder chicks would perform fewer aggressive interactions than control chicks as this has been shown to be the case in brooded chicks compared to non-brooded chicks [17,18]. However, the hypothesis could not be proven due to infrequent observations of aggressive interactions in all treatments, both in Phase I and II.

Brooder chicks spent less time on comfort behavior compared to control chicks during the brooding period (Days 1, 2, 3, 4, 9, 16, 23 and 30). However, on the last observation day (Day 42) where the brooder chicks no longer had access to brooders, they showed more comfort behavior than the control chicks (except chicks from the brooder treatment SM). Comfort behavior was defined as stretching legs and/or wings in addition to wing flapping. Often, these types of behavior occur at the start of renewed activity after a period of rest [34]. We therefore suggest that, in addition to resting, some comfort behavior may have been performed under the brooders and that the lower frequencies found in the brooder groups are an artefact of the impeded observations of behavior under the brooders. As the defined behavioral elements of comfort behavior are quickly performed time-wise, the impact on the estimated time spent resting is thought to be negligible.

4.1.1. Phase I

Overall, a lower level of activity was found during Days 1–4 of age (Phase I) in brooder chicks compared to control chicks; e.g., control chicks spent almost twice as much time on feeding and locomotion compared to the brooder chicks. The brooder chicks spent more time resting, including the time spent under the brooder, where they were assumed to be resting. From what was possible to observe, chicks settled down and rested when entering under the brooder, probably promoted by the darkness and warmth. Similar observations were done in a previous study using brooders [3]. Previous studies have shown limited activity in chicks during the dark periods [35,36,37], unless the difference in light intensity between the light and dark periods is minimal [36,37]. The increased time spent resting in the brooder chicks may be explained by the chicks staying under the brooders to maintain the normal body temperature. In a study of red jungle fowl, Sherry [4] found that the brooding time of 2–5 day-old chicks was 40%–50% longer in conditions with an ambient temperature of 19 °C compared to 28 °C. Behavioral thermoregulation may also at least partly explain the increased time spent on locomotion in control chicks. Control chicks may have been searching for areas with a microclimate within the pen matching their needs at the time. Other explanations of the increased time spent on locomotion in control chicks may be more disturbances of sleep and searches for places to hide [38].

The increased time spent resting in brooder chicks was taken from time spent on locomotion, feeding and drinking. The reduced time spent drinking may be explained by the presence of the brooders allowing the chicks to thermoregulate more efficiently which would result in less respiratory evaporation [39]. A reduced need for water intake may also result from more rest and less locomotion. Higher humidity reduces the respiratory evaporation from the chicks and thus the need for water consumption [40,41]. Not only the temperature but also the humidity under the dark brooders differed from the surroundings. During Phase I, a lower humidity was found under the brooders compared to outside the brooders, but it was still higher than the humidity in the control room (data not shown). Finally, the lower water consumption could also be directly related to reduced food intake, but although brooder chicks spent less time feeding during the first 2–3 days of age, the amount consumed did not differ between treatments [42]. Malleau *et al.* [35] reared layer chicks on two different lighting programs: continuous lighting from 0400–2320 h *versus* alternating light and dark periods of 40 min duration during the same hours as the continuous light was provided in the other treatment group. They found that although the amount of time spent on active behavior decreased, and the amount of time spent resting increased during Days 1 and 4 in the alternating light/dark treatment, no difference was found on feed intake during the period 1–4 days of age. Chicks may, therefore, compensate for the reduced time spent feeding by an increased rate of feed intake.

4.1.2. Phase II

Whereas time spent on feather pecking, locomotion and fleeing remained stable during Days 9, 16, 23, 30 and 42 of age (Phase II) in brooder chicks, these behavioral categories occurred more frequently (except on Day 9) and continued to increase in frequency in control chicks during Phase II. The increase in feather pecking in control chicks is likely to have been linked to their lower level of foraging and dust bathing activities. Feather pecking has been shown to be negatively correlated with ground pecking [43,44,45,46,47,48]. Ground pecking is part of foraging behavior, and it is also an important behavioral element of dust bathing [49,50]. It seems likely that from Day 16, the higher occurrence of locomotion was associated with the increase in feather pecking which also may have instigated the increased emotional reactivity of the chicks making them more prone to show fleeing. Previous studies have found an association between anxiety and feather pecking. For example, Hansen [51] compared laying hens in different housing systems and found that hens in a tiered wire floor aviary system expressed more feather pecking and were more restless, measured as the number of transitions between activities, than hens in other types of aviary systems. The laying hens from the tiered wire floor aviary system also spent more time on locomotion, thus expressing the same positive association between locomotion and feather pecking as found in the present study. Other studies have found chicks showing high anxiety in an open-field test to have stronger tendencies to perform feather pecking [52,53]. Higher levels of locomotion, feather pecking and fleeing can be expected to cause more disturbances of rest. However, the amount of resting in control chicks only seemed to be affected on Day 16 where control chicks rested less than brooder chicks. O'Connor [54] reported that laying hens spent more time resting when sleep was disturbed by continuous high background noise. The apparent discrepancy in reaction to disturbances of rest found between the present study and the study by O'Connor may be attributed to the fact that the sleeping patterns are likely to differ between adult hens and chicks [35]. The factors causing the disturbances differ as well, which may also have an impact on the effect.

4.2. Effects of Dark Brooders on Time Budget: Differences between the Brooder Treatments

We expected to find more drinking, feeding and foraging during Phase I in the brooder treatments where the brooders were raised for short periods during the first four days, compared to the brooder treatments, where the brooders were maintained at the same height during the entire brooding period. However, no such differences were found. Thus, raising the brooders in regular intervals during the first days did not turn out to be effective in stimulating maintenance activities (*i.e.*, drinking and feeding) in the chicks. No clear trends in the differences in activity were found between the different brooder treatments, except that on Day 9, a higher percentage of chicks were resting under the brooders in the two treatments with a smaller available area under the brooder per chick. This observation may indicate that space was starting to become limited, forcing chicks closer to the borders of the brooders where the temperature was not as high as in the middle, *i.e.*, chicks had to stay under the brooders longer to maintain normal body temperature. However, this speculation was not supported by the observations on percentage of chicks resting outside the brooders, as we did not find any clear differences between brooder treatments on this behavioral category that could indicate competition over space under the brooders.

4.3. Effect of Brooders on Fearfulness: Control vs. Brooder Birds

Our hypothesis that birds from the brooder treatments would be less fearful was supported as the results from the TI test clearly showed that birds from the brooder treatments were less fearful, irrespective of which parameter was measured and the age of the birds when tested (with the exception of minor differences in the number of inductions). Also, the results from the NO and OF tests supported the hypothesis, as the overall outcome was that control birds were either the most fearful or among the most fearful birds during the fear tests conducted when the birds were either pullets or adults. The reducing effect on fearfulness of brooders, when the birds were chicks, is similar to what has been found in chicks brooded by mother hens [13]. However, the effect of broody hens on fearfulness has been found to disappear with age when fear tests outside the home pens have been conducted [14,17]. In contrast, we found an increasing discrepancy with age between brooder and control birds in fearfulness with control birds being most fearful. Thus, our hypothesis that the reducing effect of dark brooders would be long-lasting was supported as well. We speculate whether the long-lasting effects on fearfulness of brooders, unlike broody hens, may be associated with the exclusion of emotionality of the brooding element. It has been shown that the fearfulness of the mother hen herself affects the fearfulness of her chicks complicating investigations of fearfulness in brooded chicks [55,56,57]. This factor is omitted when using dark brooders during the brooding phase. Furthermore, most studies show that being brooded in the brooding period by a mother hen, in contrast to dark brooders, does not have a reducing effect on feather pecking activity in chicks, pullets, or young adult hens [15,58]. In contrast, we found a reducing effect of brooders on feather pecking behavior during the observation period, *i.e.*, the first six weeks of age. Feather pecking is known to be positively associated with fearfulness in layers [47,59,60]. Older studies suggest that the pecked birds become more fearful due to the discomfort and pain experienced when being pecked [61,62]. A more recent study has demonstrated that fearful birds have a higher risk of developing severe feather pecking later in life [53]. We therefore suggest that the increase in fearfulness in control birds compared to brooder birds, occurring between the first and second fear tests (chicks and pullets, respectively), may at least partly be linked to the higher level of feather pecking activity observed in the control chicks.

4.4. Effect of Brooders on Fearfulness: Differences between Brooder Treatments

Differences in fearfulness between birds from the four brooder treatments were also found although they were minor and not as clear as between brooder and control birds. However, these differences are more difficult to explain. Birds from SM (Small and Movable) were expressing more fearfulness than birds from the other brooder treatments in minimum one parameter of at least one of the tests at each age. If the increased fearfulness of birds from SM is to be explained by the treatment, it appears that only the combination of brooder size and management of height could be attributed as causing the effect. Birds from the other brooder treatment with the smaller brooder size were among the least fearful in the majority of the tests conducted. Also, birds from the other brooder treatment, where the brooder was raised on a regular schedule during the first four days, were inconsistent in their fear responses with them being most fearful in the NO test conducted when they were chicks and least fearful in the open-field test when they were pullets and adults. Thus, neither the size of the brooder nor whether it was raised regularly seems to result in differences in fearfulness, but apparently the combination of a small brooder raised at a regular schedule during the first four days does.

5. CONCLUSIONS

In conclusion, the present study demonstrates that providing layer chicks with dark brooders during the brooding period results in changes of their time budget in terms of calmer chicks expressing less abnormal behavior during the first six weeks of life and in long-term reduction of the level of fearfulness. The knowledge of the effects of dark brooders gained in the present and previous studies [2,3,8,19,42] provides firm evidence that dark brooders can be a successful method of reducing or preventing some of the major welfare problems in layers, e.g., feather pecking, cannibalism and fear.

ACKNOWLEDGMENTS

We would like to thank MHJ Agroteknik A/S for the development of the water-heated brooders and the technical support during the installation. We also thank Laura Alicia Gonzalez for her statistical advice. The work was financed by the Green Development and Demonstration Programme (GUDP), established under the Ministry of Food, Agriculture, and Fisheries, Denmark. Consejo Nacional de Investigaciones Científicas y Técnicas (CONICET), Argentina, funded Diego A. Guzman's postdoctoral stay in Denmark.

REFERENCES

1. Weekstein, D.R.; Zolman, J.F. Ontogeny of heat production in chicks. *FASEB J.* **1969**, *28*, 1023–1027.
2. Gilani, A.M.; Knowles, T.G.; Nicol, C.J. The effect of dark brooders on feather pecking on commercial farms. *Appl. Anim. Behav. Sci.* **2012**, *142*, 42–50.
3. Jensen, A.B.; Palme, R.; Forkman, B. Effect of brooders on feather pecking and cannibalism in domestic fowl (*gallus gallus domesticus*). *Appl. Anim. Behav. Sci.* **2006**, *99*, 287–300.
4. Sherry, D.F. Parental care and the development of thermoregulation in red junglefowl. *Behaviour* **1981**, *76*, 250–279.
5. Cherfas, J.J. Simultaneous color discrimination in chicks is improved by brief exposure to light. *Anim. Behav.* **1978**, *26*, 1–5.
6. Zolman, J.F.; Pursley, D.G.; Hall, J.A.; Sahley, C.L. Form preferences in successive discrimination-learning of young chicks. *J. Comp. Physiol. Psychol.* **1975**, *89*, 1180–1191.
7. Wasserman, E.A.; Hunter, N.B.; Gutowski, K.A.; Bader, S.A. Autoshaping chicks with heat reinforcement—Role of stimulus-reinforcer and response-reinforcer relations. *J. Exp. Psychol. Anim. Behav. Process.* **1975**, *1*, 158–169.
8. Riber, A.B.; Nielsen, B.L.; Ritz, C.; Forkman, B. Diurnal activity cycles and synchrony in layer hen chicks (*gallus gallus domesticus*). *Appl. Anim. Behav. Sci.* **2007**, *108*, 276–287.
9. Sherry, D.F. Parental food-calling and the role of the young in burmese red junglefowl (*gallus gallus spadiceus*). *Anim. Behav.* **1977**, *25*, 594–601.
10. Stokes, A.W. Parental and courtship feeding in red jungle fowl. *Auk* **1971**, *88*, 21–29.
11. Wauters, A.M.; Richard-Yris, M.A.; Talec, N. Maternal influences on feeding and general activity in domestic chicks. *Ethology* **2002**, *108*, 529–540.
12. Riber, A.B.; Wichman, A.; Braastad, B.O.; Forkman, B. Effects of broody hens on perch use, ground pecking, feather pecking and cannibalism in domestic fowl (*gallus gallus domesticus*). *Appl. Anim. Behav. Sci.* **2007**, *106*, 39–51.
13. Shimmura, T.; Kamimura, E.; Azuma, T.; Kansaku, N.; Uetake, K.; Tanaka, T. Effect of broody hens on behaviour of chicks. *Appl. Anim. Behav. Sci.* **2010**, *126*, 125–133.
14. Perre, Y.; Wauters, A.M.; Richard-Yris, M.A. Influence of mothering on emotional and social reactivity of domestic pullets. *Appl. Anim. Behav. Sci.* **2002**, *75*, 133–146.
15. Roden, C.; Wechsler, B. A comparison of the behaviour of domestic chicks reared with or without a hen in enriched pens. *Appl. Anim. Behav. Sci.* **1998**, *55*, 317–326.
16. Rodenburg, T.B.; Uitdehaag, K.A.; Ellen, E.D.; Komen, J. The effects of selection on low mortality and brooding by a mother hen on open-field response, feather pecking and cannibalism in laying hens. *Anim. Welf.* **2009**, *18*, 427–432.

17. Shimmura, T.; Maruyama, Y.; Fujino, S.; Kamimura, E.; Uetake, K.; Tanaka, T. Persistent effect of broody hens on behaviour of chickens. *Anim. Sci. J.* **2015**, *86*, 214–220.
18. Fält, B. Differences in agressiveness between brooded and non-brooded domestic chicks. *Appl. Anim. Ethol.* **1978**, *4*, 211–221.
19. Johnsen, P.F.; Kristensen, H.H. Effect of Brooder Quality on the Early Development of Feather Pecking Behaviour in Domestic Chicks. In Proceedings of the 6th European Symposium on Poultry Welfare, Zollikofen, Switzerland, 2001; Oester, H., Wyss, C., Eds.; Swiss Branch of the World's Poultry Science Association (WPSA): Zollikofen, Switzerland, 2001; pp. 209–212.
20. Riber, A.B. *Effect of Dark Brooders on Fearfulness in Layers*; Aarhus University: Aarhus, Denmark, 2015.
21. Altmann, J. Observational study of behavior: Sampling methods. *Behaviour* **1974**, *49*, 227–267.
22. Daigle, C.L.; Siegford, J.M. When continuous observations just won't do: Developing accurate and efficient sampling strategies for the laying hen. *Behav. Process.* **2014**, *103*, 58–66.
23. Hocking, P.M.; Channing, C.E.; Waddington, D.; Jones, R.B. Age-related changes in fear, sociality and pecking behaviours in two strains of laying hen. *Br. Poult. Sci.* **2001**, *42*, 414–423.
24. Jones, R.B. Fear and adaptability in poultry: Insights, implications and imperatives. *Worlds Poult. Sci. J.* **1996**, *52*, 131–174.
25. Forkman, B.; Boissy, A.; Meunier-Salaün, M.C.; Canali, E.; Jones, R.B. A critical review of fear tests used on cattle, pigs, sheep, poultry and horses. *Physiol. Behav.* **2007**, *92*, 340–374.
26. Di Rienzo, J.A.; Casanoves, F.; Balzarini, M.G.; Gonzalez, L.; Tablada, M.; Robledo, C.W. *Infostat, InfoStat versión 2015*; InfoStat Group, National University of Córdoba: Córdoba, Argentina, 2015.
27. Di Rienzo, J.A.; Guzman, A.W.; Casanoves, F. A multiple-comparisons method based on the distribution of the root node distance of a binary tree. *JABES* **2002**, *7*, 129–142.
28. Riber, A.B. Ontogeny of Behaviour in Domestic Fowl—With Emphasis on Feather Pecking. Ph.D. Thesis, University of Copenhagen, København, Denmark, 2007; p. 55.
29. Brown, L.T. A critical period in learning of motionless stimulus properties in chicks. *Anim. Behav.* **1964**, *12*, 353–361.
30. Hess, E.H. Imprinting in birds—Research has borne out concept of imprinting as type of learning different from association learning. *Science* **1964**, *146*, 1128–1139.
31. Sanotra, G.S.; Vestergaard, K.S.; Agger, J.F.; Lawson, L.G. The relative preferences for feathers, straw, wood-shavings and sand for dustbathing, pecking and scratching in domestic chicks. *Appl. Anim. Behav. Sci.* **1995**, *43*, 263–277.
32. Vestergaard, K.S.; Baranyiova, E. Pecking and scratching in the development of dust perception in young chicks. *Acta Vet. Brno* **1996**, *65*, 133–142.
33. Jones, R.B.; Satterlee, D.G.; Ryder, F.H. Fear of humans in japanese quail selected for low or high adrenocortical response. *Physiol. Behav.* **1994**, *56*, 379–383.
34. Kruijt, J.P. Ontogeny of social behaviour in burmase red junglefowl (*gallus gallus spadiceus*) bonnaterre. *Behaviour***1964**, *Suppl XII*, 1–201.
35. Malleau, A.E.; Duncan, I.J.H.; Widowski, T.M.; Atkinson, J.L. The importance of rest in young domestic fowl. *Appl. Anim. Behav. Sci.* **2007**, *106*, 52–69.
36. Alvino, G.M.; Archer, G.S.; Mench, J.A. Behavioural time budgets of broiler chickens reared in varying light intensities. *Appl. Anim. Behav. Sci.* **2009**, *118*, 54–61.
37. Blatchford, R.A.; Klasing, K.C.; Shivaprasad, H.L.; Wakenell, P.S.; Archer, G.S.; Mench, J.A. The effect of light intensity on the behavior, eye and leg health, and immune function of broiler chickens. *Poult. Sci.* **2009**, *88*, 20–28.

38. Cornetto, T.; Estevez, I. Behavior of the domestic fowl in the presence of vertical panels. *Poult. Sci.* **2001**, *80*, 1455–1462.
39. Yahav, S. Domestic fowl - strategies to confront environmental conditions. *Avian Poult. Biol. Rev.* **2000**, *11*, 81–95.
40. Lin, H.; Zhang, H.F.; Jiao, H.C.; Zhao, T.; Sui, S.J.; Gu, X.H.; Zhang, Z.Y.; Buyse, J.; Decuypere, E. Thermoregulation responses of broiler chickens to humidity at different ambient temperatures. I. One week of age. *Poult. Sci.* **2005**, *84*, 1166–1172.
41. Nichelmann, M.; Tzschentke, B.; Burmeister, A. Evaporative heat-loss at high relative air humidity in poultry. *Arch. Geflugelkd.* **1991**, *55*, 111–115.
42. Riber, A.B.; Guzman, D.A. Effects of rearing with dark brooders on welfare and production in layers. Manuscript in preparation.
43. Blokhuis, H.J. Feather-pecking in poultry: Its relation with ground-pecking. *Appl. Anim. Behav. Sci.* **1986**, *16*, 63–67.
44. Blokhuis, H.J. The effect of a sudden change in floor type on pecking behaviour in chicks. *Appl. Anim. Behav. Sci.***1989**, *22*, 65–73.
45. Huber-Eicher, B.; Wechsler, B. Feather pecking in domestic chicks: Its relation to dustbathing and foraging. *Anim. Behav.* **1997**, *54*, 757–768.
46. Huber-Eicher, B.; Wechsler, B. The effect of quality and availability of foraging materials on feather pecking in laying hen chicks. *Anim. Behav.* **1998**, *55*, 861–873.
47. Vestergaard, K.S.; Kruijt, J.P.; Hogan, J.A. Feather pecking and chronic fear in groups of red junglefowl: Their relation to dustbathing, rearing environment and social status. *Anim. Behav.* **1993**, *45*, 1127–1140.
48. Vestergaard, K.S.; Lisborg, L. A model of feather pecking development which relates to dustbathing in the fowl. *Behaviour* **1993**, *126*, 89–105.
49. Vestergaard, K. Dust-bathing in the domestic fowl—Diurnal rhytm and dust deprivation. *Appl. Anim. Ethol.* **1982**, *8*, 487–495.
50. Schütz, K.E.; Jensen, P. Effects of resource allocation on behavioural strategies: A comparison of red junglefowl (*gallus gallus*) and two domesticated breeds of poultry. *Ethology* **2001**, *107*, 753–765.
51. Hansen, I. Behavioral expression of laying hens in aviaries and cages—Frequencies, time budgets and facility utilization. *Br. Poult. Sci.* **1994**, *35*, 491–508.
52. Jones, R.B.; Blokhuis, H.J.; Beuving, G. Open-field and tonic immobility responses in domestic chicks of two genetic lines differing in their propensity to feather peck. *Br. Poult. Sci.* **1995**, *36*, 525–530.
53. Rodenburg, T.B.; Buitenhuis, A.J.; Ask, B.; Uitdehaag, K.A.; Koene, P.; van der Poel, J.J.; van Arendonk, J.A.M.; Bovenhuis, H. Genetic and phenotypic correlations between feather pecking and open-field response in laying hens at two different ages. *Behav. Genet.* **2004**, *34*, 407–415.
54. O'Connor, E.A.; Parker, M.O.; Davey, E.L.; Grist, H.; Owen, R.C.; Szladovits, B.; Demmers, T.G.M.; Wathes, C.M.; Abeyesinghe, S.M. Effect of low light and high noise on behavioural activity, physiological indicators of stress and production in laying hens. *Br. Poult. Sci.* **2011**, *52*, 666–674.
55. Bertin, A.; Richard-Yris, M.-A. Mothers'fear of human affects the emotional reactivity of young in domestic japanese quail. *Appl. Anim. Behav. Sci.* **2004**, *89*, 215–231.
56. Houdelier, C.; Lumineau, S.; Bertin, A.; Guibert, F.; de Margerie, E.; Augery, M.; Richard-Yris, M.A. Development of fearfulness in birds: Genetic factors modulate non-genetic maternal influences. *PLoS ONE* **2011**.
57. Pittet, F.; Coignard, M.; Houdelier, C.; Richard-Yris, M.A.; Lumineau, S. Effects of maternal experience on fearfulness and maternal behaviour in a precocial bird. *Anim. Behav.* **2013**, *85*, 797–805.

58. Angevaare, M.J.; Prins, S.; van der Staay, F.J.; Nordquist, R.E. The effect of maternal care and infrared beak trimming on development, performance and behavior of silver nick hens. *Appl. Anim. Behav. Sci.* **2012**, *140*, 70–84.
59. Rodenburg, T.B.; de Haas, E.N.; Nielsen, B.L.; Buitenhuis, A.J. Fearfulness and feather damage in laying hens divergently selected for high and low feather pecking. *Appl. Anim. Behav. Sci.* **2010**, *128*, 91–96.
60. Uitdehaag, K.A.; Komen, H.; Rodenburg, T.B.; Kemp, B.; von Arendonk, J. The novel object test as predictor of feather damage in cage-housed Rhode Island Red and White Leghorn laying hens. *Appl. Anim. Behav. Sci.* **2008**, *109*, 292–305.
61. Hughes, B.O.; Duncan, I.J.H. Influence of strain and environmental factors upon feather pecking and cannibalism in fowls. *Br. Poult. Sci.* **1972**, *13*, 525–547.
62. Ouart, M.D.; Adams, A.W. Effects of cage design and bird density on layers. 1: Productivity, feathering, and nervousness. *Poult. Sci.* **1982**, *61*, 1606–1613.

CHAPTER 16

Distribution, Numbers, and Diversity of ESBL-Producing *E. coli*in the Poultry Farm Environment

Hetty Blaak[1*] ,Angela H. A. M. van Hoek[1],Raditijo A. Hamidjaja[1],Rozemarijn Q. J. van der Plaats[1],Lianne Kerkhof-de Heer[1],Ana Maria de Roda Husman[1,2],Franciska M. Schets[1]

[1] *Centre for Zoonoses and Environmental Microbiology, National Institute for Public Health and the Environment (RIVM), Bilthoven, The Netherlands,*

[2] *Institute for Risk Assessment Sciences, Utrecht University, Utrecht, The Netherlands*

ABSTRACT

This study aimed to discern the contribution of poultry farms to the contamination of the environment with ESBL-producing *Escherichia coli* and therewith, potentially to the spread of these bacteria to humans and other animals. ESBL-producing *E. coli* were detected at all investigated laying hen farms (n = 5) and broiler farms (n = 3) in 65% (46/71) and 81% (57/70) of poultry faeces samples, respectively. They were detected in rinse water and run-off water (21/26; 81%), other farm animals (11/14; 79%), dust (21/35; 60%), surface water adjacent to farms (20/35; 57%), soil (48/87; 55%), on flies (11/73; 15%), and in barn air (2/33; 6%). The highest prevalence and concentrations in the outdoor environment were observed in soil of free-range areas at laying hen farms (100% of samples positive, geometric mean concentration 2.4×10^4 cfu/kg), and surface waters adjacent to broiler farms during, or shortly after, cleaning between production rounds (91% of samples positive, geometric mean concentration 1.9×10^2cfu/l). The diversity of ESBL-producing *E. coli* variants with respect to sequence type, phylogenetic group, ESBL-genotype and antibiotic resistance profile was high, especially on broiler farms where on average 16 different variants were detected, and the average Simpson's Indices of diversity (SID; 1–D) were 0.93 and 0.94 among flock and environmental isolates respectively. At laying hen farms on average nine variants were detected, with SIDs of 0.63 (flock isolates) and 0.77 (environmental isolates). Sixty percent of environmental isolates were identical to flock isolates at the same farm. The highest proportions of 'flock variants' were observed in dust (94%), run-off gullies (82%), and barn air (67%), followed by surface water (57%), soil (56%), flies (50%) and other farm animals (35%).The introduction of ESBL-producing *E. coli* from poultry farms to the environment may pose a health risk if these bacteria reach places where people may become exposed.

INTRODUCTION

The use of antibiotics in human and animal health care has resulted in the widespread prevalence of antibiotic resistant (ABR) bacteria not only in humans and animals, but also in the environment, e.g. in surface water and soil [1–4]. As a consequence, the probability of getting exposed to ABR bacteria outside a health care setting has increased.

A specific type of antibiotic resistance that currently represents a major public health concern is the 3rd generation cephalosporin resistance induced by extended spectrum beta-lactamase (ESBL)-production [5]. ESBL-producing bacteria are resistant to almost all beta-lactam antibiotics, and often to other classes of antibiotics as well. This results in difficult to treat infections, and additionally compels the use of so-called 'last-resort antibiotics', e.g. carbapenems, resulting in increased resistance to these types of antibiotics as well [6]. Initially, ESBL-production was mainly observed in hospital infections caused by *Klebsiella pneumonia*, but today it is also frequently associated with community-acquired infections, mostly urinary tract infections caused by *E. coli* [7,8]. In the Netherlands, 1% of urinary tract infections among selected primary care centers in 2009 were associated with ESBL-producing *E. coli* [9]; the prevalence of ESBL-producing *E. coli* among patients with blood stream infections ranges between 4%- 6% [10]. In community patients and healthy individuals, a prevalence of ESBL-producing Enterobacteriaceae of 5%- 10% has been described [11,12] which, in the study on community patients where species were identified, were shown to be primarily *E. coli* [12]. The future threat of increased occurrence of untreatable infections necessitates mitigation of dissemination of ESBL-producing bacteria, and hence the identification of their possible dissemination routes. Spread of ESBL-producing *E. coli* in the community may be facilitated by direct contact with human carriers, but alternatively, may also be livestock-related. In the Netherlands, ESBL-producing *E. coli* are highly prevalent in poultry: in recent years (2009–2011), ESBL- (and/or AmpC-) producing *E. coli* were detected on 100% of Dutch broiler farms studied [13,14]. The high prevalence of ESBL-producing *E. coli* on Dutch retail chicken meat, and overlap between ESBL-genotypes from chicken meat and clinical *E. coli* isolates, has led to the suggestion of chicken meat as a source of ESBL-producing *E. coli* [15,16], although a more recent study demonstrated no evidence for recent events of clonal transmission from poultry to humans using next generation sequencing [17]. The presence of ESBL-producing *E. coli* in the environment, including Dutch surface water [18–22], suggests that dissemination through the environment should also be considered.

ABR intestinal bacteria end up in the environment with animal and human faeces. A major human contamination source is wastewater, either discharged onto surface water after treatment by wastewater treatment plants or discharged untreated through sewage overflows during heavy rain fall [13,19,20]. Examples of animal environmental contamination sources are animal manure used for field application and livestock farms [23–28]. At livestock farms, bacteria may enter the natural environment (i.e. ambient air, soil, surface water) directly with droppings of pasture animals and free-range animals, or indirectly from barns, for instance through air and dust, with hands or feet of farm workers, or with rinse water during cleaning practices. Once in the outside environment of farms, the bacteria may spread further away from farms with motile environmental compartments such as air and surface water, where people may get exposed to them, for instance through inhalation of air during outdoor activities, recreation in down-stream located surface water, or when downstream-located water is used for irrigation of crops [18,29,30]. An additional route of dissemination of ESBL-producing *E.coli* from farms may be with pest animals, e.g. flies, which have been recognized as transmitters of infectious diseases for some time [31]. Flies may move from farms where they were bred in, and have fed on, faeces and carcasses [32], to next feed on food meant for human consumption, including prepared food. Indeed, carriage of ESBL-producing *E. coli* by flies on livestock farms has been described previously [23,33,34].

The current study was aimed at determining the extent of contamination of the natural (or outside) environment at Dutch poultry farms with ESBL-producing *E. coli*. For this purpose, ESBL-producing *E. coli* were quantified in ambient air, soil, surface water and flies at laying hen and broiler farms. For comparison, the bacteria were simultaneously quantified in the suspected sources of contamination of the natural environment, e.g. poultry faeces, wastewater, barn dust and barn air. Additionally, isolates from different matrices were characterised with respect to ESBL-genotype, antibiotic resistance profile, phylogenetic group, and sequence type, to determine the proportion of environmental isolates identical to those observed in poultry. The current study strengthens the knowledge base on the contribution of animal husbandry to the dissemination of ESBL-producing bacteria to the environment and therewith, to the potential spread of ABR zoonoses that cause hard to treat human infections.

MATERIAL AND METHODS

Ethics Statement

The field samples were obtained at private properties, permission was granted by the collaborating farmers.

Poultry Farms

During 2011 and 2012 three broiler (Br1 to Br3) and five laying hen (Lh1 to Lh5) farms were visited multiple times (Fig 1). Each first visit had an orienting character, during which farms and their immediate surroundings were mapped, suitable sites for sampling were identified, and poultry faeces was sampled to establish the presence or absence of ESBL-producing *E. coli*. After this, laying hen farms were visited once more during the same production round, and broiler farms were visited during a subsequent production round: once while broilers were present, and once just after these broilers were removed and stables had just been cleaned or were being cleaned. Sampling strategies were different between farm types due to the high turnover of broiler flocks (every six to seven weeks) and the low turnover of laying hens (approximately once in every one and a half year). Farm Br1 was sampled during three different production rounds, two in 2011 and one in 2012, i.e. in total on six occasions after the first orientating visit. All broiler farms were conventional farms, with capacities of 38,000 (Br3), 87,000 (Br1) and 150,000 (Br2) broilers. Two of the laying hen farms were conventional farms with capacities of 78,000 (Lh2) and 80,000 (Lh4) chickens, and the three other ones were free range farms with capacities of 30,000 (Lh1, Lh5) and 43,000 (Lh3) chickens. During the main sampling occasions (i.e. after the orienting visit), the age of broiler flocks was four to five weeks, the age of laying hen flocks was on average 59 weeks (range 26–73 weeks).

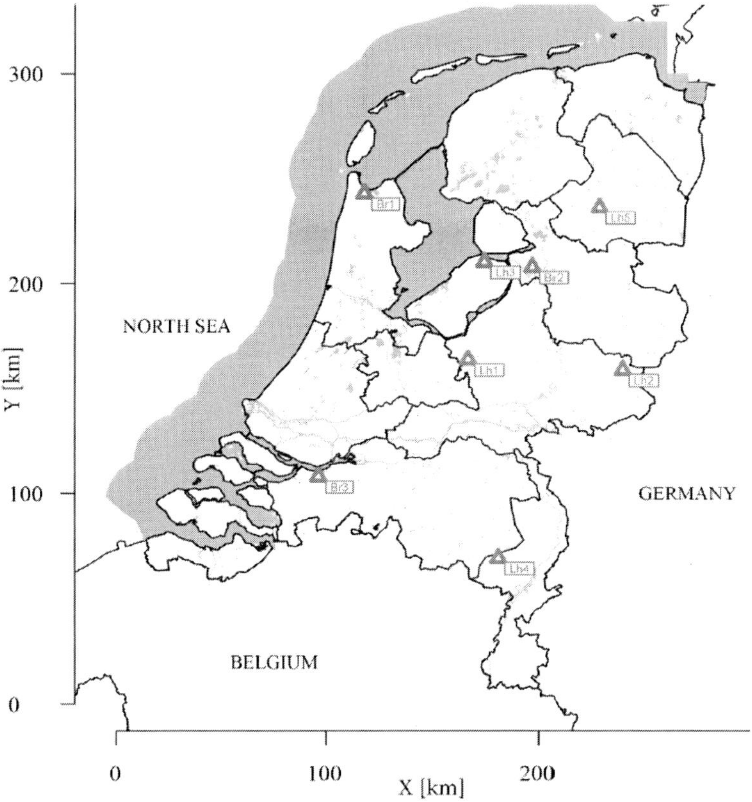

Figure 1. Map of the Netherlands showing locations of the sampled poultry farms.

Description of Sampled Matrices

Overall, 471 samples were taken, 276 from broiler farms and 195 from laying hen farms: 290 (62%) from environmental compartments (surface water, soil, dust, air, flies), 141 (30%) from poultry faeces, 26 (5.5%) from wastewater, 14 (3.0%) from faeces from other animals kept at the farms (S1 Table). Fresh (i.e. still soft and warm) and semi-fresh (i.e. generally still identifiable as individual droppings, but no longer soft and warm) poultry faeces was sampled from barns and free-range areas. At laying hen farms, poultry faeces was additionally sampled from manure belts and at different stages during air-drying. At broiler farms, when present, faeces was additionally sampled from dung-heaps. The reason for additional sampling of faeces from manure belts, during various stages of air drying and from dung-heaps was that it was anticipated that bacterial concentrations might change after prolonged time outside the body. Soil was sampled in free-range areas where applicable, and from various other sites at the premises, i.e. in the vicinity (1–5 m) of barns, manure storage sheds, wastewater storage basins, or manure belts, as well as at sites not visibly under direct influence of such faecal contamination sources. At three of the farms, soil samples were taken from dried-up ditches. When present, surface water was sampled from ditches bordering on, or within 50 m distance of farm premises, as well as from more remote (50 m– 1 km) water bodies or sampling sites. Wastewater was sampled at broiler farms from drains, collection pits, storage basins and run-off gullies. Wastewater in drains, pits and basins consisted of barn rinse water (either or not diluted with rain water), and was therefore considered to reflect poultry faeces contents. This in contrast to water from run-off gullies, which consisted of run-off from farm premises and might therefore contain input from other, undefined contamination sources. Sometimes, the water level in run-off gullies was low, and sediment was sampled instead. Only one of the laying hen farms had a wastewater pit filled with wastewater at the time of sampling, and this was sampled as well. Other farm animals included hobby laying hens at broiler farms (Br1, Br2), a dog (Br1), cattle (Br2, Lh2), horses (Lh3), and swallows (Lh2). Flies were caught inside barns, from other indoor environments (manure storage sheds, egg-sorting areas, homes, break rooms or changing rooms, stables of other animals) and outdoors. Caught flies included *Musca domesticus* (common house fly), *Stomoxys calcitrans* (stable fly), *Fannia canicularis* (lesser housefly), *Calliphoridae* (blowflies), Hydrotaea, *Muscina stabulans* (false stable fly), *Sarcophagidae* (flesh flies) and *Tachinidae* (tachina flies). Dust was sampled from surfaces inside barns, from barn ventilation fans, from a manure storage shed, and, at Br1, outside one of the barns from a fence positioned within 5 m range of a ventilation fan. Most air samples were collected inside barns. At Br1 outdoor air was also sampled (within 5 m of ventilator fans) as well as air from a break room adjoining one of the barns. All samples were transported to the laboratory and stored at 5±3°C (except for flies which were stored at room temperature), and analysed within 24 hours after sampling. The procedures of sample collection and subsequent preparation is described in S1 Materials and Methods).

Isolation and Enumeration of *E. coli* and ESBL-Producing *E. coli*

Multiple dilutions of each sample were streaked on ChromID ESBL agar (Biomerieux, Boxtel, the Netherlands) for the isolation of ESBL-producing *E. coli* and on tryptone bile x-glucuronide agar (TBX) for the isolation of *E. coli*, in accordance with ISO 16649–2 [35] (for details on sample preparation, see S1 Material and Methods). For water, samples were filtered through 0.45 μm pore size membrane filters, which were next placed on these media. Cultures were incubated for 4 to 5 hours at 36±1°C, followed by 21±3 h at 44±0.5°C[35]. For the isolation of ESBL-producing *E. coli*, environmental samples (with the exception of surface water) were additionally pre-enriched in buffered peptone water supplemented with 1 μg/ml cefotaxime (CTX), using the same incubation conditions. Pre-enriched suspensions were streaked onto ChromID ESBL agar and incubated for another 21±3 h at 44±0.5°C. Bacterial concentrations were calculated from the colony counts obtained from the direct cultures using Mathematica software 9.0.1 (WolframResearch, Champaign, IL, USA). The pre-enriched cultures were assessed for presence or absence of suspected colonies, which was used as supplementary data in case of negative direct cultures. Because of the skewed distribution of concentrations of ESBL-producing *E. coli* within matrices and the relatively high numbers of negative samples for most environmental matrices, group mean values were expressed as geometric means (i.e. based on the log-transformed concentrations). Samples yielding ESBL-producing *E. coli* after enrichment but not in direct culture were included for estimation of the geometric means, by setting the concentration in these samples to $\frac{1}{x}$ cfu/g, in which *x* is the amount of matrix that was cultured.

Phenotypic Confirmation of ESBL-Production

Suspected ESBL-producing *E. coli* isolates (n = 1,140) were confirmed to be indole-positive using BBL Dry Slide (BD), and subsequently tested for ESBL-production by disk diffusion following CLSI guidelines [36], using Sensi-Discs (BD, Breda, the Netherlands). Zone diameters were determined for cefotaxime (30μg) ± clavulanic acid (10μg), ceftazidime (30μg) ± clavulanic acid (10μg), and cefoxitin (30 μg). ESBL-producing isolates were defined as strains resistant to cefotaxime (zone diameter ≤ 22 mm) and/or ceftazidime (zone diameter ≤ 17 mm), and an increase in zone diameter of ≥ 5 mm with the disks containing clavulanic acid [36]. ESBL-producing *E. coli* concentrations were calculated from the numbers of β-glucuronidase-positive colonies, and the fraction of isolates confirmed to be indole-positive and ESBL-producing. Some isolates (n = 22) did not appear to be resistant to either cefotaxime or ceftazidime, and then ESBL-production was confirmed using an alternative AmpC and ESBL detection test, which is based on cefpodoxime (Mastgroup Ltd., Bootle, UK). Third-generation cephalosporin-resistant isolates with no significant inhibitory effect of clavulanic acid (as defined by CLSI) were defined as non-ESBL-producers and excluded from further analyses (n = 32).

Phylogenetic Group Analysis

Of each isolate, material from one single colony was suspended in Tris EDTA buffer (pH 8.0, Sigma-Aldrich, Zwijndrecht, the Netherlands), followed by incubation at 99°C for 5 min. The resulting cell lysates were stored at -20°C. PCRs were targeted to *chuA* and *yjaA* genes, and TspE4.C2 DNA fragment using primers and PCR conditions as described by Clermont at al. [37]. Strains were sub-grouped according to Escobar-Páramo et al. [38]: subgroup A_0, chuA-, yjaA-, TspE4.C2-; subgroup A_1, chuA-, yjaA+, TspE4.C2-; group B1, chuA-, yjaA-, TspE4.C2+; subgroup $B2_2$, chuA+, yjaA+, TspE4.C2-; subgroup $B2_3$, chuA+, yjaA+, TspE4.C2+; subgroup D_1, chuA+, yjaA-, TspE4.C2-; subgroup D_2, chuA+, yjaA-, TspE4.C2+. To confirm the *E. coli* identity of A_0 isolates, a PCR was performed targeted to the β-glucuronidase gene *uidA* [39]. UidA-negative isolates (n = 1) were excluded from further analysis.

ESBL-Genotyping

The presence of CTX-M-group 1, CTX-M-group 2, CTX-M-group 9, and OXA-, SHV- and TEM-genes was determined by multiplex PCR using primers and PCR conditions described by Dallenne et al. [40]. The ESBL-genotypes were based on partial gene sequences. For this purpose, PCR-products of the expected size were treated with ExoSAP-IT (GE Healthcare, Hoevelaken, the Netherlands) and sequenced using the same primers used to generate the PCR-products. To confirm the results based on partial gene sequences, 54% of the major genotypes CTX-M-1, SHV-12 and TEM-52, was sequenced full-length. For this purpose, DNA of the selected isolates was subjected to PCR analysis using primers and conditions described by Dierikx et al. [41]. Obtained sequences were compared with ESBL-gene sequences in the GenBank database and on the Lahey website (www.lahey.org/Studies).

Multilocus Sequence Typing (MLST)

Seven house-keeping genes, *adk, fumC, gyrB, icd, mdh, purA* and *recA*, were amplified and sequenced as described by Wirth et al.[42]. Primer sequences were obtained from the *E. coli* MLST database website http://mlst.warwick.ac.uk/mlst/mlst/dbs/Ecoli and used at 0.2 μM per reaction. Amplification conditions were: 15 min 95°C, followed by 35 cycles of 30s 95°C, 30s 65°C (*adk, fumC, gyrB, icd, purA, recA*) or 30s 60°C (*mdh*), 45s 72°C, and a final elongation step of 10 min 72°C. PCR-products were analysed on agarose gel and PCR-products of the expected size were treated with ExoSAP-IT (GE Healthcare, Hoevelaken, the Netherlands) followed by sequencing using the same primers used to generate PCR-products.

Sequence types were identified against the online database available at the *E. coli* MLST website (http://mlst.warwick.ac.uk/mlst/mlst/dbs/Ecoli) using Bionumerics software (version 7.1; Applied Maths NV, Sint Martens-Latem, Belgium). The same software was used for the construction of minimal spanning trees based on the concatenated sequences of the seven alleles.

Antibiotic Resistance Profiling

Isolates were screened for antibiotic susceptibility to a panel of 14 antibiotics of human and veterinary clinical relevance, using micro broth dilution using the Sensititre SensiTouch system (MCS Diagnostics, Swalmen, the Netherlands), according to the manufacturers' instructions and CLSI guidelines [36]. Included were one or two antibiotic representatives from seven classes of antibiotics: ampicillin and co-amoxiclav (penicillins), cefotaxime and ceftazidime (3rd generation cephalosporins), tetracycline (tetracyclines), ciprofloxacin and nalidixic acid ((fluoro)quinolones), gentamycin and streptomycin (aminoglycosides), sulfamethoxazole and trimethoprim (folate pathway inhibitors), chloramphenicol (phenicols), and imipenem and meropenem (carbapenems). Resistance was defined as having a minimal inhibitory concentration (MIC) above the ecological cut-off (ECOFF) value available at the EUCAST website [43]. For amoxicillin/ clavulanic acid a break-point of ≥16/8 was used [44], because an ECOFF value was not available at the EUCAST website at the time of data analysis. Multi-drug resistance was defined as resistance to 3 or more different classes of antibiotics [45].

Selection of Isolates

Overall, 1,107 confirmed ESBL-producing *E. coli* isolates were obtained from 240 positive samples. Per sample, a maximum of two isolates with similar ESBL-phenotype based on inhibition zones and the effect of clavulanic acid for cefotaxime, ceftazidime and, if applicable, cefpodoxime, were included for further analyses. The resulting 686 isolates were analysed with respect to phylogenetic group. Subsequently, maximally two isolates with similar ESBL-phenotype and identical phylogenetic group per sample (or set of samples derived from the exact same site at the same time), were characterized at the level of ESBL-gene family using PCR (n = 488, from 230 samples). These data were used to identify 'sets' of isolates, i.e. isolates obtained from different matrices at the same farm, with the same identity with respect to phylogenetic group and ESBL-gene family (n = 434), as well as 'unique variants', i.e. variants that were detected in only one of the matrices (n = 54). Within each set of isolates, at least one isolate per matrix (n = 275), as well as one of each unique variant (n = 41) were further characterized with respect to ESBL-genotype and antibiotic resistance profiles (n = 316). Eighteen of the 316 isolates were retrospectively identified as 'copy-isolates', i.e. isolates with the same characteristics obtained from the same sample; these were excluded from data analyses, leaving 298 isolates: 115 from 87 faeces and rinse water samples, and 183 from 114 environmental samples (Table 1). Of these, 288 were characterized using MLST (excluded from MLST analysis were 10 of 28 faeces isolates that, with respect to phylogenetic group, ESBL-genotype and ABR profile, had no counterparts in any of the environmental samples).

Table 1. Numbers of characterized ESBL-producing *E. coli* isolates per matrix and farm type.

Matrix	Laying hens	Broilers	Total
Faeces and wastewater			
Poultry faeces*	41 (4)	52 (6)	93
Rinse water from barns	1	21	22
Farm environment			
Air	0	3	3
Dust	5	12	17
Faeces of other animals	4	13	17
Flies	7	7	14
Run-off gullies	0	17	17
Soil	38	24	62
Surface water	11	42	53
Total	*107*	*191*	*298*

* Indicated between brackets are the numbers of isolates that were not characterized using MLST.

Statistics

The Pearson Chi square Test was used to test differences in the proportions of positive samples (e.g. prevalence in adjacent vs. remote surface waters, prevalence in soil at laying hen and broiler farms), and differences in the proportions of isolates identical to flock isolates between broiler and laying hen farms. Non-parametric tests were used to test for differences in bacterial concentrations between groups. The rationale for this was the skewed distribution of ESBL-producing *E. coli* concentrations, the inclusion of semi-quantitative pre-enrichment data (because of which also log-transformed values were not always normally distributed), and the loss of information when excluding negative samples from analysis, as is the case when using log-transformed values. The Mann-Whitney U Test and the

Kruskal-Wallis Test were used for comparison between two and multiple groups, respectively.

To express the diversity of ESBL-producing *E. coli* variants in laying hen and broiler farms, the Simpson's Index of Diversity (1-D) was calculated for poultry faeces and barn rinse water isolates, for each farm. The Simpson's Index of Diversity (1-D) was calculated from: $D = \frac{\sum_{i=1}^{N} n_i(n_i - 1)}{N(N-1)}$ where n_i represents the number of variants with the ith pheno-/genotype, and N the total number of isolates. Br1 was sampled during three different production rounds, yielding three slightly different diversity indices, which were averaged for purpose of statistical analysis. The difference in diversity indices between laying hen and broiler farms was tested using the Mann-Whitney U test, the difference between faeces and environmental samples was tested in a paired fashion using the Wilcoxon Signed Ranks Test (non-parametric tests were used because of low numbers of farms).

RESULTS

Prevalence of ESBL-Producing *E. coli* at Poultry Farms

ESBL-producing *E. coli* were present at each poultry farm, in 65% and 81% of poultry faeces samples at laying hen and broiler farms, respectively (Fig 2). The bacteria were also highly prevalent in wastewater and sediments from run-off gullies, in faeces from other animals present at the farms, in dust in the interior of barns at broiler farms, and in soil at laying hen farms. Also surface water in the direct vicinity of the farms frequently contained ESBL-producing *E. coli*, and so did, to a lesser extent, flies and barn air (Fig 2).

White bars represent broiler farms, black bars represent laying hen farms. In between brackets (n = a; b) are indicated the numbers of analysed samples at broiler (a) and laying hen farms (b) respectively; ns = not sampled; *includes water from pits and storage basins as well as run-off water and sediment from gullies.

At laying hen farms, ESBL-producing *E. coli* were less frequently detected in dust and air in barns (40% and 0% resp.) than at the broiler farms (81% and 7.7% resp.). Overall, ESBL-producing *E. coli* were detected in two of 33 indoor air samples which were both obtained at the same broiler farm (Br1), at two different sampling points. At the same broiler farm, dust and air were also sampled near ventilator exhausts outside barns; ESBL-producing *E. coli* were detected in 50% of the outside dust but none of the outside air samples.

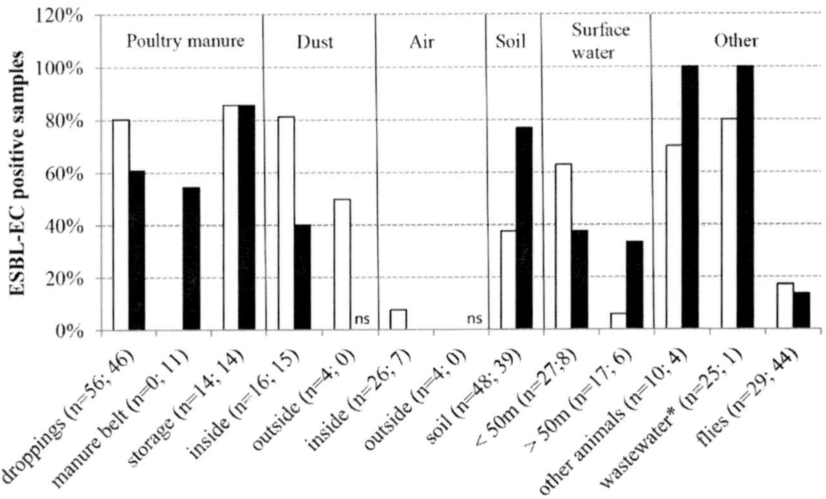

Figure 2. Prevalence of ESBL-producing *E. coli* in poultry faeces and environmental samples at laying hen and broiler farms.

In the outdoor environment, ESBL-producing *E. coli* were most frequently detected in soil at laying hen farms (77%), followed by surface water adjacent to broiler farms (63%). The prevalence in soil was higher at laying hen farms compared to broiler farms ($P<0.0001$), while the prevalence in adjacent surface water was higher at broiler farms, although this latter difference was not statistically significant ($P>0.1$) (Fig 2). The prevalence of ESBL-producing *E. coli* in surface waters adjacent to broiler farms was ten times higher than that at more remote sampling sites ($P<0.0001$). Moreover, in adjacent surface waters, the bacteria were detected more frequently during or shortly after cleaning compared to when broilers were present (91% vs. 44% of samples positive; $P = 0.01$). In surface waters adjacent to laying hen farms, the prevalence of ESBL-producing *E. coli* was similar to that at more remote sampling sites ($P>0.1$) (Fig 2).

ESBL-producing *E. coli* were detected in 14% and 17% of all fly samples at laying hen and broiler farms, respectively (Fig 2). All ESBL-positive flies belonged to four of the eight different fly species/families that were caught at the farms: blow flies (*Calliphoridae*), common house flies (*Musca domestica*), stable flies (*Stomoxys calcitrans*) and flesh flies (*Sarcophagidae)*. Among the 52 pooled samples that consisted of flies from these four species/families, the overall prevalence was 21%. Total *E. coli* were detected in 85% of these samples. Although ESBL-producers were not detected in samples consisting of the other four fly species/families (*Fannia canicularis, Hydrotea, Muscina stabulans, Tachinidae*), these flies did carry *E. coli* (53%). For the most abundant fly species that were caught at both types of farms, *M. domestica* (n = 32 pools) and *S. calcitrans* (n = 11 pools), a two-fold higher prevalence of ESBL-producing *E. coli* was observed among flies from broiler farms compared to flies from laying hen farms: 33% vs. 16% ESBL-positive, respectively (P>0.1).

Concentrations of ESBL-*E. coli* and *E. coli* in Poultry Farm Environment

Concentrations of ESBL-producing *E. coli* and total *E. coli* were determined in poultry faeces, wastewater, soil, surface water, flies, and a subset of dust samples (Table 2). In fresh and semi-fresh droppings collected in barns and free-range areas (the latter being only applicable for Lh1, Lh3, and Lh5), the concentrations of ESBL-producing *E. coli* were approximately 2-log higher on broiler farms compared to laying hens farms (geometric means 5.0×10^5 and 2.3×10^7cfu/kg, respectively; P = 0.001). The proportion of ESBL-producing *E. coli* was slightly higher among broilers compared to laying hens: on average one in 3.1×10^3 *E. coli* were ESBL-producing, compared to one in 1.8×10^4, respectively, although this difference was not statistically significant (P>0.1). Five barn dust samples, three at broiler farm Br1 and one each at laying hen farms Lh4 and Lh5, were analysed quantitatively. Concentrations of ESBL-producing *E. coli* and total *E. coli* were 1- to 3-$\log_{10}$ units lower than that in poultry faeces from barns (Table 2).

In soil, the concentrations of ESBL-producing *E. coli* were significantly higher in free-range areas at laying hen farms compared to other sites at laying hen and broiler farms (geometric means 2.4×10^4 vs. 3.6×10^2 and 3.9×10^2 cfu/kg, respectively; P = 0.001). A relative high concentration was also observed at sites that were visibly influenced by rinse water (1.1×10^4cfu/kg), although this value is based on only two (out of three) positive samples with substantially different concentrations (Table 2). In surface water bodies adjacent to poultry farms, the concentrations of ESBL-producing *E. coli* were higher than in further removed water bodies (geometric means 66 and 3.9 cfu/l, respectively; P<0.0001). Moreover, at broiler farms, ESBL-producing *E. coli* were present in higher concentrations during or shortly after cleaning compared to when broilers were present (geometric mean of 1.9×10^2 vs. 25 cfu/l, respectively; P = 0.02). An explanation might be contamination of surface water with rinse water, given the high concentrations of ESBL-producing *E. coli* in rinse water (geometric mean of 2.4×10^6cfu/l). No significant difference in ESBL-producing *E. coli* concentrations was observed between surface waters surrounding laying hen farms and broiler farms when flocks were present at both types of farms (Table 2).

Table 2. ESBL-producing *E. coli* and *E. coli* concentrations in poultry manure and the environment.

Matrix (concentration unit).and matrix subgroups (no. samples[a])	ESBL-producing *E. coli*			*E. coli*		
	% pos[b]	Geo-mean	Range[c]	% pos[b]	Geo-mean	Range[c]
Poultry faeces (cfu/kg)						
Laying hen_barns (n = 18)	56	6.9×10^5	$\geq10^2$–1.7×10^9	100	7.8×10^9	1.4×10^9–5.5×10^{10}
Laying hen_ree range (n = 9)	78	3.2×10^6	$\geq10^2$–1.2×10^7	100	1.4×10^{10}	4.5×10^9–3.7×10^7
Laying hen_ manure belts (n = 9)	56	1.4×10^6	3.1×10^4–4.5×10^7	100	9.0×10^{10}	1.7×10^{10}–1.8×10^{12}
Laying hen_ storage containers (n = 14)	88	4.4×10^6	$\geq10^2$–1.0×10^9	94	1.6×10^{11}	1.2×10^9–1.8×10^{12}
Broiler_ barns* (n = 36;28)	83	2.3×10^7	$\geq10^2$–8.3×10^9	100	4.4×10^{10}	1.4×10^9–2.2×10^{11}
Broiler_dung heaps (n = 14)	93	2.5×10^3	$\geq10^2$–7.1×10^6	86	2.5×10^7	2.7×10^5–1.8×10^9
Wastewater (cfu/l)						
Pits/basins/drains at cleaning* (n = 15;10)	80	2.4×10^6	9.0×10^3–8.3×10^7	80	3.2×10^6	1.3×10^5–4.6×10^9
Run-off gullies* (n = 5;4)	80	7.8×10^3	4.3×10^2–2.2×10^5	100	1.5×10^5	3.6×10^3–6.0×10^6
Wastewater (cfu/kg)						
Sediment from run-off gullies* (n = 4;2)	100	2.4×10^5	4.5×10^5–1.0×10^8	100	1.5×10^7	1.2×10^7–2.0×10^7
Dust (cfu/kg)						
Laying hen_barns (n = 2)	50	$\geq2\times10^2$	n.a.	100	4.1×10^8	1.3×10^8–1.4×10^9
Broiler_barns (n = 3)	100	5.0×10^5	2.5×10^5–8.8×10^5	100	6.4×10^7	5.7×10^7–7.7×10^7
Soil (cfu/kg)						
Free-range (n = 14)	100	2.4×10^4	$\geq10^2$–3.3×10^7	100	1.1×10^9	1.8×10^7–5.4×10^9
Near barns* (n = 23;21)	43	1.7×10^2	$\geq10^2$–2.0×10^4	100	7.5×10^4	8.2×10^3–2.6×10^8
Near manure storage* (n = 19;18)	68	4.4×10^2	$\geq10^2$–2.0×10^4	100	1.2×10^5	$\geq10^2$–1.8×10^8
Near rinse water (n = 3)	67	1.1×10^4	$\geq10^2$–1.2×10^6	100	1.1×10^5	$\geq10^2$–6.8×10^7
Dried-up ditches* (n = 11;10)	45	3.8×10^2	$\geq10^2$–7.8×10^4	100	6.9×10^5	9.0×10^3–1.7×10^8
Premises other* (n = 17;15)	24	3.2×10^2	$\geq10^2$–1.0×10^4	100	1.1×10^4	$\geq10^2$–1.4×10^6
Surface water (cfu/l)						
At > 50m distance (n = 23)	13	4	1.4–10	100	1.8×10^3	31–8.0×10^4
Laying hen, < 50m (n = 8)	38	27	11–1.4×10^2	100	5.0×10^3	90–1.4×10^6
Broiler_flocks present *, < 50m (n = 16;13)	50	25	1.4–2.3×10^3	100	3.6×10^3	3.1×10^2–1.2×10^5
Broiler_at cleaning, <50m (n = 11)	91	1.9×10^2	2.8–5.0×10^5	100	2.5×10^4	6.3×10^2–3.1×10^7
Flies (cfu/fly pool)[d]						
Calliphoridae (n = 8, 20)	13	2.5×10^4	n.a.	50	6.3×10^2	≥3.3–2.4×10^5
Musca domestica (n = 32; 212)	22	5.2×10^2	≥3.3–8.0×10^4	97	3.0×10^5	≥3.3–1.2×10^8
Sarcophagidae (n = 1; 2)	100	4.1×10^3	n.a.	100	5.1×10^6	n.a.
Stomoxys calcitrans (n = 11; 39)	18	1.5×10^3	1.2×10^3–2.0×10^3	82	9.4×10^2	≥3.3–5.9×10^7
Other (n = 21; 53)	0	0	n.a.	53	1.6×10^4	≥3.3–1.5×10^6

[a]For some of the samples total *E. coli* was not determined (*): in these cases the number of samples analyzed for ESBL-producing *E. coli* (a) and *E. coli* (b) are indicated separately (n = a; b).

[b]Percentages of samples positive for the indicated bacteria.

[c] Concentration ranges observed in the positive samples

[d] Indicated after fly species/family (n = x, y) is the number of pools (x) and the number of flies (y)

cfu = colony forming units; Geo-mean = geometric mean; n.a. = not applicable.

In fly samples containing flies from the four species/families shown to carry ESBL-producing *E. coli*, the geometric mean concentrations of ESBL-producing *E. coli* and total *E. coli* was 1.1×10^3 cfu/pool and 6.3×10^4 cfu/pool, respectively. Even though ESBL-producing *E. coli* were not detected in the other four species/families, the *E. coli* concentrations in these fly samples were in the same order of magnitude: 1.6×10^4 cfu/pool) (Table 2).

Diversity among ESBL-Producing *E. coli* at Poultry Farms

Among 298 isolates from poultry faeces and farm environment, all phylogenetic subgroups were represented, the most prevalent being subgroup B1 (42%), followed by A_1 (24%), A_0(11%), D_2 (10%), D_1 (7.7%), $B2_2$ (4.7%) and $B2_3$ (0.7%). The vast majority (98%) of all isolates carried $bla_{CTX\text{-}M\text{-}1}$ (41%), $bla_{SHV\text{-}12}$ (29%) and $bla_{TEM\text{-}52}$ (28%). Additional detected genotypes were $bla_{CTX\text{-}M\text{-}2}$ (0.34%), $bla_{CTX\text{-}M\text{-}14}$ (0.34%), $bla_{CTX\text{-}M\text{-}15}$ (0.34%), and $bla_{CTX\text{-}M\text{-}27}$ (0.34%). Two isolates (0.67%) contained an unidentified ESBL-gene other than $bla_{CTX\text{-}M}$, bla_{TEM}, bla_{SHV}, and bla_{OXA}. In total, 25 different variants were observed based on phylogenetic subgroup and ESBL-genotype. Eighteen of these were found in poultry faeces and rinse water from barns, and therewith considered to reflect the population of the flocks (Fig 3). With the exception of one of the isolates from broiler faeces that carried $bla_{CTX\text{-}M\text{-}2}$, all faeces and rinse water isolates (99%) carried $bla_{CTX\text{-}M\text{-}1}$, $bla_{SHV\text{-}12}$ or $bla_{TEM\text{-}52}$. The distribution of these three genotypes differed between the two farm types: $bla_{CTX\text{-}M\text{-}1}$was present in 32% and 50% of the isolates on broiler and laying hen farms respectively ($P = 0.049$), $bla_{SHV\text{-}12}$ in 45% compared to 0% ($P<0.0001$), and $bla_{TEM\text{-}52}$ in 22% and 50% of the isolates ($P = 0.002$). The diversity among ESBL-producing *E. coli* from poultry faeces and rinse water with respect to phylogenetic subgroup and ESBL-genotype combination was higher on broiler farms compared to laying hen farms (Fig 3). This was confirmed by the respective average Simpson's Indices of Diversity (SID) of 0.89 and 0.51 ($P = 0.04$, Table 3).

Shown are the proportions of ESBL-producing isolates with the indicated phylogenetic group and ESBL-genotype combinations, among the isolates from poultry faeces and barn rinse water sampled from drains and pits. Lh = laying hen farms, Br = broiler farms.

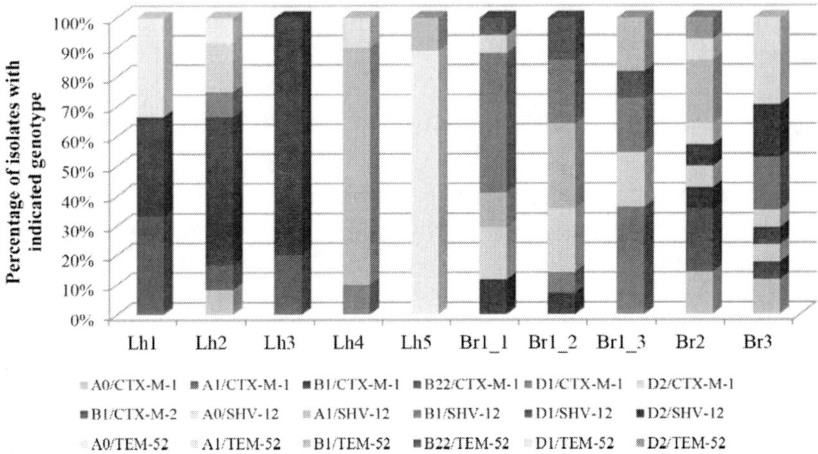

Figure 3. Diversity among ESBL-producing *E. coli* variants in poultry faeces and barn rinse water.

Table 3. Simpson's Index of Diversity (1-D) among isolates from poultry faeces and farm environment.

		Average Simpson's Index of Diversity [range]	
Origin of isolates[a]	Isolate characteristics[b]	Broilers (n = 3)	Laying hens (n = 5)
Flocks	Phylo/ESBL	0,89 [0.82–0.92]	0,51 [0.22–0.76]*
Flocks	ST/Phylo/ESBL/ABR	0.93 [0.91–0.96]	0,63 [0.22–0.92]
Flocks	ESBL/ABR	0,90 [0.83–0.93]	0,37 [0.00–0.56]*
Environment	Phylo/ESBL	0,88 [0.83–0.92]	0,60 [0.40–0.90]
Environment	ST/Phylo/ESBL/ABR	0,94 [0.92–0.94]	0,77 [0.44–1.00][c]
Environment	ESBL/ABR	0,91 [0.88–0.93]	0,57 [0.17–0.90][c]

[a] Isolates from faeces and rinse water ('Flocks) or isolates from the farm environmental ('Environment')

[b] Diversity was analysed for three combinations of isolate characteristics: phylogenetic subgroup and ESBL-genotype; (Phylo/ESBL); ST, phylogenetic subgroup, ESBL-genotype and ABR profile (ST/Phylo/ESBL/ABR) and ESBL-genotype and ABR profile (ESBL/ABR)

* = At the same characteristics level, the difference between broiler and laying hen farms is statistically significant ($P<0.05$) using the Mann-Whitney U test

[c] = Within farm type and at the same characteristics level, the difference between 'Flocks' and 'Environment' is statistically significant ($P<0.05$) using the Wilcoxon Signed Ranks test.

Overall, 65 different ESBL-producing *E. coli* sequence types (STs) were detected at poultry farms: 45 at broiler farms, of which 27 (60%) in poultry faeces and rinse water, and 29 at laying hen farms, of which ten (34%) in poultry faeces and rinse water. Among these were seven new STs (ST4976 –ST4981, ST4994). Thirteen (20%) of the STs were observed at multiple farms, and nine of these were detected at both types of farms (ST10, ST48, ST58, ST155, ST162, ST212, ST746, ST1276, ST3249). The most widespread

STs were ST48 and ST155, which were both detected at six of the eight farms.

Inclusion of sequence types (STs), phylogenetic group, ESBL-genotype and antibiotic resistance profiles in variant analysis demonstrated on average 16 and 9.0 different ESBL-producing *E. coli* variants at broiler farms and laying hen farms respectively, of which 10.0 and 4.2 were detected in faeces and rinse water (Figs 4 and 5). The diversity among ESBL-producing isolates from flocks was higher on broiler farms (average SID: 0.93) compared to laying hen farms (average SID: 0.63), although this difference was not statistically significant ($P = 0.071$). The lack of statistical significance may at least partially be due to the large variation between laying hen farms and small number of farms analysed. Four of five laying hen farms had SIDs that were considerably lower than for each of the three broiler farms. The only exception was Lh2 which, compared to the other laying hen farms, had an uncharacteristically high diversity of ESBL-producing variants (SID = 0.92, see also Fig 4). Despite the large variation at this latter farm, the variation at the level of antibiotic resistance appeared to be much more limited: eight different STs carried *bla*$_{CTX\text{-}M\text{-}1}$ in combination with sulfamethoxazole and tetracycline resistance (Fig 4). Indeed, at laying hen farms, a lower diversity was observed at the level of antibiotic resistance properties than at the level of total isolate characteristics (average SIDs 0.37 vs. 0.63, respectively). Although this difference was not statistically significant ($P = 0.068$), it was clearly more pronounced than that at broiler farms (average SIDs 0.90 vs. 0.93, respectively). Based on antibiotic resistance properties alone, the diversity was significantly higher at broiler farms compared to laying hen farms ($P = 0.036$). At laying hen farms, but not at broiler farms, the diversity among ESBL-producing *E. coli* from environmental matrices was significantly higher compared to ESBL-producing *E. coli* from flocks (Table 3).

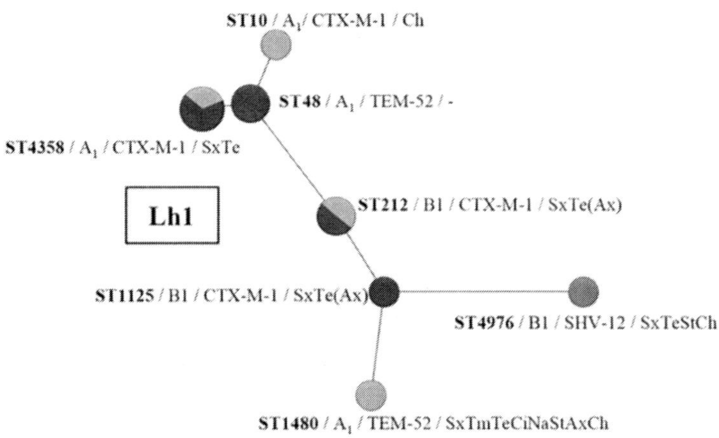

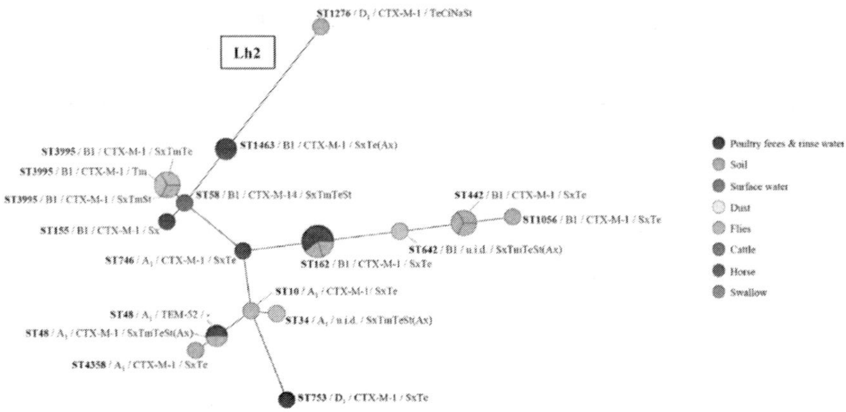
Lh2
ST1276 / D_1 / CTX-M-1 / TeCiNaSt
ST1463 / B1 / CTX-M-1 / SxTe(Ax)
ST3995 / B1 / CTX-M-1 / SxTmTe
ST3995 / B1 / CTX-M-1 / Tm
ST3995 / B1 / CTX-M-1 / SxTmSt
ST58 / B1 / CTX-M-14 / SxTmTeSt
ST155 / B1 / CTX-M-1 / Sx
ST746 / A_1 / CTX-M-1 / SxTe
ST162 / B1 / CTX-M-1 / SxTe
ST642 / B1 / u.i.d. / SxTmTeSt(Ax)
ST442 / B1 / CTX-M-1 / SxTe
ST1056 / B1 / CTX-M-1 / SxTe
ST10 / A_1 / CTX-M-1/ SxTe
ST34 / A_1 / u.i.d. / SxTmTeSt(Ax)
ST48 / A_1 / TEM-52 / -
ST48 / A_1 / CTX-M-1 / SxTmTeSt(Ax)
ST4358 / A_1 / CTX-M-1 / SxTe
ST753 / D_1 / CTX-M-1 / SxTe
Poultry feces & rinse water
Soil
Surface water
Dust
Flies
Cattle
Horse
Swallow

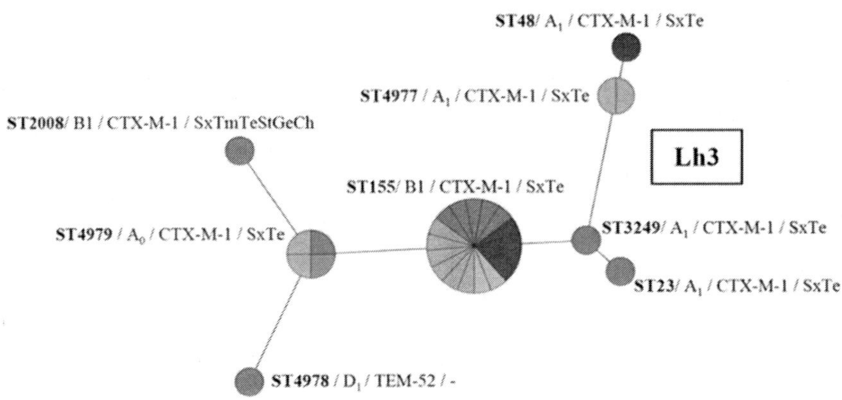
ST48/ A_1 / CTX-M-1 / SxTe
ST4977 / A_1 / CTX-M-1 / SxTe
ST2008/ B1 / CTX-M-1 / SxTmTeStGeCh
Lh3
ST155/ B1 / CTX-M-1 / SxTe
ST4979 / A_0 / CTX-M-1 / SxTe
ST3249/ A_1 / CTX-M-1 / SxTe
ST23/ A_1 / CTX-M-1 / SxTe
ST4978 / D_1 / TEM-52 / -

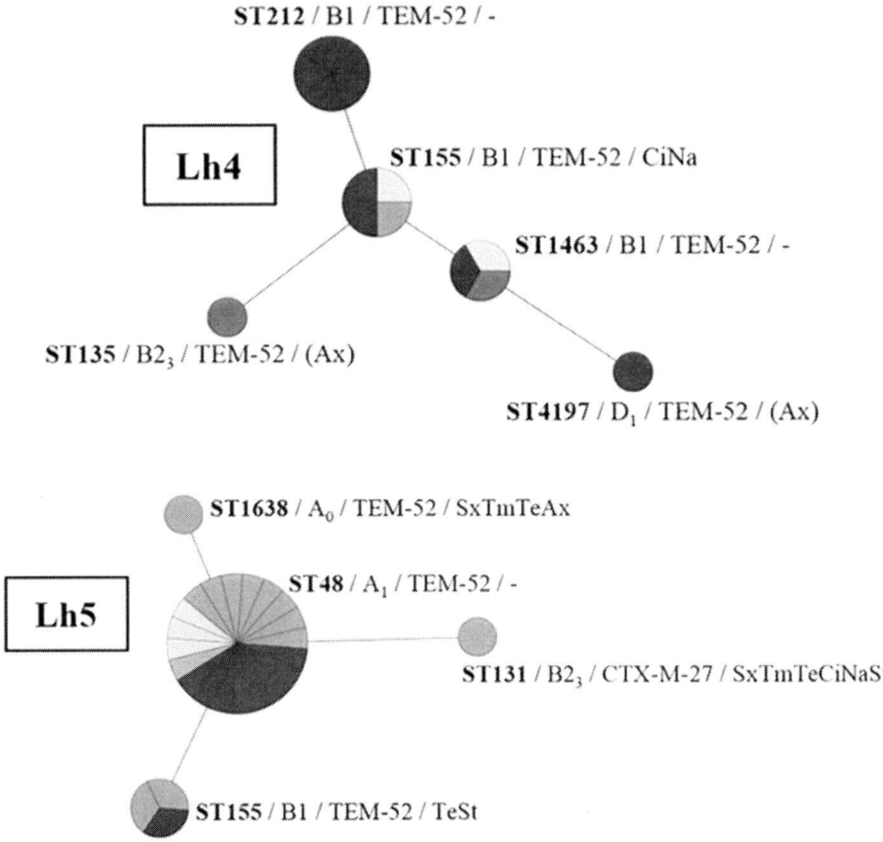

Figure 4. ESBL-producing *E. coli* variants on laying hen farms.

Maximum parsimony trees constructed based on the concatenated sequences of the seven MLST alleles, using Bionumerics 7.1 software. Node sizes reflect the number of isolates per ST and node colours represent different matrices. Additionally indicated are the phylogenetic group, ESBL-genotype and ABR profiles of isolates in every node (i.e., ST). Note that per sample maximally one isolate of each variant was included, and that multiple isolates of a specific variant reflects detection in multiple samples. ABR profiles represent antibiotics to which resistance was observed additionally to 3rd generation cephalosporins. In between brackets are antibiotics with MICs just above and just below epidemiological cut-off values (i.e. with a maximal 2-fold difference) among isolates with the same ST and/or ESBL genotype. Sx = sulfamethoxazole, Tm = trimethoprim; Te =

tetracycline, Ci = ciprofloxacin, Na = nalidixic acid, St = streptomycin, Ge = gentamycin, Ax = amoxicillin+clavulanic acid, Ch = chloramphenicol; u.i.d. = unidentified.

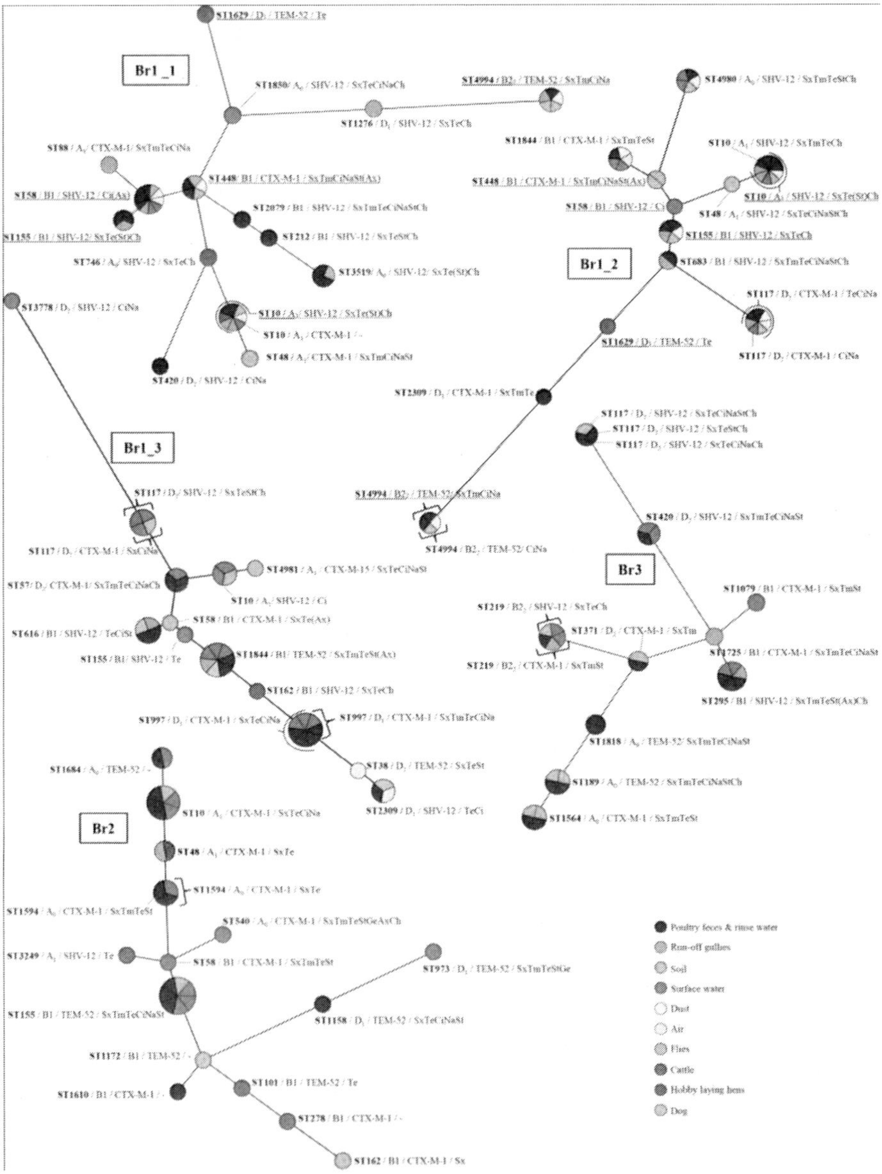

Figure 5. ESBL-producing *E. coli* variants on broiler farms.

Maximum parsimony trees constructed based on the concatenated sequences of the seven MLST alleles, using Bionumerics 7.1 software. Node sizes reflect the number of isolates per ST and node colours represent different matrices. Additionally indicated are the phylogenetic group, ESBL-genotype and ABR profiles of isolates in every node (i.e., ST). Note that per sample maximally one isolate of each variant was included, and that multiple isolates of a specific variant reflects detection in multiple samples. ABR profiles represent antibiotics to which resistance was observed additionally to 3rd generation cephalosporins. In between brackets are antibiotics with MICs just above and just below epidemiological cut-off values (i.e. with a maximal 2-fold difference) among isolates with the same ST and/or ESBL genotype. Sx = sulfamethoxazole, Tm = trimethoprim; Te = tetracycline, Ci = ciprofloxacin, Na = nalidixic acid, St = streptomycin, Ge = gentamycin, Ax = amoxicillin+clavulanic acid, Ch = chloramphenicol. Br1 was sampled during three production rounds: August/September 2011 (Br1_1), November 2011 (Br1_2), and August/September 2012 (Br1_3); underlined are variants that were detected at multiple time-points.

Relation between ESBL-Producing Isolates from Poultry Faeces and Farm Environment

Overall, 60% of all isolates from farm environment had a variant type that was identical with respect to ST, phylogenetic group, ESBL-genotype and ABR profile, to variants observed in poultry faeces or rinse water at the same farm (i.e. 'flock' variants). An additional 7.1% of isolates from farm environments had identical counterparts in other environmental matrices at the same farm (i.e. 'ubiquitous' variants) (Fig 6). The proportion of environmental ESBL-producing isolates with antibiotic properties (i.e. ESBL genotype and ABR profile) identical to those in flocks and/or observed in multiple matrices was 74% and 4.4%, respectively.

Indicated are the proportions of isolates from the different environmental matrices, with identical counterparts in manure and rinse water and/or other environmental matrices at the same farm, based on phylogenetic group, ESBL-genotype, ABR profile and ST. White bars represent broiler farms, black bars represent laying hen farms, and hatched bars represent the results for all farms combined.

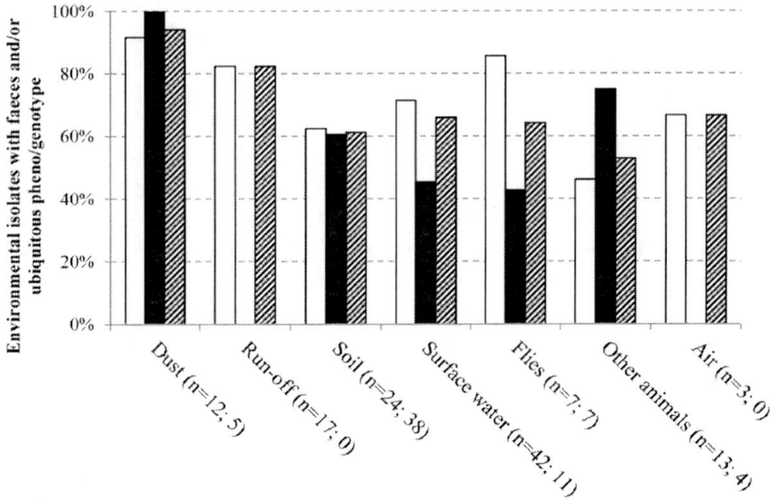

Figure 6. Relationship between ESBL-producing variants from different matrices.

The highest proportion of flock and/or ubiquitous variants was observed in dust (94%), run-off gullies (82%) and barn air (67%) (Fig 6). Among surface water, soil and fly variants, 61–66% were identical to those observed in flocks and other environmental matrices; in other farm animals approximately half of the isolates were identical to those in flocks and farm environment. The proportion of flock and ubiquitous variants in the farm environment was slightly higher on broiler (71%) compared to laying hen farms (60%) (*P>0.1*). The two farm types differed particularly with respect to the proportions of flock and ubiquitous variants in surface water and on flies (both higher at broiler farms), and that in other farm animals (higher at laying hen farms) (Fig 6), however, none of these differences were statistically significant. Farm animals harbouring ESBL-producing variants identical to those in flocks or in other environmental matrices at the same farm consisted of cattle (Lh3, Br2), swallows (Lh3), dogs (Br1), and hobby laying hens (Br1_1) (Fig 4). An isolate obtained from horse faeces at Lh2 had an ESBL-genotype that was not detected in any of the poultry isolates, *bla*$_{\text{CTX-M-14}}$.

At Br1, six of the variants detected at the first time-point (August/September 2011), were still detected a couple of production rounds later, in November of the same year. One of these variants was persistent in hobby laying hens only, three were detected in poultry faeces as well as several environmental matrices at both time points, and two were detected in poultry faeces and farm environment at the first time-point, but only in farm environment (soil and surface water) at the latter time-point (Fig 5). In August/September 2012, the variants present in 2011 were no longer detected.

DISCUSSION

ESBL-producing *E. coli* were detected in flocks at all eight investigated poultry farms, i.e. in broiler and laying hen flocks. A high prevalence of ESBL-producing *E. coli* in Dutch broiler flocks has previously been described [13,14], and was shown to be at least partially traceable to breeding chickens and parent stocks [46]. Although a limited number (five) of laying hen farms was included in the present study, our findings demonstrate a high prevalence of ESBL-producing *E. coli* in Dutch laying hen flocks as well. The prevalence of ESBL-producing *E. coli* in poultry faeces as well as the barn environment (air, dust), was higher at broiler farms compared to laying hen farms. In the Netherlands, the total amount of antimicrobials used in broilers is approximately twenty times higher than that in laying hens: in 2013, 18.1 daily dosages per animal year (dd/ay) were registered for conventionally held broilers, compared to 0.5–1.2 dd/ay for battery and free-range laying hens respectively[47]. The lower use of antibiotics means there is a lower selection pressure on ESBL-producing *E. coli* in laying hens as compared with broilers. Assuming that part of the broiler and laying hen flocks already carry ESBL-producing *E. coli* at entrance of the production farms, the low selection pressure in laying hens might result in a decrease in ESBL-producing bacteria in flocks over time, while they remain present for a longer time among broilers. This, in combination with the older age of the laying hen flocks compared to the broiler flocks at the time of sampling might explain the observed differences in prevalence.

In the outdoor farm environment, ESBL-producing *E. coli* were frequently detected in soil and surface water. Other farm animals when present at the poultry farms, as well as flies, were also shown to carry ESBL-producing *E. coli*. Overall, the prevalence in soil was higher at sites that were visibly influenced by poultry faeces, e.g. free-range areas, sites near manure transport belts, near manure storage sheds, near dung heaps and near a rinse water storage container. Both the detection frequency and the average concentrations in soil were higher at laying hen farms compared to broiler farms, which was largely attributable to the inclusion of free-range areas at laying hen farms (which were not present at the broiler farms). Broiler but not laying hen farms significantly contributed to the contamination of surface water, as evidenced from the statistically significant higher prevalence and average concentrations in water adjacent to broiler farms compared to remote sampling sites. This difference appeared largely associated with the cleaning of broiler farms between two production rounds (i.e. every six to seven weeks). The effect of cleaning of barns on the environmental load was not studied at the laying hen farms, due to the long lifespan of laying hens (approximately one and a half year). It is conceivable however, that the observed difference between the two farm types with respect to surface water contamination is mainly associated with the frequency of clearing and cleaning. The high prevalence of ESBL-producing *E. coli* in free-range areas suggests that run-off from such areas represents a source of surface water contamination as well. This is supported by the observation that one of the three positive surface water samples obtained in the vicinity of laying hen farms was adjacent to a free-range area. In this water, ESBL-producing *E. coli* was detected with the same identity with respect to phylogenetic subgroup, sequence type, ESBL-genotype, and ABR profile, as isolates detected in, amongst others, free-range soil and poultry faeces (Lh3: ST155/B1/CTX-M-1/SxTe). The other two positive surface water sites were in the proximity of a barn ventilation fan (Lh3) and a manure storage shed (Lh4), and also these waters contained isolates that were present in poultry faeces and other environmental matrices at the corresponding farms (Lh3: ST155/B1/CTX-M-1/SxTe, Lh4: ST1463/B1/TEM-52/-).

Comparison of isolates from flocks and farm environment indicated the poultry flocks as major source of environmental contamination. At both types of farms, a very high diversity of ESBL-producing variants with respect to phylogenetic subgroup, sequence type, ESBL-genotype and ABR profile was observed in poultry faeces and barn rinse water (matrices assumed to reflect flock carriage). A small majority (62%) of all variants obtained from the farm environment had the exact same identity as isolates observed in flocks present at the same time at the same farm. It should be emphasised that this percentage presumably reflects an underrepresentation of the proportional load of variants in the farm environment that are identical to 'flock' variants. This is the result of sequential rounds of selection of isolates after each test, in which a maximum of one or two isolates with specific characteristics per sample was included for the next round (i.e. selection was aimed to include a maximum of diversity). Based on the current data, the contribution of each variant (including the 'flock' variants and the 'unique' variants) to the total environmental load cannot be calculated.

The detection of variants in the farm environment not detected in flocks may have multiple explanations. Firstly, part of the variants in the farm environment may originate from other sources than the poultry flocks at the farms. This is most probable for 'motile' environmental compartments such as surface water and flies. Surface water, may contain variants introduced upstream of the investigated farms (e.g. from other farms or wastewater treatment plants). Flies may move between habitats [48], and possibly carry ESBL-producing *E. coli* from one farm to the next. Other farm animals, when they are not bred at the poultry farms, could also be considered as motile. In the case of other farm animals, the presence of non-poultry isolates might additionally be explained by separate treatment regimens. For barn air and soil on farm premises, the presence of ESBL-producing *E. coli* from another origin than the flocks is less easily envisioned. One explanation may be the persistence of bacteria from previous flocks in for instance dust, soil, or water. *E. coli* may survive for weeks to months in soil and the aquatic environment, depending on physical, chemical and biological properties of the matrix, such as temperature, pH, sunlight, soil moisture and presence of nutrients [49,50]. This possibility was supported by our own findings at Br1. Following the same line of reasoning, the significantly higher turn-over of flocks at broiler farms as compared to laying hen farms, might contribute to the higher diversity observed on the latter farm type. A second explanation for discrepancies between matrices may simply be chance, related to the probability of omitting variants during the process of colony selection, given the high level of variation and a limited number of selected colonies. In other words, the fact that certain variants were not detected in poultry faeces does not rule out the possibility that they were actually present in faeces. That this may indeed explain at least part of the discrepancies was suggested by the presence of variants in rinse water directly derived from barns, which were not detected in poultry faeces from the same barns (S2 Table). The same explanation might underlie the occasional observation of variants being abundantly present in the farm environment (as indicated by their presence in a variety of different matrices) while not detected in poultry faeces or barn rinse water. A third explanation could be the development of new variants in the environment itself through the exchange of plasmids containing ESBL-genes or other antibiotic resistance genes [17,51–53]. The genes encoding ESBLs are generally located on plasmids that may contain other resistance genes as well [7,8,54,55]. Transfer of mobile elements could explain the observations of isolates with the same ESBL-genotype and/or other ABR profiles in different *E. coli* backgrounds (i.e. sequence types) at the same farm. This process may particularly contribute to the variation observed in the other farm animals, assuming intestinal tracts to be an optimal site for horizontal gene transfer [56,57], but possibly also in soil, surface water and sediments of run-off-gullies. *E. coli* has been shown to be able to act both as donor and recipient of mobile elements in soil and surface water. The efficiency of the gene transfer among *E. coli* in soil was shown to be dependent on its physicochemical properties, such as the presence of nutrients (e.g.

amended soil, or non-sterile agricultural soil), temperature (e.g. at 15°C—30°C), and humidity (e.g. at 20%-60%) [58–60]. These conditions may, at least occasionally, be met in soil and surface water at poultry farms. Given the fact that genes responsible for antibiotic resistance are often located on plasmids, the current observation of different STs with identical ESBL-genotype and antibiotic resistance profiles at the same farm, supports the occurrence of horizontal gene transfer. In line with this, the proportion of isolates in the farm environment with antibiotic resistance properties identical to those in the poultry flocks was higher than the proportion of isolates that was exactly the same (74% vs. 62%). Typing of the plasmids that carry ESBL and the other ABR genes could confirm that gene transfer indeed contributes to the diversity observed at farms and in the farm environment [61,62], but this was not part of the aim of the current study.

Our study is one of few studies investigating ESBL producing *E. coli* in the natural (i.e. outside) environment at poultry farms [24–28]. The most thoroughly studied exterior environmental matrices so far were ambient air and soil or ground surfaces, whereas surface water was included only once [24]. Altogether, previous and current data demonstrate that dissemination of ESBL-producing *E. coli* from livestock to the natural environment occurs. From a public health perspective, the introduction of ESBL-producing *E. coli* (and ABR zoonotic bacteria in general) into the environment may pose a risk if these bacteria are transported to places where the general public may become exposed. The environmental compartments that should be considered as likely vehicles for dissemination are water, air, and flies, because of the distance that may be covered (as opposed to e.g. soil). The risk of human exposure through the environment depends on the bacterial concentrations at the source (i.e. farm environment), and the reduction in these concentrations during the travel-time to places where people may become exposed. For exposure through water, factors involved are for example dilution and bacterial die-off in water; human exposure to bacteria carried by flies depends for example on the proportion of flies that migrate from farms, and the numbers of bacteria that survive during this process and are subsequently transferred from flies to food meant for human consumption. The role of water and flies in human exposure to ESBL-producing *E. coli* is currently being assessed using quantitative microbial risk assessment (QMRA) (Schijven et al. submitted, Evers et al. in press [63]). Even though our findings did not designate air as a major contaminated compartment at poultry farms, air has previously been implicated as source of dissemination from broiler as well as fattening pig farms [26,28]. However, even though in these studies the prevalence of 3rd generation-resistant *E. coli* in barn air was higher (16% in both studies) compared to the current findings (8% at broiler farms), in these studies, the prevalence in ambient air was still relatively low (7.5% and 6%, respectively). In particular surface water may prove to be a relevant route of environmental exposure, because of its use for recreation, irrigation and production of drinking water. Indeed, waterborne transmission has been implicated as an important route of transmission of intestinal bacteria in general [64–66]. Besides animal husbandry farms, other sources likely contribute to the contamination of the aquatic environment, such as arable farming (run-off of animal manure) and discharge of partially treated human wastewater by wastewater treatment plants [18];. To globally reduce the prevalence of ABR bacteria, not only reduced antibiotic use (in humans and animals) but also reduced dissemination of resistant bacteria is important. Further study is required to gain insight in the relative contribution of different contamination sources to the load of clinically relevant ABR bacteria in the (aquatic) environment, which is imperative for the development of effective intervention strategies.

ACKNOWLEDGMENTS

The authors would like to thank Wilma Jacobs-Reitsma and Arieke Docters van Leeuwen for their pleasant participation during sampling days, Christiaan Veenman for technical assistance, and Jeroen van Leuken for supplying Fig 1. We are very grateful to the eight poultry farmers who cooperated by their own volition and gave us unlimited access to every inch of their premises.

REFERENCES

1. **Kümmerer K. Resistance in the environment. J Antimicrob Chemother. 2004;54: 311–320. pmid:15215223**
2. Wright GD. Antibiotic resistance in the environment: a link to the clinic? Curr Opin Microbiol. 2010;13: 589–594. pmid:20850375
3. Zhang XX, Zhang T, Fang HHP. Antibiotic resistance genes in water environment. Appl Microbiol Biotechnol. 2009;82: 397–414. pmid:19130050
4. Baquero F, Martínez JL, Cantón R. Antibiotics and antibiotic resistance in water environments. Curr Opin Biotechnol. 2008;19: 260–265. pmid:18534838
5. Cantón R, Novais A, Valverde A, Machado E, Peixe L, Baquero F, et al. Prevalence and spread of extended-spectrum ß-lactamase-producing Enterobacteriaceae in Europe. Clin Microbiol Infect. 2008;14 Suppl. 1: 144–153. pmid:18154538
6. Cantón R, Akóva M, Carmeli Y, Giske CG, Glupczynski Y, Gniadkowski M, et al. Rapid evolution and spread of carbapenemases among Enterobacteriaceae in Europe. Clin Microbiol Infect. 2012;18: 413–431. pmid:22507109

7. Livermore DM, Canton R, Gniadkowski M, Nordmann P, Rossolini GM, Arlet G, et al. CTX-M: changing the face of ESBLs in Europe. J Antimicrob Chemother. 2007;59: 165–174. pmid:17158117
8. Paterson DL,Bonomo RA. Extended-spectrum beta-lactamases: a clinical update. Clin Microbiol Rev. 2005;18: 657–686. pmid:16223952
9. den Heijer CD, Donker GA, Maes J, Stobberingh EE. Antibiotic susceptibility of unselected uropathogenic Escherichia coli from female Dutch general practice patients: a comparison of two surveys with a 5 year interval. J Antimicrob Chemother. 2010;65: 2128–2133. pmid:20682565
10. EARSNet (2011) European Centre for Disease Prevention and Control. Antimicrobial resistance surveillance in Europe 2012. Annual report of the European Antimicrobial Resistance Surveillance Network (EARS-Net). Available: http://www.ecdc.europa.eu/en/publications/publications/antimicrobial-resistance-surveillance-europe-2011.pdf. Accessed 27 May 2015.
11. Huijbers PM, de Kraker M, Graat EA, van Hoek AH, van Santen MG, de Jong MC, et al. Prevalence of extended-spectrum beta-lactamase-producing Enterobacteriaceae in humans living in municipalities with high and low broiler density. Clin Microbiol Infect. 2013;19: E256–259. pmid:23397953
12. Reuland EA, Overdevest IT, Al Naiemi N, Kalpoe JS, Rijnsburger MC, Raadsen SA, et al. High prevalence of ESBL-producing Enterobacteriaceae carriage in Dutch community patients with gastrointestinal complaints. Clin Microbiol Infect. 2013;19: 542–549. pmid:22757622
13. Dierikx C, van der Goot J, Fabri T, van Essen-Zandbergen A, Smith H, Mevius D. Extended-spectrum-beta-lactamase- and AmpC-beta-lactamase-producing *Escherichia coli* in Dutch broilers and broiler farmers. J Antimicrob Chemother. 2013;68: 60–67. pmid:22949623
14. Huijbers PM, Graat EA, Haenen AP, van Santen MG, van Essen-Zandbergen A, Mevius DJ, et al. Extended-spectrum and AmpC beta-lactamase-producing Escherichia coli in broilers and people living and/or working on broiler farms: prevalence, risk factors and molecular characteristics. J Antimicrob Chemother. 2014;69: 2669–2675. pmid:24879667
15. Overdevest I, Willemsen I, Rijnsburger M, Eustace A, Xu L, Hawkey P, et al. Extended-spectrum beta-lactamase genes of *Escherichia coli* in chicken meat and humans, The Netherlands. Emerg Infect Dis. 2011;17: 1216–1222. pmid:21762575
16. Leverstein-van Hall MA, Dierikx CM, Cohen Stuart J, Voets GM, van den Munckhof MP, van Essen-Zandbergen A, et al. Dutch patients, retail chicken meat and poultry share the same ESBL genes, plasmids and strains. Clin Microbiol Infect. 2011;17: 873–880. pmid:21463397
17. de Been M, Lanza VF, de Toro M, Scharringa J, Dohmen W, Du Y, et al. Dissemination of Cephalosporin Resistance Genes between Escherichia coli Strains from Farm Animals and Humans by Specific Plasmid Lineages. PLoS Genet. 2014;10: e1004776. pmid:25522320
18. Blaak H, de Kruijf P, Hamidjaja RA, van Hoek AHAM, de Roda Husman AM, Schets FM. Prevalence and characteristics of ESBL-producing *E. coli* in Dutch recreational waters influenced by wastewater treatment plants. Vet Microbiol. 2014;171: 448–459. pmid:24690376
19. Chen H, Shu W, Chang X, Chen JA, Guo Y, Tan Y. The profile of antibiotics resistance and integrons of extended-spectrum beta-lactamase producing thermotolerant coliforms isolated from the Yangtze River basin in Chongqing. Environ Pollut. 2010;158: 2459–2464. pmid:20447743
20. Dhanji H, Murphy NM, Akhigbe C, Doumith M, Hope R, Livermore DM, et al. Isolation of fluoroquinolone-resistant O25b:H4-ST131 *Escherichia coli* with CTX-M-14 extended-spectrum ß-lactamase from UK river water. J Antimicrob Chemother. 2011;66: 512–516. pmid:21172781

21. Hong H, Chun J, Lee Y. Detection of extended-spectrum β-lacamase-producing, multidrug-resistant environmental isolated of *Escherichia coli* that bind to human bladder cells. Microb Drug Resist. 2004;10: 184–189. pmid:15256035
22. Zurfluh K, Hächler H, Nüesch-Inderbinen M, Stephan R. Characteristics of extended-spectrum β-lactamase- and carbapenemase-producing Enterobacteriaceae isolates from rivers and lakes in Switzerland. Appl Environ Microbiol. 2013;79: 3021–3026. pmid:23455339
23. Blaak H, Hamidjaja RA, Van Hoek AHAM, De Heer L, De Roda Husman AM, Schets FM. Detection of extended-spectrum beta-lactamase (ESBL)-producing *Escherichia coli* on flies at poultry farms. Appl Environ Microbiol. 2014;80: 239–246. pmid:24162567
24. Gao L, Tan Y, Zhang X, Hu J, Miao Z, Wei L, et al. Emissions of Escherichia coli carrying extended-spectrum beta-lactamase resistance from pig farms to the surrounding environment. Int J Environ Res Public Health. 2015;12: 4203–4213. pmid:25893997
25. Hartmann A, Locatelli A, Amoureux L, Depret G, Jolivet C, Gueneau E, et al. Occurrence of CTX-M producing *Escherichia coli* in soils, cattle, and farm environment in France (Burgundy region). Front Microbiol. 2012;3: 83. pmid:22408639
26. Laube H, Friese A, von Salviati C, Guerra B, Rösler U. Transmission of ESBL/AmpC-producing *Escherichia coli* from broiler chicken farms to surrounding areas. Vet Microbiol. 2014;172: 519–527. pmid:25035165
27. Ma J, Liu JH, Lv L, Zong Z, Sun Y, Zheng H, et al. Characterization of extended-spectrum beta-lactamase genes found among Escherichia coli isolates from duck and environmental samples obtained on a duck farm. Appl Environ Microbiol. 2012;78: 3668–3673. pmid:22407683
28. von Salviati C, Laube H, Guerra B, Roesler U, Friese A. Emission of ESBL/AmpC-producing Escherichia coli from pig fattening farms to surrounding areas. Vet Microbiol. 2015;175: 77–84. pmid:25465658
29. Soraas A, Sundsfjord A, Sandven I, Brunborg C, Jenum PA. Risk factors for community-acquired urinary tract infections caused by ESBL-producing enterobacteriaceae—a case-control study in a low prevalence country. PloS one. 2013;8: e69581. pmid:23936052
30. Njage PM,Buys EM. Pathogenic and commensal Escherichia coli from irrigation water show potential in transmission of extended spectrum and AmpC beta-lactamases determinants to isolates from lettuce. Microb Biotechnol. 2015;8: 462–473. pmid:25488608
31. Greenberg B (1973) Flies and disease vol II, biology and disease transmission; Greenberg B, editor. Princeton, NJ: Princeton University Press.
32. Nazni WA, Luke H, Wan Rozita WM, Abdullah AG, Sa'diyah I, Azahari AH, et al. Determination of the flight range and dispersal of the house fly, *Musca domestica* (L.) using mark release recapture technique. Trop Biomed. 2005;22: 53–61. pmid:16880754
33. Sola-Gines M, Gonzalez-Lopez JJ, Cameron-Veas K, Piedra-Carrasco N, Cerda-Cuellar M, Migura-Garcia L. Houseflies (Musca domestica) as Vectors for Extended-Spectrum beta-Lactamase-Producing Escherichia coli on Spanish Broiler Farms. Appl Environ Microbiol. 2015;81: 3604–3611. pmid:25795670
34. Usui M, Iwasa T, Fukuda A, Sato T, Okubo T, Tamura Y. The role of flies in spreading the extended-spectrum beta-lactamase gene from cattle. Microb Drug Resist. 2013;19: 415–420. pmid:23659602
35. Anonymous. ISO 16649–2 Microbiology of food and animal feeding stuffs—Horizontal method for the enumeration of beta-glucuronidase-positive *Escherichia coli*—part 2: Colony count technique at 44°C using 5-bromo-4-chloro-3-indolyl beta-D-glucuronide. 2001.

36. Clinical and Laboratory Standards Institute. Performance standards for antimicrobial susceptibility testing; twentieth informational supplement M100-S20. 2010.
37. Clermont O, Bonacorsi S, Bingen E. Rapid and simple determination of the *Escherichia coli* phylogenetic group. Appl Environ Microbiol. 2000;66: 4555–4558. pmid:11010916
38. Escobar-Páramo P, Le Menac'h A, Le Gall T, Amorin C, Gouriou S, Picard B, et al. Identification of forces shaping the commensal *Escherichia coli* genetic structure by comparing animal and human isolates. Environ Microbiol. 2006;8: 1975–1984. pmid:17014496
39. Frahm E,Obst U. Application of the fluorogenic probe technique (TaqMan PCR) to the detection of Enterococcus spp. and Escherichia coli in water samples. J Microbiol Methods. 2003;52: 123–131. pmid:12401234
40. Dallenne C, Da Costa A, Decré D, Favier C, Arlet G. Development of a set of multiplex PCR assays for the detection of genes encoding important β-lactamases in Enterobacteriaceae. J Antimicrob Chemother. 2010;65: 490–495. pmid:20071363
41. NCBIDierikx CM, van Duijkeren E, Schoormans AHW, van Essen-Zandbergen A, Veldman K, Kant A, et al. Occurrence and characteristics of extended-spectrum-beta-lactamase- and AmpC-producing clinical isolates derived from companion animals and horses. J Antimicrob Chemother. 2012;67: 1368–1374. pmid:22382469
42. Wirth T, Falush D, Lan R, Colles F, Mensa P, Wieler LH, et al. Sex and virulence in *Escherichia coli*: an evolutionary perspective. Mol Microbiol. 2006;60: 1136–1151. pmid:16689791
43. EUCAST (2014) Available: http://www.eucast.org/. Accessed February 2015.
44. Clinical and Laboratory Standards Institute (2015) Performance standards for antimicrobial susceptibility testing; twentieth informational supplement M100-S25. CLSI, Wayne, PA, USA.
45. Magiorakos AP, Srinivasan A, Carey RB, Carmeli Y, Falagas ME, Giske CG, et al. Multidrug-resistant, extensively drug-resistant and pandrug-resistant bacteria: an international expert proposal for interim standard definitions for acquired resistance. Clin Microbiol Infect. 2012;18: 268–281. pmid:21793988
46. Dierikx CM, van der Goot JA, Smith HE, Kant A, Mevius DJ. Presence of ESBL/AmpC-producing Escherichia coli in the broiler production pyramid: a descriptive study. PLoS One. 2013;8: e79005. pmid:24244401
47. Gezondheidsdienst voor Dieren en Produktschap Pluimvee en Eieren (2014) Antibioticumgebruik pluimveesector in 2013 en de trends van de afgelopen jaren. Available: http://www.pve.nl/wdocs/dbedrijfsnet/up1/ZsiuljcJI_2014-453-M0057-rapportage_2013.pdf.
48. Lysyk TJ,Axtell RC. Movement and distribution of house flies (Diptera: Muscidae) between habitats in two livestock farms. J Econ Entomol. 1986;79: 993–998. pmid:3745634
49. Crane SR,Moore JA. Modeling enteric bacterial die-off: a review. Water Air Soil Poll. 1986;27: 411–439.
50. Franz E, Schijven JF, De Roda Husman AM, Blaak H. Meta-regression analysis of commensal and pathogenic Escherichia coli survival in soil and water. Environ Sci Technol. 2014;48: 6763–6771. pmid:24839874
51. Coughter JP,Stewart GJ. Genetic exchange in the environment. Antonie van Leeuwenhoek. 1989;55: 15–22. pmid:2662898
52. Davison J. Genetic exchange between bacteria in the environment. Plasmid. 1999;42: 73–91. pmid:10489325
53. Trevors JT, Barkay T, Bourquin AW. Gene transfer among bacteria in soil and aquatic environments: a review. Can J Microbiol. 1987;33: 191–198.
54. Wang J, Stephan R, Karczmarczyk M, Yan Q, Hachler H, Fanning S. Molecular characterization of bla ESBL-harboring conjugative plasmids identified in multi-drug resistant Escherichia coli isolated from food-producing animals and healthy humans. Front Microbiol. 2013;4: 188. pmid:23874325

55. Poirel L, Naas T, Nordmann P. Genetic support of extended-spectrum beta-lactamases. Clin Microbiol Infect. 2008;14 Suppl 1: 75–81. pmid:18154530
56. Salyers AA, Gupta A, Wang Y. Human intestinal bacteria as reservoirs for antibiotic resistance genes. Trends Microbiol. 2004;12: 412–416. pmid:15337162
57. Smith HW. Transfer of antibiotic resistance from animal and human strains of Escherichia coli to resident E. coli in the alimentary tract of man. Lancet. 1969;1: 1174–1176. pmid:4181837
58. Trevors JT. Survival of *Escherichia coli* donor, recipient and transconjugant cells in soil. Water Air Soil Poll. 1987;34: 409–414.
59. Trevors JT,Oddie KM. R-plasmid transfer in soil and water. Can J Microbiol. 1986;32: 610–613. pmid:3527386
60. Trevors JT,Starodub ME. R-plasmid transfer in non-sterile agricultural soil. System Appl Microbiol. 1987;9: 312–315.
61. Carattoli A, Bertini A, Villa L, Falbo V, Hopkins KL, Threlfall EJ. Identification of plasmids by PCR-based replicon typing. J Microbiol Methods. 2005;63: 219–228. pmid:15935499
62. Google ScholarDierikx C, van Essen-Zandbergen A, Veldman K, Smith H, Mevius D. Increased detection of extended spectrum beta-lactamase producing Salmonella enterica and Escherichia coli isolates from poultry. Vet Microbiol. 2010;145: 273–278. pmid:20395076
63. Evers EG, Blaak H, Hamidjaja RA, de Jonge R, Schets FM. A QMRA for the transmission of ESBL-producing *Escherichia coli* and *Campylobacter* from poultry farms to humans through flies. Risk Anal. 2015;In Press.
64. Domingues AR, Pires SM, Halasa T, Hald T. Source attribution of human campylobacteriosis using a meta-analysis of case-control studies of sporadic infections. Epidemiol Infect. 2012;140: 970–981. pmid:22214729
65. Domingues AR, Pires SM, Halasa T, Hald T. Source attribution of human salmonellosis using a meta-analysis of case-control studies of sporadic infections. Epidemiol Infect. 2012;140: 959–969. pmid:22152439
66. Rangel JM, Sparling PH, Crowe C, Griffin PM, Swerdlow DL. Epidemiology of *Escherichia coli* O157:H7 outbreaks, United States, 1982–2002. Emerg Infect Dis. 2005;11: 603–609. pmid:15829201

CHAPTER 17

Assessment of The Effectiveness of Head Only and Back-of-The-Head Electrical Stunning of Chickens

T. J. Gibson, A. H. Taylor[1], And N. G. Gregory[1]

Department of Production and Population Health, Royal Veterinary College, University of London, Hatfield, UK

1Department of Clinical Sciences and Services, Royal Veterinary College, University of London, Hatfield, UK

ABSTRACT

1. The study assesses the effectiveness of reversible head-only and back-of-the-head electrical stunning of chickens using 130–950 mA per bird at 50 Hz AC.
2. Three trials were conducted to compare both stunning systems: (a) behavioural assessment of return of consciousness, (b) insensibility to thermal pain, and (c) assessment of return of brain activity with visually evoked potentials (VEPs).
3. Assessment of behaviour suggested that the period of unconsciousness following head-only electrical stunning was shorter in hens compared to roilers.
4. Stunning across the back-of-the-head delayed the time to return of brainstem function compared to stunning with standard head-only

electrodes. Additionally, back-of-the-head stunning produced a more prolonged period of electroanalgesia compared to head-only.

5. Based on examination of return of brain function with VEPs in hens, back-of-the-head stunning produced a shorter-lasting stun than standard head-only. However, even for standard head-only, the stun was notably shorter than previously reported. In some birds, brain function had returned within 9 s after the end of stunning.

6. The results suggest that some birds may recover consciousness prior to or during the neck cut. Based on these findings, back-of-the-head stunning and standard head-only stunning of hens should not be recommended without further development.

INTRODUCTION

Waterbath stunning of poultry is used in a number of countries. There have been calls for this system to be phased out in the United Kingdom (UK) and Europe, because of the risk of pre-stun shocks, suboptimum stuns, suspension of conscious birds upside down (which is stressful and, in some cases, painful) and variations in electrical current delivered to birds in multi-bird waterbath stunners (FAWC, 2009; Hindle et al., 2010; EFSA, 2014). In large-scale poultry processing plants, this system is being replaced by controlled atmosphere stunning (CAS) systems, which do not involve shackling before stunning. Presently, CAS systems are not feasible for small-scale poultry processing plants because of their relatively high capital and running costs. Instead, there is a need to develop cost-effective alternatives for stunning chickens in small-scale plants. Potential options include systems based on head-only stunning. Head-only stunning is already used for small-scale and seasonal poultry production and there has been recent development of cone restraint/head-only (Lambooij et al., 2010, 2014) and head-only waterbath (Lines et al., 2011) stunning systems.

Head-only electrical stunning, like some forms of high-frequency head-to-body and headto- shackle stunning, produces a reversible stun. Consequently, it is important with head-only electrical stunning that the duration of induced unconsciousness should last longer than the time it would take for neck cutting to be performed and brain death to occur through bleeding. In mammals, electrical stunning can induce analgesia (electroanalgesia), which lasts longer than the period of unconsciousness. This is mediated by the neurotransmitter GABA (Cook et al., 1992, 1993), but has not been demonstrated in poultry. In other words, after delivering an electric current the animal is stunned and then it appears to regain consciousness but is still unresponsive to painful stimuli. This state can last for 10 or more min (Gregory and Wotton, 1988).

Commercially available head-only stunning equipment for poultry generally consists of either a pair of single tapered copper electrodes or more commonly two rosettes of pointed needle (pin) electrodes (up to 12 needles on each rosette) placed on either side of the head. During head-only stunning, often only two or three needles on each electrode actually contact the surface of the head. The reduced area of contact can result in localised heating effects and enhance the build-up of carbon, which increases the electrical impedance (Sparrey andWotton, 1997).When this occurs, the operators are often required to press the electrodes with more force against the surface of the head to improve electrical contact. This can result in bending of the pins, which can further reduce electrical current flow and cause pain to the bird associated with increased pressure and tissue damage to the head prior to the onset of unconsciousness (T. J. Gibson, unpublished observation). An alternative novel system is making electrical contact across the back-ofhead (corresponding to the occipital bone) with smooth-surfaced electrodes, which can provide a relatively large area of contact. Unpublished work by the authors suggests that stunning chickens in this position could be effective. The potential advantages of the back-of-the head position are that it is relativity simple to apply, and that a single operator can perform both manual restraint and the stun.

Consciousness/unconsciousness following head-only electrical stunning has previously been assessed indirectly with a number of indicators, including: coordinated behavioural responses (Gregory and Wotton, 1990a, 1994; Lines et al., 2011); response to noxious stimulus (Lambooij et al., 2010); induction of epileptic/polyspike waveform or a quiescent electroencephalogram (EEG) (Richards and Sykes, 1967; Wormuth et al., 1981; Gregory and Wotton, 1990a; Schütt-Abraham, 1998; Raj and O'Callaghan, 2004; Lambooij et al., 2010, 2014; Lines et al., 2011), spectral analysis of the EEG (including correlation dimension analysis) (Raj and O'Callaghan, 2004; Lambooij et al., 2010, 2014; Lines et al., 2011), and recovery/absence of evoked potentials (Gregory and Wotton, 1990a).

The aim of the study was to evaluate the effectiveness of back-of-the-head compared to head-only electrical stunning, in terms of peak current delivered, duration of induced unconsciousness (behaviour evoked responses and return of brain function) and insensibility to noxious thermal laser stimulation.

MATERIALS AND METHODS

All birds were sourced from either commercial indoor or free-range units. Birds were kept in accordance with normal husbandry practices. All procedures were carried out under the provisions of the Animals (Scientific Procedures) Act 1986 and with the approval of the institute's Ethical Review Panel.

In all experiments, birds were stunned with a 50 Hz sine wave alternating current (AC) delivered using a constant voltage stunner (Whitehead Engineering Ltd, Bath, UK). Voltage and current in all experiments were recorded with the 199C Fluke Scope meter and 179 Fluke multimeter (Fluke Corporation, Everett, WA, USA). All current and voltage recordings in the paper are root mean squared (RMS) values.

Birds were stunned in the head-only or backof- the-head positions. Standard commercially available needle pin electrodes were used for head-only stunning. These consisted of 12, pointed copper pins (10 mm high × 1 mm diameter) spaced apart in a circular rosette arrangement with an overall diameter of 15 mm. The electrodes were placed on either side of the head spanning the brain between the eyes and the ears. This corresponded to the lateral aspects of the frontal bone and extended to the squamosal bone. The back-of-the-head electrodes were custom built (Solutions for Research Ltd., Silsoe, Bedford, UK), consisting of two adjustable stainless steel plate electrodes (190 mm high × 100 mm wide × 5 mm thick) arranged in a V shape. The bird's neck was placed in the V, with backward pressure applied so the caudal aspect of the cranium contacted the electrodes on either side of the head. This was caudal to the ears and corresponded to the occipital bone.

Experiment 1. Behavioural assessment of return of consciousness

The duration of unconsciousness following electrical stunning was examined in broilers and exlayer hens from the return of: rhythmic breathing, neck tension and balance (Gregory and Wotton, 1990a). Apnoea has been associated with induced epileptiform activity in the brain, while the termination of this activity has been shown to coincide with recovery of spontaneous breathing. The recovery and maintenance of neck tension relates to CNS control of muscle tone, which is absent during electrical stunning-induced unconsciousness. The recovery of and maintenance of balance indicates recovery of the vestibular system and cerebellar motor control. This reflects recovery of higher order brain function, which could be associated with consciousness.

Birds were stunned in either the head-only (broilers $n = 24$, hens $n = 23$) or back-of-the-head (hens $n = 25$) positions for 7 s; this duration was selected based on previous studies and government guidelines. As soon as rhythmic breathing returned, birds were repeatedly assessed in sequence for the time to return of neck tension and balance. Return of breathing was assessed by rhythmic movement of the vent and body related to respiration. While each bird was lying on its side, fingers were positioned under the head and used to repeatedly lift up the head and upper neck. Neck tension was recorded as returning when the birds were able to hold their head/neck up away from the supporting hand with obvious tension in the neck muscles. Immediately after testing for neck tension, birds were placed on their feet and shanks and gently pushed from the left or right sides of the body to assess for return of balance. This was conducted for a maximum of 5 s before retesting neck tension; if the bird fell onto its side it was said to not have control over its balance. Once breathing, neck tension, and balance had returned, all birds were immediately restunned and slaughtered with a ventral neck incision.

Experiment 2. Insensibility to thermal pain

Insensibility to pain (IP) was assessed using physical reactions to stimulation of the comb of hens with laser heat from a CO_2 laser (48-1S Synrad, Inc., Mukilteo, WA, USA) before and after electrical stunning with head-only ($n = 28$) and back-ofthe- head electrodes ($n = 9$). This technique has been used previously for nociceptive threshold research in pigs (Herskin et al., 2009), cattle (Veissier et al., 2000), and lambs (Guesgen et al., 2011). The same trained operator individually restrained each bird. A hood was placed over the head of the birds with an opening for exposure of the comb. This was to minimise distress and stress associated with handling and to protect the eyes and the rest of the head from the laser. Hens were placed 380 mm in front of the laser, a visible low power cold laser beam (1.25 diode pointer with ZnSe beam combiner, Synrad, Inc., Mukilteo, WA, USA) was introduced into the path of the CO_2 laser to allow for targeting of specific areas of the comb. The wavelength of the CO_2 laser was 10.57–10.63 μm, with the power set to 4.5%, this equated to 0.3 watts as measured with a power meter (Flash-500–55, Gentec Electro-optics, Inc., Quebec, Canada). Laser beam diameter was 3 mm and was focused with a 5.0" focal lens (LaserMech, Inc., Novi, MI, USA). Surface temperature of the comb was measured in a sample of birds using a thermal imaging camera (P620, FLIR Systems, West Malling, United Kingdom) or a remote infrared thermometer (LaserSight, Optris GmbH, Berlin, Germany). Prior to stimulation the comb was 31.8 ± 1.0°C. During laser stimulation, the mean temperature on the target area was 42.4 ± 0.7°C.

Prior to electrical stunning, the time to physical response (latency) to a laser beam focused on the bird's comb was assessed. This formed a baseline for time to physical response. The baseline laser stimulus was repeated no more than three times. Prior to electrical stunning, the hood was repositioned to allow placement of the stunning electrodes. Birds were then stunned with either a back-of-the-head or head-only electrical stunner with a minimum current of 240 mA. Immediately after the completion of convulsive wing flapping, the birds received a thermal laser stimulus on a fresh section of the comb. The stimulus was repeated every 30 s for up to 5 min after stunning. For each stimulus, the laser was directed to a fresh section of the comb (Figure 1). When there was no physical reaction, the laser stimulation lasted no longer than 10 s. The physical reaction consisted of recoil of the head and neck, with or without head shaking. Stimulation was discontinued as soon as a physical response occurred. The latency in the time to return of physical responses to the laser stimulus before and after stunning was assessed. Immediately on completion of experimentation, the birds were killed by electrical stunning followed by ventral neck incision.

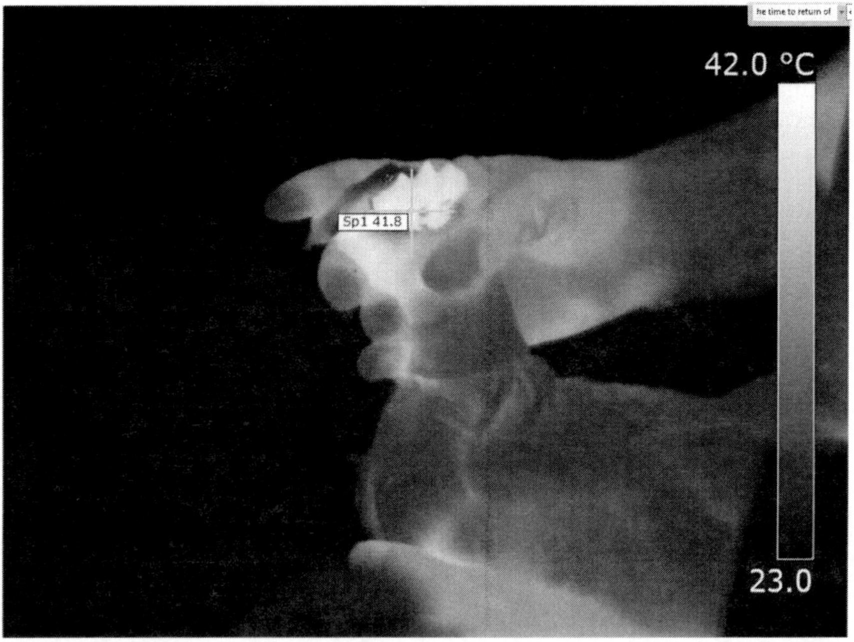

Figure 1. Experiment 2: thermal camera image of the site of laser stimulation of the comb.

Experiment 3.

Assessment of return of brain activity with visually evoked potentials (VEPs)

Thirty end-of-lay hens were randomly allocated into two stunning treatment groups. Birds either received a head-only (n = 14) or a back-of-thehead (n = 16) stun with a 50 Hz constant voltage stunner. Prior to placement of the electrodes, the birds were restrained in a cat restraint bag (Four Flags Over Aspen, Inc., St. Clair, MN, USA), with the wings, body, and legs restrained and the head and neck exposed. The birds were restrained to minimise movement artefact during VEP recording. One channel of VEPs was recorded from a three-electrode montage using three 28 gauge stainless steel subdermal electrodes (F-E3-48, Grass Technologies, Natus Neurology Inc., Warwick, RI, USA). The tips of each electrode were placed as follows: active (non-inverting) 4 mm right of midline, ≈3 mm rostral of bregma over the right optic lobe; reference (inverting), over the right rostral aspect of the forebrain 4 mm right of midline, ≈16 mm rostral of bregma; and ground electrode caudal to the back-of-thehead, respectively. Interelectrode impedance ranged between 1.7 and 3.1 kΩ (MkIII Checktrode, UFI, Morro Bay, CA, USA). VEPs were amplified, digitalised (1 kHz), and recorded with a 4/20 PowerLab (ADInstruments Ltd, Sydney, Australia) digital to analogue converter. All signals were filtered with an analogue filter (Bio Amp, ADInstruments Ltd., Sydney, Australia) with low and high pass filters of 100 and 0.1 Hz, respectively. A Grass photic stimulator (PS33-PLUS, Grass Technologies, Natus Neurology Inc., Warwick, RI, USA) was positioned 200 mm from the bird's head and was triggered and synchronized with the VEP recordings by the data acquisition software package Chart Pro 7 (ADInstruments Ltd., Sydney, Australia). All recordings were performed in a darkened room with the birds manually restrained. A 100 ms prestimulus (light) followed by a 200 ms post-stimulus period was recorded (flash every 300 ms). VEPs were assessed for a 1-min pretreatment period and for the first 5 min following the end of the stun for both treatments. After stunning, birds were monitored for return of rhythmic breathing and head shaking/movement that could be associated with movement artefact in the recorded waveforms. These waveforms were excluded from further analysis. On completion of the recording of VEPs, all birds were immediately killed with head-only electrical stunning followed by ventral neck incision.

Analysis of VEPs was conducted offline after the completion of experimentation. Initial averages were calculated from 15 consecutive sweeps (4.5 s). Where possible, the number of sweeps averaged was reduced to refine the time to return of VEP activity. Individual sweeps contaminated by noise were removed from analysis. The presence of genuine evoked potentials was assessed by comparison of positive and negative potential latencies with baseline (pre-stunning) and subsequent averaged evoked potentials. Waveforms with a signal to noise ratio < 2, based on baseline recordings were excluded from further analysis. All birds acted as their own controls. Signal averaging was performed with the Chart software, with subsequent analysis in Excel 2011 (Mac, Microsoft Corporation, Redmond, Washington, USA).

Statistical analysis

All analysis was performed using Prism 6.0e (GraphPad Software Inc., San Diego, CA, USA) and SPSS 20.0 (IBM Corporation, Chicago, IL, USA). The D'Agostino & Pearson omnibus normality test was used to determine the distribution of the data. Comparisons between head-only and back-of-the-head stunning were analysed with either an unpaired t-test with Welch's correction or Mann–Whitney test where appropriate. The relationships between peak current, return of VEPs, and stunning electrode impedance were analysed with linear regression. Return of VEPs and rhythmic breathing were analysed with a Spearman correlation. Mean values are displayed ± standard error of the mean (SE). The level of significance for all tests was $P < 0.05$.

RESULTS

Experiment 1. Behavioural assessment of return of consciousness

In broilers that received a conventional head-only stun of 618 ± 43 mA (range 130–950) and 133 ± 1 V (range 123–142), breathing, neck tension, and balance returned after 33 ± 2, 113 ± 14, and 182 ± 22 s, respectively (Table 1). Hens that received a conventional head-only stun of 469 ± 28 mA (range 240–780) and 139 ± 1 V (range 125–142), had a shorter time to return of breathing (24 ± 1 vs. 33 ± 2 s, $P < 0.05$), neck tension (76 ± 4 vs. 113 ± 14 s, $P < 0.05$), and balance (111 ± 6 s vs. 182 ± 22 s, $P < 0.05$) compared to the broilers. Hens that received a back-of-the-head stun of 476 ± 23 mA (range 240–680) with 132 ± 1 V (range 125–140), had a longer time to recovery of breathing (34 ± 2 s) compared to conventionally head-only stunned hens ($P < 0.05$) (Table 1). The time to return of neck tension was similar between back-of-the-head stunned hens and conventionally head-only stunned hens. Time to return of balance was longer with the back-of-the-head compared to conventionally head-only stunned hens (P < 0.05). One bird after back-of-the-head stunning did not recover and died.

Table 1. Experiment 1: effect of stunning method (mean ± SE (range)) on time to return of rhythmic breathing, neck tension, and balance in broilers and laying hens.

Stunning method	Bird type	Mean ± SE (range) time to return of rhythmic breathing (s)	Mean ± SE (range) time to return of neck tension (s)	Mean ± SE (range) time to return of balance (s)
Head-only	Broilers (n = 24)	33 ± 2 (17–59)^{a}	113 ± 14 (39–306)	182 ± 22 (66–512)^{a}
Head-only	End of lay hens (n = 23)	24 ± 1 (19–39)^{b}	76 ± 4 (45–138)	111 ± 6 (75–176)^{b}
Back-of-the-head	End of lay hens (n = 25)	34 ± 2 (20–53)^{a}	76 ± 4 (52–115)	260 ± 36 (80–974)^{a}

Means in a column with no common superscript letter differ significantly at $P < 0.05$.

Table 2. Experiment 2: effect of stunning method (head-only or back-of-the-head electrodes) on insensibility to thermal laser pain, behavioural indices, current and voltage.

	Stun method		
	Head-only	Back-of-the-head	Significance
Number of birds	28	9	-
Birds with physical response within 5 min, %	36% (n = 10)	0% (n = 0)	P = 0.079
Mean (± SE) time to return of physical response (range), s	83 ± 30 (9–32)	0	-
Mean (± SE) time to return of rhythmic breathing (range)[1], s	33 ± 4 (20–70)	40 ± 4 (26–58)	P = 0.107
Mean (± SE) time of return of neck tension (range)[1], s	79 ± 10 (45–132)	100 ± 10 (69–156)	P = 0.098
Mean (± SE) time of return of balance (range)[1], s	153 ± 16 (77–226)	279 ± 13 (210–306)	P = 0.002
Mean (± SE) peak current (range), mA	502 ± 48 (250–920)	589 ± 50 (420–910)	P = 0.167
Mean (± SE) peak voltage (range), V	132 ± 2 (110–143)	128 ± 1 (121–130)	P = 0.091

[1] Behavioural data only available for 19 birds (10 head-only; 9 back-of-the-head).

Experiment 2. Insensibility to thermal pain

Prior to electrical stunning, the latency in time to physical response to laser stimulation ranged between 1 and 5 (mean 1.5 ± 0.3) s. Mean peak current in the head-only and back-of-the-head groups were 502 ± 48 mA (range 250–920) and 589 ± 50 mA (range 420–910), respectively (P = 0.167). The mean peak voltage was 132 ± 2 V (range 110–143) and 128 ± 1 V (range 121–130) for head-only and back-of-the-head groups, respectively (P = 0.091) (Table 2). In the head-only electrical stunning group, 36% (n = 10) of birds responded to the laser after stunning, with the mean time for return of physical response in those birds of 83 ± 30 s. During the 5-min recording period, no birds in the back-ofthe- head group had a physical response to the laser stimulus. There was an association between stunning methods and return of physical response that approached significance (P = 0.079). The mean time to return of breathing, neck tension, and balance was recorded in 19 birds (Table 2). There was no significant difference in return of breathing or neck tension between the two treatment groups. However, there was a significant difference in the time of return of balance (P =

0.002), with animals in the back-of-the-head group having a mean time of return of balance of 279 ± 13 s compared to head-only with 153 ± 16 s.

Experiment 3. Assessment of return of brain activity with visually evoked potentials (VEPs)

There was no significant difference in the mean peak current delivered to hens in the head-only (453 ± 34 mA (range 280–650)) or back-of-the head (520 ± 16 mA (range 360–620)) stunning groups (P = 0.086) (Table 3). The mean peak voltage was 138 ± 1 V (range 131–146) and 138 ± 1 V (range 130–142) for head-only and back-of-the-head groups, respectively (P = 0.754). Initially, following electrical stunning, VEP was abolished in all hens from both treatment groups. However, VEP returned within 13 ± 1 s (range 7–26) in the backof- the-head group, compared to 20 ± 2 s (range 9–32) for birds in the head-only group ($P < 0.01$) after the end of the electrical stun. Examples of VEPs from two hens that received either the headonly or back-of-the-head stuns are shown in Figure 2. The bird that received the head-only stun recovered VEP activity within 17.5–21.0 s, whereas the one that received the back-of-the-head stun was showing signs of VEP activity starting to return at 5.0–9.5 s, with full recovery within 10.5–14.0 s. There were no significant differences in the mean time to return of breathing (P = 0.262), or peak voltage (P = 0.754) between the treatment groups. In addition, there was no significant relationship between current and time to return of VEPs (head-only R2 = 0.008, P = 0.755; back-of-the-head R^2 = 0.051, P = 0.397) (Figure 3).

Table 3. Experiment 3: effect of stunning method (head-only or back-of-the-head electrodes) on brain function, rhythmic breathing, peak current, and voltage in hens assessed for visually evoked potentials (VEPs).

	Stun method		
	Head-only	Back-of-the-head	Significance
Number of birds	14	16	-
Mean (± SE) time to return of VEPs (range), s	20 ± 2 (9–32)	13 ± 1 (7–26)	$P < 0.01$
Mean (± SE) time to return of rhythmic breathing (range), s	26 ± 1 (20–36)[1]	25 ± 1 (19–36)	$P = 0.262$
Mean (± SE) peak current (range), mA	453 ± 34 (280–650)	520 ± 16 (360–620)	$P = 0.086$
Mean (± SE) peak voltage (range), V	138 ± 1 (131–146)	138 ± 1 (130–142)	$P = 0.754$
Mean (± SE) weight (range), kg	4.0 ± 0.1 (3.2–4.7)	4.2 ± 0.1 (3.6–4.8)	$P = 0.310$

[1] No data from one animal, which died after recovery of visually evoked potentials (VEPs).

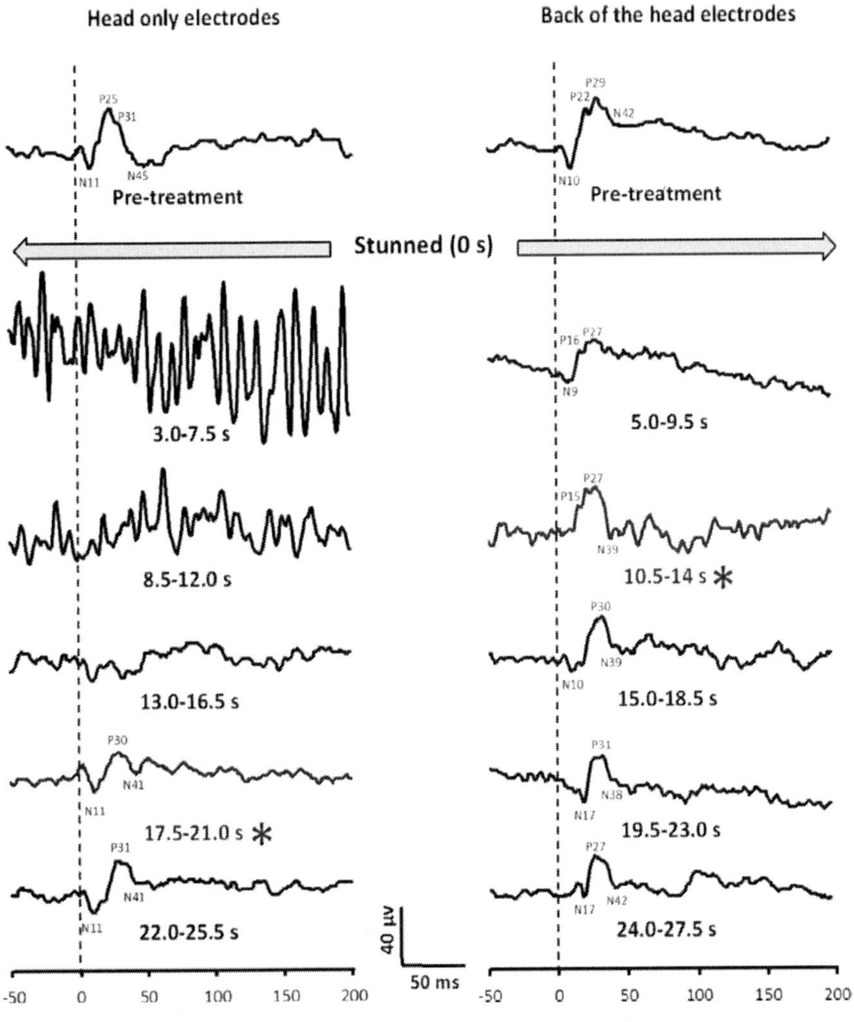

Figure 2. Experiment 3: examples of visual evoked potentials (VEP) from two hens after stunning with either standard head-only or backof- the-head electrodes. The * (grey) waveforms are where VEP activity first returned. Each waveform is the average of 15 traces (4.5 s). There was a stimulus delay of 100 ms (50 ms shown) with the stimulus delivered at 0 ms. Time after treatment is from the end of the stun. The dashed line indicates the time of the visual stimulus.

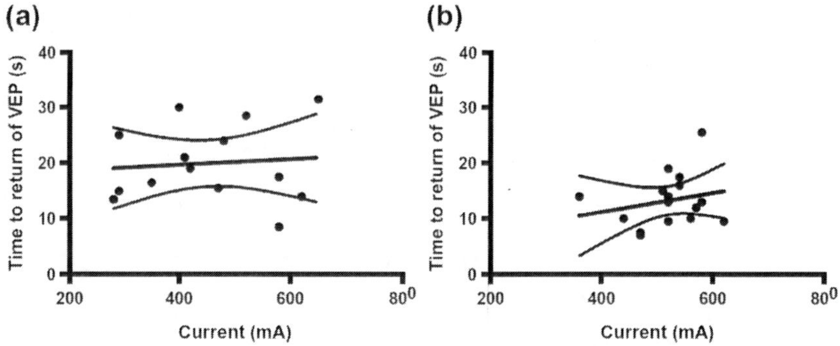

Figure 3. Experiment 3: linear regression with 95% CI for current (mA) compared to time to return of visual evoked potentials (VEPs) in hens stunned with (a) head-only ($R2 = 0.008$, $P = 0.755$) and (b) back-of-the-head ($R2 = 0.051$, $P = 0.397$) electrodes.

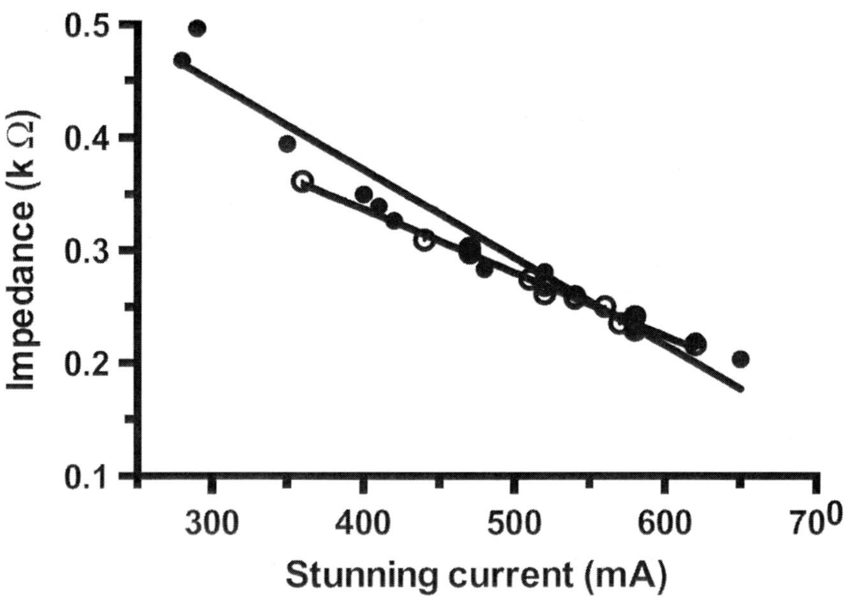

Figure 4. Experiment 3: linear regression for stunning current compared to impedance in hens stunned with (•) head-only ($R2 = 0.943$, $P < 0.001$) or (◯) back-of-the-head electrodes ($R2 = 0.987$, $P < 0.001$) in the visual evoked potentials (VEP) experiment.

The return of VEPs preceded the return of rhythmic breathing. There was no correlation between time to return of breathing and VEPs in birds stunned with the head-only electrodes (P = 0.449). However, in birds stunned with the back-of-the-head electrodes, there was a significant negative correlation between time to return of rhythmic breathing and time to return of VEP (r = −0.504, P = 0.046). There was a significant relationship between stunning current and impedance in both head-only (R^2 = 0.943, P < 0.001) and back-of-the-head (R^2 = 0.987, P < 0.001) stunning groups (Figure 4). There was a significant difference in the slopes of the regression lines between the two treatments (P = 0.013).

DISCUSSION

Indicators of effectiveness of electrical stunning Head-only electrical stunning involves the application of an electrical current across the head sufficient to induce brain dysfunction resulting in recoverable unconsciousness. This is in terms of generalised grand mal epileptic activity or an unresponsive isoelectric or quiescent EEG waveform (Raj and O'Callaghan, 2004; EFSA, 2014). Unlike stun to kill methods, the stun is reversible and does not induce cardiac arrest. In terms of animal welfare, the stun must induce near immediate unconsciousness and the duration of unconsciousness should last longer than the time it would take for neck cutting to be performed and brain death to occur through bleeding.

In broilers that received a conventional headonly stun, the time to return of breathing was similar to that previously reported by Gregory and Wotton (1990a), however the time to return of neck tension was greater in the current study. This could be due to the higher current levels delivered to broilers in the present study (618 vs. 336 mA). Meanwhile, hens stunned with the needle head-only electrodes, had a shorter time to return of breathing, neck tension and balance compared to broilers. These findings suggest that the duration of unconsciousness following electrical stunning is shorter in hens than in broilers, and it shows that this applies also to head-only stunning, whereas previously this was only evaluated for waterbath stunning (Gregory and Wotton, 1994). The reported differences between broilers and hens may be due to: (a) the increased feather coverage in end-of-lay hens, which could have increased the impedance of the head, (b) broilers compared to hens have increased soft tissue over the skull, which could have improved the electrical contact with the pin electrodes, reducing the impedance of the head, (c) hens are generally more reactive, agile, and coordinated than broilers, this may allow them to more readily express behavioural signs of recovery than broilers, (d) there may be differences in skull bone density between broilers and hens, affecting the electrical resistance of the head, and (e) there may be differences between broilers and hens in the duration and severity of bradycardia induced during head-only electrical stunning. This may time to return of consciousness.

To distinguish between the effects of current on brain function independent of the effects on spinal cord and brainstem function, VEPs were recorded from chickens prior to and after headonly and back-of-the-head stunning. VEPs are time-locked responses of the brain (optic lobe) to a light stimulus delivered in front of the eyes. They do not represent conscious awareness of the stimulus since they can occur in both the conscious and unconscious state. Rather they represent the rudimentary brain processing of the external stimulus below the level of conscious perception (Daly et al., 1988). In terms of stunning and slaughter, the absence of VEPs represents a degree of brain dysfunction that is inconsistent with the maintenance of consciousness.

From assessment of VEPs, the study found that brain dysfunction following head-only stunning on average only lasted 20 ± 2 s and ranged between 9–32 s. Compared to previous studies with head-only electrical stunning, the duration of brain dysfunction which was incompatible with consciousness is notably shorter, but it must be noted that the present finding was in hens rather than broilers. Richards and Sykes (1967) previously reported a period of unconsciousness that lasted between 30 and 60 s (ECoG) in chickens. Gregory and Wotton (1990a) reported 26–108 s in broilers (behaviour), and Lambooij et al., (2010) reported a range of 30–65 s (EEG) in broilers. A shorter period of induced unconsciousness could result in some birds recovering from the stun prior to or during exsanguination. Based on the current results, it is suggested that head-only electrical stunning (with the minimum recommended stunning parameters of 240 mA) of hens should not be recommended as a reliable stunning method.

Previous work with broilers has shown that increasing current delivered to the head increased the duration of the stun (Gregory and Wotton, 1990b; Raj and O'Callaghan, 2004). This relationship was not present in the current experiment and could be due to the low sample size and small current ranges in the study. However, work by Gregory and Wotton (1994) also found no effect of stunning current on the duration of unconsciousness in hens.

Electrical stunning and electroanalgesia

The results for noxious thermal laser stimulation demonstrate that in hens after electrical stunning there is a period of electroanalgesia that outlasts the period of induced unconsciousness. This was longest in hens stunned with the back-of-the-head electrodes (> 5 min). Previously electroanaglesia has been reported in mammals after electrical stunning; however this is the first report of it in birds and the first use of the thermal laser methodology in chickens. Potentially, the localized heating during electrical stunning could have produced desensitisation of the comb. However, this is considered unlikely as: (a) the stunning electrodes did not contact the comb, (b) they were sufficiently distant to avoid direct stimulation, (c) direct burning/cauterisation may cause localized deafferentation (however heating is not thought to cause widespread desensitisation, (d) the comb was still sensitive to non-noxious pressure in the form of touch, suggesting there was no localized numbing of the sensory nerves, and finally (e) the laser was used in multiple locations never stimulating the same area twice, reducing the potential for repeatedly stimulating desensitising regions or producing laser-induced desensitisation. The advantages of using lasers compared with physical tests such as a comb pinch are that: (a) the delivery of a laser stimulus is more reproducible because the stimulus is more uniform, (b) administering the laser does not involve a visual threat to the bird, whereas moving the hand to the comb does, (c) it is suspected that the bird would be less prone to becoming habituated to a laser stimulus compared with a pinch, (d) there are marked differences between regions of the comb in sensitivity to a pinch, and (e) an acute noxious laser stimulus does not involve other non-pain sensory modalities, which could complicate assessment of return of sensibility to pain. When doing repeated comb pinches, it is possible to run out of testable sensitive regions on the comb. This is less of an issue with laser stimulation as the area of the comb that is stimulated by the laser for each stimulus is relatively small (3 mm in diameter), leaving sufficient space for further measures.

Prior to experimentation on hens, the thermal laser stimulus was tested on the medial aspect of the forearm of the researchers. This confirmed that in humans the sensations provoked by thermal stimulation using this type of laser were comparable to those described by Willer et al., (1979) and Bromm and Treede (1987). Initially, there was a short-lasting pain, which was comparable to a localised pinprick. This was followed by a burning wave, which was a less severe pain, less localised, longer lasting and instilled was a sense of tissue injury. The initial pain is probably due to Aδ plus C-fibre activation, and the later pain to C-fibre activation alone (Bromm and Treede, 1987; Bromm and Lorenz, 1998; Ringkamp et al., 2013). In the chicken trials during baseline recordings the birds responded within 1.5 s, hich would have corresponded to the initial pricking pain experienced by humans.

There are several regions in the brain, which are thought to mediate the analgesic effect produced by electrical stimulation of the brain (Mayer, 1984; White et al., 2001). Three important regions are the periaqueductal grey matter in the midbrain, the raphe nuclei of the midbrain, and raphe nuclei of the medulla of the brainstem (Liebeskind et al., 1973; Mayer and Liebeskind, 1974). Cook et al., (1992, 1993) reported that electroanalgesia following electrical stunning is mediated by the neurotransmitter GABA in sheep. Furthermore, it is well recognised from experience in rats, cats, sheep, and humans that under specific circumstances when analgesia is produced by electrical stimulation of deep brain structures, the subject can be conscious (Urca et al., 1981; Mayer, 1984; Hosobuchi, 1986; Young and Brechner, 1986; Gregory and Wotton, 1988; Kumar et al., 1997). The analgesia can apply to ongoing intractable pain associated with diseases and disorders, or pain provoked by noxious heat applied to the tail, pinching the ear, foot or tail, or electrical stimulation of a tooth.

Although a period of protracted electroanalgesia that outlasts the duration of unconsciousness provided by the stun is beneficial in terms of welfare, the purpose of electrical stunning should be to induce complete unconsciousness, not just insensibility to pain. Furthermore, this period of induced electroanalgesia could inhibit some brainstem/spinal indices that are used to assess stun performance (e.g. response to noxious stimulus: comb pinch).

Advantages and disadvantages of back-of-the-head and head-only electrical stunning

The findings from the behavioural and laser studies indicated that back-of-the-head compared to standard head-only electrical stunning of chickens can be effective in inducing unconsciousness based on time to return of behavioural responses. Furthermore, the advantages of backof- the-head compared to other electrode positions is that it is relativity simple to apply, birds are not shackled before stunning, and the stun is reversible, potentially allowing its use for halal slaughter. However, based on examination of brain function in hens using VEPs, it was found that back-of-head stunning produced a shorterlasting period of brain dysfunction than the standard head-only electrodes and that VEP activity returned sooner than behavioural indices. This suggests that the brainstem/ cerebellummediated behavioural responses of return of breathing, neck tension, and balance underrepresent the true time to return of consciousness following head-only electrical stunning of hens. Similarly, Lines et al. (2011) reported in broilers following head-only waterbath stunning that the suppression of breathing (22 s) and neck muscle (42 s) tone was longer than the period of induced EEG suppression (12 s).

Based on the time of return of VEPs, the back-of-the-head stun produces a shorter-lasting stun than the standard head-only electrodes. It is possible that higher current levels could result in a longer lasting stun in this position. When sheep (Gilbert et al., 1991) and pigs (Hoenderken, 1978: Anil and Mckinstry, 1998) are stunned in the high neck region, an effective stun can be produced, depending on the current level and the proximity of the electrodes to the atlanto-occipital axis. Until further work is conducted, which establishes a longer lasting stun, it is suggested that back-ofthe- head stunning of chickens should not be used.

Head-only and back-of-the-head electrical stunning are reversible methods, where if the animals are not bled consciousness will return. When the duration of induced unconsciousness is short, or there are delays in the stun/cut interval, some animals could start recovering consciousness during the bleeding process prior to the onset of cerebral hypoxaemia from exsanguination. In the EU regulation 1099/2009, the requirement is that the loss of consciousness and sensibility from the stun shall be maintained until the death of the animal (Anon, 2009). Therefore, the duration of unconsciousness must be longer than the sum of the time from end of stunning and onset of death, including any delays in the stun-to-stick interval (Raj, 2006). To ensure this, the maximum stun/ cut interval (SCmax) is used to calculate the nterval in which the neck cut needs to be conducted to prevent recovery of consciousness prior to death. This is calculated by subtraction of the longest time to onset of unconsciousness following neck cutting without stunning (26 s based on Barnett et al. (2007)) from the shortest period of stun-induced unconsciousness.

Using return of neck tension (head-only stunned broilers: 39 s) as an indicator of return of consciousness, the SC_{max} to prevent recovery of consciousness during bleeding would be 13 s. This means that provided both carotid arteries are cut within the SC_{max} interval, birds will not resume consciousness during the bleeding period. However, in reality there can be considerable delays between stunning and the neck cut (transferring between cones and shackles, line speed, breakdowns etc.). Gregory and Wotton (1990b) reported that the average interval between stunning and the neck cut in UK broiler plants was 21 s (range 14–37). Therefore, if there are any delays that are longer in duration than the SC_{max}, then there is a risk that some birds will start to regain some level of consciousness and may suffer pain and distress before the onset of cerebral hypoxaemia from exsanguination. If the time to return of VEP ctivity (head-only stunned hens: 9 s) is used, it is probable that some birds would have recovered from the stun prior to or during the bleeding process and could be conscious for up to 18 s prior to the loss of consciousness from cerebral hypoxia. This would result in significant suffering from the pain and distress associated with recovery from the stun, the neck cut, and during the bleeding process.

CONCLUSIONS

In conclusion, back-of-the-head stunning delayed the time to return of brainstem function in broilers and hens, and induced a period of electroanalgesia that exceeded that of stunning with headonly electrodes. However, based on examination of brain function with VEPs in hens, it was found that back-of-the-head stunning produced a shorter-lasting stun than the head-only electrodes. This could result in birds recovering consciousness prior to or during the neck cut. Furthermore, the study found that the duration of unconsciousness with standard head-only stunning was notably shorter than previously reported. Based on these findings, back-of-the-head stunning should not be recommended without further development. The results also suggest that headonly electrical stunning of hens with a minimum current of 240 mA should not be recommended as a reliable stunning method.

ACKNOWLEDGEMENTS

The authors would like to thank the assistance of Christopher Davies, Frances Benstead, Nikolas Dadios, Natalie Chancellor, and Emily Bill in experimental work and initial data analysis authors.

FUNDING

This work was supported by the Department for Environment, Food and Rural Affairs (DEFRA) UK

REFERENCES

1. Anil, M.H. & Mckinstry, J.L. (1998) Variations in electrical stunning tong placements and relative consequences in slaughter pigs. *The Veterinary Journal*, 155: 85–90. doi:10.1016/S1090-0233(98)80042-7.
2. Anon (2009) Council Regulation (EC) No 1099/2009 of 24 September 2009 on the protection of animals at the time of killing, *Offical Journal of the European Union*. L 303/10.
3. Barnett, J.L., Cronin, G.M. & Scott, P.C. (2007) Behavioural responses of poultry during kosher slaughter and their implications for the birds' welfare. *Veterinary Record*, 160: 45–49. doi:10.1136/vr.160.2.45
4. Bromm, B. & Lorenz, J. (1998) Neurophysiological evaluation of pain. *Electroencephalography and Clinical Neurophysiology*, 107: 227–253. doi:10.1016/S0013-4694(98)00075-3

5. Bromm, B. & Treede, R.D. (1987) Human cerebral potentials-evoked by Co2-laser stimuli causing pain. *Experimental Brain Research*, 67: 153–162. doi:10.1007/BF00269463
6. Cook, C.J., Devine, C.E., Maasland, S.A. & Gilbert, K.V. (1993) Humane slaughter: an achievable goal? *Proceedings of the New Zealand Society of Animal Production*, 53: 197–199.
7. Cook, C.J., Devine, C.E., Tavener, A. & Gilbert, K.V. (1992) Contribution of amino acid transmitters to epileptiform activity and reflex suppression in electrically head stunned sheep. *Research in Veterinary Science*, 52: 48–56. doi:10.1016/0034-5288(92)90057-9
8. Daly, C.C., Kallweit, E. & Ellendorf, F. (1988) Cortical function in cattle during slaughter: conventional captive bolt stunning followed by exsanguination compared with shechita slaughter. *Veterinary Record*, 122: 325–329. doi:10.1136/vr.122.14.325
9. EFSA (2014) Scientific opinion on electrical requirements for poultry waterbath stunning equipment. *EFSA Journal*, 12: 3745.
10. FAWC (2009) *Report on the Welfare of Farmed Animals at Slaughter or Killing: Part 2. White Meat Animals.* (London, Farm Animal Welfare Council).
11. Gilbert, K.V., Cook, C.J. & Devine, C.E. (1991) Electrical stunning in cattle and sheep: electrode placement and effectiveness. Proceedings of the International Congress of Meat Science Technology, Kulmback, Germany, pp. 245–248.
12. Gregory, N.G. & Wotton, S.B. (1988) Sheep slaughtering procedures V. Responsiveness to potentially painful stimuli following electrical stunning. *British Veterinary Journal*, 144: 573–580. doi:10.1016/0007-1935(88)90027-9
13. Gregory, N.G. & Wotton, S.B. (1990a) An evaluation of the effectiveness of hand-held stunners for stunning chickens. *Veterinary Record*, 126: 290–291.
14. Gregory, N.G. & Wotton, S.B. (1990b) Effect of stunning on spontaneous physical activity and evoked activity in the brain. *British Poultry Science*, 31: 215–220. doi:10.1080/00071669008417248
15. Gregory, N.G. & Wotton, S.B. (1994) Effect of electrical stunning current on the duration of insensibility in hens. *British Poultry Science*, 35: 463–465. doi:10.1080/00071669408417711
16. Guesgen, M.J., Beausoleil, N.J., Minot, E.O., Stewart, M., Jones, G. & Stafford, K.J.(2011) The effects of age and sex on pain sensitivity in young lambs. *Applied Animal Behaviour Science*, 135: 51–56. doi:10.1016/j.applanim.2011.09.008
17. Herskin, M.S., Ladewig, J. & Arendt-Nielsen, L. (2009) Measuring cutaneous thermal nociception in group-housed pigs using laser technique—effects of laser power output. *Applied Animal Behaviour Science*, 118: 144–151. doi:10.1016/j.applanim.2009.02.016
18. Hindle, V.A., Lambooij, E., Reimert, H.G., Workel, L.D. & Gerritzen, M.A. (2010)Animal welfare concerns during the use of the water bath for stunning broilers, hens, and ducks. *Poultry Science*, 89: 401–412. doi:10.3382/ps.2009-00297
19. Hoenderken, R. (1978) *Elektrische Bedwelming Van Eslachvarkens.* (Utrecht, The Netherlands, State University).
20. Hosobuchi, Y. (1986) Subcortical electrical-stimulation for control of intractable pain in humans – report of 122 Cases (1970-1984). *Journal of Neurosurgery*, 64:543–553. doi:10.3171/jns.1986.64.4.0543
21. Kumar, K., Toth, C. & Nath, R.K. (1997) Deep brain stimulation for intractable pain: a 15-year experience. *Neurosurgery*, 40: 736–747. doi:10.1097/00006123-1997 04000 -00015
22. Lambooij, E., Reimert, H.G.M. & Hindle, V.A. (2010) Evaluation of head-only electrical stunning for practical application: assessment of neural and meat quality parameters. *Poultry Science*, 89: 2551–2558. doi:10.3382/ps.2010-00815
23. Lambooij, E., Reimert, H.G.M., Verhoeven, M.T.W. & Hindle, V.A. (2014) Cone restraining and head-only electrical stunning in broilers: effects on physiological

responses and meat quality. *Poultry Science*, 93: 512–518. doi:10.3382/ps.2013-03318

24. Liebeskind, J.C., Guilbaud, G., Besson, J.-M. & Oliveras, J.-L. (1973) Analgesia from electrical stimulation of the periaqueductal gray matter in the cat: behavioral observations and inhibitory effects on spinal cord interneurons. *Brain Research*, 50: 441–446. doi:10.1016/0006-8993(73)90748-8
25. Lines, J.A., Raj, A.B.M., Wotton, S.B., O'callaghan, M. & Knowles, T.G. (2011) Head-only electrical stunning of poultry using a waterbath: a feasibility study. *British Poultry Science*, 52: 432–438. doi:10.1080/00071668.2011.587180
26. Mayer, D.J. (1984) Analgesia produced by electrical-stimulation of the brain. *Progress in Neuro-Psychopharmacology & Biological Psychiatry*, 8: 557–564. doi:10.1016/0278-5846(84)90015-0
27. Mayer, D.J. & Liebeskind, J.C. (1974) Pain reduction by focal electrical stimulation of the brain: an anatomical and behavioral analysis. *Brain Research*, 68: 73–93. doi:10.1016/0006-8993(74)90534-4
28. Raj, A.B.M. (2006) Recent developments in stunning and slaughter of poultry. *Worlds Poultry Science Journal*, 62: 467–484. doi:10.1079/WPS2005109
29. Raj, A.B.M. & O'Callaghan, M. (2004) Effect of amount and frequency of head-only stunning currents on the electroencephalogram and somatosensory evoked potentials in broilers. *Animal Welfare*, 13: 159–170.
30. Richards, S.A. & Sykes, A.H. (1967) Physiological effects of electrical stunning and venesection in the fowl. *Research in Veterinary Science*, 8: 361–368.
31. Ringkamp, M., Raja, S.N., Campbell, J.N. & Meyer, R.A. (2013) Peripheral mechanisms of cutaneous nociception, in: McMahon, S.B., Koltzenbury, M., Tracey, I. & Turk, D.C. (Eds) *Wall and Melzack's Textbook of Pain*, pp. 1–11(Philadelphia, Elsevier).
32. Schütt-Abraham, I. (1998) Stunning methods for poultry: influence on birds' welfare and prospects for future EC regulation. Proceedings of the COST Action 97, 2. Poultry Productions Microbiology, European Regulations and Quality Assurance Systems, Berlin, Germany, pp. 333–344.
33. Sparrey, J.M. & Wotton, S.B. (1997) The design of pig stunning tong electrodes - a review. *Meat Science*, 47: 125–133. doi:10.1016/S0309-1740(97)00047-8
34. Urca, G., Yitzhaky, J. & Frenk, H. (1981) Different opioid systems may participate in post-electroconvulsive shock (Ecs) analgesia and catalepsy. *Brain Research*, 219: 385–396. doi:10.1016/0006-8993(81)90301-2
35. Veissier, I., Rushen, J., Colwell, D. & de Passillé, A.M. (2000) A laser-based method for measuring thermal nociception of cattle. *Applied Animal Behaviour Science*, 66: 289–304. doi:10.1016/S0168-1591(99)00099-4
36. White, P.F., Li, S.T. & Chiu, J.W. (2001) Electroanalgesia: its role in acute and chronic pain management. *Anesthesia and Analgesia*, 92: 505–513. doi:10.1213/00000539-200102000-00042
37. Willer, J.C., Boureau, F. & Berny, J. (1979) Nociceptive flexion reflexes elicited by noxious laser radiant-heat in man. *Pain*, 7: 15–20. doi:10.1016/0304-3959(79)90103-9
38. Wormuth, H.J., Schutt, I. & Fessel, J. (1981) *Tierschutzgerechte Elektrische Betäubung Von Schlachtgeflügel. VetMed Berichte 2/1981*. (Berlin, Dietrich Reimer Verlag).
39. Young, R.F. & Brechner, T. (1986) Electrical-stimulation of the brain for relief of intractable pain due to cancer. *Cancer*, 57: 1266–1272. doi:10.1002/(ISSN)1097-0142

Index

A

animal welfare, 36, 38, 51, 52, 57, 79, 277, 278, 280, 283, 294, 298, 301, 306, 397

Animal welfare, 34, 302, 324, 403

Aspergillus, 135, 136, 137, 139, 151, 195

B

broiler production, 34, 37, 38, 56, 57, 248, 382

brooders, 327, 328, 329, 330, 342, 343, 344, 345, 346, 347, 348, 350

C

Chick, 102

E

ecological survey, 308

economic performance, 33, 34, 35, 36, 43, 46, 51, 52, 54

edible components, 81, 82, 93, 95, 96, 99

F

farmers' perception, 34, 35

fertility functions, 119, 123, 126, 127, 129, 130

Fusarium, 17, 135, 136, 141, 143, 152, 153, 154

L

lameness, 33, 34, 35, 36, 37, 48, 49, 50, 51, 52, 53, 54, 55, 56, 57, 60, 80, 290, 303, 305

lipopolysaccharide, 101, 102, 108, 109, 112, 116, 117, 118, 144

M

metabolism, 120, 128, 138, 254, 274, 277, 288

mycotoxins,, 135, 136, 144, 146, 148

N

non-edible components, 81
nutritive value, 1, 3, 8, 20, 29, 31, 202, 203, 204

P

Penicillium, 135, 136, 139
phosphorus, 5, 13, 157, 162, 172, 173, 178, 203, 235, 236, 255, 256, 259, 273, 275
poultry, 1, 2, 3, 4, 5, 6, 11, 14, 379, 380, 381, 383, 386, 387, 402, 403, 404
prevention, 56, 124, 135, 136, 137, 146, 149, 150, 151, 285
prostaglandin E2, 101, 102, 111, 117
protein quality, 1, 2, 9, 10, 11, 16, 17, 18, 19, 21, 26, 28, 30, 31, 182, 198, 203, 275

R

requirements, 12, 21, 28, 121, 147, 148, 158, 172, 191, 209, 321, 403
rock phosphate, 236, 246

S

selenium, 120, 121, 125, 129, 130, 132, 209, 212, 213, 216, 229, 230, 231, 232, 233
selenoamino acids, 209, 232

T

technology, 277, 278, 279, 280, 281, 283, 284, 286, 287, 288, 289, 290, 293, 294, 295, 298, 299, 300, 301, 302, 303, 306
tissue components, 81, 82, 86, 95, 96, 99
tonic immobility, 327, 329, 334, 350

V

vitamin E, 119, 120, 121, 122, 123, 125, 126, 127, 128, 129, 130, 131, 132, 133, 229, 232

Z

Zea mays, 1, 2, 3